New Frontiers in Sustainable Geotechnics

New Frontiers in Sustainable Geotechnics

Editors

Małgorzata Jastrzębska
Krystyna Kazimierowicz-Frankowska
Gabriele Chiaro
Jaroslaw Rybak

MDPI • Basel • Beijing • Wuhan • Barcelona • Belgrade • Manchester • Tokyo • Cluj • Tianjin

Editors
Małgorzata Jastrzębska
Silesian University of Technology
Poland

Krystyna Kazimierowicz-Frankowska
Institute of Hydro-Engineering of the Polish Academy of Sciences
Poland

Gabriele Chiaro
University of Canterbury
New Zealand

Jaroslaw Rybak
Wroclaw University of Science and Technology
Poland

Editorial Office
MDPI
St. Alban-Anlage 66
4052 Basel, Switzerland

This is a reprint of articles from the Special Issue published online in the open access journal *Applied Sciences* (ISSN 2076-3417) (available at: https://www.mdpi.com/journal/applsci/special_issues/NewFrontiers_SustainableGeotechnics).

For citation purposes, cite each article independently as indicated on the article page online and as indicated below:

LastName, A.A.; LastName, B.B.; LastName, C.C. Article Title. *Journal Name* **Year**, *Volume Number*, Page Range.

ISBN 978-3-0365-6527-9 (Hbk)
ISBN 978-3-0365-6528-6 (PDF)

Cover image courtesy of Marcin Ćwirko

Contents

About the Editors

Małgorzata Jastrzębska

Małgorzata Jastrzębska is the Associate Professor of Geotechnics at the Silesian University of Technology in Gliwice, Poland. Her research activities include the calibration and verification of soil models, the tests of weak cohesive soils and their mixtures with waste rubber material, the methodology of the triaxial tests and the problems of maintaining the homogeneity of the state of stress and strain, the sustainable geotechnics in the field of Renewable Energy Sources, the impact of geotechnical works on the behavior of the subsoil, the securing slopes and strengthening the soil, the studies on the impact of microwave radiation on the physical and mechanical parameters of soils. She has experience in planning and implementing advanced experimental research on weak, swelling soils. Author and co-author of over 60 publications and 3 monographs on experimental geotechnics. She completed two scientific-research internships at the "Soils, Solides, Structures—Risques" Laboratory at the Fourier University in Grenoble, France. She has an experience in training Ph.D., masters and undergraduate students. As part of the Erasmus program, she gave lectures related to advanced soil mechanics in experimental terms at the J. Fourier University in Grenoble and at the University 2 in Montpellier.

ORCID iD: 0000-0003-0080-5784

Web of Science profile: https://www.webofscience.com/wos/author/record/796221

Scopus profile: https://www.scopus.com/authid/detail.uri?authorId=57197324744

ResearchGate website: https://www.researchgate.net/profile/Malgorzata_Jastrzebska

Krystyna Kazimierowicz-Frankowska

Krystyna Kazimierowicz-Frankowska, PhD, DSc, is Associate Professor at the Institute of Hydro-Engineering of the Polish Academy of Sciences. She is the Deputy Head in the Department of Geomechanics. Author and co-author of over 80 publications and 4 books on marine and inland civil engineering specifically focused on mechanics of reinforced soil, experimental mechanics and investigation of liquefaction of saturated and unsaturated soils, soil-water-structure interaction, modelling of engineering structure behaviour under different types of loading. As part of the implementation of international projects (such as: Numerical modelling of liquefaction around marine structures—http://nulimas.info), she cooperates with foreign partners from international research groups and companies. She also has extensive experience in carrying out expert works commissioned by many international and national institutions such as: European Research Executive Agency, National Centre for Research and Development, Polish Agency for Enterprise Development and others. Her expert work mainly concerns the assessment of various innovative solutions in the field of civil and marine engineering.

ORCID iD: 0000-0002-6648-6600

https://www.ibwpan.gda.pl/employee/82

https://scholar.google.es/citations?user=rmeHFXIAAAAJ&hl=es&oi=ao

Gabriele Chiaro

Gabriele Chiaro is Associate Professor of Geotechnical Engineering at the University of Canterbury, New Zealand. His research focuses on Geotechnical Engineering for Resilience and Sustainability with special interests on: Earthquake Geotechnical Engineering; Geoenvironmental Engineering; Ground Improvement; Experimental and Computational Geotechnics. He holds a BSc and MSc in Civil Engineering (Italy) and a PhD (Tokyo, Japan). His career involves 15 years of work in the academy (5 years in Japan; 3 years in Australia; 7 years in New Zealand). To date, he has secured over NZ$ 2.2 million in research funding as Principal Investigator. He has authored/co-authored over 120 peer-reviewed scientific publications, and his excellence in research has been recognized by many accolades, including the JGS Best Paper Award (2022), IABSE Best Scientific Paper Award (2021), NZSEE Otto Glogau Award (2020) and a JSPS Research Fellowship (2014). He served as Team Leader of the 2016 NZSEE LfE Mission, Kumamoto, Japan, and Team Co-Leader of the 2015 JGS/JSCE LfE Mission, Gorkha, Nepal. Currently, he represents the New Zealand Geotechnical Society in the ISSMGE TC101 and AsRTC1.

ORCID iD: 0000-0001-7328-2932

https://www.canterbury.ac.nz/engineering/contact-us/people/gabriele-chiaro.html

https://scholar.google.com.au/citations?user=yITGRvIAAAAJ&hl=en

Jaroslaw Rybak

Jaroslaw Rybak is Assistant Professor at Wroclaw University of Science and Technology, Poland. He works in Geotechnical Engineering, especially on deep excavations, foundation piles, soil improvement methods and environmental impact of geotechnical and mining technologies, and is responsible of many research projects as well as contracts with various companies. He is the author of 173 publications (63 articles in periodicals, 94 conference papers, 7 papers in monographic works) and of manifold unpublished works. In total, 52 of his papers are indexed in Web of Science (h-index=11), 61 of their papers are indexed in Scopus (h-index=15). He did over 200 reviews of journal and conference papers (verified in Web of Science) including over 150 reviews for journals indexed in Journal Citation Reports (Clarivate Analytics). He is a member of the Reviewer Boards for journals: *Studia Geotechnica et Mechanica*, and *Applied Sciences*, and he is currently involved in many editorial works, including managing Special Issues (as Academic Editor) in: *Energies, Sensors, Applied Sciences* and *Buildings*. He actively collaborates with numerous international conferences (Poland, Slovakia, Czechia, Russia, Uzbekistan, Vietnam), working in Scientific Committees and reviewing contributions. Since 2022 he joined Editorial Board of *Journal of the Mechanical Behavior of Materials*.

ORCID iD: 0000-0002-7267-1044

Web of Science profile: https://www.webofscience.com/wos/author/record/F-8558-2018

Scopus profile https://www.scopus.com/authid/detail.uri?authorId=14024816900

personal or ResearchGate website: https://www.researchgate.net/profile/Jaroslaw-Rybak

Preface to "New Frontiers in Sustainable Geotechnics"

Sustainable Geotechnics are a relatively new, but very dynamically developing area of ground engineering. Its development is a response to the currently emerging problems and challenges related to the growing world population (which is expected to reach almost 10,000 million by 2050) and the need to reconcile this with the development of a low-carbon economy addressing global climate change challenges, dwindling fossil fuel reserves, and environmental governance. A wide variety of sustainable approaches is already used in geotechnics, and the proposed solutions are becoming more and more advanced and effective. A sustainable engineered system should be efficient, reliable, resilient, and adaptive. Combining all these elements is not easy and requires knowledge from various fields (such as construction, economy, ecology, and others) and poses a challenge for many scientists and engineers.

The aim of this book is to present to the readers different solutions in the field of sustainable geotechnics and to bring together the state of the art in modern geotechnics. The main attention was paid to such key elements as sustainable design, construction, and monitoring of different kinds of engineering structures. An overview of the different research studies in this area was carried out by researchers from different countries and with different professional backgrounds. The reader can find examples of the use of ecofriendly solutions and new materials in geotechnics and description on interesting case studies regarding advanced monitoring systems. The book also includes a description of the various mathematical tools used to develop new solutions in the field of sustainable geotechnics. The effectiveness of their use to simulate the behavior of real engineering structures is examined. The presented research results indicate growing awareness of the importance of sustainable geotechnics in the professional community, both for researchers and engineers. There is no turning back from the possibility/necessity of seeking new solutions in line with this spirit. This, in turn, creates new challenges that must be met during the design and construction of new engineering structures. Such challenges will be easier to address if the knowledge about the currently used solutions in this area is more complete and congregated. That is what this book intends to achieve.

Małgorzata Jastrzębska, Krystyna Kazimierowicz-Frankowska, Gabriele Chiaro, and Jaroslaw Rybak

Editors

applied sciences

MDPI

Editorial

New Frontiers in Sustainable Geotechnics

Małgorzata Jastrzębska [1], Krystyna Kazimierowicz-Frankowska [2], Gabriele Chiaro [3] and Jarosław Rybak [4,*]

1 Department of Geotechnics and Roads, Faculty of Civil Engineering, Silesian University of Technology, Akademicka 5, 44-100 Gliwice, Poland
2 Institute of Hydro-Engineering, Polish Academy of Sciences, Kościerska 7, 80-328 Gdańsk, Poland
3 Department of Civil and Natural Resources Engineering, University of Canterbury, Christchurch 8041, New Zealand
4 Department of Geotechnology, Hydro Technology, and Underground and Hydro Engineering, Faculty of Civil Engineering, Wroclaw University of Science and Technology, Wyspiańskiego 27, 50-370 Wrocław, Poland
* Correspondence: jaroslaw.rybak@pwr.edu.pl

Citation: Jastrzębska, M.; Kazimierowicz-Frankowska, K.; Chiaro, G.; Rybak, J. New Frontiers in Sustainable Geotechnics. *Appl. Sci.* **2023**, *13*, 562. https://doi.org/10.3390/app13010562

Received: 21 November 2022
Accepted: 9 December 2022
Published: 31 December 2022

1. Introduction

With increasing ecological awareness, the idea of balanced development has become more popular. Meeting the socio-economic needs of humans in the context of maintaining the ecological balance is incredibly important. Geotechnical engineers are also faced with this difficult task. Currently, their work is focused on three wide areas: sustainable ground improvement, sustainable foundation engineering, and sustainable geotechnical design. All these areas create a raising demand for a precise evaluation of geotechnical parameters in field and laboratory testing, together with a cautious quality control of the geotechnical works. A social awareness should also be considered as one of the crucial factors in the planning of future research.

"What are the hottest topics in Geotechnical Engineering right now and in the next 10 years?" [1] This question asked in 2018 on ResearchGate by Mousa Bani Baker from Al-Zaytoonah University of Jordan still seems to be more relevant and is worth considering. Surprisingly, the most recommended answer to this question in ResearchGate was not focused to very sophisticated models and methods that are the domain of scientific world at the universities but rather to "the current situation in the developing countries, where there is a big need for *infrastructure*, *food* and *energy* for the growing populations".

Geotechnical Engineering can significantly influence the sustainability of infrastructure development because of its early position in the construction process [2]. A key focus should be the use of sustainable geotechnical materials, technologies, and applications, which are those that have very little negative impact on the environment, are easily available and affordable, are environmentally friendly and do not constitute any health hazard, and finally meet the design specifications in terms of durability and maintainability [3]. In this context, the reuse and recycling of industrial by-products, commercial wastes, and construction and demolition materials in geotechnical engineering applications, as well as development of new construction techniques and ground improvement methods, are progressively sought as they provide important benefits in terms of increased sustainability and reduced environmental impacts [4].

A collection of articles gathered under the title "New Frontiers in Sustainable Geotechnics" was intended to share various experiences in the following fields: (1) the use of alternate environmentally friendly materials in geotechnical constructions (such as embankments, slopes, and dams) and the reuse of waste materials (such as rubber waste, fly ash, natural, or artificial fibers) for soil improvement and stabilization; (2) innovative and energy-efficient ground improvement techniques; (3) bio-slope engineering; (4) the efficient use of geosynthetics; (5) the retrofitting and reuse of foundations, and foundations for energy extraction (e.g., termopile); (6) the use of underground space for the storage of energy; (7) the mining of shallow and deep geothermal energy; (8) making geo-structures reliable and resilient in the case of natural or man-made hazards; (9) design and modeling on the basis of geotechnical parameters derived; and (10) experiences with solving geotechnical challenges in marine engineering (such as predicting seabed liquefaction, designing marine foundations, and using anchors). Finally, in total, 18 contributions from various countries and diverse areas of research expertise were gathered to form the basis for comments listed in proceeding sections. Every listed contribution is supplemented with information taken directly from the articles' abstracts.

2. Laboratory Testing for Geotechnical Design

It must be underlined that the majority of contributions related to laboratory testing refer to "material science" applied in geotechnical engineering as opposed to the testing of geotechnical construction models. In the Editors' opinion, this is due to the higher universality of obtained results compared to the time and high costs of laboratory models of geotechnical structures (related to a singular case study). Nonetheless, all contributions listed below have a significant scientific and educational value.

2.1. Laboratory Testing and Theoretical Modeling of Deformations of Reinforced Soil Wall [5]

From a scientific point of view, an important element of this work was analyzing the effect of friction between the backfill and the side walls of the test box on the measured displacements. For the investigated case, it was shown that the impact of this element caused a reduction in the value of external loading of more than 60%. The final results may be particularly useful in the design process of structures used in transportation engineering (bridge abutments), where deformation limit values cannot be exceeded [5].

2.2. Influence of Laboratory Compaction Method on the Compaction and Strength Characteristics of Unbound and Cement-Bound Mixtures [6]

For mixtures with a higher cement content, the optimal moisture content difference depending on the laboratory compaction method used can be significant, so the laboratory compaction method should be chosen carefully, particularly for moisture-susceptible materials. This paper also reveals that by increasing the proportion of rubber in the mixture, the compaction and strength characteristics differ significantly when the compaction method is used. Therefore, when using alternative and insufficiently researched materials, the compaction method should also be chosen carefully [6].

2.3. Effect of Chitosan Solution on Low-Cohesivity Soil's Shear Modulus G Determined through Resonant Column and Torsional Shearing Tests [7]

In this study, the effect of using a biopolymer soil stabilizer on soil stiffness characteristics was investigated. It was shown that chitosan solution added to medium-grained materials substantially improves their shear modulus G (up to 3 times), even for solutions with relatively low chitosan concentrations (1.5 g of chitosan per 1 kg of dry silica sand). The results obtained in this study and the known chitosan properties suggest that chitosan solutions can be a very effective and eco-friendly short-term stabilizer for temporary geotechnical structures, e.g., working platforms [7].

2.4. The Effect of MICP on Physical and Mechanical Properties of Silt with Different Fine Particle Contents and Pore Ratios [8]

Microbial-induced calcium carbonate precipitation (MICP) is a new soil remediation technology which can improve the physical and mechanical properties of soil by transporting bacterial solution and cementation solution to loose soil and precipitating calcium carbonate precipitation between soil particles through microbial mineralization. Based on this technique, the effects of different fine particle contents and pore ratios on the physical and chemical properties of silt after reinforcement were studied. The content of calcium carbonate, the ability of silt to fixed bacteria, unconfined compressive strength (UCS), permeability coefficient, and microstructure of samples were determined. The results showed that the permeability coefficient of cured silt can be reduced by 1 to 4 orders of magnitude compared with that of untreated silt. In particular, the permeability of MICP-treated silt A is almost impermeable [8].

2.5. The Effects of Hydroxypropyl Methylcellulose (HPMC) on the Reinforcement of Sand by Microbial-Induced Calcium Carbonate Precipitation (MICP) [9]

Existing MICP treatment technology seems to be more suitable for deeper soils due to its high permeability. In this study, HPMC, a cohesive material combined with Sporosarcina pasteurii-induced calcium carbonate precipitation, was used to improve the surface layer of the soil. The addition of HPMC effectively improved the ammonia absorption rate and reduced the release of ammonia in the process of MICP technology, which is of great significance for environmental protection. The microstructure showed that the addition of HPMC and the increase in the number of treatments using MICP technology can make the surface structure of the specimens more compact, and the calcium carbonate can more effectively fill the pores and cement the soil particles, while the addition of HPMC may not change the calcium carbonate crystal type [9].

2.6. Geotechnical Evaluation of Diesel-Contaminated Clayey Soil [10]

Soil contamination by different hydrocarbons has rapidly expanded worldwide, surpassing the self-purification capacity of soils and increasing the number of contaminated sites. Although considerable effort has been devoted to studying the effects of diesel contamination on the geotechnical properties of soil, there is still limited available information about it. Presented results showed that the soil could only retain 12.6% of the added diesel and the excess was expulsed. At such a diesel concentration, the saturation rate of the soil was lower than 80%. Diesel contamination increased the plasticity and the internal friction angle of the soil, while its cohesion was considerably decreased. It should be noted that the matric suction of contaminated soil was lower than the one obtained for natural soil. However, its osmotic suction was considerably higher. This indicates that osmotic suction must be considered to evaluate the shear strength of contaminated soils [10].

2.7. Evaluating Biosedimentation for Strength Improvements in Acidic Soil [11]

Marine clay soils are problematic soils in the construction industry when they are subjected to construction loads. In order to stabilize the marine soil, new methods for soil improvement were built with biogrouting, and physical, biological, and chemical treatments were incorporated into the soil. To address issues with marine clay soil, this study aimed to minimize the high cost of a special foundation system and the use of non-environmentally friendly materials such as calcium-based binders, aside from the reduction of deformations caused by loading. The findings of this study can be used for acidic soils and to improve the soil's geotechnical behavior in general [11].

2.8. Strength and Compressibility of Ammonia–Soda Residue from the Solvay Sodium Plant [12]

This work discussed the results of an experimental study conducted to characterize the mechanical behavior of ammonia–soda residue (ASR). According to the physical state of ASR and the depth of sampling, two different evolutions of the critical state in the stress–strain space were observed. In light of the assessed stress–strain–strength behaviors, key design engineering parameters of ASR were calculated [12].

2.9. Modelling of Static Liquefaction of Partially Saturated Non-Cohesive Soils [13]

This study applied a semi-empirical model to predict the response of partially saturated soils under undrained conditions. The model proposed is based on an incremental equation describing the pre-failure undrained response of partially saturated non-cohesive soils during monotonic shearing in a standard triaxial test. Model predictions were confronted with the results of triaxial tests for two types of non-cohesive soils (quartz medium sand and copper ore post-flotation industrial tailings). Good agreement between experimental data and theoretical predictions was achieved [13].

3. Execution of Geotechnical Works and Quality Control Procedures

Contributions presented in this section provide a concise and precise description of the experimental results derived from field (real-life) experiments or different interpretations, as well as conclusions that can be drawn from the experimental testing of geomaterials.

3.1. Measurement-While-Drilling-Based Estimation of Dynamic Penetrometer Values Using Decision Trees and Random Forests [14]

Machine learning is a branch of artificial intelligence (AI) that involves the application of various algorithms to obtain information from large data sets. The objective of this study was to correlate the drilling parameters of deep foundation machinery (measurement-while-drilling, MWD) with a number of blows in the dynamic penetrometer test. Therefore, the drilling logs could be equated with said tests, providing information that can be easily interpreted by a geotechnical engineer and that can help to validate the design hypotheses. Decision trees and random forest algorithms have been used for this purpose. The ability of these algorithms to replicate the complex relationships between drilling parameters and terrain characteristics has allowed the penetrometric profile of the traversed soil to be a reliably reproduced [14].

3.2. An Experimental Study of Nailed Soil Slope Models: Effects of Building Foundation and Soil Characteristics [15]

A soil nailing system is a proven effective and economic method used to stabilize earth slopes from the external (factors increasing the shear stress) and internal (factors decreasing material strength) failure causes. It is found that the increase in soil density reduces both slopes facing displacement and building foundation settlements. The slope face displacement and footing settlement will increase with an increase in the width of the foundation and foundation position near the crest of the slope [15].

3.3. Behaviour of the Steel Welded Grid during a Simplified Pullout Test in Fine Sand [16]

This study considered the possibility of using steel gabion baskets made of welded mesh for a soil-strengthening function. It was unequivocally stated that as the stiffness of the steel grid itself increases, its strength increases during the pullout test, which is not so obvious in the case of popular steel woven meshes. In addition, it has been shown that steel-welded meshes with wire diameters less than 6 mm are suitable for soil reinforcement in structures with gabion facing, and the determined apparent friction coefficient (μ_k = 0.39–1.47) takes values similar to the friction coefficient given in references for welded meshes with larger diameters. This is a positive premise for starting further research on the use of wires of smaller diameters for welded mesh production used as soil reinforcement [16].

3.4. Recycling of End-of-Life Tires (ELTs) for Sustainable Geotechnical Applications: A New Zealand Perspective [17]

In this study, gravel and recycled granulated rubber were mixed to explore the possibility of obtaining synthetic granular geomaterials (with adequate geotechnical and environmental characteristics) that are suitable to act as structural fills for geotechnical applications including foundation systems for low-rise light-weight residential buildings. Moreover, an original framework with a set of geo-environmental criteria is proposed to demonstrate the use of gravel–rubber mixtures (GRMs) as structural fills [17].

3.5. The Influences of Local Glacitectonic Disturbance on Overconsolidated Clays for Upland Slope Stability Conditions: A Case Study [18]

This study used numerous calculation model analyses of the optional clay position in the context of slope stability conditions. A wide range of variable soil properties was taken into account, resulting from both lithogenesis and subsequent processes which disintegrate the original soil structure. Regarding the geological conditions of the slip surface, the use of classical computational methods and numerical modelling (FEM) was considered for comparative purposes. The results indicated that local changes in equilibrium conditions were affected by the different morphologies of the clay roof surface of the slope and alterations in strength characteristics on the slip surfaces. The findings of the study contribute to the sustainable spatial planning of near-slope regions [18].

3.6. Improving Mudstone Materials in Badland in Southwestern Taiwan by Increasing Density and Low-Cement Amount [19]

This study aimed to offer a new solution with the use of compaction techniques and also the addition of a small amount of cement used as soil amendment. The feasibility of this concept was examined by performing a series of tests, including the basic physical property test, the compaction test, the unconfined compression test, the static triaxial CU and UU tests, the consolidation test, the California bearing ratio (CBR) test, and the triaxial permeability test. The results of consolidation and the CBR test showed that improvement is possible by using low amounts of cement. Overall, the present method not only adheres to low-carbon and environmental protection requirements, but also verifies the feasibility of using compacted mudstone as an engineering material [19].

3.7. Probabilistic Analysis as a Method for Ground Freezing Depth Estimation [20]

This paper presented a probabilistic method to assess the depth of soil freezing. The obtained results are not the same as those given in the older Polish Standard which was based on simplified and limited data. The results confirm the impact of climate change on ground freezing depth [20].

3.8. Numerical Analysis of Pressure Profiles and Energy Dissipation across Stepped Spillways with Curved Risers [21]

Stepped spillway models with curved risers based on the increasing suspension angles were tested to check for improvements in energy dissipation and pressure distributions. It was estimated that curving the risers increases the energy dissipation up to three percent for lower flow rates, whereas this has no significant impact on energy dissipation for higher flow rates. It was found that in simply stepped spillways, lower steps dissipate more energy as compared to stepped spillways with curved risers where energy dissipation is shifted to higher steps. It was also observed that a higher energy dissipating step as experienced more negative pressures as compared to the lower energy dissipating step [21].

3.9. Monitoring the Impact of the Large Building Investments on the Neighborhood [22]

This article proposed the concept of monitoring buildings and infrastructure elements located near large construction investments The monitoring results allow corrections to be made in the technology of works (e.g., reductions in vibration amplitudes, the application of additional protections at excavations, etc.) or allow additional safety measures to be used. Currently, there are also monitoring systems used during the operation of completed facilities [22].

4. Summary

A brief overview of the articles submitted for this Special Issue "New Frontiers in Sustainable Geotechnics" allows us to draw the following general conclusions.

- By bearing in mind the geographical spread of the research presented in this Special Issue, it is clear that the topic of sustainable geotechnics is highly relevant for geotechnical engineers and researchers working in many different places and conditions.
- The ways in which individual issues and problems are presented by the authors indicate that, currently, in order to operate effectively in the field of geotechnics, knowledge from various fields should be combined. For example, performing laboratory tests, carrying out geotechnical works, and frequently overseeing quality control procedures require knowledge of the principles of modelling and a sound grasp of theory. Examples of such a comprehensive approach can be seen in papers

such as [5,13,20,21]. It should also be noted that various mathematical tools are used to solve complex geotechnical problems. In this Special Issue, the reader can find examples of the use of analytical and numerical methods for theoretical modelling. The reader also has an opportunity to become acquainted with the use of such tools to verify the accuracy of predicting experimental data.

– The sustainability in geotechnical engineering covers different aspects, allowing researchers to contribute in a wide variety of ways suitable for their particular interests. The reader can find different examples of the use of pro-ecological solutions and materials in geotechnics [5,15,17] and a description of an effective monitoring system for the analysis of large building investments' impact on the neighbourhood [22].

We would like to thank all the authors for their contributions and encourage all colleagues (both researchers and practical engineers) who are interested in the subject of sustainable development in geotechnics to read the articles presented in this Special Issue. Given the variety of topics, we hope that everyone finds items that interest them. We encourage everyone to join the growing community that is dedicated to sustainability in geotechnical engineering.

Author Contributions: Conceptualization, J.R.; writing—original draft preparation, J.R., K.K.-F. and G.C.; writing—review and editing, M.J.; visualization, M.J.; supervision, M.J.; project administration, M.J. All authors have read and agreed to the published version of the manuscript.

Acknowledgments: The authors would like to express their gratitude to Tracy Yu who was our Contact Editor and who helped us a lot with their competence and devotion in the editorial process.

Conflicts of Interest: The authors declare no conflict of interest.

References

1. Baker, M.B. What Are the Hottest Topics in Geotechnical Engineering Right Now and in the Next 10 Years? Available online: https://www.researchgate.net/post/What-are-the-hottest-topics-in-Geotechnical-Engineering-right-now-and-in-the-next-10-years (accessed on 15 November 2022).
2. Basu, D.; Misra, A.; Puppala, A.J. Sustainability and geotechnical engineering: Perspectives and review. *Can. Geotech. J.* **2015**, *52*, 96–113. [CrossRef]
3. Cunningham, W.P.; Cunningham, M.A.; Saigo, B.W. *Environmental Science: A Global Concern*; McGraw-Hill: Boston, MA, USA, 2001.
4. Arulrajah, A.; Narsilio, G.; Kodikara, J.; Orense, R.P. Key Issues in Environmental Geotechnics: Australia-New Zealand. *Environ. Geotech.* **2015**, *6*, 326–330. [CrossRef]
5. Kazimierowicz-Frankowska, K.; Kulczykowski, M. Laboratory Testing and Theoretical Modeling of Deformations of Reinforced Soil Wall. *Appl. Sci.* **2022**, *12*, 6895. [CrossRef]
6. Zvonarić, M.; Barišić, I.; Galić, M.; Minažek, K. Influence of Laboratory Compaction Method on Compaction and Strength Characteristics of Unbound and Cement-Bound Mixtures. *Appl. Sci.* **2021**, *11*, 4750. [CrossRef]
7. Bocheńska, M.; Bujko, M.; Dyka, I.; Srokosz, P.; Ossowski, R. Effect of Chitosan Solution on Low-Cohesive Soil's Shear Modulus *G* Determined through Resonant Column and Torsional Shearing Tests. *Appl. Sci.* **2022**, *12*, 5332. [CrossRef]
8. Zhao, Y.; Wang, Q.; Yuan, M.; Chen, X.; Xiao, Z.; Hao, X.; Zhang, J.; Tang, Q. The Effect of MICP on Physical and Mechanical Properties of Silt with Different Fine Particle Content and Pore Ratio. *Appl. Sci.* **2022**, *12*, 139. [CrossRef]
9. Zhu, W.; Yuan, M.; He, F.; Zhao, Y.; Xiao, Z.; Wang, Q.; Meng, F.; Tang, Q. Effects of Hydroxypropyl Methylcellulose (HPMC) on the Reinforcement of Sand by Microbial-Induced Calcium Carbonate Precipitation (MICP). *Appl. Sci.* **2022**, *12*, 5360. [CrossRef]
10. Hernández-Mendoza, C.E.; García Ramírez, P.; Chávez Alegría, O. Geotechnical Evaluation of Diesel Contaminated Clayey Soil. *Appl. Sci.* **2021**, *11*, 6451. [CrossRef]
11. Saad, A.H.; Nahazanan, H.; Yusoff, Z.B.M.; Mustafa, M.; Elseknidy, M.H.; Mohammed, A.A. Evaluating Biosedimentation for Strength Improvement in Acidic Soil. *Appl. Sci.* **2021**, *11*, 10817. [CrossRef]
12. Zięba, J.; Rzepka, P.; Olek, B.S. Strength and Compressibility of Ammonia-Soda Residue from the Solvay Sodium Plant. *Appl. Sci.* **2021**, *11*, 11305. [CrossRef]
13. Świdziński, W.; Smyczyński, M. Modelling of Static Liquefaction of Partially Saturated Non-Cohesive Soils. *Appl. Sci.* **2022**, *12*, 2076. [CrossRef]
14. García, E.M.; Alberti, M.G.; Arcos Álvarez, A.A. Measurement-While-Drilling Based Estimation of Dynamic Penetrometer Values Using Decision Trees and Random Forests. *Appl. Sci.* **2022**, *12*, 4565. [CrossRef]
15. Mohamed, M.H.; Ahmed, M.; Mallick, J. An Experimental Study of Nailed Soil Slope Models: Effects of Building Foundation and Soil Characteristics. *Appl. Sci.* **2021**, *11*, 7735. [CrossRef]

16. Ćwirko, M.; Jastrzębska, M. Behaviour of the Steel Welded Grid during a Simplified Pullout Test in Fine Sand. *Appl. Sci.* **2021**, *11*, 9147. [CrossRef]
17. Tasalloti, A.; Chiaro, G.; Murali, A.; Banasiak, L.; Palermo, A.; Granello, G. Recycling of End-of-Life Tires (ELTs) for Sustainable Geotechnical Applications: A New Zealand Perspective. *Appl. Sci.* **2021**, *11*, 7824. [CrossRef]
18. Kiełbasiński, K.; Dobak, P.; Kaczmarek, Ł.; Kowalczyk, S. The Influences of Local Glacitectonic Disturbance on Overconsolidated Clays for Upland Slope Stability Conditions: A Case Study. *Appl. Sci.* **2021**, *11*, 10718. [CrossRef]
19. Hsiao, D.; Hsieh, C. Improving Mudstone Materials in Badland in Southwestern Taiwan by Increasing Density and Low-Cement Amount. *Appl. Sci.* **2022**, *12*, 2290. [CrossRef]
20. Godlewski, T.; Wodzyński, Ł.; Wszędyrówny-Nast, M. Probabilistic Analysis as a Method for Ground Freezing Depth Estimation. *Appl. Sci.* **2021**, *11*, 8194. [CrossRef]
21. Saqib, N.U.; Akbar, M.; Pan, H.; Ou, G.; Mohsin, M.; Ali, A.; Amin, A. Numerical Analysis of Pressure Profiles and Energy Dissipation across Stepped Spillways Having Curved Risers. *Appl. Sci.* **2022**, *12*, 448. [CrossRef]
22. Łupieżowiec, M. Monitoring the Impact of the Large Building Investments on the Neighborhood. *Appl. Sci.* **2021**, *11*, 6537. [CrossRef]

applied sciences

MDPI

Article

Laboratory Testing and Theoretical Modeling of Deformations of Reinforced Soil Wall

Krystyna Kazimierowicz-Frankowska * and Marek Kulczykowski

Department of Geomechanics, Institute of Hydro-Engineering of Polish Academy of Sciences, ul. Kościerska 7, 80-328 Gdansk, Poland; marek@ibwpan.gda.pl
* Correspondence: krystyna@ibwpan.gda.pl; Tel.: +48-58-552-29-62

Abstract: This paper presents the results of an experimental investigation of a vertical reinforced soil (RS) wall. The structure was built on a laboratory scale. Horizontal displacements on the surface of the model wall were monitored at the end of construction and during surcharge application (as post-construction displacements). The experimental results were compared with their theoretical predictions. The accuracy of the selected analytical approach was examined to predict deformations of the RS structure under external loading. It was shown that the proposed original and relatively simple analytical method for estimating structural deformation can be successfully used in practice (the average difference between the recorded and calculated values of deformation did not exceed 25%). From a scientific point of view, an important element of this work was the analysis of the effect of friction between the backfill and the side walls of the test box on the measured displacements. For the investigated case, it was shown that the impact of this element caused a reduction in the value of external loading of more than 60%. The final results may be particularly useful in the design process of structures used in transportation engineering (bridge abutments), where deformation limit values cannot be exceeded.

Keywords: reinforced soil walls; experimental tests; facing displacements; analytical methods

Citation: Kazimierowicz-Frankowska, K.; Kulczykowski, M. Laboratory Testing and Theoretical Modeling of Deformations of Reinforced Soil Wall. *Appl. Sci.* **2022**, *12*, 6895. https://doi.org/10.3390/app12146895

Academic Editors: Stefano Invernizzi and Daniel Dias

Received: 15 May 2022
Accepted: 3 July 2022
Published: 7 July 2022

1. Introduction

The use of reinforced soil (RS) walls, treated as a new, more economical proposition compared to conventional retaining walls, has increased significantly across the world during the last few decades. In recent periods of time, a large number of such structures has been built because of their numerous advantages, including reliability, aesthetics, cost effectiveness, simplicity of construction, tolerance of differential settlements, and good seismic performance [1–13].

Predicting the behavior of reinforced soil walls, especially under external loading, is complicated by the varied properties of their component materials (such as soil, reinforcement, wall elements), the complexity of their immediate interactions, different boundary conditions, variable structural geometry, and the different construction methods used in particular cases. Several investigations have been performed in order to recognize how the behavior of reinforced soil walls is influenced by particular design factors, such as wall geometry, the types of loading, foundation quality, the facing type and its connection with reinforcement, the reinforcement type and tensile parameters, the reinforcement layout in horizontal and vertical directions, and backfill parameters.

The available results show that the effectiveness of RS wall technology is significantly dependent on the interaction between two essential materials used to build this type of structure: the soil that is used as the backfill and the reinforcements. In a correctly constructed RS wall, this interaction substantially increases the tensile strength inside the soil mass, allowing the overall structure to behave in a way similar to that of a monolithic body, supporting not only its own weight, but also external loads [14–18].

To improve knowledge in this area, many researchers have studied both the static performance of reinforced soil walls [19–26] and the dynamic behavior of such structures. They have performed both experimental tests and theoretical modeling. The analysis of typical RS walls is still often carried out using analytical methods, but in many (especially more complex) cases, a numerical simulation by FEM may also be useful and is included in the design process. In addition, it is often worth comparing the results obtained using different methods. The most commonly used approach takes into account the assumption that the plane strain conditions and calculations are carried out in a two-dimensional (2D) framework [27–35].

The different design concepts for RS walls using particular types of materials as reinforcement have been extensively analyzed and clearly described in the literature [36–38]. However, it should be noted that the methods of designing such structures have focused mainly on their ultimate limit states, such as pull-out and global and local stability failure. Although many studies have investigated the influence of selected reinforcement parameters on the behavior of reinforced soil foundations, research on the serviceability limit state has been scarce [39,40]. Deformation in RS structures is becoming an ever more important design consideration, as such structures are built with increasingly tight tolerances. It is crucial, for example, when they are used as elements of infrastructure in road engineering. The performance of such structures as geosynthetic reinforced soil (GRS) bridge abutments under service loads and working stress has been reported as satisfactory from engineering point of view [41–49]. That is why the problem of deformation in such structures needs further investigation.

The main aim of this study is to present the results of our experimental investigations into deformations in a reinforced soil wall, and to compare these results with the theoretical predictions of these deformations. For this purpose, we used an analytical method involving the calculation of displacements of reinforced soil structures. The method, elaborated by [40], is an extension of a method proposed by [18,50,51].

The final results may be particularly useful in the design process of structures used in transportation engineering (bridge abutments), where deformation limit values cannot be exceeded.

2. Methods

2.1. Background

Several studies have dealt with the behavior of footings constructed on stabilized sand walls [52–70]. Some authors are interested in both experimental and theoretical investigations of RS structures. For example, Adams et al. [71] built a few large-scale mini-pier tests using silty gravel as back-fill material, concrete masonry units as facing elements, and woven geotextiles of different tensile strengths and vertical spacing as reinforcement layers. The authors concluded that the performance of the RS mass and its behavior under loading was more dependent on reinforcement spacing than on geotextile strength. Pham [72] carried out large-scale generic soil-geosynthetic composite tests under plane strain conditions. The gravel was used as back-fill material and woven geotextiles of different tensile strengths and vertical spacing were used as reinforcement layers. During the experiments, Pham [72] applied different confining pressures to the two sides of the reinforced soil mass. The author [72] pointed that vertical spacing of the reinforcement had a more significant influence on the performance of the RS mass than its tensile strength.

Based on published results [72,73], it can be concluded that large-scale loading tests are very useful to investigate the real behavior of RS structures, considering their aggregate size and reinforcement spacing [74,75]. However, it is worth bearing in mind that this kind of investigation also has some disadvantages. Nicks et al. [76] paid attention to two major problematic items associated with large-scale RS performance tests. Firstly, these tests require specialized equipment which is not always available. Secondly, a performance test gives results only for a specific case (for a specific RS structure model characterized by an individually selected combination of parameters). Therefore, the test results obtained cannot be considered representative of other structures—without additional checking—when the geometry and/or materials will be changed. Additionally, it should be noted that this is a very expensive type of research that requires significant funding. Moreover, in some cases, many interesting results can be obtained by limiting activities to small-scale experiments. This is the approach that was used during this research, the results of which are presented in the article.

In this study, special attention was paid to the theoretical analysis of the behavior of RS walls under serviceable loading. The main part of this investigation focused on recognizing the scope and character of these deformations. It should be noted that only limited studies concerning this aspect of the behavior of RS structures are presented in the literature. However, the deformation of RS walls is becoming an ever more important design consideration since they are built with increasingly tight tolerances. The deformation performance of these composite structures, which combine the benefits of strong soil and reinforcement, can be influenced by a large number of factors. The most important are the geometrical properties of RS structures, the time-dependent characteristics of the reinforcement, and the main parameters of the soil used as backfill and subgrade/foundation materials.

During their professional activities, engineers are interested mainly in accurately predicting two types of deformations of RS structures: horizontal displacements of the RS facing and vertical displacements of the crest (Figure 1). The scope of this paper is limited to the analysis the former. It should be noted that additional information regarding the problems associated with the deformation of RS structures is available from other published articles [50,51,77–80].

There are two general mathematical approaches that can be used to solve the problem. The first one is based on analytical models for the calculation of deformations of RS structures. This approach depends on accurate and precise assumptions regarding the workings of RS structures and their particular elements. Modeling errors cause significant inaccuracies in the output results. Thus, such methods require good theoretical background regarding the mechanisms of deformations in RS structures in order to be useful.

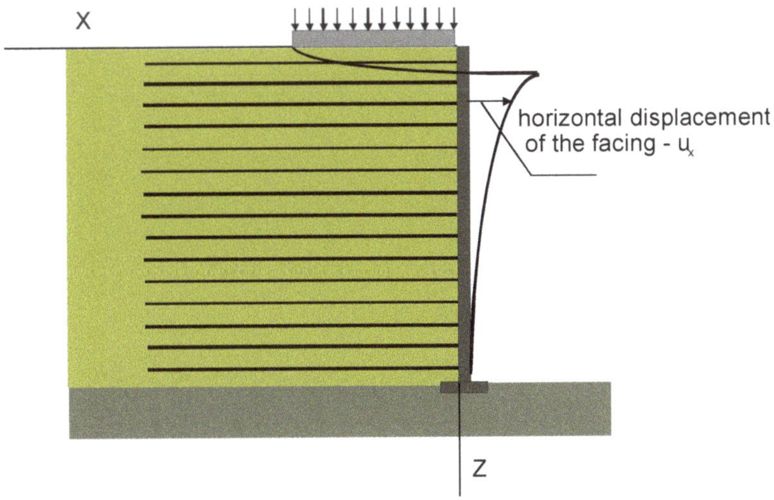

Figure 1. Deformations of RS structure under serviceable loading.

Although the number of studies has increased, so far there is no single accepted method which could be used in the design process to predict deformations in RS structures. The calculation methods that are available can be classified into a few groups (Figure 2), all of which have advantages and disadvantages [51].

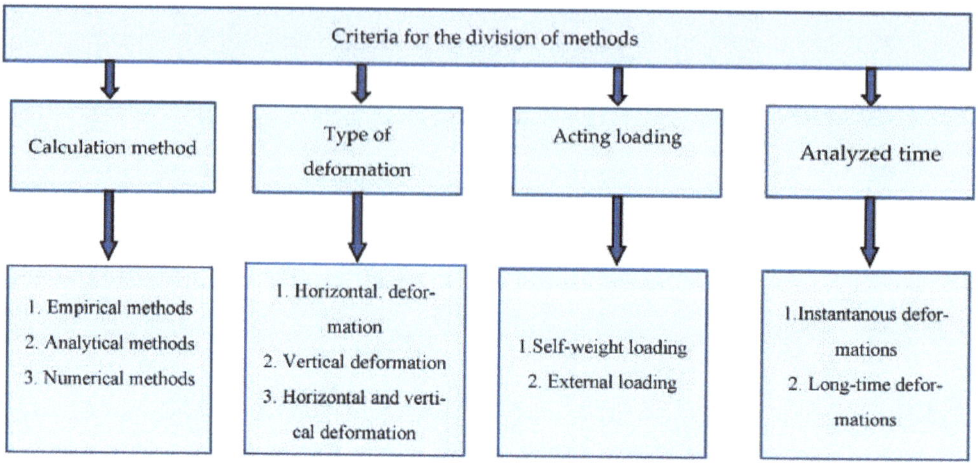

Figure 2. Methods used to investigate deformations in RS walls.

The second approach utilizes numerical methods to investigate the scope and magnitude of the deformations. Such calculations allow the assignment of material parameters that would be difficult to recognize in experimental studies. It should also be noted that the development of typical numerical procedures led to some important idealizations of the problem with regard to the geometry of the RS structure, the type of loading, the properties of the materials, and the boundary conditions. Therefore, the correctness of the proposed models and assumptions should be verified against extensive experimental data.

The methods used to solve particular cases (engineering or scientific problems) should be carefully selected. A short overview of the main information regarding analytical and numerical methods is presented below.

2.2. Analytical Methods

The most popular methods used for calculating the horizontal displacements of typical reinforced soil structures are shown in Table 1. The solutions that are commonly used are not very new, with most having been developed in the 1990s. In the 1990s, the level of professional knowledge was different and more limited. All of the factors/parameters that strongly influence the behavior of RS structures were not considered. Typically, they neglect the effects of the time-dependent properties of the materials used as reinforcement. Thus, the results obtained using these methods are usually not very accurate [77]. More detailed information about these analytical methods is presented by Kazimierowicz-Frankowska and Kulczykowski [40].

Table 1. Short description of the most popular analytical methods used to calculate the horizontal deformations of reinforced soil walls.

Reference	Basic Correlations	Assumptions	Notations
[81]	$\delta_h = \frac{\varepsilon_d L}{2}$	δ_h—horizontal deformation of RS walls ε_d—assumed strain limit	the strain limit is established as less than 10%
[73]	$\Delta_h = \left(\frac{1}{2}\right)\left(\frac{P_{rm}}{K_{reinf}}\right)(H - z_i)\cdot$ $\cdot\left[\tan\left(45^0 - \frac{\psi}{2}\right) + \tan\left(90^0 - \phi_{ds}\right)\right]$	Δ_h—deformations of RS wall P_{rm}—max. force Ψ—dilatation angle of soil used as backfill	$\frac{L}{H} \geq 0.7$;
[82]	$D_L = \frac{2b_{q,vol} D_v}{H}$ $\varepsilon_L = \frac{D_L}{b_{q,vol}} = \frac{2D_v}{H} = 2\varepsilon_v$	D_L—horizontal deformation D_V—vertical deformation ε_L—horizontal strain ε_v—vertical strain	the same values of soil and reinforcement horizontal deformations are assumed

2.3. Numerical Methods

Numerical modeling has been successfully used by many researchers to evaluate the behavior of reinforced soil walls [20,23,25,83–88]. Typically, numerical approaches are divided into finite element methods (FEM) and finite difference methods (FDM). In order to perform numerical calculations, it is possible to use many different finite element numerical tools [89]. The most popular are DYNA3D, Plaxis, FLAC, Abaqus, and Ansys.

Table 2 presents information about the typical assumptions made in the numerical modeling of the behavior of RS structures.

Table 2. Typical assumptions in the numerical modeling of the behavior of RS walls.

Reference	Code	Facing Model	Reinforcing Model	Soil Model	Soil/Reinforcement Interfaces	Soil/Wall Interfaces
[43]	FLAC	linear elastic element	elastic-plastic (two-node cable element)	Mohr–Coulomb with hyperbolic stress–strain model	grout material with zero thickness	slip element
[90]	Abaqus	linear elastic manner; eight-node plane strain elements	linear elastic manner; three-node truss elements having no significant compressive or bending strength	Mohr–Coulomb failure criterion and non-associated flow	grout material with zero thickness	slip element
[91]	PLAXIS	linear elastic element	linear elastic geogrid element	stress-dependent hyperbolic Hardening Soil (HS) model	grout material with zero thickness	slip element
[88]	FLAC	linear elastic element	linearly elastic-plastic strip element	Cap-Yield (CY) soil model	linearly elastic-perfectly plastic model with the Mohr–Coulomb failure criterion	linearly elastic-perfectly plastic model with the Mohr–Coulomb failure criterion
[92]	ABAQUS	linearly elastic element	1D bar element, elastoplastic viscoplastic bounding surface model	Drucker–Prager creep model modified with nonlinear and cyclic hysteric behavior	thin layer elements follow Mohr–Coulomb failure criterion	thin layer elements follow Mohr–Coulomb failure criterion
[89]	FLAC3D	linearly elastic element	three-node shell elements; isotropic linear-elastic material	Mohr–Coulomb with hyperbolic stress–strain model	linear spring-slider system	linear spring-slider system
[93]	PLAXIS	three-node beam elements with flexural rigidity and normal stiffness	elastic material	elastoplastic Mohr–Coulomb material	Interface elements with three pairs of nodes, characterized by zero thickness	Interface elements with three pairs of nodes, characterized by zero thickness

3. Laboratory Model Tests

3.1. Experimental Set-Up

The vertical retaining wall of reinforced soil was constructed and loaded at laboratory conditions. The experiment was carried out in a rigid test box with inside dimensions of 100 cm (height) × 37 cm (width) × 180 cm (length). The front and back walls of the box were made of glass (both 2 cm thick) in order to observe deformations of the model during loading. These walls were not lubricated, ensuring that good images of the soil deformations could be captured. The RS model consisted of 10 layers of sand with fabric reinforcement laid between them. The model was characterized by the following dimensions: 50 cm (height) × 60 cm (length) × 37 cm (width). The model of the footing used for the test was made of a 2.5 cm thick rigid steel plate with a rough base of 29 cm × 37 cm. The length of the plate was equal to the width of the tank to maintain plane strain conditions. The footing was connected to a vertical pull-out loading arm with a load cell, which in turn was connected to an electro-mechanical actuator. The footing displacement was measured by displacement transducers. Five horizontal displacement transducers were located along the height of the RS model's facing in order to measure lateral displacements during loading. A schematic view of the test configuration is presented in Figure 3.

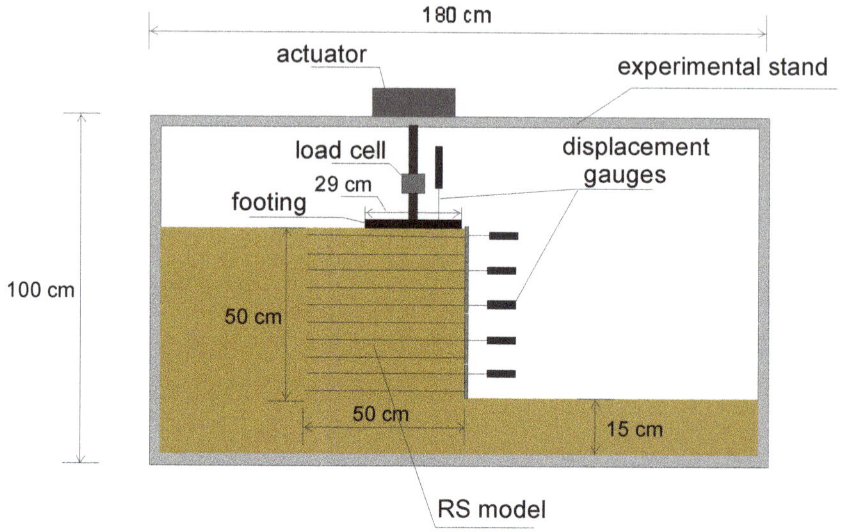

Figure 3. Cross-section of the experimental stand showing the positions of transducers.

3.2. *Characteristics of Materials Used in the Experimental Procedure*

3.2.1. Backfill

Dry silicon sand was selected as the backfill soil in the experimental stand. A sieve analysis was performed in order to determine the particle size distribution curve. Based on the results obtained, the grain size distribution curve of the material used as backfill was plotted (Figure 4). The coefficient of uniformity was 2.0 and the average particle size was 0.15 mm. The unit weight was $\gamma = 18.5 \times 10^3$ N/m^3, the relative density = 0.73, and the void ratio = 0.47 (Table 3). The friction angle of the sand obtained from standard triaxial compression tests was $\phi = 34.5°$ and the cohesion of the soil was $c = 0$ kPa.

Table 3. Properties of the main materials used for construction of the RS wall.

Parameters	Unit	Value
	Soil: sand	
Unit weight γ	kN/m^3	18.5
Friction Angle ϕ	degrees	34.5
Cohesion c	kN/m^2	0
Relative density	-	0.73
Void ratio	-	0.47
Average particle size	mm	0.15
Uniformity coefficient	-	2.0
	Reinforcement: aluminum foil	
thickness	m	18×10^{-6}
ultimate tensile strength R	N/m^2	61×10^6
stiffness E	N/m^2	3533×10^6
elongation at failure	%	3.7
coefficient of friction between the soil and the reinforcement	-	0.05

Table 3. *Cont.*

Parameters	Unit	Value
	Parameters of RS model	
height of the structure H	m	0.5
length of the reinforcement strips L	m	0.5
vertical spacing of the reinforcement ΔH	m	0.05
horizontal spacing of the reinforcement ΔB	m	0.123
width of a single reinforcement strip B	m	0.01

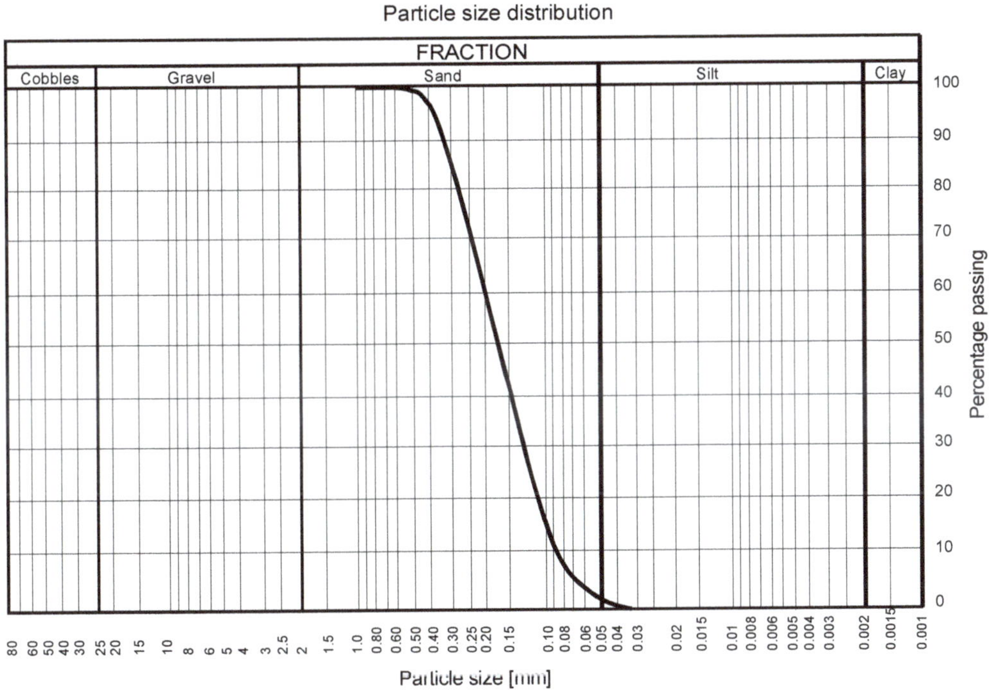

Figure 4. Grain size distribution curve of the material used as backfill.

3.2.2. Reinforcement

Aluminum foil with a thickness of 18×10^{-6} m was used as the reinforcement. The load–elongation behavior of this foil was determined from a standard tension test (Figure 5).

The ultimate tensile strength measured was R = 61×10^6 N/m^2. Reinforcement stiffness reached the value of E = 3533×10^6 N/m^2 and elongation at failure was equal to 3.7%. The coefficient of friction between the backfill and the reinforcement, calculated from a direct shear test, was 0.18. The model reinforcement in each layer consisted of three independent fabric strips (5.0 cm wide and 50 cm long) with a center-to-center spacing of 12.3 cm (Figure 6b).

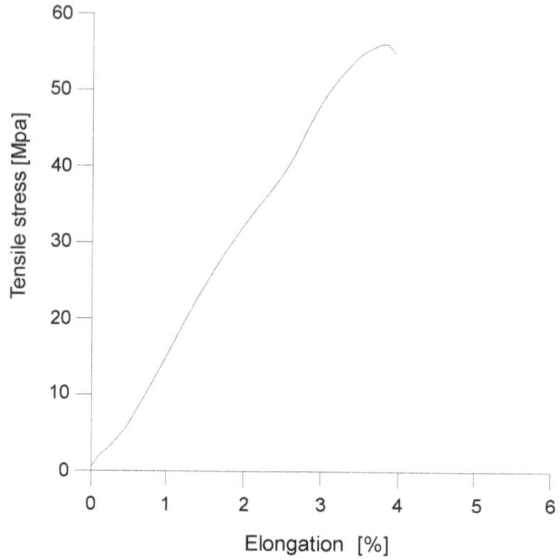

Figure 5. Stress-elongation behavior of the material used as reinforcement.

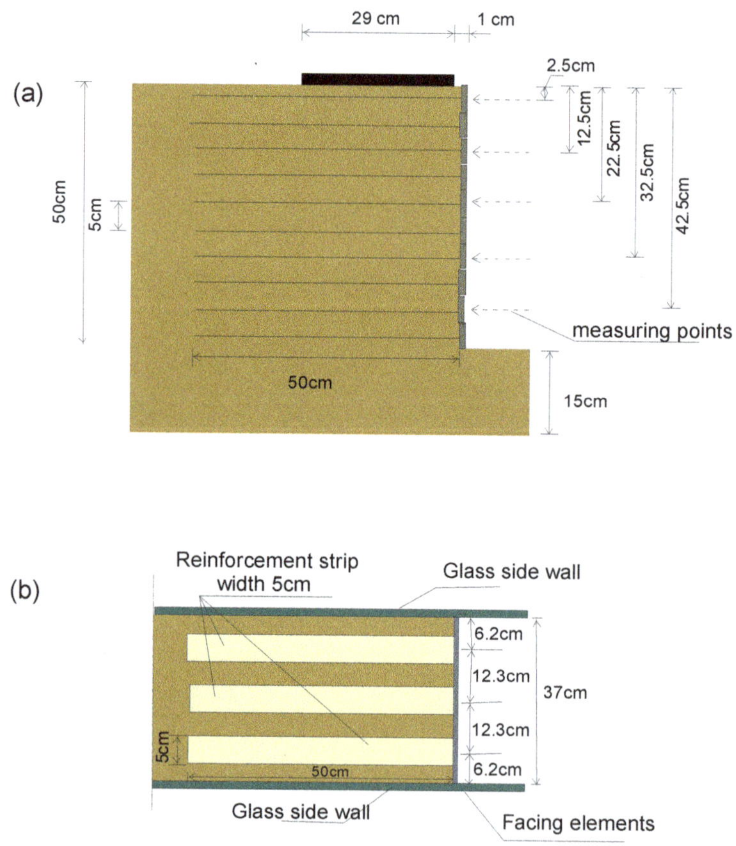

Figure 6. Cross-section of the RS model (**a**); layout of reinforcing strips (**b**).

3.2.3. Facing

The model reinforcement was connected to the elements used as wall facings. The timber panels were 5 cm high, 37 cm long, and 0.7 cm thick. The panels were made of pine wood with a unit weight of $\gamma_w = 4 \text{ kN/m}^3$, a Young's modulus of $E_w = 12 \times 10^6 \text{ kN/m}^2$, and a Poisson ratio of $\nu_w = 0.05$.

3.3. Testing Procedure

The sand was placed into the test box using pluviation technique. The height of fall was 100 cm to obtained the assumed unit weight of the backfill ($\gamma = 18.5 \times 10^3 \text{ N/m}^3$). First, a 15 cm layer of soil foundation was formed at the bottom of test box. Then, a temporary support for the model, in the form of a wooden element, was positioned in front of the wall face in order to keep the facing surface in its initial place during construction. The model was built layer by layer from the bottom to the top, with 5 cm spacing between the soil and reinforcement layers. Each layer of soil was flattened with a grader. After this, a layer of reinforcement (consisting of three strips connected to the facing element) was placed on the sand, and then another sand layer was set down. This procedure was repeated until the structure was complete. Next, after the temporary support was removed, five displacement transducers were fixed on the wall facing at 5 points with regular intervals (Figure 6a). The actuator with the load cell and footing was then placed over the top of the model. The displacement transducers and the load cell were connected to a digital data logger. Finally, loading was applied on top of the structure through a footing that was pushed downwards at a constant displacement rate of 0.22 cm/min. The model was loaded continuously until failure was achieved with continuous measurements of the loading force, footing displacement, and facing deformation. The strip footings were loaded at a rate equal to those employed by other authors [94].

3.4. Experimental Results

The model failed rapidly at an applied vertical load of 3450 N and a vertical footing displacement of 0.5 cm. The relationship between vertical load and footing displacement is presented in Figure 7.

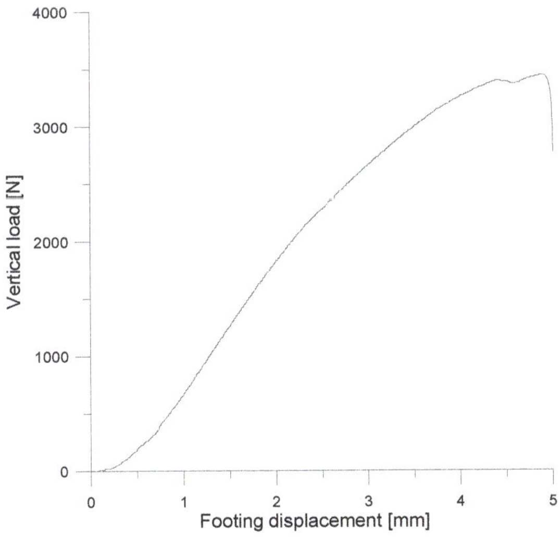

Figure 7. Load vs. footing displacement.

Figure 8 presents horizontal displacements of the facing measured at different loading levels. The same tendency was observed for all cases. The largest values of displacement were measured in the top parts of the wall and diminished as the height of the structure decreased. The minimum horizontal movement was measured at the base of the structure. The measured values of deformations increased with increasing values of external loading.

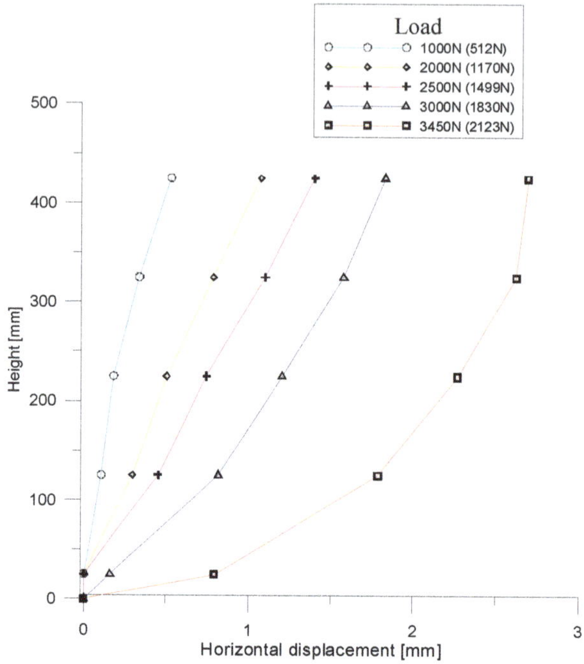

Figure 8. Horizontal deformation of the facing recorded at different loading levels.

The failure surface of the structure determined by the location of the points of tensile rupture of the reinforcement in each layer is presented in Figure 9. The theoretical Rankine failure surface (inclined to the horizontal plane at $\pi/4 + \phi/2$) is also plotted in this figure using a dotted line. It should be noted that the experimental slip surface corresponded reasonably well to this theoretical surface, and the experimental failure zone resembled a triangular wedge.

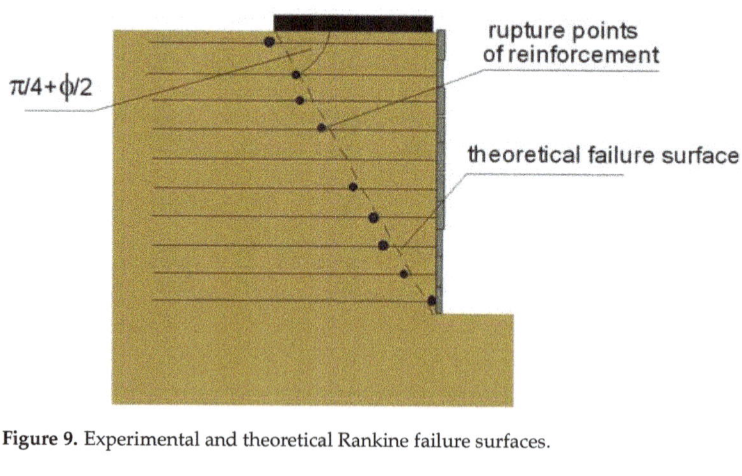

Figure 9. Experimental and theoretical Rankine failure surfaces.

3.5. The Effect of Friction between the Backfill and Side Walls of the Test Box

As previously noted, the front and back walls of the box were not lubricated. Therefore, the effect of friction between the backfill and the side walls on the load force should be taken into account. In order to estimate the increase in load due to such frictional resistance, a simple approach proposed by Kulczykowski [95] was used.

The resistance force (P_F) caused by the side wall friction can be estimated using the following formula, see [95]:

$$P_F = a^2 \tan\alpha \left[\tan\delta(1-\sin\phi)\left(\frac{a\gamma\tan\alpha}{3} + \frac{P_{E+F}}{aL}\right) + c_\delta \right], \quad (1)$$

where

$$a = \frac{H_P}{\tan\alpha}, \quad (2)$$

α is the angle of inclination of the slip surface to the horizontal plane, ϕ is the friction angle of the sand, δ is the angle of friction between the backfill and the side wall surface, and c_δ is the resistance resulting from adhesion (in N/m^2). P_{E+F} is the load force recorded during the experiment (in N), H_E is the height of the failure zone, and ΔL denotes the width of the test box (both in m) (Figure 10).

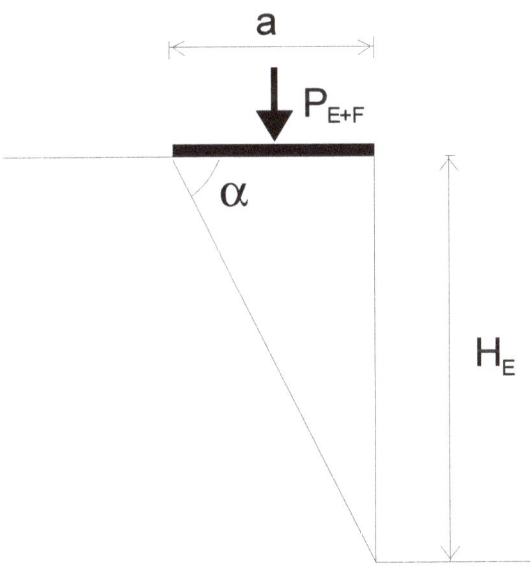

Figure 10. Triangular contact area between the sand and the side wall in the failure zone.

The value of the experimental load (P_E) reduced by the friction effect can be calculated using the following equation:

$$P_E = P_{E+F}\left[1 - \frac{a}{\Delta L}\tan\delta\tan\alpha(1-\sin\phi)\right] - a^2\tan\alpha\left[\frac{1}{3}(1-\sin\phi)a\gamma\tan\delta\tan\alpha + c_\delta\right] \quad (3)$$

The value of δ was determined by the method presented in [96] (a "tipping experiment"). A layer of sand was placed on the surface of sheet glass, which was the same as that used for the side walls of the test box. Then the sheet glass was slowly tilted. The angle of tipping was gradually increased until the soil mass began to slide. This angle of inclination was taken as the angle of friction δ between the sand and the glass surface.

The analysis of the digital images captured during a similar test and processed by particle image velocimetry (PIV), presented in [96], showed that a deformation zone with a shape similar to that of the failure zone formed even at an early loading stage. Therefore, it was assumed that the above relationship can be used to calculate the value of P_E at any stage of loading.

The following data, corresponding to the experimental conditions, were used to calculate the value of the load (P_E) at seven different stages of experimental loading: P_{E+F}: $H_E = 0.5$ m, $\phi = 34.5°$, $\gamma = 18.5 \times 103$ N/m^3, $\delta = 26°$, and $c_\delta = 0$ N/m^2. The results of these calculations are presented in Table 4.

Table 4. Recorded reduced external load.

P_{E+F} [N]	P_E [N]
500	184
1000	512
1500	841
2000	1170
2500	1499
3000	1830
3450	2123

4. Theoretical Analysis: Analytical Approach

4.1. Basic Assumptions

The following initial assumptions were made [18,40,50]:

- The typical cross-section of the RS retaining wall was taken into consideration (Figure 11).

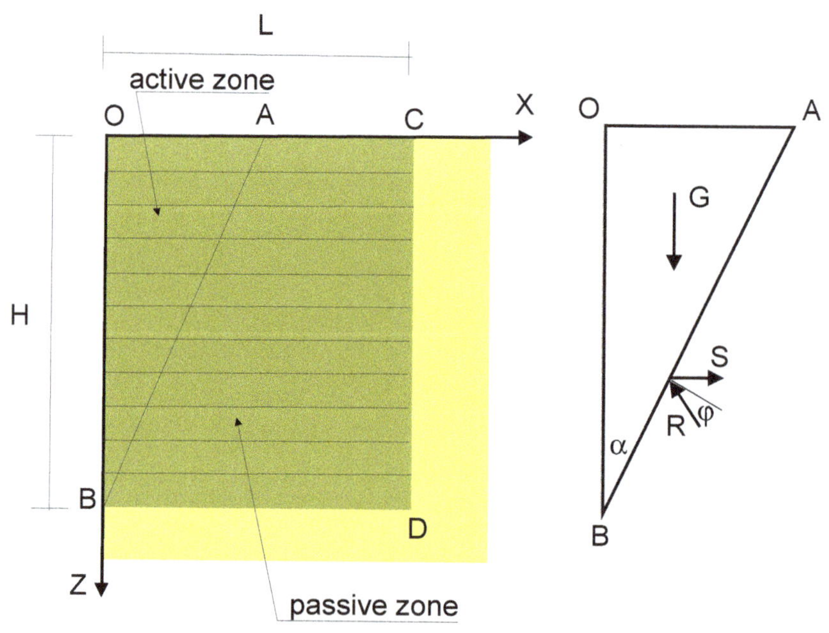

(**a**) (**b**)

Figure 11. General model of the RS wall (**a**) and equilibrium of the failure wedge (**b**).

- The soil was in a plastic state within the potential failure wedge *ABO* (active zone) and was rigid outside this area (passive zone).

- Perfect bonding between the soil and the reinforcement was assumed (slippage of the reinforcement did not occur).
- The increasing external load (acting on the top of the structure) was, at first, smaller than, and eventually equal to, the collapse load.
- The slippage of the wedge occurred along a planar failure surface that passed through the toe of the structure.
- The horizontal displacement (u_x) of the facing of the model RS wall consisted of two parts:

$$u_x = u_{act} + u_{pass} \tag{4}$$

where u_{act} is the displacement resulting from the deformation of the reinforcement in the active zone, and u_{pass} is the displacement resulting from the deformation in the rigid (passive) zone (caused by pull-out, hence $u_{pass} = u_{pullout}$).

4.2. Deformation in the Plastic Zone

The following constitutive relationship between tensile force and strain can be assumed in the elastic range (Hooke's law):

$$F = E\varepsilon_{act} \tag{5}$$

where $F = A_r\sigma_X^r$ is the force in the reinforcement, $E = A_rE_r$ is the elastic stiffness of the reinforcement, and A_r is the cross-sectional area of the reinforcement.

The horizontal displacement of the RS facing caused by the elastic deformation of the reinforcement in the active zone is determined by the integration of strains:

$$u_{act} = \int_0^{x^*} \varepsilon_{act}dx \tag{6}$$

where x^* denotes the length of the reinforcement in the active zone (Figure 12)

$$x^* = (H - z)\tan\alpha \tag{7}$$

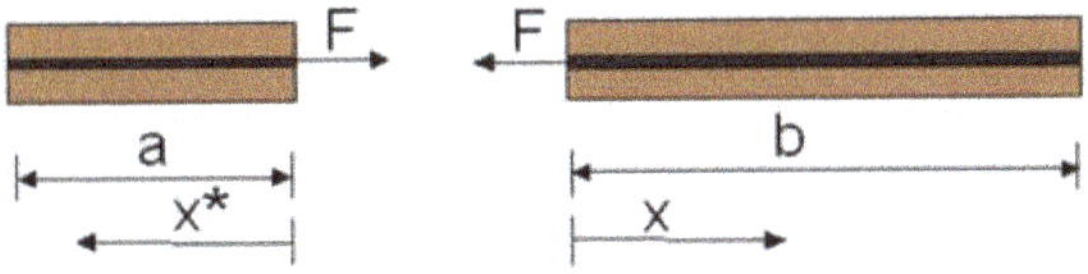

Figure 12. Overview of forces acting on the reinforcing element.

4.3. Deformation in the Rigid Zone

Displacements in this zone are caused by pull-out of the reinforcement. Sawicki [18], Kazimierowicz-Frankowska [50], and Kazimierowicz-Frankowska and Kulczykowski [40] presented detailed formulas which can be used to calculate these kinds of deformations.

For this purpose, the following governing equations can be used:

(1) The differential equation derived from the equilibrium of the reinforcing element (Figure 13) [18,50]:

$$\frac{dF}{dx} = -2B\tau, \tag{8}$$

where F denotes the tensile strength of the reinforcement, and B is the width of the reinforcement strip.

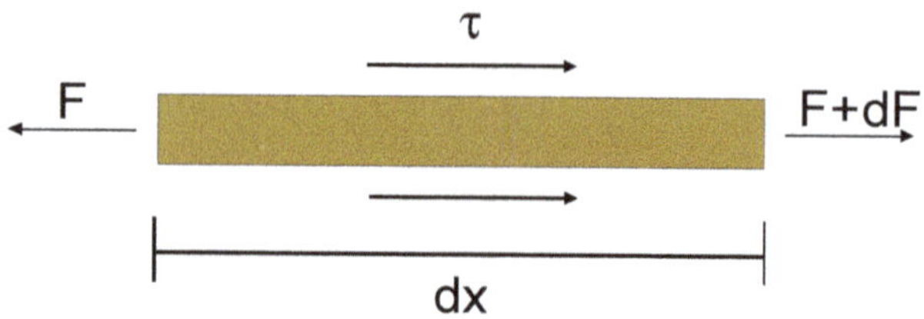

Figure 13. Reinforcing strip.

(2) The equation describing strain in the reinforcement strip:

$$\varepsilon = {}^{du}/_{dx}, \tag{9}$$

where u is displacement.

(3) The correlation between sheer stress (τ) and displacement (u):

$$\tau = -Gu, \tag{10}$$

where G denotes a coefficient of proportionality.

(4) The relationship between tensile strength and strain:

$$F = E\varepsilon, \tag{11}$$

where E is the stiffness of the reinforcement strip.

Equations (8)–(11) give rise to the differential equation, which shows the distribution of forces along the reinforcement strip:

$$\frac{d^2F}{dx^2} - \beta^2 F = 0 \tag{12}$$

where $\beta = \sqrt{\frac{2BG}{E}}$. The solution for Equation (8) is the following function:

$$F = C_1 e^{-\beta x} + C_2 e^{\beta x}, \tag{13}$$

where C_1 and C_2 are coefficients determined from the corresponding boundary conditions:

$$F(x = 0) = F \text{ and } F(x = b) = 0 \tag{14}$$

It is possible to estimate the displacement ($u_{pullout}$) of the reinforcement strip using Equations (8), (10), (13), and (14).

$$u_{pullout} = -\frac{F}{\beta E[1 - \exp(-2\beta b)]} \cdot [\exp(-\beta x) + \exp(-2\beta b)\exp(\beta x)] \tag{15}$$

where b is the length of the reinforcement strip in the rigid zone (b = L − a), compared to Figures 11 and 12. The coefficient G (Equation (10)) can be calculated using $\beta = \sqrt{\frac{2BG}{E}}$. β can be determined using the following equation:

$$C = -\frac{\beta E[1 - \exp(-2\beta b)]}{1 + \exp(-2\beta b)} \tag{16}$$

5. Comparison of Experimental and Theoretical Results

Verification of the Accurate Prediction of Experimental Results

Firstly, the experimental values of the horizontal displacements were compared with their analytical predictions. The experimental results took into account the effect of friction between the backfill and the side walls of the test box. It should be also noted that the horizontal displacement caused only by external loading was measured during the experiment.

The results obtained are listed in Table 5 and presented in Figure 14. It can be seen that the experimentally recorded horizontal displacements are lower than the predicted horizontal displacements. The largest differences between the recorded and calculated values of deformation were observed at the top of the model. This was due to friction between the sand and the lower surface of the footing, which significantly reduced the horizontal deformation in this zone. However, compatibility between the experimental results and their theoretical predictions was considered satisfactory (i.e., acceptable for engineering practice). The average difference between the measured and calculated values of horizontal deformation was not greater than twenty five percent. Differences between the model results and the realistic behavior of the RS model structure may result from errors and inaccuracies that arose when the key material parameters used for the construction of the model wall were determined, or over the course of the experiments.

Table 5. Calculated (u_T) and experimental (u_E) displacement of the facing at subsequent load levels.

Dist. from the Top [m]	Displacement of the Facing at Subsequent Load Levels [mm]													
	184 N		512 N		841 N		1170 N		1499 N		1830 N		2123 N	
	u_T	u_E	u_T	u_E	u_T	u_E	u_T	u_E	u_T	u_E	u_T	u_E	u_T	u_E
0.025	0.31	0.25	0.86	0.53	1.42	0.79	1.97	1.07	2.53	1.40	3.09	1.83	3.58	2.69
0.125	0.25	0.14	0.70	0.34	1.15	0.54	1.61	0.79	2.06	1.10	2.51	1.58	2.91	2.62
0.225	0.19	0.05	0.54	0.18	0.89	0.32	1.24	0.50	1.59	0.74	1.94	1.20	2.25	2.27
0.325	0.14	0.03	0.38	0.11	0.62	0.19	0.87	0.30	1.11	0.45	1.36	0.82	1.58	1.79
0.425	0.08	0.00	0.22	0.00	0.36	0.00	0.50	0.00	0.64	0.00	0.78	0.17	0.91	0.80

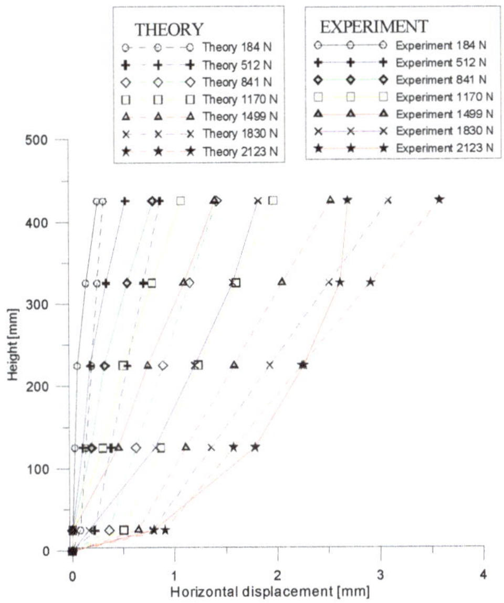

Figure 14. Comparison of experimental results with their analytical predictions.

The predicted horizontal displacements were larger than those that were experimentally recorded. From the point of view of design standards, there is a tendency for this to occur. Our results confirm that the proposed analytical method upholds the principles of safe design, which includes safety factors related to the potential occurrence of unexpected events.

6. Conclusions and Discussion

This paper aimed to investigate the accuracy with which horizontal deformations in the RS wall are modelled. This was achieved by comparing predictions of horizontal deformations with the experimental results obtained after constructing an instrumented RS wall on a laboratory scale. The displacements were recorded under external loading on top of the structure, and the theoretical analysis was carried out using an analytical approach.

The main findings of this study are as follows:

- The experimental investigations of the reinforced soil wall made it possible to observe the magnitude and pattern of horizontal displacements under a sustained external load at the top of the model. The largest displacements occurred in the top layers of the RS structure and gradually decreased as the height of the model wall decreased. Our results were similar to those observed by other researchers [40,50,51].
- An important and new insight that this paper introduces is the effect of friction between the backfill and the side walls of the test box on displacement. The effects of this factor tend to be omitted, or only generally discussed, by the authors of other experiments. However, our results show that friction has a significant impact on the value of external loading actually acting on the structure. In the analyzed case, even with the highest value of external loading on top of the structure, minimal reductions in the value of external loading actually acting on the structure were above 60% (Table 4). For example, the minimal reduction was 62.5% for an external load of 3450 N, 63.9% for an external load of 3000 N, and 66.8% for an external load of 2500 N. Therefore, the results clearly show that friction between the backfill and the side walls of the test box significantly influenced the horizontal displacement of the RS wall during the tests.
- Our results suggest that the proposed analytical method can be used as an alternative approach to other analytical and numerical methods used to model the deformations of RS wall structures. Although far from exhaustive, the first verification of the accuracy of the IBW PAN method produced promising results.
- The average difference between the recorded and calculated values of deformation did not exceed twenty five percent. Discrepancies between the model predictions and the experimental results may have resulted from inaccurate soil and reinforcement data.
- The theoretical horizontal displacements that were analytically predicted for the model wall were larger than the experimental horizontal displacements. This tendency is correct and shows that the proposed analytical method can be used without breaching safety rules.
- More complex constitutive soil models are investigated in the literature, but these models require input properties that are seldom available to design engineers. Moreover, greater accuracy in the predictions for more advanced models may not be assured [92]. Considering the accuracy of the results obtained, it should be stated that the proposed approach (and the models selected) are adequate from an engineering point of view.
- The final results may be particularly useful when designing the reinforced soil structures used in transport engineering (e.g., bridge abutments), where the deformation limit values cannot be exceeded. In such cases, the key factor determining the permissible durability of the structures is the second limit state of use (related to the size of the deformations occurring).

Author Contributions: Conceptualization, K.K.-F. and M.K.; Investigation, M.K.; Methodology, K.K.-F.; Resources, K.K.-F.; Supervision, K.K.-F.; Visualization, M.K.; Writing—original draft, K.K.-F.; Writing—review & editing, M.K. All authors have read and agreed to the published version of the manuscript.

Funding: This research received no external funding.

Institutional Review Board Statement: Not applicable.

Informed Consent Statement: Not applicable.

Data Availability Statement: The data presented in this study are available on request from the authors.

Conflicts of Interest: The authors declare no conflict of interest.

References

1. Allen, T.; Bathurst, R. Soil reinforcement loads in geosynthetic walls at working stress conditions. *Geosynth. Int.* **2002**, *9*, 525–566. [CrossRef]
2. Lavasan, A.A.; Ghazavi, M. Behavior of closely spaced square and circular footings on reinforced sand. *Soils Founduation* **2012**, *52*, 160–167. [CrossRef]
3. Mehrjardi, G.T.; Ghanbari, A.; Mehdizadeh, H. Experimental study on the behaviour of geogrid-reinforced slopes with respect to aggregate size. *Geotext. Geomembr.* **2016**, *44*, 862–871. [CrossRef]
4. Sun, X.; Han, J.; Corey, R. Equivalent modulus of geogrid-stabilized granular base back-calculated using permanent deformation. *J. Geotech. Geoenviron. Eng.* **2017**, *143*, 06017012. [CrossRef]
5. Weerasekara, L.; Hall, B.; Wijewickreme, D. A new approach for estimating internal stability of reinforced soil structures. *Geosynth. Int.* **2017**, *24*, 419–434. [CrossRef]
6. Javankhoshdel, S.; Bathurst, R.J. Deterministic and probabilistic failure analysis of simple geosynthetic reinforced soil slopes. *Geosynth. Int.* **2017**, *24*, 14–29. [CrossRef]
7. Mehrjardi, G.T.; Motarjemi, F. Interfacial properties of geocell-reinforced granular soils. *Geotext. Geomembr.* **2018**, *46*, 384–395. [CrossRef]
8. Abd, A.H.; Utili, S. Design of geosynthetic-reinforced slopes in cohesive backfills. *Geotext. Geomembr.* **2017**, *45*, 627–641. [CrossRef]
9. Wang, J.Q.; Zhang, L.L.; Xue, J.F. Load-settlement response of shallow square footings on geogrid-reinforced sand under cyclic loading. *Geotext. Geomembr.* **2018**, *46*, 586–596. [CrossRef]
10. Xu, C.; Liang, C.; Shen, P. Experimental and theoretical studies on the ultimate bearing capacity of geogrid-reinforced sand. *Geotext. Geomembr.* **2019**, *47*, 417–428. [CrossRef]
11. Rowe, R.K.; Yu, Y. Magnitude and significance of tensile strains in geomembrane landfill liners. *Geotext. Geomembr.* **2019**, *47*, 439–458. [CrossRef]
12. Chehade, H.A.; Dias, D.; Sadek, M. Seismic analysis of geosynthetic- reinforced retaining wall in cohesive soils. *Geotext. Geomembr.* **2019**, *47*, 315–326. [CrossRef]
13. Bildik, S.; Laman, M. Effect of geogrid reinforcement on soil—Structure–pipe interaction in terms of bearing capacity, settlement and stress distribution. *Geotext. Geomembr.* **2020**, *48*, 844–853. [CrossRef]
14. Vidal, H. The principle of reinforced earth. *Highw. Res. Rec.* **1969**, 1–16. Available online: https://onlinepubs.trb.org/Onlinepubs/hrr/1969/282/282-001.pdf (accessed on 14 May 2022).
15. Lee, K.L.; Adams, B.D.; Vagneron, J.M.J. Reinforced earth retaining walls. *J. Soil Mech. Found. Div.* **1973**, *99*, 745–764. [CrossRef]
16. Schlosser, F. Mechanically stabilized earth retaining structures in Europe. In *Design and Performance of Earth Retaining Structures*; Lambe, P., Hansen, L.A., Eds.; Geotechnical Special Publication; ASCE: Reston, VA, USA, 1990; pp. 347–378.
17. Jones, C.J.F.B. *Earth Reinforcement and Soil Structures (Butterworth's Advanced Series in Geotechnical Engineering)*; Butterworth-Heinemann: London, UK, 1996; p. 379.
18. Sawicki, A. *Mechanics of Reinforced Soil*; CRC Press: Brookfield, VT, USA, 2000.
19. Karpurapu, R.; Bathurst, R.J. Behavior of geosynthetic reinforced soil retaining walls using the finite element analysis. *Comput. Geotech.* **1995**, *17*, 179–299. [CrossRef]
20. Ho, S.K.; Rowe, R.K. Effect of wall geometry on the behaviour of reinforced soil walls. *J. Geotext. Geomembr.* **1996**, *14*, 521–541. [CrossRef]
21. Rowe, R.K.; Ho, S.K. Continuous panel reinforced soil walls on rigid foundations. *J. Geotech. Geoenviron. Eng.* **1997**, *123*, 912–920. [CrossRef]
22. Rowe, R.K.; Skinner, G.D. Numerical analysis of geosynthetic reinforced retaining wall constructed on a layered soil foundation. *Geotext. Geomembr.* **2001**, *19*, 387–412. [CrossRef]
23. Ling, H.I.; Leshchinsky, D. Finite element parametric study of the behavior of segmental block reinforced-soil retaining walls. *Geosynth. Int.* **2003**, *10*, 77–94. [CrossRef]
24. Hatami, K.; Bathurst, R.J. Development and verification of a numerical model for the analysis of geosynthethic-reinforced soil segmental walls under working stress conditions. *Can. Geotech. J.* **2005**, *42*, 1066–1085. [CrossRef]

25. Hatami, K.; Bathurst, R.J. Numerical model for reinforced soil segmental walls under surcharge loading. *J. Geotech. Geoenviron. Eng.* **2006**, *132*, 673–684. [CrossRef]
26. Skinner, G.D.; Rowe, R.K. Design and Behaviour of a Geosynthetic Reinforced Retaining Wall and Bridge Abutment on a Yielding Foundation. *Geotext. Geomembr.* **2005**, *23*, 234–260. [CrossRef]
27. Chungsik, Y.; Kim, S.-B. Performance of a two-tier geosynthetic reinforced segmental retaining wall under a surcharge load: Full-scale load test and 3D finite element analysis. *Geotext. Geomembr.* **2008**, *26*, 460–472.
28. Mendonça, A.; Lopes, M.; Pinho-Lopes, M. Construction and post-construction behaviour of a geogrid-reinforced steep slope. *Geotech. Geol. Eng.* **2003**, *21*, 129–147. [CrossRef]
29. Bathurst, R.J.; Nernheim, A.; Walters, D.L.; Allen, T.M.; Burgess, P.; Saunders, D.D. Influence of reinforcement stiffness and compaction on the performance of four geosynthetic-reinforced soil walls. *Geosynth. Int.* **2009**, *16*, 43–59. [CrossRef]
30. Kongkitkul, W.; Tatsuoka, F.; Kawahata, S.; Hirakawa, D.; Sugimoto, T.; Ito, M. Time histories of tensile force in geogrid arranged in two full-scale high walls. *Geosynth. Int.* **2010**, *17*, 12–32. [CrossRef]
31. Horpibulsuk, S.; Suksiripattanapong, C.; Niramitkornburee, A.; Chinkulkijniwat, A.; Tangsutthinon, T. Performance of an earth wall stabilized with bearing reinforcements. *Geotext. Geomembr.* **2011**, *29*, 514–524. [CrossRef]
32. Vieira, C.S.; Lopes, M.D.L.; Caldeira, L.M. Earth pressure coefficients for design of geosynthetic reinforced soil structures. *Geotext. Geomembr.* **2011**, *29*, 491–501. [CrossRef]
33. Yang, G.; Liu, H.; Lv, P.; Zhang, B. Geogrid-reinforced lime-treated cohesive soil retaining wall: Case study and implications. *Geotext. Geomembr.* **2012**, *35*, 112–118. [CrossRef]
34. Yang, G.; Liu, H.; Zhou, Y.T.; Xiong, B.L. Post-construction performance of a two-tiered geogrid reinforced soil wall backfilled with soil-rock mixture. *Geotext. Geomembr.* **2014**, *42*, 91–97. [CrossRef]
35. Riccio, M.; Ehrlich, M.; Dias, D. Field monitoring and analyses of the response of a block-faced geogrid wall using fine-grained tropical soils. *Geotext. Geomembr.* **2014**, *42*, 127–138. [CrossRef]
36. Benjamim, C.V.S.; Bueno, B.S.; Zornberg, J.G. Field monitoring evaluation of geotextile-reinforcedsoil-retaining walls. *Geosynth. Int.* **2007**, *14*, 100–118.
37. Liu, J.; Liu, M.; Zhu, Z. Sand deformation around an uplift plate anchor. *J. Geotech. Geoenviron. Eng.* **2011**, *138*, 728–737. [CrossRef]
38. Portelinha, F.H.M.; Zornberg, J.G.; Pimentel, V. Field performance of retaining walls reinforced with woven and nonwoven geotextiles. *Geosynth. Int.* **2014**, *21*, 270–284. [CrossRef]
39. Scotland, I.; Dixon, N.; Frost, M.; Fowmes, G.; Horgan, G. Modelling deformation during the construction of wrapped geogrid-reinforced structures. *Geosynth. Int.* **2016**, *23*, 219–232. [CrossRef]
40. Kazimierowicz-Frankowska & Kulczykowski, 2021 Deformation of model reinforced soil structures: Comparison of theoretical and experimental results. *Geotext. Geomembr.* **2021**, *49*, 1176–1191.
41. Adams, M.T.; Ooi, P.S.K.; Nicks, J.E. Mini-pier testing to estimate performance of full-scale geosynthetic reinforced soil bridge abutments. *Geotech. Test. J. ASTM Int.* **2014**, *37*, 884–894. [CrossRef]
42. Eliahu, U.; Watt, S. Geogrid-Reinforced Wall Withstands Earthquake. In *Geotechnical Fabrics Report*; IFAI: Saint Paul, MN, USA, 1991; Volume 9, pp. 8–13.
43. Fakharian, K.; Attar, I. Static and seismic numerical modeling of geosynthetic- reinforced soil segmental bridge abutments. *Geosynth. Int.* **2007**, *14*, 228–243. [CrossRef]
44. Helwany, S.M.; Wu, J.T.; Froessl, B. GRS bridge abutments—An effective means to alleviate bridge approach settlement. *Geotext. Geomembr.* **2003**, *21*, 177–196. [CrossRef]
45. Lee, K.Z.; Wu, J.T. A synthesis of case histories on GRS bridge-supporting structures with flexible facing. *Geotext. Geomembr.* **2004**, *22*, 181–204. [CrossRef]
46. Won, G.; Hull, T.; De Ambrosis, L. *Performance of a Geosynthetic Segmental Block Wall Structure to Support Bridge Abutments*; Ochiai, H., Yasufuku, N., Omine, K., Eds.; Earth Reinforcement: Rotterdam, The Netherlands, 1996; pp. 543–548.
47. Xiao, C.; Han, J.; Zhang, Z. Experimental study on performance of geosynthetic- reinforced soil model walls on rigid foundations subjected to static footing loading. *Geotext. Geomembr.* **2016**, *44*, 81–94. [CrossRef]
48. Zheng, Y.; Fox, P.J. Numerical investigation of geosynthetic-reinforced soil bridge abutments under static loading. *J. Geotech. Geoenviron. Eng.* **2016**, *142*, 1–18. [CrossRef]
49. Zheng, Y.; Fox, P.; Shing, P.B.; McCartney, J.S. Physical model tests of half- scale geosynthetic reinforced soil bridge abutments. I: Static loading. *J. Geotech. Geoenviron. Eng.* **2019**, *145*, 04019094. [CrossRef]
50. Kazimierowicz-Frankowska, K. Deformation of model RS retaining walls due to creep and reinforcement pull-out. *Geosynth. Int.* **2003**, *10*, 153–164. [CrossRef]
51. Kazimierowicz-Frankowska, K. Deformations of reinforced-soil retaining walls. In Proceedings of the 11th International Conference on Geosynthetics, Seoul, Korea, 16–21 September 2018.
52. Selvadurai, A.; Gnanendran, C. An experimental study of a footing located on a sloped fill: Influence of a soil reinforcement layer. *Can. Geotech. J.* **1989**, *26*, 467–473. [CrossRef]
53. Huang, C.; Tatsuoka, F.; Sato, Y. Failure mechanism of reinforced sand slopes loaded with footing. *Soil Found.* **1994**, *24*, 27–40. [CrossRef]
54. Bathurst, R.J. Lessons learned from full scale model tests of reinforced walls and slopes: Keynote paper. In Proceedings of the 2nd Asian Regional Conference on Geosynthetics, Kuala Lumpur, Malaysia, 29–31 May 2000; Volume 1, pp. 1–22.

55. Lee, K.M.; Manjunath, V.R. Experimental and numerical studies of geosynthetic reinforced sand slopes loaded with footing. *Can. Geotech. J.* **2000**, *37*, 828–842. [CrossRef]
56. Yoo, C. Laboratory investigation of bearing capacity behavior of strip footing on geogrid reinforced sand slope. *Geotext. Geomembr.* **2001**, *19*, 298. [CrossRef]
57. Bathurst, R.J.; Jones, C.J.F.P. Earth retaining structures and reinforced slopes. In *Geotechnical and Geoenvironmental Engineering Handbook*; Rowe, R.K., Ed.; Kluwer Academic Publishing: Norwell, MA, USA, 2001; Chapter 17.
58. Bathurst, R.J.; Blatz, J.A.; Burger, M.H. Performance of instrumented large-scale unreinforced and reinforced embankments loaded by a strip footing to failure. *Can. Geotech. J.* **2003**, *40*, 1067–1083. [CrossRef]
59. El Sawwaf, M. Behaviour of strip footing on geogrid reinforced sand over a soft clay slope. *Geotext. Geomembr.* **2007**, *25*, 50–60. [CrossRef]
60. Alamshahi, S.; Hataf, N. Bearing capacity of strip footings on sand slopes reinforced with geogrid and grid anchors. *Geotext. Geomembr.* **2009**, *27*, 217–226. [CrossRef]
61. Mittal, S.; Shah, M.Y.; Verma, N.K. Experimental study of footing on reinforced earth slope. *Int. J. Geotech. Eng.* **2009**, *3*, 251–260. [CrossRef]
62. Shukla, S.K.; Sivakugan, N.; Das, B.M. A state of the art review of geosynthetic reinforced slopes. *Int. J. Geotech. Eng.* **2011**, *5*, 17–32. [CrossRef]
63. Bathurst, R.J.; Hatami, K.; Alfaro, M.C. Geosyntheticreinforced soil walls and slopes-seismic aspects. In *Handbook of Geosynthetic Engineering*; Shukla, S.K., Ed.; ICE Publishing: London, UK, 2012; pp. 317–369.
64. Gill, K.S.; Choudhary, A.K.; Jha, J.N.; Shukla, S.K. Experimental and numerical studies of loaded strip footing resting on reinforced fly ash slope. *Geosynth. Int.* **2013**, *20*, 13–25. [CrossRef]
65. Gawali, S.L.; LGKalurkar, L.G. Deformation Behaviour of Geogrid Reinforced Soil Walls under Strip Loading. *Int. J. Eng. Res. Technol. (IJERT)* **2020**, *9*, 23–30.
66. Riccio, M.; Ehrlich, M. *Engineered Fabrics*; Singh, M.K., Ed.; IntechOpen: London, UK, 2018.
67. Lu, L.; Lin, H.; Wang, Z.; Xiao, L.; Ma, S.; Arai, K. Experimental and Numerical Investigations of Reinforced Soil Wall Subjected to Impact Loading. *Rock Mech. Rock Eng.* **2021**, *54*, 5651–5666. [CrossRef]
68. Jia, X.; Xu, J.; Sun, Y. Deformation Analysis of Reinforced Retaining Wall Using Separate Finite Element. *Discret. Dyn. Nat. Soc. (DDNS)* **2018**, *2018*, 6946492. [CrossRef]
69. Wu, J.T.H. *Geosynthetic Reinforced Soil (GRS) Walls*, 2nd ed.; Wiley Blackwell: Hoboken, NJ, USA, 2020.
70. Yoo, C.; Tabish, A.; Yang, J.W.; Abbas, Q.; Song, J.S. Effect of internal drainage on deformation behavior of GRS wall during rainfall. *Geosynth. Int.* **2022**, *29*, 137–150. [CrossRef]
71. Adams, M.T.; Ketchart, K.; Wu, J.T.H. Mini Pier Experiments: Geosynthetic Reinforcement Spacing and Strength as Related to Performance. *Geosynth. Reinf. Hydraul. Appl.* **2007**, 165.
72. Pham, T.Q. Investigating Composite Behavior of Geosynthetic Reinforced Soil (GRS). Ph.D. Thesis, University of Colorado, Denver, CO, USA, 2009.
73. Wu, J.T.H.; Pham, T.Q. Load-carrying capacity and required reinforcement strength of closely spaced soil-geosynthetic composites. *J. Geotech. Geoenviron. Eng.* **2013**, *139*, 1468–1476. [CrossRef]
74. Elton, D.J.; Patawaran, M.A.B. Mechanically stabilized earth reinforcement tensile trength from tests of geotextile-reinforced soil. *Transport. Res. Rec. J. Transp. Res. Board* **2004**, *1868*, 81–88. [CrossRef]
75. Shen, P.; Jie Hanb Zornberg, J.G.; Morsy, A.M.; Leshchinsky, D.; Tanyu, B.F.; Xu, C. Two and three-dimensional numerical analyses of geosynthetic-reinforced soil (GRS) piers. *Geotext. Geomembr.* **2019**, *47*, 352–368. [CrossRef]
76. Nicks, J.E.; Esmaili, D.; Adams, M.T. Deformations of geosynthetic reinforced soil under bridge service loads. *Geotext. Geomembr.* **2016**, *44*, 641–653. [CrossRef]
77. Khosrojerdi, M.; Xiao, M.; Qiu, T.; Nicks, J. Evaluation of prediction methods for lateral deformation of GRS walls and abutments. *J. Geotech. Geoenviron. Eng.* **2017**, *143*, 1–8. [CrossRef]
78. Ozturk, T.E. Artificial neural networks approach for earthquake deformation determination of geosynthetic reinforced retaining walls. *Int. J. Intell. Syst. Appl. Eng.* **2014**, *2*, 1–9. [CrossRef]
79. Wu, J.T.; Pham, T.Q.; Adams, M.T. *Composite Behaviour of Geosynthetic Reinforced Soil Mass*; Report No. FHWA-HRT-10-077; Office of Infrastructure Research and Development: McLean, VA, USA, 2013.
80. Kazimierowicz-Frankowska, K. A case study of geosynthetic reinforced wall with wrap-around facing. *Geotext. Geomembr.* **2005**, *23*, 107–115. [CrossRef]
81. Giroud, J.P.; Fluet, J.E., Jr.; Gross, B.A. *FHWA-HI-89-002: Geotextile Engineering Workshop-Design Examples*; Federal Highway Administration: Washington, DC, USA, 1989.
82. Adams, M.T.; Lillis, C.P.; Wu, J.T.H.; Ketchart, K. Vegas Mini Pier experiment and postulate of zero volume change. In *Proceedings of the 7th International Conference Geosynthetics*; Swets and Zeitinger: Lisse, The Netherlands, 2002; pp. 389–394.
83. Christopher, B.R.; Gill, S.A.; Giroud, J.P.; Juran, I.; Scholsser, F.; Mitchell, J.K.; Dunnicliff, J. Summary of Research and Systems Information. In *Reinforced Soil Structures*; Report No. FHWA-RD-89-043; Federal Highway Administration: Washington, DC, USA, 1989; Volume 2, p. 287.
84. Guler, E.; Hamderi, M.; Demirkan, M.M. Numerical analysis of reinforced soilretaining wall structures with cohesive and granular backfills. *Geosynth. Int.* **2007**, *14*, 330–345. [CrossRef]

85. Huang, B.Q.; Bathurst, R.J.; Hatami, K. Numerical study of reinforced soil segmental walls using three different constitutive soil models. *J. Geotech. Geoenviron. Eng.* **2009**, *135*, 1486–1498. [CrossRef]
86. Huang, J.; Parsons, R.L.; Han, J.; Pierson, M.C. Numerical analysis of a laterally loaded shaft constructed within an MSE wall. *Geotext. Geomembr.* **2011**, *29*, 233–241. [CrossRef]
87. Huang, J.; Han, J.; Parsons, R.L.; Pierson, M.C. Refined numerical modeling of a laterally-loaded drilled shaft in an MSE wall. *Geotext. Geomembr.* **2013**, *37*, 61–73. [CrossRef]
88. Jiang, Y.; Han, J.; Parsons, R.L. Numerical evaluation of secondary reinforcement effect on geosyntheticreinforced retaining walls. *Geotext. Geomembr.* **2020**, *48*, 98–109. [CrossRef]
89. Bhattacharjee, A.; Krishna, A.M. Development of numerical model of wrap-faced walls subjected to seismic excitation. *Geosynth. Int.* **2012**, *19*, 354–369. [CrossRef]
90. Yoo, C. Serviceability state deformation behaviour of two-tiered geosynthetic reinforced soil walls. *Geosynth. Int.* **2018**, *25*, 12–25. [CrossRef]
91. Chen, J.; Zhang, W.; Xue, J. Zoning of reinforcement forces in geosynthetic reinforced cohesionless soil slopes. *Geosynth. Int.* **2017**, *24*, 565–574. [CrossRef]
92. Liu, H.; Wang, X.; Song, E. Reinforcement load and deformation mode of geosynthetic-reinforced soil walls subject to seismic loading during service life. *Geotext. Geomembr.* **2011**, *29*, 1–16. [CrossRef]
93. Cristelo, N.; Félix, C.; Lopes, M.L.; Dias, M. Monitoring and numerical modelling of an instrumented mechanically stabilised earth wall. *Geosynth. Int.* **2016**, *23*, 48–61. [CrossRef]
94. Roy, S.; Deb, K. Bearing capacity of rectangular footings on multilayer geosynthetic-reinforced granular fill over soft soil. *Int. J. Geomech.* **2017**, *17*. [CrossRef]
95. Kulczykowski, M. Determination of the effect of sidewall friction in reinforced soil retaining wall experiments. *Arch. Hydro-Eng. Environ. Mech.* **2021**, *68*, 137–158. [CrossRef]
96. Lambe, T.W.; Whitman, R.V. *Soil Mechanics. SI Version*; Wiley: New York, NY, USA, 1979.

applied sciences

MDPI

Article

Influence of Laboratory Compaction Method on Compaction and Strength Characteristics of Unbound and Cement-Bound Mixtures

Matija Zvonarić *, Ivana Barišić, Mario Galić and Krunoslav Minažek

Faculty of Civil Engineering and Architecture Osijek, Josip Juraj Strossmayer University of Osijek, Vladimira Preloga 3, 31000 Osijek, Croatia; ivana@gfos.hr (I.B.); mgalic@gfos.hr (M.G.); krumin@gfos.hr (K.M.)
* Correspondence: mzvonaric@gfos.hr

Abstract: During road construction, granular materials for the unbound base course (UBC) and cement-bound base course (CBC) are mostly compacted by vibratory rollers. A widespread laboratory test for determining the optimal moisture content (OMC) and maximum dry density (MDD) of the mixture for installation in UBC and CBC is the Proctor test. Considering that the Proctor test does not produce any vibrations during compaction, this paper compares the Proctor test and the vibrating hammer test. The examination was conducted on UBC and CBC with varying cement content and aggregate types. All mixtures were compacted by both methods with the aim of determining the compaction and strength characteristics. The results indicated the high comparability of the two test methods for mixtures with natural aggregate in terms of MDD, OMC, density and strength characteristics (California bearing ratio (CBR) for UBC and 28-day compressive strength for CBC). For mixtures with higher cement content, the OMC difference depending on the laboratory compaction method used can be significant, so the laboratory compaction method should be chosen carefully, particularly for moisture-susceptible materials. This paper also reveals that by increasing the proportion of rubber in the mixture, the compaction and strength characteristics differ significantly due to the compaction method. Therefore, when using alternative and insufficiently researched materials, the compaction method should also be chosen carefully.

Keywords: cement bound base course; unbound base course; compaction technology; Proctor test; vibrating hammer test; recycled rubber

Citation: Zvonarić, M.; Barišić, I.; Galić, M.; Minažek, K. Influence of Laboratory Compaction Method on Compaction and Strength Characteristics of Unbound and Cement-Bound Mixtures. *Appl. Sci.* **2021**, *11*, 4750. https://doi.org/10.3390/app11114750

Academic Editor: Malgorzata Jastrzebska

Received: 12 April 2021
Accepted: 17 May 2021
Published: 21 May 2021

1. Introduction

Compaction is a significant process in pavement construction. For pavement installation, all the installed layers must be compacted to the target degree of compaction and density. As a result of the application of the required compaction effort, a certain targeted porosity and shear strength of the compacted material is achieved. Also, compaction has a great impact on the pavement load-bearing capacity. Good compaction is a guarantee of the road service level [1]. The compaction of pavement courses is performed by various types of roller or plate compactors [2] and the selection of the type of compaction machine has a significant impact on the compaction characteristics achieved. An overview of the importance and historical development of road compaction technology is presented in [3]. Commonly for road construction, several different types of roller are used, i.e., smooth-wheel rollers, pneumatic rollers, and sheepsfoot rollers. Their dead weight, drum width, and diameter are common characteristics for all rollers, but all rollers differ in certain properties. Smooth-wheel rollers appear as static, vibratory, and oscillating. Static rollers can consist of one, two, or three drums and they rely on dead weight, speed and the number of passes. On the other hand, the vibratory rollers' efficiency greatly depends on their weight, speed, amplitude, and the frequency of the drum. Namely, a drum consists of one or two eccentric and rotating masses that produce vibrations. By compacting with vibratory rollers, the friction of the material during compaction reduces, and aggregate grains nest more easily into the cavities in the mixture. Dynamic compaction gives the best effect in compacting cohesionless materials [4–7]. Fathi et al. [8] tried to examine the depth of influence of intelligent compaction rollers using simulated and field data. Research has shown an increasing depth of influence for granular materials, while the influence decreases with increase in the cohesion of the compacted material. In road construction, large areas have to be uniformly compacted to the targeted degree of compaction and density. Taking into account the aforementioned, the simultaneous control of all the parameters for effective compaction tends to be a challenging task. Nowadays the continuous compaction control method (CCC), presented in more detail in [9], is widespread. CCC allows control of the compaction effect achieved in real time, based on the dynamic response of the interaction between the drum and the layer being installed [10,11]. Most recently, published studies are presenting the developing approach of quality control and acceptance (QC/QA) based on intelligent compaction (IC) technology in road construction, which is arguably dependent on and determined by the development of sensor technology and robotic operations [3,12–14]. In [1], the main drawbacks of existing compaction assessment methods are underlined, based on discrete manual measurements, such as the inability to provide results on the quality of the compaction of the entire road, as well as being potentially destructive of the layers and time-consuming. The authors presented a QC/QA-based IC technology with an alternative measurement based on a changing pattern of compaction meter values (CMV) in regard to the number of compaction roller passes. They reported that this is more likely to achieve the desired compaction quality under the optimal moisture content (OMC) situation. A similar approach is presented in [15], comprising the implementation of roller-integrated CCC technology on granite residual subgrades. The authors underlined in their conclusions that the roller operation parameters and lift thickness should be kept constant during further quality assurance using the roller-integrated CCC technology, since the roller measurement values rely strongly on the roller operation parameters and lift thickness, especially for cement-treated layers. As for further research, the authors suggest investigating the effect of the moisture content on the roller measurement values, as well as the correlations between the roller measurement values and further location-specific in situ point measurements. The results of a geostatistical analysis of IC technology showed that this technology can provide detailed information about the uniformity of the spatial compaction and that it could be adopted to monitor the changes in spatial uniformity during the compaction [16].

Finally, for effective compaction, appropriate material needs to be used, with proper aggregate gradation and binder content depending on the constructed bearing layer. The recipe for the reference mixture is obtained by laboratory testing. For a proper reference mixture definition in laboratory conditions, realistic on-site conditions need to be simulated through proper simulation of the in-situ compaction energy and other conditions. The standard laboratory tests for determining the maximum dry density (MDD) and OMC are the Proctor test, vibrating hammer, and vibrating table [17,18].

The unbound base layer is an indispensable part of every pavement, made of a mixture of unbound granular stone materials. The main characteristics of this layer are the degree of compaction and the modulus of compressibility. In pavements with a heavy traffic load, a hydraulically bound base layer is installed on the unbound base layer, which is a mixture of well-graded aggregate, hydraulic binder and an appropriate amount of water. The main feature of this layer is the degree of compaction and compressive strength. To determine the compaction parameters in laboratory conditions, the most widespread test is the Proctor test, while other test methods can also be applied. Since granular soils are mostly compacted by vibrating rollers, with vibration compaction, it can be inferred that the Proctor test does not give sufficiently realistic results, because it does not use any vibrations during specimen compaction. As an alternative to sample preparation by the Proctor compaction method, vibratory hammer compaction can be used. Izaquiredo et al. [19] concluded that when using the Proctor compaction method, grains tend to pre-shred and be unevenly distributed, which leads to unrealistically high optimal moisture content and maximum dry density values which are hard to achieve in real compaction. Sengün et al. [18] compared four compaction methods on roller-compacted concrete mixtures to determine the compressive strength. The results proved the better efficiency of the vibrating hammer test than the modified Proctor test for mixtures that contain lower amounts of cement, while for mixtures with higher cement content, smaller differences between the test methods are observed. Also, Drnevich and Evans [20] claim that the vibrating hammer represents the field compaction conditions of granular soils better than the vibrating table test. Hoff et al. [21] studied the influence of different laboratory compaction methods on the resistance to permanent deformation of unbound granular materials. Comparisons were made for samples compacted with four different methods: gyratory compactor, modified Proctor, vibratory table and vibratory hammer, and showed that the modified Proctor method shows less resistance to permanent deformation than methods based on vibration. McLachlan and Bagshaw [22] addressed the standardization of laboratory compaction energies for different compaction methods, giving the description and comparison of three compaction methods; vibratory hammer, vibrating table and gyratory compactor, considering, among other parameters, costs, ease of use, consistent energy input as well as correlation with field compaction. As the alternative methods were weighted on the vibrating hammer test till such a decision is made, the recommendation on the vibrating hammer standard should be revised and made more precise, e.g., requiring calibration. Investigations of the effects of different laboratory compaction methods on the properties of cement-stabilized materials were done by Wu and Houben [23]. The mechanical performance of cement-stabilized samples of three types of soil—sand, sandy clay and clay—were compared, defining the appropriate technique for compaction of each of the soil types used. It was observed that for sand materials vibratory compaction gives substantially higher densities compared to the Proctor test (resulting in two times higher compressive and indirect tensile strength), and also that there is an increase in densities for stabilized sandy clay, as opposed to clay and stabilized clay, where no obvious improvement is noticed. A new soil compaction vibratory machine which can produce densities and optimal moisture contents that would mimic the Proctor test is described in [17], with the conclusion that soil densities increased with the increasing amplitude, time, and frequency of the machine, and that the vibratory soil compactor would work best at a frequency of 17 Hz and amplitude of 1.17 mm for 5 min. Field test results presented in [24] on the energy transferred to the ground compacted by a 4-sided impact roller confirmed that the towing speed influences the pressure imparted to the ground and hence the compaction effort, and that the current practice of using either kinetic energy or gravitational potential energy should be avoided, as neither can accurately quantify the rolling dynamic compaction when the towing speed is varied.

This research aimed to determine the applicability of the vibrating hammer method compared to the modified Proctor test in the testing of unbound and hydraulically bound mixtures for pavement purposes using standard and some non-standard materials (waste rubber). The main purpose of the research conducted was to compare the compaction test conditions and their influence on the material behaviour through compaction and strength characterization. Also, the appropriateness of the two compaction methods on an alternative, waste material—waste rubber—was investigated. Namely, the use of waste rubber in the cement-bound base course (CBC) can reduce the occurrence of reflective cracks in the asphalt wearing course, resulting in natural resource preservation and extended service life of a pavement structure [25,26]. Little research has been done on effect of crumb rubber on cement bound mixtures, while extensive research has been conducted on rubberized concrete. Natural aggregate replacement by crumb rubber inevitably contributes to the reduction of compressive strength, for which some researches have established a prediction model [27]. Crumb rubber addition to concrete changes failure pattern from brittle to ductile at the microscale observation and shows better performance under severe dynamic loads at the macroscale observation [28]. According to [29] rubber addition in concrete improves its resistance to frost which, besides brittleness, is a major problem in

CBC mixtures. Encouraged by positive impact of rubber on concrete, authors decide to examine the influence of crumb rubber on CBC. Due to their long decomposition time, waste tyres cause severe environmental issue. This problem is growing every day, as the number of vehicles on roads increase all over the world. Therefore, application of waste tyres in engineering structures, especially the big one as road infrastructure may have a great impact on environment.

2. Materials and Methods

Within this research, for the unbound bearing course and cement-bound course, natural aggregates (sand and gravel) with different gradations (i.e., 0–2 mm, 0–4 mm, 4–8 mm, and 8–16 mm) were used (Figure 1). Due to its similar grain size distribution, as an alternative to the 0–2 mm fraction, a fine fraction of recycled rubber was used (Figure 2). The chosen material was prepared according to EN 933-1 [30] with a grain size distribution curve to fit between the upper and lower limits according to EN 13285 [31] for the unbound base course (UBC) material (mixture named 0%C), and EN 14227-1 [32] for the cement-bound material (CBC). For CBC, Portland cement of grade 32.5 (CEM II B/M (P-S) 32.5R) was used as a binder, comprising 3%, 5%, and 7% of the total aggregate mass. Solid particle densities were determined in a pycnometer according to standard EN 1097-6 [33]. The density of the 0–2, 0–4, 4–8, and 8–16 mm natural aggregate fractions used was 2.86, 2.96, 2.63 and 2.70 g/cm^3 respectively, while the waste rubber density of 1.12 g/cm^3 was determined using ethanol instead of water due to its low density (similar to water). The density of the cement used was 2.92 g/cm^3, determined by standard EN 196-6 [34]. Three additional mixtures for the CBC were designed, within which the binder content of 5% was selected as an optimal share according to previous research using the same aggregate type [35] in order to preliminarily determine the influence of the waste rubber on the CBC mixture characteristics. In those mixtures, the 0–2 mm fraction was replaced with granular rubber (due to a similar grain size distribution) by volumetric percent (due to high-density differences between these two materials). The replacement was in amounts of 20% (mixture 5%C + 20%R), 30% (mixture 5%C + 30%R) and 60% (mixture 5%C + 60%R) of the 0–2 mm fraction volume in the mix. When recalculated, the 0–2 mm fraction replacement by rubber in amounts of 20%, 30% and 60% represents 4.64%, 6.95% and 13.91% of the total aggregate volume of the mixture respectively, and 1.92%, 2.93% and 6.13% of the total aggregate mass of the mixture. Table 1 presents the proportions of all the constituents in the mixtures.

Figure 1. Natural aggregate and waste rubber used.

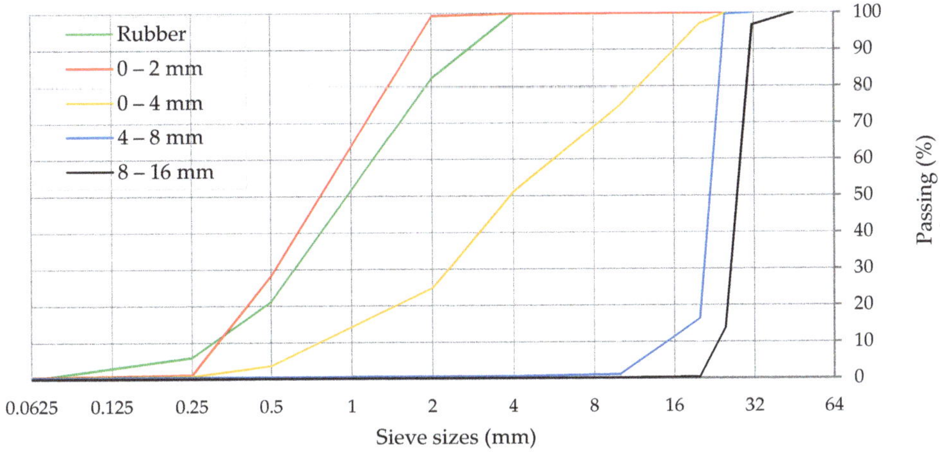

Figure 2. Grain size distribution curve of materials used.

Table 1. Unbound base course (UBC) and cement-bound base course (CBC) mixture compositions.

Mix	0%C	3%C	5%C	7%C	5%C + 20%R	5%C + 30%R	5%C + 60%R
Cement [vol.%]	0	2.78	4.56	6.26	4.56	4.56	4.56
0–2 mm [vol.%]	19.60	23.61	23.18	22.77	18.54	16.23	9.27
0–4 mm [vol.%]	38.00	22.87	22.46	22.06	22.46	22.46	22.46
4–8 mm [vol.%]	32.00	25.69	25.21	24.76	25.21	25.21	25.21
8–16 mm [vol.%]	10.40	25.05	24.59	24.15	24.59	24.59	24.59
Rubber [vol.%]	0	0	0	0	4.64	6.95	13.91
Mix volume [%]	100	100	100	100	100	100	100

In order to compare the results from the Proctor and vibrating hammer tests, the tests were conducted on both the unbound base course mixture and cement-bound base course mixture. The modified Proctor test was done using an automatic Proctor compaction machine and the operator (human) factor on compaction is minimized. On the other hand, as described in standard EN 13286-51 [36], the vibrating hammer is greatly affected by human influence.

To find the OMC and MDD, five test specimens with five different amounts of water were prepared for each test method and each tested mixture, following the recommendations of EN 13286-2 and 13286-4 [37,38]. Both test methods require the use of the Proctor mould B (150 mm diameter and 120 mm height) for the chosen aggregate mixture, which makes the test results comparable (Figure 3). Specimen preparation using the Proctor hammer consisted of specimen compaction in 5 layers, with 56 blows per layer with a 4.5 kg rammer falling from 457 mm height, which is equal to a modified Proctor energy of 2.66 MJ/m^3. The amount of material, placed in each of 5 layers, must occupy one fifth (1/5) of the mould after layer compaction. For specimen preparation using the vibrating hammer, standard [38] prescribes the compaction of a specimen in 3 layers with 60 s of vibration per layer. Analogous to the previously explained method, the amount of the material, placed in each of 3 layers, must occupy one third (1/3) of the mould after compaction of the layer. A vibrating hammer with a weight of 6.4 kg and power of 1100 W is placed on a supporting frame and has a 146 mm diameter tamping foot as an extension. To prevent the hammer bouncing, a surcharge load during compaction was entered by an operator, whereby the total weight of the hammer with the additional load was between 300 and 400 N. Prior to the start of the test, the operator used a balance to determine how much pressure on the hammer corresponds to a force input of approximately 300 N so that the applied force with the hammer weight seemed a force between 300 N and 400 N. Operator experience is required to estimate the appropriate amount of material for each layer. Considering that the duration of specimen compaction by the Proctor hammer was longer than the cement's initial setting time, the mixtures prepared for compaction by the vibrating hammer were left to cure for 10-15 min in order to enable cement hydration. Calculation of the OMC and MDD for both methods used was undertaken following the specifications given in standards EN 13286-2 [37] and 13286-4 [38].

(a) (b)

Figure 3. (**a**) Specimen preparation by Proctor method; (**b**) specimen preparation by vibrating hammer (operator entering surcharge during compaction).

For mechanical characterization, the California bearing ratio (CBR) test and compressive strength test were conducted according to the standards EN 13286-47 [39] and EN 13286-41 [40] respectively. CBR is used to characterize the bearing capacity of UBC after 4 days under a full soaking curing regime and with 2 kg surcharge discs. The compressive strength determined as a stress at failure of a specimen under uniaxial unconfined compression is used to characterize the bearing capacity of CBC. Both tests were conducted on samples made at their OMC according to both methods, the Proctor compaction [41] and vibrating hammer [36]. For compressive strength and CBR determination, three test specimens for each mixture by both methods were prepared in order to establish the validity of the research. Average values of the obtained results are presented as the relevant value for each mechanical characteristic. The presented average value is calculated between results which vary less than 20% according to the propositions of EN 14227-1 [32].

For CBC mixtures, density ρ as a ratio of the mass to the volume of the sample after a 28-day curing period is calculated in order to analyse the effectiveness of the compaction methods.

3. Results and Discussion

On samples prepared using both (Proctor compaction and vibrating hammer) methods, mechanical characteristics were determined and Table 2 presents the average results for three samples.

Table 2. Results obtained by two test methods, vibrating hammer (marked green) method and Proctor (marked red), respectively.

	0%C	3%C	5%C	7%C	5%C + 20%R	5%C + 30%R	5%C + 60%R
OMC [%]	4.79/4.72	4.91/5.15	5.35/5.87	5.53/6.10	5.32/5.53	5.21/5.13	4.94/4.94
MDD [g/cm^3]	2.01/1.97	2.08/2.09	2.12/2.12	2.15/2.15	2.10/2.06	2.03/2.03	1.87/1.89
ρ [g/cm^3]	not applicable	2.17/2.28	2.21/2.29	1.96/2.37	2.18/2.23	2.07/2.18	1.96/2.05
CBR [%]	101/98	not applicable	not applicable	not applicable	not applicable	not applicable	not applicable
f_c [MPa]	not applicable	4.11/4.06	7.59/7.60	12.99/12.92	1.89/3.13	1.33/2.48	0.49/0.94

3.1. Compaction Characteristics

The results of the tests conducted are presented in Table 2 and Figure 4. For mixtures with natural aggregate (0%C, 3%C, 5%C, 7%C), similar results are obtained with both compaction methods. The difference between the MDD values obtained is less than those of OMC. For cement-bound mixtures, similar MDD values (third decimal difference) are obtained, but Proctor compaction required a higher water content. Generally, the vibrating hammer compaction method provides a lower OMC for the same MDD values. The similarity of the results obtained leads to the conclusion of the high comparability of the two test methods for mixtures with natural aggregate. It can be concluded that the vibrating hammer compaction method can be used as an alternative to the modified Proctor compaction method since it is more practical and less time-consuming. A similar observation is presented in [20], where the authors also state that, compared with the standard Proctor method, higher OMC and MDD values should be expected when using the vibrating hammer compaction method.

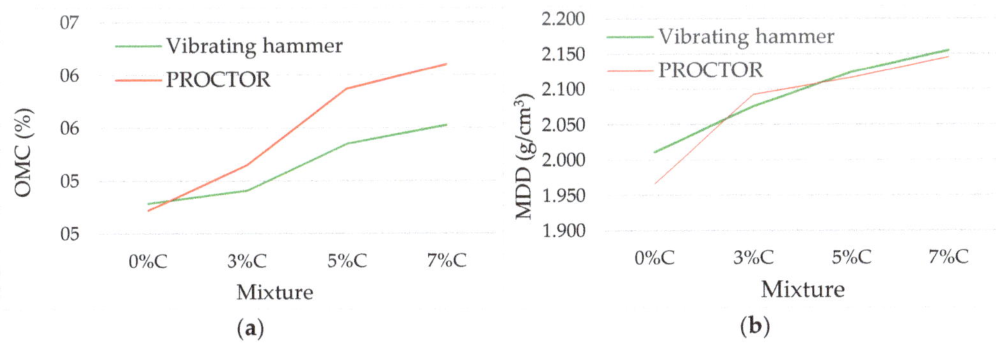

Figure 4. (**a**) Optimal moisture content (OMC) and (**b**) maximum dry density (MDD) for UBC and CBC mixtures.

Also, an expected increase in OMC and MDD is observed with the increase in cement content for both the compaction methods used. The cement hydration reaction and cement paste creation require more water for a higher cement content within the mixture [42]. The vibrating hammer is a less time-consuming method, considering compaction in 3 layers for 60 s per layer and time taken for filling the mould with material, gives about 5 min per specimen. So the lower OMC could be the result of less time being available for the cement's initial setting time and less water being used for initial hydration. To be more precise, cement has more time available for hydration in the Proctor compaction method. Unlike the 60 s for the vibrating hammer method, 56 blows per layer take longer. With a longer compaction of the one layer, the Proctor method still requires a larger number of layers (5), which makes this process longer still and takes approximately 15 min per sample. From this discussion, it can be concluded that the specimen compacted by the Proctor method has about 3 times more time for cement hydration which requires more water, compared to vibrating hammer specimens. However, similar strength characteristics are obtained and high comparability is observed. For higher cement content mixtures, the OMC difference depending on the laboratory compaction method used can reach 0.6%, so the laboratory compaction method should be chosen carefully, particularly if moisture-susceptible materials are used.

As expected, the addition of waste rubber results in a decrease in both the OMC and MDD values compared to the natural aggregate mixture with the same binder content. This is a result of rubber's hydrophobic nature, rough surface (Figure 5), and low density, but also its high elasticity that absorbs some of the compaction energy. As can be seen from Figure 5, rubber particles have more irregular and indented surface which captures air molecules thus taking up more space in the structure of the mixture which reduce the volume of the mixture that hydrates with cement. Furthermore, for irregularly shaped particles it is more difficult to adhere to the particles with flat surfaces, such as sand particles, and thus create larger gaps in the mixture. The results are in accordance with some previous research dealing with waste rubber applications in the pavement system [26,43,44]. From Table 2, it can be noted that there is a larger difference in OMC than MDD between waste rubber and non-waste rubber mixtures for the Proctor compaction method. This greater drop in the compaction characteristics for the Proctor compaction method means that the choice of laboratory compaction method should be made carefully if alternative materials are used.

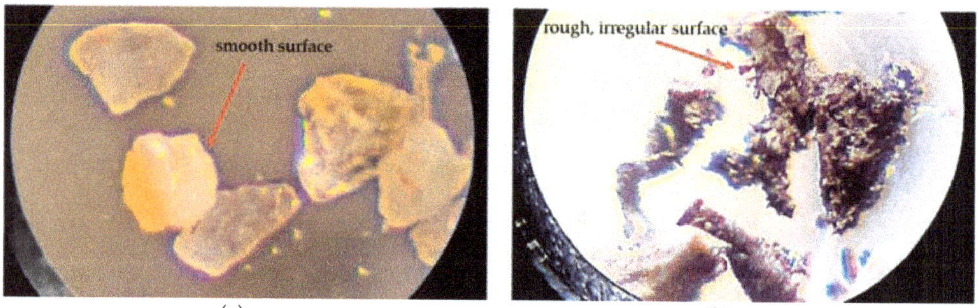

(a) (b)

Figure 5. Optical microscopy image of sand grain (**a**) and waste rubber grain (**b**).

As presented in Table 2, a lower density (ρ) is observed for the vibrating hammer method compared to the modified Proctor compaction method, for which an increasing trend is observed with an increase of the cement content. However, for the vibrating hammer method, with a high cement content a decrease in density occurred. During compaction, a cement paste layer formed on the sample surface at the end of each compaction layer (Figure 6), resulting in a decrease of the apparent density. Under high vibrations and a high cement paste content within the mixture, segregation occurred, influencing the apparent density decrease. The same phenomenon can explain the high compressive strength difference for higher values of $f_{c,28}$. Namely, cement paste segregation on the sample surface caused uneven density of the sample and a higher $f_{c,28}$ difference when comparing the two compaction methods, as presented in Figure 7b.

Figure 6. Vibrating hammer compaction of mixture with high cement content.

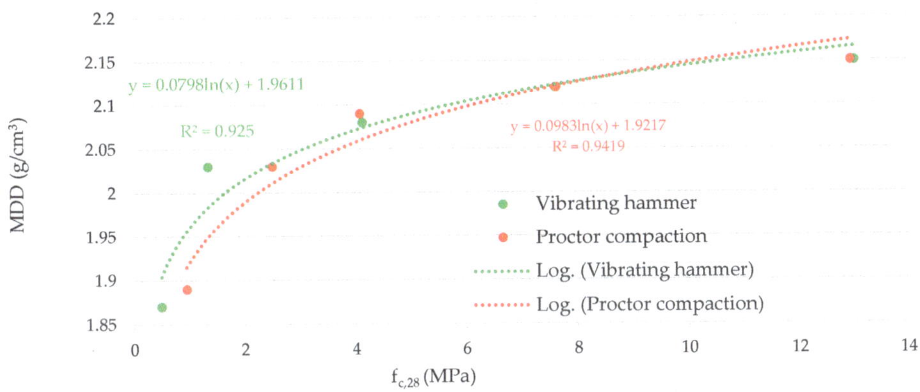

(**a**)

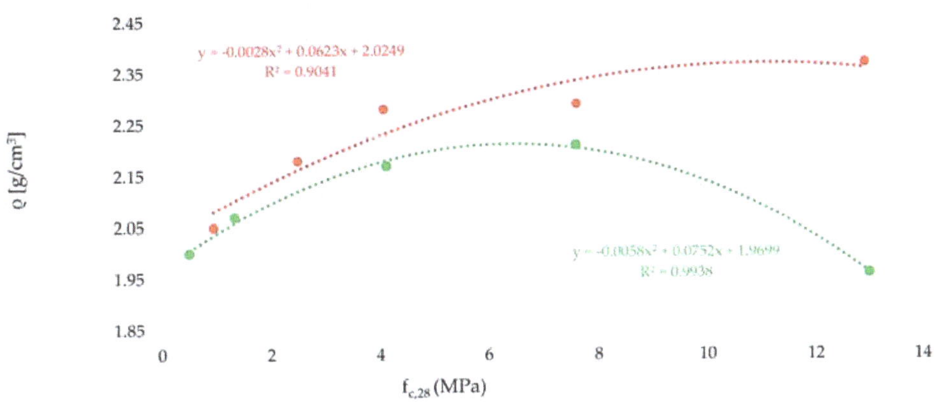

(**b**)

Figure 7. Correlation between compressive strength and MDD (**a**) and density (**b**).

The addition of waste rubber also resulted in a density decrease due to the lower rubber density compared to the replaced sand. The rubber's high elasticity and compaction energy absorption resulted in a greater difference in the measured density. For the vibrating hammer method a 1.63%, 6.3% and 11.31% decrease in density occurred for 20%, 30% and 60% sand replacement by rubber respectively, while for the Proctor compaction method the density reduction was 2.62%, 4.8% and 10.53% compared to the reference mixture (5%C).

3.2. Strength Characteristics

Strength characterization of the UBC was undertaken by measuring the CBR. As presented in Table 2, similar results were obtained for both the compaction methods used. Therefore, again the high comparability of the test results can be observed, but, as indicated earlier, due to variation in the OMC, the compaction method for moisture-susceptible materials should be chosen carefully.

Strength characterization of CBC was undertaken by measuring the 28-day compressive strength and the results are presented in Table 2 and Figure 7. Small differences can be observed between the two compaction methods for natural aggregate mixtures. As expected, an increase in $f_{c,28}$ is observed for the cement content due to the formation of calcium silicate hydrates (C-S-H gel), which is mostly responsible for the development of compressive strength. A decrease in compressive strength is observed with the addition of waste rubber, which is in accordance with some previous research [44]. For mixtures with waste rubber, a higher compressive strength is recorded for the Proctor compaction method. The differences between the two compaction methods used are 39.62%, 46.4% and 47.9% for 20%, 30% and 60% sand replacement by rubber respectively. To analyse this high compressive strength difference between the two compaction methods used, the correlation of $f_{c,28}$ and compaction characteristics (MDD and density) is presented in Figure 7.

There is a high correlation between the compressive strength and MDD regardless of the compaction method. Also, the same conclusion can be drawn for the correlation between $f_{c,28}$ and ρ. Since the MDD and ρ are influenced by the lower density of rubber, that is one reason for the lower $f_{c,28}$ of rubber mixtures and the high difference between the compaction methods. However, for mixtures with waste rubber, the difference between MDD and ρ is much smaller compared to the $f_{c,28}$ difference obtained by the two compaction methods. The reason could be the great difference in the behaviour of the rubber mixtures when different compaction methods are used but also when the compression test is conducted. This finding presents a good basis for further research. Finally, the optimal waste rubber content is expected to be up to 30% sand replacement, since a higher rubber content could lead to a drop in the 28-day compressive strength under 2 MPa, as is usually required for most pavement purposes.

4. Conclusions

Within this paper, the compaction characteristics of unbound and cement-bound mixtures for application as a pavement base course are investigated on two aggregate types using two laboratory compaction methods. The purpose of these tests was to compare the efficiency and difference in the compaction and strength characteristics achieved in the well-known Proctor test and, less commonly used in practice, the vibrating hammer test. The conclusions presented are drawn from the research conducted on a limited number of samples and, to reach general conclusions, more tests are to be conducted. Based on the laboratory research results, several conclusions can be drawn:

- There is strong correlation between the compaction and strength characteristics obtained in laboratory conditions using the modified Proctor and vibrating hammer compaction methods for natural aggregate mixtures.
- A greater drop in the compaction characteristics for rubber mixtures and the use of the Proctor compaction method lead to the conclusion that the laboratory compaction method should be chosen carefully if alternative materials such as waste rubber are used.
- For higher cement content mixtures, a higher OMC difference depending on the laboratory compaction method used is observed and the laboratory compaction method should be chosen carefully, particularly if moisture-susceptible materials are used.

- Comparing the compaction effects (MDD, OMC and density), it can be observed that approximately the same compaction energy was applied using both compaction methods. However, calculation of the compaction energy using a vibrating hammer is more difficult than for the Proctor test, and it should be investigated in more detail. Further research should provide an additional insight into the comparability of the vibrating hammer and standard Proctor tests, applying less compaction energy than the modified Proctor test used in this research.
- The choice of laboratory compaction method should be made carefully depending on the specific behaviour of materials used and the expected field conditions.

Author Contributions: Conceptualization, I.B. and M.Z.; methodology, I.B.; validation, I.B., M.G. and K.M.; investigation, I.B., M.G., K.M. and M.Z.; writing—original draft preparation, M.Z.; writing—review and editing, I.B., M.G. and K.M. All authors have read and agreed to the published version of the manuscript.

Funding: This research was supported by Croatian Science Foundation under the project UIP-2019-04-8195 Cement stabilized base courses with waste rubber for sustainable pavements—RubSuPave.

Institutional Review Boards Statement: Not applicable.

Informed Consent Statement: Not applicable.

Conflicts of Interest: The authors declare no conflict of interest.

References

1. Zhu, X.; Bai, S.; Xue, G.; Yang, J.; Cai, Y.; Hu, W.; Jia, X.; Huang, B. Assessment of compaction quality of multi-layer pavement structure based on intelligent compaction technology. *Constr. Build. Mater.* **2018**, *161*, 316–329. [CrossRef]
2. Bezak, S.; Bezak, S.; Ereš, M. (Eds.) *General Technical Conditions for Road Works*; Institut građevinarstva Hrvatske: Zagreb, Croatia, 2001.
3. Xu, Q.; Chang, G.K. Adaptive quality control and acceptance of pavement material density for intelligent road construction. *Autom. Constr.* **2016**, *62*, 78–88. [CrossRef]
4. Chen, L.; Qiao, L.; Li, Q. Study on dynamic compaction characteristics of gravelly soils with crushing effect. *Soil Dyn. Earthq. Eng.* **2019**, *120*, 158–169. [CrossRef]
5. Triantafyllidis, T.; Kimmig, I. A simplified model for vibro compaction of granular soils. *Soil Dyn. Earthq. Eng.* **2019**, *122*, 261–273. [CrossRef]
6. Chilukwa, N.N. *Vibratory Hammer Compaction of Granular Materials*; Stellenbosch University: Stellenbosch, South Africa, 2013.
7. Kodikara, J.; Islam, T.; Sounthararajah, A. Review of soil compaction: History and recent developments. *Transp. Geotech.* **2018**, *17*, 24–34. [CrossRef]
8. Fathi, A.; Tirado, C.; Rocha, S.; Mazari, M.; Nazarian, S. Assessing depth of influence of intelligent compaction rollers by integrating laboratory testing and field measurements. *Transp. Geotech.* **2021**, *28*, 100509. [CrossRef]
9. Pistrol, J.; Adam, D. Fundamentals of roller integrated compaction control for oscillatory rollers and comparison with conventional testing methods. *Transp. Geotech.* **2018**, *17*, 75–84. [CrossRef]
10. Pistrol, J.; Villwock, S.; Völkel, W.; Kopf, F.; Adam, D. Continuous Compaction Control (CCC) with Oscillating Rollers. *Procedia Eng.* **2016**, *143*, 514–521. [CrossRef]
11. Paulmichl, I.; Adam, C.; Adam, D. Analytical modeling of the stick-slip motion of an oscillation drum. *Acta Mech.* **2019**, *230*, 3103–3126. [CrossRef]
12. Hu, W.; Shu, X.; Huang, B.; Woods, M.E. An examination of compaction meter value for asphalt pavement compaction evaluation. *Int. J. Pavement Eng.* **2017**, *19*, 447–455. [CrossRef]
13. Hu, W.; Huang, B.; Shu, X.; Woods, M. Utilising intelligent compaction meter values to evaluate construction quality of asphalt pavement layers. *Road Mater. Pavement Des.* **2016**, *18*, 980–991. [CrossRef]
14. Hu, W.; Jia, X.; Zhu, X.; Gong, H.; Xue, G.; Huang, B. Investigating key factors of intelligent compaction for asphalt paving: A comparative case study. *Constr. Build. Mater.* **2019**, *229*, 116876. [CrossRef]
15. Ling, J.; Lin, S.; Qian, J.; Zhang, J.; Han, B.; Liu, M. Continuous Compaction Control Technology for Granite Residual Subgrade Compaction. *J. Mater. Civ. Eng.* **2018**, *30*, 04018316. [CrossRef]
16. Hu, W.; Shu, X.; Jia, X.; Huang, B. Geostatistical analysis of intelligent compaction measurements for asphalt pavement compaction. *Autom. Constr.* **2018**, *89*, 162–169. [CrossRef]
17. Leonard, L.; Ekwue, E.I.; Taylor, A.; Birch, R. Evaluation of a machine to determine maximum bulk density of soils using the vibratory method. *Biosyst. Eng.* **2019**, *178*, 109–117. [CrossRef]
18. Sengun, E.; Alam, B.; Shabani, R.; Yaman, I. The effects of compaction methods and mix parameters on the properties of roller compacted concrete mixtures. *Constr. Build. Mater.* **2019**, *228*, 116807. [CrossRef]

19. Izquierdo, M.; Querol, X.; Vazquez, E. Procedural uncertainties of Proctor compaction tests applied on MSWI bottom ash. *J. Hazard. Mater.* **2011**, *186*, 1639–1644. [CrossRef]
20. Drnevich, V.; Evans, A.; Prochaska, A. A Study of Effective Soil Compaction Control of Granular Soils. *Jt. Transp. Res. Program* **2007**. [CrossRef]
21. Hoff, I.; Bakløkk, L.J.; Aurstad, J. Influence of Laboratory Compaction Method on Unbound Granular Materials. In *International Symposium on Pavements Unbound*, 6th ed.; Dawson, A., Ed.; CRC Press: Nottingham, England, 2004.
22. McLachlan, R.; Bagshaw, S. *Standardisation of Laboratory Compaction Energies*; NZ Transport Agency: Wellington, New Zeland, 2017.
23. Wu, P.; Houben, L.J.M. Investigation of the Effects of Different Laboratory Compacting Methods on the Properties of Cement Stabilized Materials. In Proceedings of the Ninth International Conference on the Bearing Capacity of Roads, Railways and Airfields, Trondheim, Norway, 25–27 June 2013; Hoff, I., Mork, H., Garba Saba, R., Eds.; CRC Press: Boca Raton, FL, USA, 2018; Volume 2.
24. Scott, B.T.; Jaksa, M.B.; Mitchell, P. Influence of towing speed on effectiveness of rolling dynamic compaction. *J. Rock Mech. Geotech. Eng.* **2020**, *12*, 126–134. [CrossRef]
25. Farhan, A.H.; Dawson, A.R.; Thom, N.H. Characterization of rubberized cement bound aggregate mixtures using indirect tensile testing and fractal analysis. *Constr. Build. Mater.* **2016**, *105*, 94–102. [CrossRef]
26. Farhan, A.H.; Dawson, A.R.; Thom, N.H. Effect of cementation level on performance of rubberized cement-stabilized aggregate mixtures. *Mater. Des.* **2016**, *97*, 98–107. [CrossRef]
27. Huang, W.; Huang, X.; Xing, Q.; Zhou, Z. Strength reduction factor of crumb rubber as fine aggregate replacement in concrete. *J. Build. Eng.* **2020**, *32*, 101346. [CrossRef]
28. Habib, A.; Yildirim, U.; Eren, O. Mechanical and dynamic properties of high strength concrete with well graded coarse and fine tire rubber. *Constr. Build. Mater.* **2020**, *246*, 118502. [CrossRef]
29. Gou, Y.; Zhang, L.; Liu, C.; Zhang, H.; Wei, C.; Cai, X.; Yang, H.; Guan, Q.; Zhai, S.; Liu, L. Investigation of freeze-thaw mechanism for crumb rubber concrete by the online strain sensor. *Measurement* **2021**, *174*, 109080. [CrossRef]
30. European Commitee for Standardization. *HRN EN 933-1: 2012 Test for Geometrical Properties of Aggregates—Part 1: Determination of Particle size Distribution—Sievin Method*; CEN: Brussels, Belgium, 2012.
31. European Commitee for Standardization. *HRN EN 13285:2018 Unbound Mixtures—Specifications*; CEN: Brussels, Belgium, 2018.
32. European Commitee for Standardization. *HRN EN 14227-1:2013 Hydraulically bound mixtures—Specifications—Part 1: Cement Bound Granular Mixtures*; CEN: Brussels, Belgium, 2013.
33. European Commitee for Standardization. *HRN EN 1097-6:2013 Test for Mechanical and Physical Properties of Aggregates—Part 6: Determination of Particle Density and Water Absorption*; CEN: Brussels, Belgium, 2013.
34. European Commitee for Standardization. *HRN EN 196-6:2018 Methods of Testing Cement—Part 6: Determination of Fineness*; CEN: Brussels, Belgium, 2018.
35. Barišić, I.; Dimter, S.; Rukavina, T. Characterization of cement stabilized pavement layers with ultrasound testing. *Teh. Vjesn. Tech. Gaz.* **2016**, *23*, 447–453. [CrossRef]
36. European Commitee for Standardization. *HRN EN 13286-51:2004 Unbound and Hydraulically Bound Mixtures—Part 51: Methods for the Manufacture of Test Specimens of Hydraulically Bound Mixtures Using Vibrating Hammer Compaction*; CEN: Brussels, Belgium, 2004.
37. European Commitee for Standardization. *HRN EN 13286-2:2010 Unbound and Hydraulically Bound Mixtures—Part 2: Test Method for Laboratory Dry Density and Water Content—Proctor Compaction*; CEN: Brussels, Belgium, 2010.
38. European Commitee for Standardization. *HRN EN 13286-4:2003 Unbound and Hydraulically Bound Mixtures—Part 4: Test Methods for Laboratory Reference Density and Water Content—Vibrating Hammer*; CEN: Brussels, Belgium, 2003.
39. European Commitee for Standardization. *HRN EN 13286-47:2012 Unbound and Hydraullically Bound Mixtures—Part 47: Test Method for the Determination of California Bearing Ratio, Immediate Bearing Index and Linear Swelling*; CEN: Brussels, Belgium, 2012.
40. European Commitee for Standardization. *HRN EN 13286-41:2003 Unbound and Hydraulically Bound Mixtures-Pari 41: Test Method for the Determination of the Compressive Strenght of Hydraulically Bound Mixtures*; CEN: Brussels, Belgium, 2003.
41. European Commitee for Standardization. *EN 13286-50 Unbound and hydraulicylly Bound Mixtures—Part 50: Method for the Manufacture of Test Specimens of Hydraulically Bound Mixtures Using Proctor Equipment or Vibrating Table Compaction*; CEN: Brussels, Belgium, 2004.
42. Ovad, E.; Barišić, I.; Zagvozda, M. Comparison of Croatian and European standards for Determining the Frost Suspectibility of Hydraulically Bound Mixtures. *e-GFOS* **2014**, *5*, 41–48. [CrossRef]
43. Farhan, A.H.; Dawson, A.R.; Thom, N.H.; Adam, S.; Smith, M.J. Flexural characteristics of rubberized cement-stabilized crushed aggregate for pavement structure. *Mater. Des.* **2015**, *88*, 897–905. [CrossRef]
44. Sun, X.; Wu, S.; Yang, J.; Yang, R. Mechanical properties and crack resistance of crumb rubber modified cement-stabilized macadam. *Constr. Build. Mater.* **2020**, *259*, 119708. [CrossRef]

applied sciences

MDPI

Article

Effect of Chitosan Solution on Low-Cohesive Soil's Shear Modulus *G* Determined through Resonant Column and Torsional Shearing Tests

Marta Bocheńska [1,*], Marcin Bujko [1], Ireneusz Dyka [1], Piotr Srokosz [1] and Rafał Ossowski [2]

[1] Faculty of Geoengineering, University of Warmia and Mazury, 10-957 Olsztyn, Poland; marcin.bujko@uwm.edu.pl (M.B.); i.dyka@uwm.edu.pl (I.D.); psrok@uwm.edu.pl (P.S.)
[2] Faculty of Civil and Environmental Engineering, Gdansk University of Technology, 80-233 Gdansk, Poland; rafal.ossowski@pg.edu.pl
* Correspondence: marta.baginska@uwm.edu.pl

Abstract: In this study the effect of using a biopolymer soil stabilizer on soil stiffness characteristics was investigated. Chitosan is a bio-waste material that is obtained by chemical treatment of chitin (a chemical component of fungi or crustaceans' shells). Using chitosan solution as a soil stabilizer is based on the assumption that the biopolymer forms temporary bonds with soil particles. What is important is that these bonds are biodegradable, so the product does not leave any harmful waste and has high eco-compatibility. The biopolymer itself is a by-product of many industrial chemical processes, so its application is compliant with the goals of sustainable geotechnical engineering. The effect of chitosan on soil shear strength, permeability or surface erosion has already been investigated in several different studies. In this study specimens of low-cohesive soil stabilized with two different chitosan solutions were subject to cyclic loading (torsional shearing test) and dynamic loading (resonant column) to obtain soil shear modulus *G* as a function of strain values. It has been shown that chitosan solution added to medium-grained materials improves their shear modulus *G* substantially (up to 3 times) even for relatively low chitosan concentration solutions (1.5 g of chitosan per 1 kg of dry silica sand). The results obtained in this study and the known chitosan properties suggest that chitosan solutions can be a very effective and eco-friendly short-term stabilizer for temporary geotechnical structures, e.g., working platforms.

Keywords: soil stabilization; chitosan; non-traditional additives; soil shear modulus; temporary geotechnical structures

Citation: Bocheńska, M.; Bujko, M.; Dyka, I.; Srokosz, P.; Ossowski, R. Effect of Chitosan Solution on Low-Cohesive Soil's Shear Modulus G Determined through Resonant Column and Torsional Shearing Tests. *Appl. Sci.* **2022**, *12*, 5332. https://doi.org/10.3390/app12115332

Academic Editors: Jaroslaw Rybak, Gabriele Chiaro, Małgorzata Jastrzębska and Krystyna Kazimierowicz-Frankowska

Received: 30 April 2022
Accepted: 23 May 2022
Published: 25 May 2022

1. Introduction

Geotechnical engineering is a branch of civil engineering that focuses on the behavior of soil and rock materials near the surface of the earth [1]. It is unquestionable that civil engineering has a significant impact on the natural environment as it rapidly consumes natural resources and usually interferes with the local ecosystem functioning. Therefore, it is of vital importance to promote sustainable development in this branch of industry.

Development of geotechnical engineering is stimulated, in a large part, by advancement of technology and new material solutions that enable realization of construction projects in difficult geotechnical conditions. One of the main aspects of sustainable geotechnical engineering is safe and effective use of soil stabilizers. Soil stabilization is a process of enhancing mechanical properties of soil. It can alter soil physical and chemical characteristics like strength, compressibility and permeability. It can also include actions aimed at limiting soil surface erosion or preventing air pollution by fine-dispersed dust [2].

Ground improvement methods include deep or surface soil stabilization. Surface soil stabilization usually consists of applying an admixture to the surface layer of soil and compacting the resulting mixture. It can be used not only for increasing performance of in situ subsoils but also for, e.g., soils used in road embankments.

Soil can be stabilized mechanically, chemically, thermally or electrically. Mechanical methods (compacting) are used most often as they do not alter the soil composition.

For chemical stabilization the following materials can be used

- cement,
- calcium-based stabilizers,
- fly ash,
- bitumen,
- non-standard stabilizers.

Desired performance characteristics and economic aspects often call for using (and studying) non-traditional additives. Due to ecological reasons, non-standard soil additives are becoming more and more popular. Especially for forest road construction, bike lanes, pedestrian ways on protected sites, weak soil protection from excessive water absorption or working platform construction on weak ground.

Traditional soil stabilizers have been thoroughly studied and their properties and stabilizing mechanisms are mostly identified. Various non-traditional soil additives are currently intensively developed and there is still a great need for research in this subject. Most non-traditional additives can be classified as ionic, enzymes, lignosulfonates, salts, petroleum resins, polymers or tree resins. Many products contain additional substances, e.g., surfactants. However, due to commercial confidentiality the exact composition of a given product is not available to the public.

In [3–5] non-traditional additives were tested. Series of laboratory tests were performer on stabilized coarse- and fine-grained soils and the increase in soil strength was evaluated. The stabilization mechanisms were broadly divided into two types, mechanical and chemical bonding. However, the experiments were mainly focused on stabilization effect evaluation and not on the underlying mechanisms.

Tingle et al. [6] conducted research that was specifically aimed at identifying the chemical and physical bonding mechanisms of non-traditional additives. Series of laboratory experiments included macroscopic analysis, physical characterization and chemical analysis of mineral, soil, stabilizer and soil-stabilizer composites.

The research suggests that stabilization mechanisms that occur in soil-stabilizer mixtures cannot be considered separately. The final effect results from a complex interplay of many different factors. Many chemical stabilizers (ionic, lignosulfonates, salts and enzymes) can react with soil mineral particles, so the stabilization effect depends on a particular soil's mineral composition.

Synthetic stabilizers are mostly vinyl acetates and acrylic copolymers [6]. A polymer stabilizer covers soil particles, and the bonding effect is created when the water is evaporated from the emulsion, leaving a solidified soil-polymer structure. Therefore, the increase in soil strength depend mainly on how well are soil particles covered in emulsion.

A similar mechanism is observed in natural polymer (lignosulfonates or natural resins) stabilization. Laboratory experiments [4,5,7] confirm a clearly physical mechanism of solidification. For fine-grained soils the stabilization effect is less pronounced as the coverage of the smallest particles insufficient. Stabilization with polymer emulsions is more appropriate for coarse-grained soils, also because of the larger specific surface area that allows for more efficient mixing.

Synthetic polymers have a very high tension and bending strength, and create strong physical bonding in the soil-polymer composites. Depending on the specific chemical composition of the stabilizer, an ionic exchange between polymer and soil minerals can also occur and affect the stabilizing effect. The polymer stabilization mechanisms are comprehensively described in literature (see [8]).

Due to rapid global climate changes, a strong emphasis is placed on carbon neutral material production processes. The material research is more and more focused on environmentally neutral materials and possible applications of waste materials that can be used as alternatives for traditional solutions (e.g., eggshell-derived lime as an alternative pozzolanic material for sandy soil stabilization [9] or clayey soil stabilization [10]). In addition, biopolymers (synthetically or naturally obtained) are getting more and more attention. In recent years many review articles concerning the use of biopolymers for soil stabilization were published [11–16]. Biopolymers are typically produced from gum trees, shrimp shell, milk, fermentation of glucose, algae, fungi or bacteria that consist of polysaccharides [17]. Chang et al. [18] presented a broad analysis of applications and potential benefits of using various biopolymers in geotechnical engineering. The review is focused on the results of the latest studies on biopolymers like agar gum, guar gum, gellan gum, dextran, β-glucan, xanthan gum, chitosan, starch and casein. The main benefits of using those materials are increase in soil strength, modification of consistency through increase of water absorption, erosion reduction, etc. (see Table 1). All of these ground improvement effects are obtained in accordance with the principles of sustainable development, as the biopolymers are mostly microbial hydrocarbons with low CO_2 footprints compared to conventional soil binders and promote seed germination and the growth of vegetation in soils [18]. Another aspect of sustainable development is effective and safe waste management and reuse of waste materials [19].

Table 1. Biopolymer additives used in geotechnical engineering [18].

Biopolymer	Chemical Composition	Behavior with Soils
Agar Gum	$C_{14}H_{24}O_9$	Strengthening Pore clogging Erosion reduction
β-glucan	$C_{18}H_{32}O_{16}$	Grouting Strengthening Superplasticizer in concrete
Casein	$C_{81}H_{125}N_{22}O_{39}P$	Strengthening Water resistance Hydraulic conductivity reduction
Chitosan	$C_{18}H_{35}N_3O_{13}$	Strengthening Coagulant effects Removal of heavy metals in water
Dextran	$H(C_6H_{10}O_5)_xOH$	Drilling muds Conditioners Erosion reduction
Gellan Gum	$C_{24}H_{37}O_{20}$	Strengthening Pore clogging Erosion reduction
Guar Gum	$C_{10}H_{14}N_5Na_2O_{12}P_3$	Dust control Strengthening Grouting
Starch	$C_{27}H_{48}O_{20}$	Adhesives for drilling fluids Strengthening Erosion reduction
Xanthan Gum	$C_{36}H_{58}O_{29}P_2$	Drilling mud thickener Strengthening

Most of the published studies that concern mechanisms and effects of biopolymer soil stabilization focus mainly on the composite's strength parameters, like internal friction angle, cohesion and compression strength [17,20–24]. There are few studies that focus on deformation characteristics, stiffness or damping.

Chitosan is a waste product of the sea food processing industry. This cationic polymer is obtained by deacetylation of chitin, found abundantly in crustacean, insect, arthropod exoskeletons, and molluscs [25]. Despite its numerous applications in water treatment [26], pharmaceutics [27], agriculture [28], winemaking [29] or energy storage [30,31] this natural resource hasn't been fully utilized. The biggest advantages of this polymer include availability, low cost, high biodegradability and ease of chemical modification [32–35]. Polymer blending can also be performed to improve the mechanical properties of the bioproduct [25,32].

Application of biopolymers in geotechnical and geo-environmental engineering is relatively new. There are limited studies on the application of biopolymers for soil improvement. These studies were generally related with permeability [36], erosion control [37] and shear strength of the soils [21,38]. Experimental studies have shown CU Triaxial tests performed on compacted silt mixed with 1% chitosan solution indicate shear strength improvement of up to 30 percent within a week [39]. In another study [40], fine silica sand containing dextran showed 20 times increased critical shear stress (τ_c) compared to the untreated condition. Chang and Cho [41] proved, the presence of biopolymers in soil might enhance shear stiffness (Kirchhoff modulus G).

Various applications of chitosan are still studied. In geotechnical research there are still few results available that specifically concern soil–chitosan mixtures. Recent studies focus mainly on soil–chitosan composites shear strength or soil surface erosion prevention. To the authors' best knowledge there are yet no studies that focus on soil–chitosan mixtures' stiffness and dynamic properties.

In this study the effect of chitosan on stiffness and dynamic properties of medium-grained low-cohesive soil was investigated. Soil–chitosan composited were subjected to cyclic slow-changing loading (torsional shearing test) and dynamic loading (resonant column). The specimens were prepared using two different chitosan solution concentrations. The obtained properties of soil–chitosan composites were compared with the properties of non-enhanced soil specimens with the same water content.

2. Materials and Methods

2.1. Materials

2.1.1. Silica Sand

Non-cohesive materials have proven to be suitable for soil stabilization with polymer emulsions (increased mixing efficiency resulting from their large specific surface area) [7]. Therefore, silica sand was chosen for the experiments in this study. Silica sand was originally used by Wichtmann [42] to perform experiments in resonant column. Numerous tests on this type of soil are presented in Wichtmann's dissertation [42] and other studies [43,44]. "Silica sand 1" is classified as a poorly graded sand with a uniformity coefficient C_U value of 1.6 (see Figure 1). Sand with high uniformity of soil particles is appropriate to observe influence of additives on specimen mechanical properties [45]. Geometrical and physical parameters of tested soil are presented in Table 2.

The maximum dry density and the optimum moisture content of "silica sand 1" were determined through the Standard Proctor Compaction Test and the results are presented in Figure 2. All the specimens were prepared with the optimum moisture content (3.0–3.5%) to maximize the contact surface between the soil particles, and therefore provide a better interaction between the soil and the chitosan solution.

2.1.2. Chitosan

Chitosan is biocompatible, biodegradable and non-toxic cationic polymer obtained by deacetylation of chitin [46]. Chemical structure of chitin is made up of 1–4 linked 2-acetamido-2-deoxy-β-D-glucopyranose. Chitosan, a deacetylated form of chitin to at least 50% of the free amine form, has a heterogeneous chemical structure made up of both 1–4 linked 2-acetamido-2-deoxy-β-D-glucopyranose as well as 2-amino-2-deoxy-β-D-glucopyranose [47]. Chemical structure of chitosan is presented in Figure 3, and chitosan powder particles are shown in magnification in Figure 4.

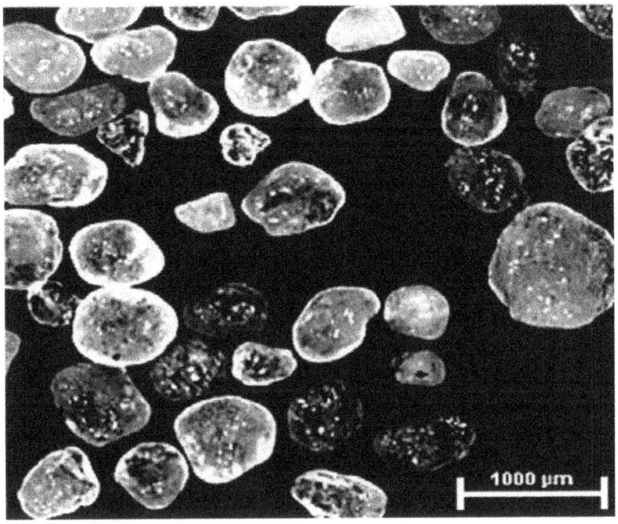

Figure 1. Grains of "silica sand 1" in magnification.

Table 2. Parameters of the tested soil ("silica sand 1") *.

Soil Type	G_s [-]	d_{50} [mm]	d_{60} [mm]	d_{10} [mm]	C_U [-]	e_{max} [-]	e_{min} [-]	ρ_{dmax} [g/cm^3]	w [%]
Silica sand 1	2.65	0.33	0.35	0.22	1.6	0.68	0.41	1.68	3.30

* d, d_{60}, d_{10}—the values of the particle diameter at 50%, 60% and 10% in the cumulative distribution; e_{max}—maximum void ratio, e_{min}—minimum void ratio, C_U—uniformity coefficient, ρ_{dmax}—maximum dry soil density, w—moisture content, G_s—specific gravity.

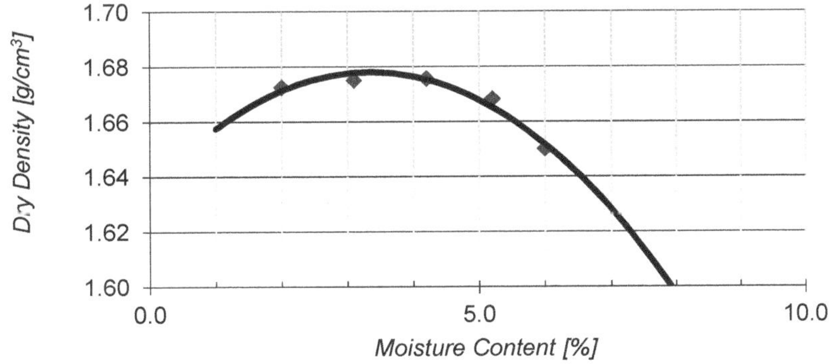

Figure 2. Determination of maximum dry density (g/cm^3) and optimum moisture content (%) for "silica sand 1".

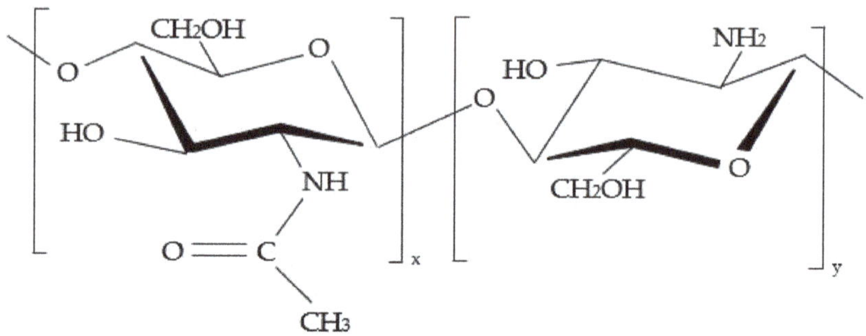

Figure 3. Chemical structure of chitosan [47].

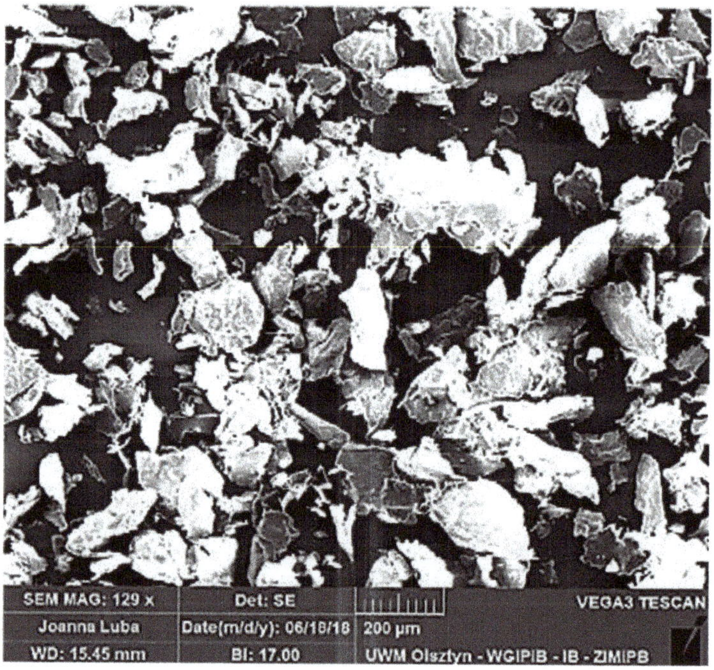

Figure 4. Grains of chitosan in magnification.

The physical parameters of the chitosan powder that was used in this study are presented in Table 3. Additionally, in Figure 5 particle size distribution of both materials (sand and chitosan) is compared.

Table 3. Parameters of the tested chitosan *.

G_s [-]	d_{50} [mm]	d_{60} [mm]	d_{10} [mm]	C_U [-]	e_{max} [-]	e_{min} [-]
1.47	0.11	0.13	0.044	2.85	3.29	1.73

* d_{50}, d_{60}, d_{10}—the values of the particle diameter at 50%, 60% and 10% in the cumulative distribution; e—void ratio, C_U—uniformity coefficient, G_s—specific gravity.

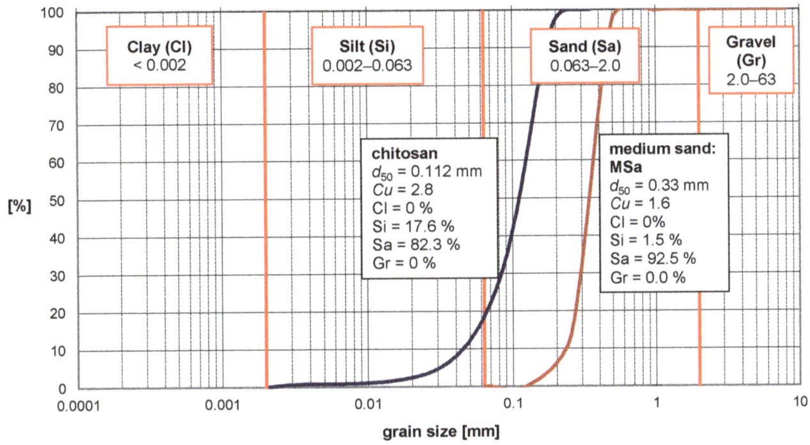

Figure 5. Particle size distribution of used materials.

It can be noted (Figure 5) that the particle size distributions make it possible to create a uniform particle mixture of both materials even when mixing dry ingredients. This aspect can be beneficial for further physical-chemical processing.

Commercial chitosan is usually a fine powder of white to slightly yellow color (Figure 6a). It is insoluble in water, but it is easy to dissolve in solutions of weak acids, e.g., in 10% acetic acid CH_3COOH (Figure 6b,c). Chitosan solutions in acetic acid are viscous fluids. The viscosity increases with the increase in chitosan concentration (in this study 4% solution in 10% acetic acid and 7% solution in 10% acetic acid were used. The increase in the liquid's viscosity was easily noticeable but has not been quantitatively measured).

(**a**) (**b**) (**c**)

Figure 6. Chitosan (**a**) powder, (**b**) 4% solution in 10% acetic acid, (**c**) 7% solution in 10% acetic acid.

Laboratory tests were performed on cylindrical specimens of oven-dry "silica sand 1" and two types of mixtures of "silica sand 1" and chitosan-water-acetic acid solutions. Specimens were compacted with respect to the ASTM standard [48]. In Table 4 the composition of prepared specimens is presented.

Table 4. Composition of specimens' mixtures [in g per 1 kg of dry silica sand].

	Dry Chitosan	Reversed Osmosis Water	10% Acetic Acid Solution
specimen 1	0	35	0
specimen 2	2.5	0	35
specimen 3	1.5	15	20

2.2. Methods

2.2.1. Resonant Column/Torsional Shearing Device

The laboratory tests were performed using modified Stokoe's fixed-free type Resonant Column apparatus with Torsional Shear mode (Wykeham Farrance RC/TS, model WF8500). The modified version of the device and new procedure of interpretation of results were presented in recent publication [49]. Standard WF8500 measurement method is based on the external measurement system of deformation in one plane of the specimen's cross-section—at its upper end. In resonant column mode (RC), the device enables dynamic loading with torque causing shear strain in the tested soil specimen. During a test, torque frequency increases abruptly or smoothly in the range of 0–300 Hz. In torsional shear mode (TS), the device generates cyclic, slowly changing torque with a constant frequency in the range of 0.01–50 Hz. The device is equipped with an aluminum base with a steel plinth equipped with a porous stone and pipes for water supply and drainage from triaxial compression cell together with valves. This allows for performing a controlled saturation process and isotropic consolidation. Pressure control is provided by sensors for cell pressure, pore pressure and backpressure. The size of plinth allows for the installation of a cylindrical soil specimen with a base diameter of 70 mm and a height of 140–150 mm. The loading is generated by an electromagnetic system consisting of four coils and neodymium magnets. The magnets are the moving part of the device and can generate a maximum torque of 1.2 Nm and maximum angle of 10 degrees. The control of the electromagnetic drive system is correlated with the automatic acquisition of measurement data from all sensors installed in the device. The device is powered by an external unit and controlled with a dedicated software "DYNATOR". Main control parameters in the RC mode are initial and final frequency, time and amplitude of the voltage that produces torsional loading (moment of electrodynamic force). In the TS mode, the control parameters are frequency, voltage amplitude and number of torsion cycles. Static values, characterizing the mechanical properties of the drive head, are altered due to changes of conditions in the laboratory and change with time (equipment components' wear). For these reasons, before each series of tests, the device is calibrated. Figure 7 presents a schematic drawing of the RC/TS device.

2.2.2. Specimen Preparation and Experimental Procedure

Dry chitosan powder was dissolved in the 10% acetic acid solution at 20 °C. Chitosan was added in small portions and mixed until completely dissolved. Two different chitosan solutions were used in this study, 4% and 7% (to obtain 1.5 g and 2.5 g of chitosan per 1 kg of dry sand respectively for soil specimens with optimum water content). The prepared solutions were stored in hermetically sealed laboratory glassware at 20 °C. The "silica sand 1" was dried to a constant weight at 105 °C, then cooled down and stored in a hermetically sealed container. The chitosan solutions were gradually added to soil portions of a known mass and mixed with the soil to a homogenous mixture. The prepared mixtures were stored in hermetically sealed containers for about 1 h.

The specimens were formed using a hollow metal cylinder, weighted and prepared for installation in the RC/TS device.

The specimen installation includes placing the specimen on the bottom porous stone and placing the drive head on the top surface of the specimen. The specimen is then covered with a latex membrane that separates the material from pressurized water in the testing chamber. For specimens of height 140–150 mm, 235 mm long membranes were used to create an additional overlap on the drive head and the bottom base. The membrane placement was secured with rubber O-rings at the top and bottom.

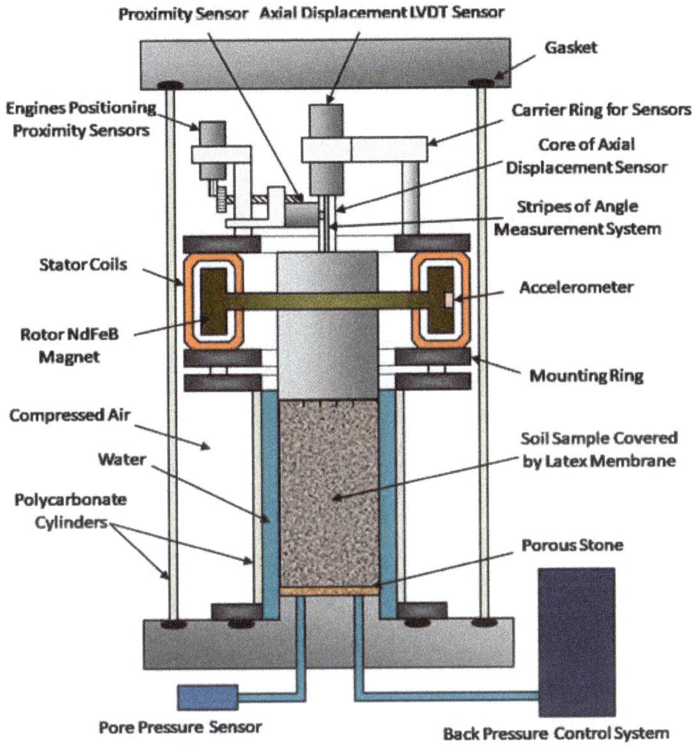

Figure 7. Schematic drawing of RC/TS apparatus.

The prepared specimen is placed inside a polycarbonate internal cylinder filled out with water just before starting the testing procedure. The final step of specimen installation is covering the entire setup with an external cylinder that separates the specimen, drive and measuring equipment from the external environment. The RC and TS tests were performed according to the ASTM standard [50]. All of the tests were performed in drained conditions (the pore water can easily drain out from the soil matrix and the liquefaction is not expected to occur).

After the tests were conducted, the fresh soil–chitosan mixtures were cured. The thermal curing was done at 60 °C for 24 h. After the curing process the same experiments were carried out on the cured soil–chitosan composite specimens.

After the tests the specimens were left at 20 °C (relative humidity 50%) and subjected to observation for 28 days. The last phase of the experiment was the observation of the composite degradation in water.

2.2.3. Method of Results Interpretation

Standard RC test is performed to measure the propagation speed V_S of a transverse (shearing) wave in a soil specimen.

For material with bulk density ρ, the shear modulus is given by the following formula

$$G = \rho \cdot V_S{}^2. \quad (1)$$

And the propagation speed V_S of a transverse wave is

$$V_S = \frac{2\pi \cdot f_r \cdot L}{\beta}, \tag{2}$$

where f_r—resonant frequency of the system (specimen + drive system), L—specimen's height. The parameter β [rad] is obtained using the formula

$$\frac{I}{I_0} = \beta \cdot \tan(\beta), \tag{3}$$

where I—mass moment of inertia of the specimen, I_0—mass moment of inertia of the drive system, determined during the device calibration process.

To find the resonant frequency f_r the specimen is subjected to torsional vibrations of gradually increasing frequency. The resonant frequency would be the frequency value corresponding to the maximum twist angle that was measured during the test.

Equation (1) takes the following form

$$G = \rho \frac{4\pi^2 f_r^2 L^2}{\beta^2}. \tag{4}$$

The amplitude of applied torque and the measured corresponding twist angle can be converted to shear stress τ and shear strain γ. The shear modulus can then be calculated using the formula

$$G = \frac{\tau(\gamma_{max})}{\gamma_{max}} \tag{5}$$

considering the fact that the graph of the $\tau(\gamma)$ function forms a characteristic hysteresis loop.

Alternatively, the shear modulus of the specimen can be found using back analysis and numerically simulated TS test results.

The goal of back analysis is to find a function that accurately describes relationship between the observed (known) effects and unknown causes. To avoid problems with solution ambiguity (there may be many sets of causes that lead to the same final result), the optimization approach is often used

$$L = \underset{X}{Min}(|\Gamma(X) - Y|), \tag{6}$$

where $\Gamma(\cdot)$—cause-and-effect relationship, X—set of unknown causes, Y—set of observed effects, L—objective function (or error function).

The optimal X parameters for the $\Gamma(\cdot)$ problem are found through minimization of the L function, which measures the difference between observed effects Y and the effects predicted by the Γ(X) model. The following objective function was used in the performed back analyses

$$L = \sum_n w\left(\tau_\gamma^e - \tau_\gamma^c\right)^2, \tag{7}$$

where τ_γ^e—measured (experimentally, during the TS test) τ for specific values of γ, τ_γ^c—numerically calculated τ for specific values of γ, w—weights of the computational nodes (set to 1), n—number of nodes.

It is assumed that Hooke's law applies, therefore

$$\tau(\gamma) = G\gamma, \tag{8}$$

where τ—average shear stress, γ—average shear strain, G—shear modulus. This ensures that the problem solution is stable and unambiguous. However, considering the soil stiffness degradation in small strain range the following nonlinear relation is used

$$\tau(\gamma) = G(\gamma)\gamma. \tag{9}$$

The degree of nonlinearity can only be determined through repetitive G evaluation for different γ values. Therefore $G(\gamma)$ can be obtained with iterative back analysis of a dataset from a single TS test. The $\tau(\gamma)$ relation in numerical simulation is successively fitted to the $\tau(\gamma)$ from the laboratory experiment. The numerical simulations were performed using the finite element method (FEM). Particular attention was paid to appropriate modeling of boundary conditions.

It should be noted that the soil stiffness degradation function is not always monotonic, although it is usually a nonincreasing function. Therefore, a complete model of soil behavior under cyclic loading requires numerical recreation of the full loading path that includes incremental stiffness changes.

This nonlinearity problem is typically addressed using, e.g., tangential stiffness method, Newton–Raphson method, modified Newton–Raphson method, aided by numerical integration techniques (Euler or Runge–Kutta method).

To perform back analysis of results obtained in this study, a concept of hysteresis loop modeling based on the Masing's rules was programed in C language (program TS.exe developed by the authors). A detailed description of the algorithm can be found in [51].

Determination of the non-linear $G(\gamma)$ relationship is done through iterative fitting of FEM-simulated $\tau(\gamma)$ to the $\tau(\gamma)$ from TS tests. The goal is to minimize the value of the objective function $L(\cdot)$, which measures fitness of the solution. In order to do that some initial form of the $G(\gamma)$ function is assumed and modified in a way that minimizes the value of the objective function $L(\cdot)$. This interpretation method can only be used when the most probable shape of the G degradation function is known. The optimization process can be limited to finding the coordinates of data points that are corresponding to the experimental dataset (computational nodes).

To minimize $L(\cdot)$ function a non-gradient optimization method was used, specifically, the Nelder–Mead algorithm [52]. The algorithm sequentially generates simplices. For an n-argument objective function the simplices are defined by $n + 1$ vertices. The Nelder–Mead algorithm is implemented in Matlab in the *fminsearch*$(\cdot)$ function. It is based on concept of sequential searching of minimum [53]. To find the solution of a back analysis problem a program developed by the authors (TS.exe) was integrated with Matlab.

Numerical model of the specimen was discretized using 167 TH_{10}^{15} elements (tetrahedral elements with 10 nodes and 15 Gauss points), 338 nodes in total (Figure 8). Geometrical parameters and boundary conditions for the numerical model were set to match the physical parameters of real specimens tested in laboratory experiments (Table 5).

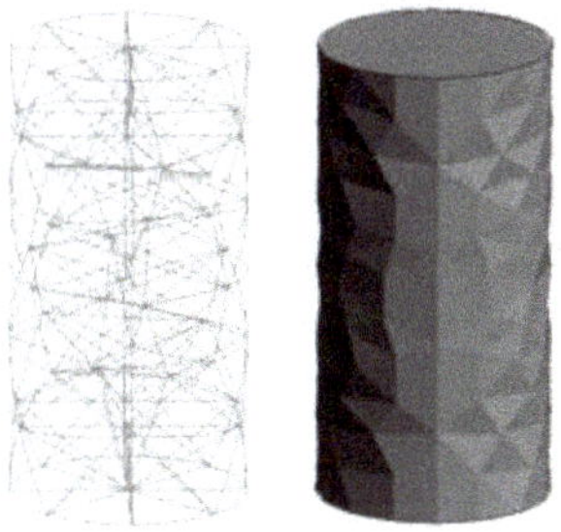

(a) (b)

Figure 8. Discretized specimen model (**a**) Finite Element Method mesh, (**b**) after rendering.

Table 5. Parameters of the numerical model of the specimens *.

Specimen	d [mm]	h [mm]	N [-]	k [-]	f [Hz]	α_{min} [mRad]	α_{max} [mRad]
F1-32551	70	140	1	80	0.05	−0.8982	1.0036
C1-30437	70	140	1	80	0.05	−0.5200	0.7550
F2-30608	70	140	1	80	0.05	−0.7751	0.9017
C2-23608	70	140	1	80	0.05	−0.3367	0.3887

* d—specimen diameter; h—specimen height; N—number of TS cycles; k—number of FEM steps per cycle; f—frequency; α_{min}, α_{max}—minimum and maximum twist angle; F, C—fresh and cured composite.

A numerical simulation of one full cycle took about 14 s (Intel Core i7 4790K@4.4 GHz, RAM 32 GB@1600 MHz). Time of the complete back analysis of one experiment did not exceed 3 h.

3. Results

The confining pressure p, applied in the RC/TS cell on the sample was equal to 50, and 100 kPa. After applying confining pressure, RC tests were performed. Specimens 1 and 2 were tested directly three times: before mixing with chitosan (pure silica sand, specimen P), after specimen preparation mixing (fresh specimens, F1 and F2) and after 7 days of chitosan solution binding (cured specimens, C1 and C2). Figures 9 and 10 present the maximum shear modulus values for each type of specimen for different confining pressure values.

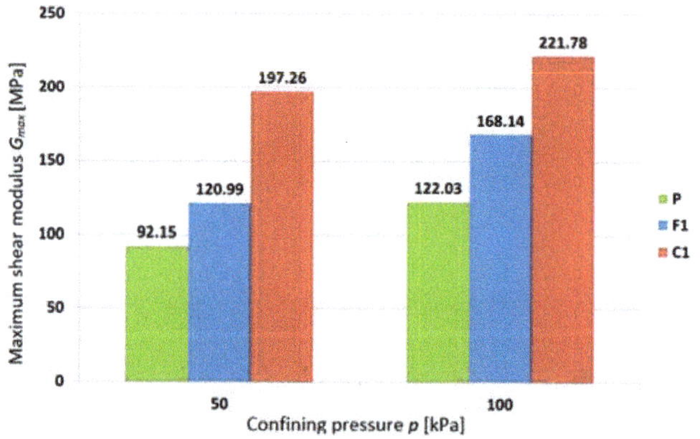

Figure 9. Comparison of RC test results. Specimens P, F1-32551 and C1-30437.

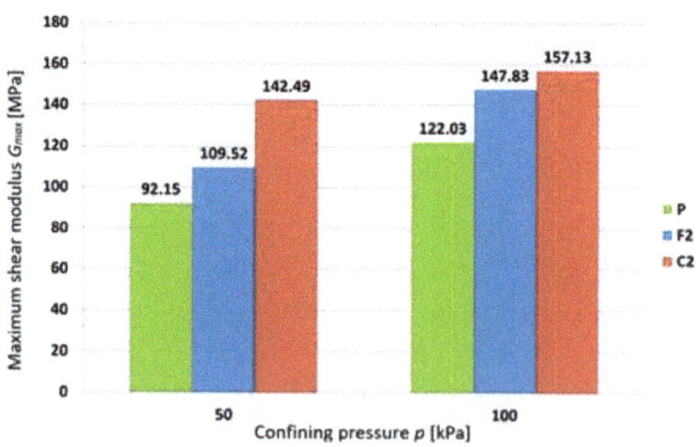

Figure 10. Comparison of RC test results. Specimens P, F2-30608 and C2-23608.

Figures 11 and 12 present comparison of RC test results performed on pure silica sand specimen (P) at different values of confining pressure p.

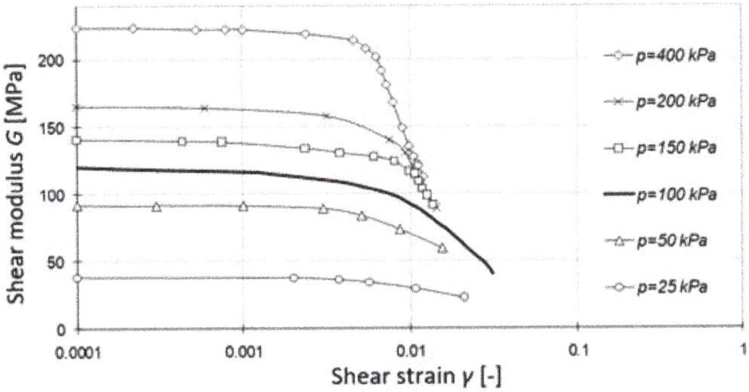

Figure 11. Comparison of RC test results of silica sand specimen (P).

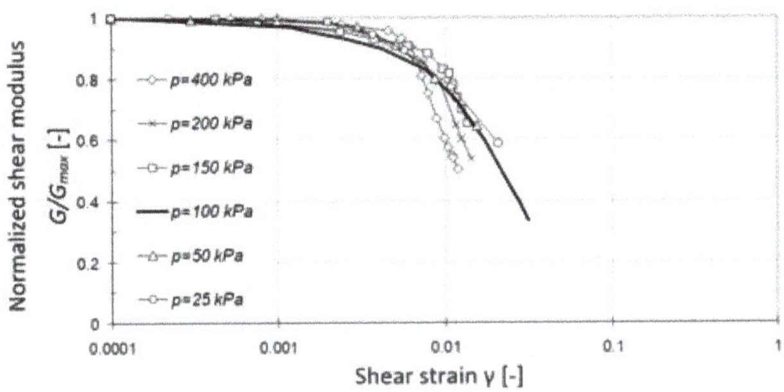

Figure 12. Comparison of RC test results of specimen P—normalized shear modulus.

Figures 13–20 presents the results of back analysis of the TS tests for fresh (F1, F2) and cured (C1, C2) composite specimens.

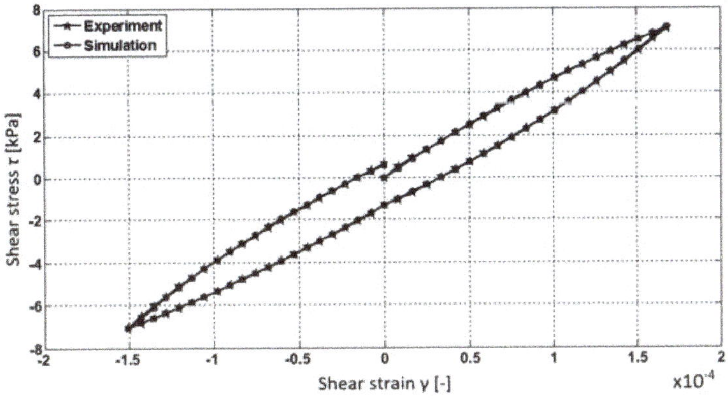

Figure 13. Comparison of back analysis results. Specimen F1-32551.

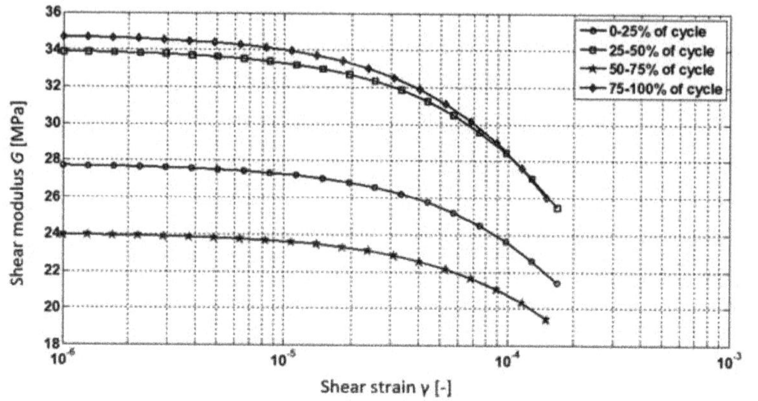

Figure 14. Comparison of back analysis results. Specimen F1-32551.

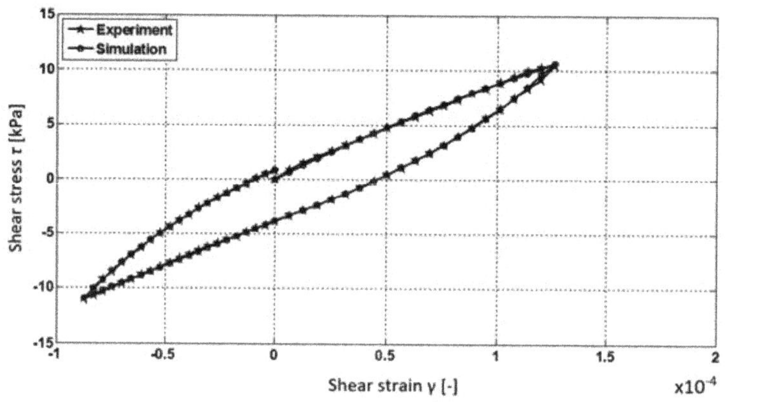

Figure 15. Comparison of back analysis results. Specimen C1-30437.

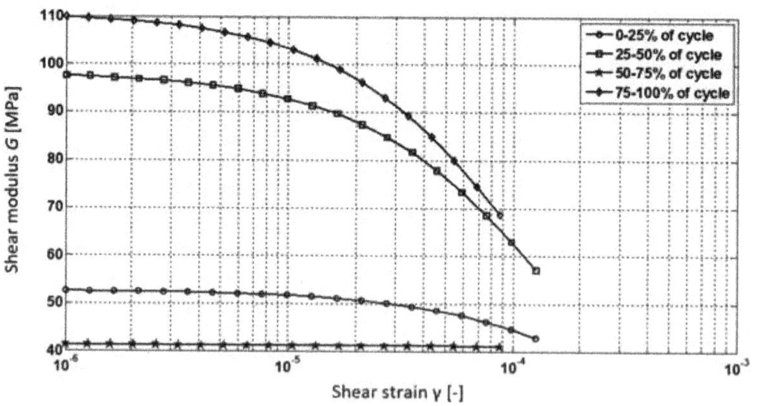

Figure 16. Comparison of back analysis results. Specimen C1-30437.

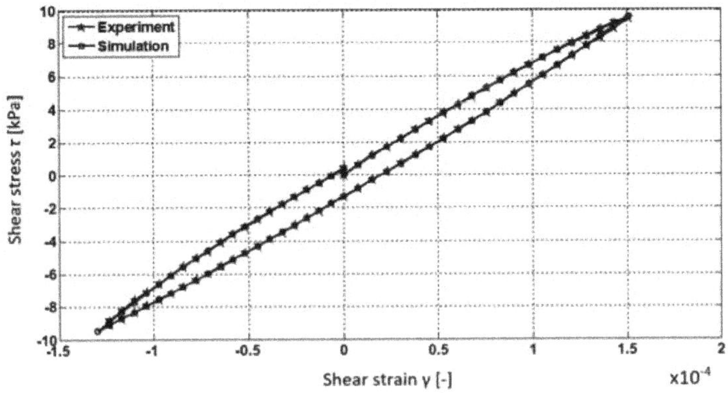

Figure 17. Comparison of back analysis results. Specimen F2-30608.

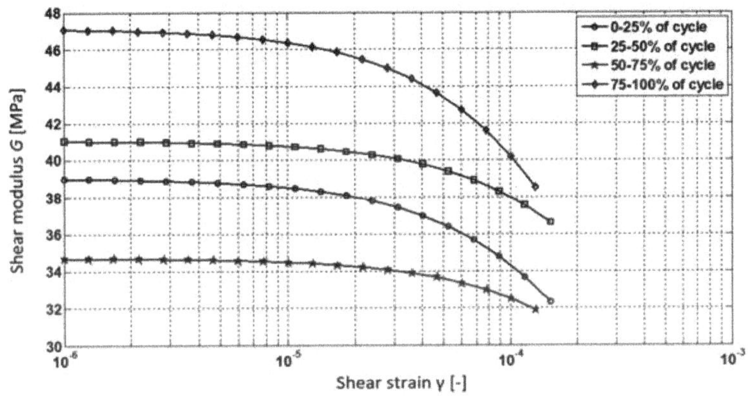

Figure 18. Comparison of back analysis results. Specimen F2-30608.

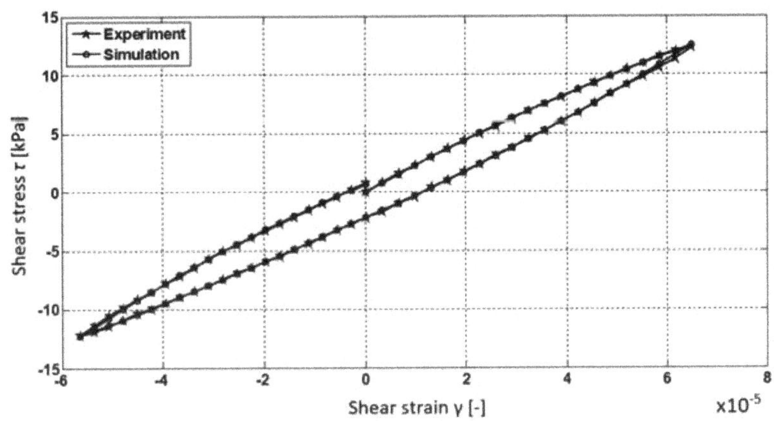

Figure 19. Comparison of back analysis results. Specimen C2-23608.

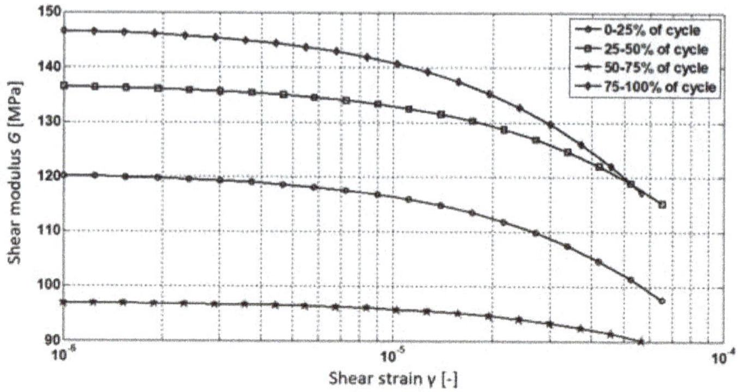

Figure 20. Comparison of back analysis results. Specimen C2-23608.

Figure 23 presents a cured composite specimen after 14 days of degradation in water (20 °C). The observation confirmed that gradual degradation of bonds between soil particles and the biopolymer.

4. Discussion

Comparing results presented in Figures 9 and 10, increments of dynamic shear modulus G_{max} are noticeable. Maximum shear modulus values of fresh composite specimens F1 and F2 are 1.2 to 1.4 times higher than corresponding values for untreated, pure silica sand specimen P. Specimens curing resulted in increasing shear modulus values from 1.1 to 1.6 higher than corresponding values for fresh F1, F2 specimens.

Influence of the chitosan binder and curing process on the stiffness of composites is significant. For cured composite specimens C1 and C2 maximum shear modulus values are 1.3 to 2.1 times higher than values for untreated, pure silica sand specimen P.

It should be stressed that adding chitosan solution to soil enhances soil dynamic properties even before specimen curing. The beneficial effect of chitosan solution before curing was only observed in dynamic (RC) tests. It is an important observation, considering the fact that optimal curing conditions are easier to provide and control in laboratory, but might not be possible to provide on a construction site. Another important issue is obtaining a homogenous soil–chitosan mixture. While it is not a problem in the laboratory conditions, it can be quite difficult at the construction site. Additional research (including in situ tests of stabilized soil) should be conducted to propose and test appropriate mixing technology.

In TS tests there was no significant difference between pure soil specimen and fresh composites' behavior.

In the TS tests (as can be seen in Figures 13–20) the effect of chitosan on composite stiffness is substantial. Initial shear modulus G_0 of cured specimens (C1 and C2) is 1.9 to 3.1 times higher compared to the fresh specimen (F1 and F2 respectively).

Considering G values at the beginning of each of the four phases in full loading cycle (see Table 6), the cured specimens' stiffness is from 1.7 up to 3.3 times higher than for fresh mixtures (see Figures 21 and 22).

Table 6. Four phases in one full cycle of the TS test.

Phase	γ	$d\gamma$	τ	$d\tau$	Load Direction [-]
1	positive	positive	positive	positive	load
2	positive	negative	positive	negative	unload
3	negative	negative	negative	negative	reload
4	negative	positive	negative	positive	re-unload

γ—shear strain; $d\gamma$—change in shear strain; τ—shear stress; $d\tau$—change in shear stress.

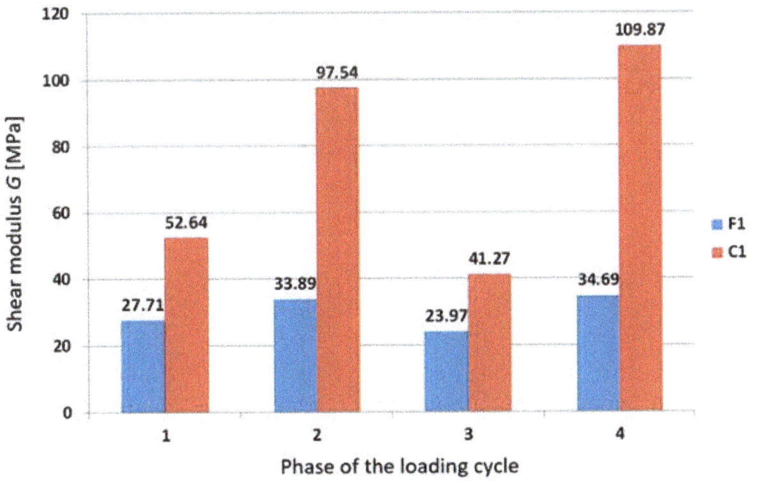

Figure 21. Comparison of back analysis results. Specimens F1-32551 and C1-30437.

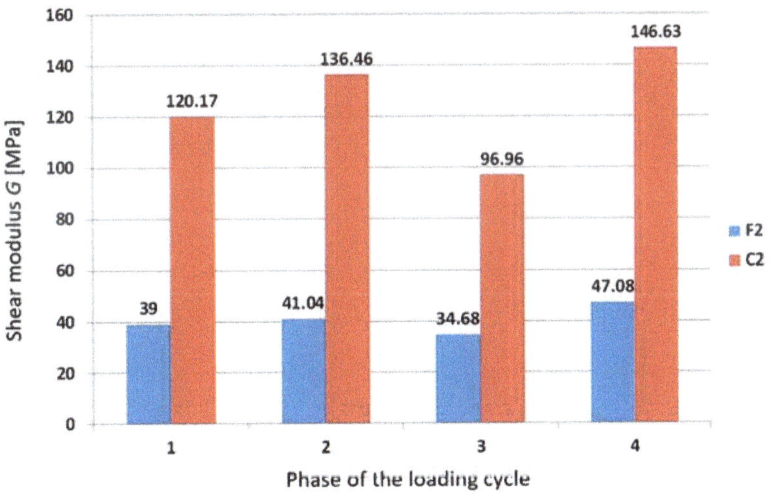

Figure 22. Comparison of back analysis results. Specimens F2-30608 and C2-23608.

Cured soil–chitosan composites exhibit better dynamic properties in response to high frequency and high amplitude loadings than non-enhanced soil. Is suggests chitosan can be considered as a soil stabilizer for temporary structures that are often in danger of stability loss due to, e.g., operation of construction machines.

Stiffness (measured by the shear modulus G) of the composite specimen is several times greater than stiffness of the non-enhanced soil specimen.

The composites exhibit slow degradation in a humid environment, especially when totally submerged in water. Some observable effects of water degradation were documented (Figure 23), but a precise determination of the rate of bond degradation is outside the scope of this study. As it is important to ensure geotechnical structures stability over time, further research is needed to assess the change in chitosan stabilization effect for large scale use of chitosan as a soil stabilizer.

Figure 23. Specimen C2-23608 after 14 days of degradation in water.

Considering the fact that both soil and chitosan are natural materials, the properties of the composite can be affected by many factors and can vary noticeably from specimen to specimen. Large number of tests should be performed to confirm the repeatability of the enhanced mechanical parameters. It should also be noted that the experiments were carried out in the controlled laboratory environment. The use of chitosan in geotechnical structures is a promising solution but it cannot be fully recommended without further studies on how soil–chitosan behavior is influenced by various environmental factors (e.g., seawater exposition). Future research should also focus on understanding the underlying binding mechanisms of chitosan stabilizer so that optimal conditions for binding can be clearly identified and recreated in engineering practice.

5. Conclusions

The following main conclusions can be drawn from this study.

1. Medium-grained low-cohesive soils can be effectively stabilized with chitosan solutions.
2. Adding chitosan to the tested soil specimens improves their shear modulus G substantially (even up to 3 times).
3. Soil–chitosan mixtures show better dynamic properties (dynamic shear modulus) even before the curing process. Therefore, the soil dynamic characteristics are noticeably enhanced even when the optimal curing conditions cannot be met.
4. Even relatively low chitosan concentration solutions (1.5 g of chitosan per 1 kg of dry silica sand) can be a very effective stabilizer.
5. Chitosan can be used as an eco-friendly short-term soil stabilizer. The specimens show signs of degradation after 14 days of being submerged in water.

The findings presented in this study suggest that chitosan solutions can be successfully used as a soil stabilizer. It positively affects soil dynamic properties. This, along with the biodegradability of chitosan makes it a possible solution for stabilizing temporary soil structures (e.g., working platforms) or for regular slope surface stabilization. Chitosan stabilization is a low-cost (as chitosan is a waste material) and easy-to-use (does not require specialized equipment to apply) method for temporary ground improvement. Considering high eco-compatibility of the material, its application is compliant with the main goals of sustainable engineering.

The authors plan to continue their research on chitosan–soil composites. Specifically, more dynamic tests will be conducted and focused on determining other dynamic properties (e.g., the damping ratio) and more in-depth analysis of the underlying binding mechanisms of chitosan solutions will be performed.

Author Contributions: Conceptualization, I.D. and P.S.; methodology, M.B. (Marcin Bujko) and I.D.; software, P.S.; validation, I.D. and P.S.; formal analysis, I.D. and R.O.; investigation, I.D., M.B. (Marcin Bujko) and M.B. (Marta Bocheńska); resources, I.D., P.S. and R.O.; writing—original draft preparation, M.B. (Marta Bocheńska) and M.B. (Marcin Bujko); writing—review and editing, M.B. (Marta Bocheńska) and P.S.; visualization, M.B. (Marta Bocheńska); supervision, I.D. All authors have read and agreed to the published version of the manuscript.

Funding: This research received no external funding.

Institutional Review Board Statement: Not applicable.

Informed Consent Statement: Not applicable.

Data Availability Statement: The data that support the findings of this study are available from the corresponding author, upon reasonable request.

Conflicts of Interest: The authors declare no conflict of interest.

References

1. Das, B.M. *Principles of Geotechnical Engineering*; PWS Engineering: Boston, MA, USA, 1985.
2. Katra, I. Soil Erosion: Dust Control and Sand Stabilization. *Appl. Sci.* **2020**, *10*, 8044. [CrossRef]
3. Santoni, R.L.; Tingle, J.S.; Nieves, M. Accelerated Strength Improvement of Silty Sand with Nontraditional Additives. *Transp. Res. Rec. J. Transp. Res. Board* **2005**, *1936*, 34–42. [CrossRef]
4. Tingle, J.S.; Santoni, R.L. Stabilization of Clay Soils with Nontraditional Additives. *Transp. Res. Rec. J. Transp. Res. Board* **2003**, *1819*, 72–84. [CrossRef]
5. Newman, J.K.; Tingle, J.S.; Gill, C.; McCaffrey, T. Stabilization of Sand Using Polymer Emulsions. *Int. J. Pavements* **2005**, *4*, 1–12.
6. Tingle, J.S.; Newman, J.; Larson, S.; Weiss, C.; Rushing, J. Stabilization mechanisms of nontraditional additives. *Transp. Res. Rec.* **2007**, *1*, 59–67. [CrossRef]
7. Santoni, R.L.; Tingle, J.S.; Webster, S.L. Stabilization of Silty Sand with Nontraditional Additives. *Transp. Res. Rec. J. Transp. Res. Board* **2002**, *1787*, 61–70. [CrossRef]
8. Huang, J.; Kogbara, R.B.; Hariharan, N.; Masad, E.A.; Little, D.N. A state-of-the-art review of polymers used in soil stabilization. *Constr. Build. Mater.* **2021**, *305*, 124685. [CrossRef]
9. Consoli, N.C.; Caicedo, A.M.L.; Beck Saldanha, R.; Filho, H.C.S.; Acosta, C.J.M. Eggshell Produced Limes: Innovative Materials for Soil Stabilization. *J. Mater. Civ. Eng.* **2020**, *32*, 06020018. [CrossRef]
10. Bensaifi, E.; Bouteldja, F.; Nouaouria, M.S.; Breul, P. Influence of crushed granulated blast furnace slag and calcined eggshell waste on mechanical properties of a compacted marl. *Transp. Geotech.* **2019**, *20*, 100244. [CrossRef]
11. Choi, S.-G.; Chang, I.; Lee, M.; Lee, Y.-H.; Han, J.-T.; Kwon, T.-H. Review on geotechnical engineering properties of sands treated by microbially induced calcium carbonate precipitation (MICO) and biopolymers. *Constr. Build. Mater.* **2020**, *246*, 118415. [CrossRef]
12. Soldo, A.; Miletić, M.; Auad, M.L. Biopolymers as a sustainable solution for the enhancement of soil mechanical properties. *Sci. Rep.* **2020**, *10*, 267. [CrossRef] [PubMed]
13. Benzerara, M.; Guihéneuf, S.; Belouettar, R.; Perrot, A. Combined and synergic effect of algerian natural fibres and biopolymers on the reinforcement of extruded raw earth. *Constr. Build. Mater.* **2021**, *289*, 123211. [CrossRef]
14. Losini, A.E.; Grillet, A.G.; Bellotto, M.; Woloszyn, M.; Dotelli, G. Natural additives and biopolymers for raw earth construction stabilization—A review. *Constr. Build. Mater.* **2021**, *304*, 124507. [CrossRef]
15. Mendonça, A.; Morais, P.V.; Pires, A.C.; Chung, A.P.; Oliveira, P.V. A Review on the Importance of Microbial Biopolymers Such as Xanthan Gum to Improve Soil Properties. *Appl. Sci.* **2021**, *11*, 170. [CrossRef]
16. Fatehi, H.; Ong, D.E.L.; Yu, J.; Chang, I. Biopolymers as Green Binders for Soil Improvement in Geotechnical Applications: A Review. *Geosciences* **2021**, *11*, 291. [CrossRef]
17. Shariatmadari, N.; Reza, M.; Tasuji, A.; Ghadir, P.; Javadi, A. Experimental Study on the Effect of Chitosan Biopolymer on Sandy Soil Stabilization. In Proceedings of the 4th European Conference on Unsaturated Soils, E3S Web of Conferences, Les Ulis, France, 19–21 October 2020; Volume 195, p. 06007. [CrossRef]
18. Chang, I.; Lee, M.; Tran, A.T.P.; Lee, S.; Kwon, Y.-M.; Im, J.; Cho, G.-C. Review on biopolymer-based soil treatment (BPST) technology in geotechnical engineering practices. *Transp. Geotech.* **2020**, *24*, 100385. [CrossRef]

19. Chang, I.; Im, J.; Cho, G.-C. Introduction of microbial biopolymers in soil treatment for future environmentally-friendly and sustainable geotechnical engineering. *Sustainability* **2016**, *8*, 251. [CrossRef]
20. Wiszniewski, M.; Skutnik, Z.; Biliniak, M.; Çabalar, A.F. Some geomechanical properties of a biopolymer treated medium sand. *Ann. Wars. Univ. Life Sci. SGGW Land Reclam.* **2017**, *49*, 201–212. [CrossRef]
21. Hataf, N.; Ghadir, P.; Ranjbar, N. Investigation of soil stabilization using chitosan biopolymer. *J. Clean. Prod.* **2018**, *170*, 1493–1500. [CrossRef]
22. Jang, J. A Review of the Application of Biopolymers on Geotechnical Engineering and the Strengthening Mechanisms between Typical Biopolymers and Soils. *Adv. Mater. Sci. Eng.* **2020**, *2020*, 1465709. [CrossRef]
23. Ni, J.; Li, S.-S.; Ma, L.; Geng, X.-Y. Performance of soils enhanced with eco-friendly biopolymers in unconfined compression strength tests and fatigue loading tests. *Constr. Build. Mater.* **2020**, *263*, 120039. [CrossRef]
24. Ni, J.; Hao, G.-L.; Chen, J.Q.; Ma, L.; Geng, X.-Y. The Optimisation Analysis of Sand-Clay Mixtures Stabilised with Xanthan Gum Biopolymers. *Sustainability* **2021**, *13*, 3732. [CrossRef]
25. Santos, V.; Marques, N.; Maia, P.; Lima, M.; Franco, L.; Campos-Takaki, G. Seafood Waste as Attractive Source of Chitin and Chitosan Production and Their Applications Int. *J. Mol. Sci.* **2020**, *21*, 4290. [CrossRef]
26. Zenga, D.; Wua, J.; Kennedy, J. Application of a chitosan flocculant to water treatment. *Carbohydr. Polym.* **2008**, *71*, 135–139. [CrossRef]
27. Şenel, S.; İkinci, G.; Kaş, S.; Yousefi-Rad, A.; Sargon, M.; Hıncal, A. Chitosan films and hydrogels of chlorhexidine gluconate for oral mucosal delivery. *Int. J. Pharm.* **2000**, *193*, 197–203. [CrossRef]
28. Paulraj, M.G.; Ignacimuthu, S.; Gandhi, M.J.; Shajahan, A.; Ganesan, P.; Packiama, S.M.; Al-Dhabib, N.A. Comparative studies of tripolyphosphate and glutaraldehyde cross-linked chitosan-botanical pesticide nanoparticles and their agricultural applications. *Int. J. Biol. Macromol.* **2017**, *104*, 1813–1819. [CrossRef] [PubMed]
29. Benucci, I.; Liburdi, K.; Cacciotti, I.; Lombardelli, C.; Zappino, M.; Nannic, F.; Estia, M. Chitosan/clay nanocomposite films as supports for enzyme immobilization: An innovative green approach for winemaking applications. *Food Hydrocoll.* **2018**, *74*, 124–131. [CrossRef]
30. Ramkumar, R.; Minakshi, M. Fabrication of Ultrathin CoMoO4 Nanosheet Modified with Chitosan and their Improved Performance in Energy Storage Device. *Dalton Trans.* **2015**, *44*, 6158–6168. [CrossRef]
31. Ramkumar, R.; Minakshi, M. A biopolymer gel-decorated cobalt molybdate nanowafer: Effective graft polymer cross-linked with an organic acid for better energy storage. *New J. Chem.* **2016**, *40*, 2863–2877. [CrossRef]
32. Li, Q.; Dunn, E.T.; Grandmaison, E.W.; Goosen, M.F.A. Applications and Properties of Chitosan. *J. Bioact. Compat. Polym.* **1992**, *7*, 370–397. [CrossRef]
33. Morin-Crini, N.; Lichtfouse, E.; Torri, G.; Crini, G. Fundamentals and Applications of Chitosan. In *Sustainable Agriculture Reviews: Chitin and Chitosan: History, Fundamentals and Innovations*; Crini, G., Lichtfouse, E., Eds.; Springer: Cham, Switzerland, 2019; Volume 35, pp. 49–124.
34. Jiménez-Gómez, C.P.; Cecilia, J.A. Chitosan: A Natural Biopolymer with a Wide and Varied Range of Applications. *Molecules* **2020**, *25*, 3981. [CrossRef] [PubMed]
35. Azmana, M.; Mahmood, S.; Hilles, A.R.; Rahman, A.; Bin Arifin, M.A.; Ahmed, S. A review on chitosan and chitosan-based bionanocomposites: Promising material for combatting global issues and its applications. *Int. J. Biol. Macromol.* **2021**, *185*, 832–848. [CrossRef] [PubMed]
36. Wan, M.-W.; Petrisor, I.G.; Lai, H.-T.; Kim, D.; Yen, T.F. Copper adsorption through chitosan immobilized on sand to demonstrate the feasibility for in situ soil decontamination. *Carbohydr. Polym.* **2004**, *55*, 249–254. [CrossRef]
37. Ham, S.; Kwon, T.; Chang, I.; Chung, M. Ultrasonic P-wave reflection monitoring of soil erosion for erosion function apparatus. *Geotech. Test. J.* **2016**, *39*, 301–314. [CrossRef]
38. Aguilar, R.; Nakamatsu, J.; Ramírez, E.; Elgegren, M.; Ayarza, J.; Kim, S.; Pando, M.A.; Ortega-San-Martin, L. The potential use of chitosan as a biopolymer additive for enhanced mechanical properties and water resistance of earthen construction. *Constr. Build. Mater.* **2016**, *114*, 625–637. [CrossRef]
39. Karimi, S. A Study of Geotechnical Applications of Biopolymer Treated Soils with an Emphasis on Silt. Ph. D. Thesis, University of Southern California, Los Angeles, CA, USA, 1998.
40. Ham, S.-M.; Chang, I.; Noh, D.-H.; Kwon, T.-H.; Muhunthan, B. Improvement of surface erosion resistance of sand by microbial biopolymer formation. *J. Geotech. Geoenvironmental. Eng.* **2018**, *144*, 06018004. [CrossRef]
41. Chang, I.; Cho, G.C. Geotechnical behavior of a beta-1,3/1,6-glucan biopolymer-treated residual soil. *Geomech. Eng.* **2014**, *7*, 633–647. [CrossRef]
42. Wichtmann, T. Soil Behaviour under Cyclic Loading—Experimental Observations, Constitutive Description and Applications. Ph.D. Thesis, Karlsruhe Institute of Technology, Karlsruhe, Germany, 2016.
43. Dyka, I.; Srokosz, P.E.; Bujko, M. Influence of grain size distribution on dynamic shear modulus of sands. *Open Eng.* **2017**, *7*, 317–329. [CrossRef]
44. Srokosz, P.E.; Bujko, M.; Bocheńska, M.; Ossowski, R. Optical flow method for measuring deformation of soil specimen subjected to torsional shearing. *Measurement* **2021**, *174*, 109064. [CrossRef]
45. Im, J.; Tran, A.T.P.; Chang, I.; Cho, G.-C. Dynamic properties of gel-type biopolymer-treated sands evaluated by Resonant Column (RC) tests. *Geomech. Eng.* **2017**, *12*, 815–830. [CrossRef]

46. Bhardwaj, N.; Kundu, S.C. Electrospinning: A fascinating fiber fabrication technique. *Biotechnol. Adv.* **2010**, *28*, 325–347. [CrossRef] [PubMed]
47. Jiang, T.; James, R.; Kumbar, S.G.; Laurencin, C.T. Chitosan as a Biomaterial: Structure, Properties, and Applications in Tissue Engineering and Drug Delivery. In *Natural and Synthetic Biomedical Polymers*; Elsevier: Amsterdam, The Netherlands, 2014; pp. 91–113.
48. *ASTM D1557–12e1*; Standard Test Methods for Laboratory Compaction Characteristics of Soil Using Modified Effort (56,000 ft-lbf/ft3 (2700 kN-m/m3)). ASTM International: West Conshohocken, PA, USA, 2012.
49. Srokosz, P.E.; Dyka, I.; Bujko, M.; Bocheńska, M. A Modified Resonant Column Device for In-Depth Analysis of Vibration in Cohesive and Cohesionless Soils. *Energies* **2021**, *14*, 6647. [CrossRef]
50. *ASTM D4015–21*; Standard Test Methods for Modulus and Damping of Soils by Fixed-Base Resonant Column Devices. ASTM International: West Conshohocken, PA, USA, 2021.
51. Srokosz, P.E.; Dyka, I.; Bujko, M. Determination of Shear Modulus of Soil in the RC/TS Apparatus for Designing Offshore Wind Power Plant Foundations. *Pol. Marit. Res.* **2018**, *25*, 69–83. [CrossRef]
52. Nelder, J.A.; Mead, R. A simplex method for function minimization. *Comput. J.* **1965**, *7*, 308–313. [CrossRef]
53. Lagarias, J.C.; Reeds, J.A.; Wright, M.H.; Wright, P.E. Convergence Properties of the Nelder-Mead Simplex Method in Low Dimensions. *SIAM J. Optim.* **1998**, *9*, 112–147. [CrossRef]

applied sciences

MDPI

Article

The Effect of MICP on Physical and Mechanical Properties of Silt with Different Fine Particle Content and Pore Ratio

Yang Zhao [1], Qian Wang [1], Mengnan Yuan [1], Xi Chen [1], Zhiyang Xiao [1], Xiaohong Hao [1], Jing Zhang [2,*] and Qiang Tang [3,*]

1 College of Geosciences and Engineering, North China University of Water Resources and Electric Power, Zhengzhou 450046, China; china.zhaoyang@hotmail.com (Y.Z.); wq13276901507@163.com (Q.W.); ymengnan1223@163.com (M.Y.); cx1055216675@163.com (X.C.); 201401519@stu.ncwu.edu.cn (Z.X.); haoxiaohong@ncwu.edu.cn (X.H.)

2 Department of Neuroimmunology, Henan Institute of Medical and Pharmaceutical Sciences, Zhengzhou University, Zhengzhou 450052, China

3 School of Rail Transportation, Soochow University, Suzhou 215131, China

* Correspondence: sgrzdy@163.com (J.Z.); tangqiang@suda.edu.cn (Q.T.)

Abstract: Microbial-induced calcium carbonate precipitation (MICP) is a new soil remediation technology, which can improve the physical and mechanical properties of soil by transporting bacterial solution and cementation solution to loose soil and precipitating calcium carbonate precipitation between soil particles through microbial mineralization. Based on this technique, the effects of different fine particle content and pore ratio on the physical and chemical properties of silt after reinforcement were studied. The content of calcium carbonate, the ability of silt to fixed bacteria, unconfined compressive strength (UCS), permeability coefficient and microstructure of the samples were determined. The results showed the following: In the process of calcium carbonate precipitation induced by microorganisms, more than 50% bacterial suspension remained on the surface of silt particles and their pores. The higher the bacterial fixation rate of silt, the more $CaCO_3$ was generated during the solidification process. The bacterial fixation rate and $CaCO_3$ content both decreased with the increase in the pore ratio and increased with the increase in the fine particle content. XRD and SEM images show that the calcium carbonate is mainly composed of spherical vaterite and acicular cluster aragonite. There is an obvious correlation between unconfined compressive strength and $CaCO_3$ content of silt. When $CaCO_3$ content accumulates to a certain extent, its strength will be significantly improved. The unconfined compressive strength of silt A with pore ratio of 0.75 and fine particle content of 75% is 2.22 MPa when the single injection amount of cementing fluid is 300 mL. The permeability coefficient of cured silt can be reduced by 1 to 4 orders of magnitude compared with that of untreated silt. In particular, the permeability of MICP-treated silt A is almost impermeable.

Keywords: calcium carbonate precipitation induced by microorganisms; silt; void ratio; fine particle content; unconfined compressive strength; permeability coefficient; XRD; SEM

Citation: Zhao, Y.; Wang, Q.; Yuan, M.; Chen, X.; Xiao, Z.; Hao, X.; Zhang, J.; Tang, Q. The Effect of MICP on Physical and Mechanical Properties of Silt with Different Fine Particle Content and Pore Ratio. *Appl. Sci.* **2022**, *12*, 139. https://doi.org/10.3390/app12010139

Academic Editor: Angeles Sanroman Braga

Received: 12 November 2021
Accepted: 15 December 2021
Published: 23 December 2021

1. Introduction

Microbial-induced calcite precipitation (MICP) is a bacteria-induced bio-mineralization process, which has been paid close attention to civil, infrastructure and environmental engineering [1–3]. Taking advantage of natural biological processes, it contributes to additional cementation at particle to particle contacts of soils, i.e., strength, stiffness or permeability improvement of soils [4–6]. In a more environmentally friendly and sustainable manner, calcium carbonate precipitation induced by microorganisms is a promising solution for soil-related engineering problems.

To date, most studies of microbial-induced calcium carbonate precipitation have focused on treating various kinds of sands [7–9]. Few studies have been done on other soil types. However, Dejong et al. [10] and Mortensen et al. [11] considered calcium carbonate precipitation induced by microorganisms has the potential to improve silt. Tropical residual soil classified as silt was tested by Soon et al. [12,13], and the greatest improvements in shear strength and reduction in hydraulic conductivity achieved are 100% and 90%, respectively. Zamani et al. [14] studied the changes of hydraulic conductivity of fine sand and silt after MICP treatment, and the test proved that the application of MICP reduced the hydraulic conductivity of soil, while improving its strength and stiffness. By adding finer particles changing the grain size distribution, Jiang et al. [15] concluded samples treated by microorganisms had better fine particle content behavior better in the erosion test. Li et al. [16] indicates that incorporation of fly ash in Biological cement reaction is an effective means of increasing the strength of soils. Moreover, void ratio (relative density) is taken into consideration on the improvement of soils. Gao et al. [17] conducted triaxial consolidation and drainage tests on sand with different relative compactness (Dr = 30%, 50%, 90%), and found that the strength and deformation control of biocemented loose and moderately compacted sand could reach or exceed that of compacted sand.

However, experimental data on microbial cemented silt are still very limited considering the variation of both grain size distribution and void ratio. Therefore, the effects of grain size distribution and void ratio on the bio-mineralization process and mechanical properties of microbial cemented silt require further investigation. In the present study, three different types of silt samples with two different void ratios were treated with biological cementation, and the yield of calcium precipitation and uniaxial compression strength (UCS) of the silt samples after biological cementation were evaluated and discussed.

2. Materials and Methods

2.1. Biological Treatment Process

In the study, urease-active strain Sporosarcina pasteurii (CGMCC 1.3687) was used, provided by the China General Microbiological Culture Collection. It is an aerobic bacterium with a diameter of 2 μm to 3 μm. Optical density was measured using a spectrophotometer at a length of 600 nm (OD_{600}), and urease activity was measured immediately after sampling at 25 °C by the conductivity method [18].

Under aerobic batch conditions at 32 °C, bacteria were cultivated in a medium containing 5 g/L soy peptone, 3 g/L beef extract, 0.02 g/L $NiCl_2$, and 20 g/L $CO(NH_2)_2$, at a pH of 8, to late exponential phase with OD_{600} of 3 $\pm$ 0.5 and urease activity of 18 $\pm$ 2 mM urea/min.

2.2. Soil Properties and Gradation

The soil used in the experiment was taken from the floodplain area of the middle and lower reaches of the Yellow River (Zhengzhou and Jinan, China), as shown in Figure 1. According to the particle size larger than 0.075 mm, using the sieving method, and less than 0.075 mm using the hydrometer method to carry out particle separation test on the soil in the two areas, the particle grading curve of the two is shown in Figure 2, in which the fine particle content of the soil in Zhengzhou area is 75%, and the fine particle content of the soil in Jinan area is 47%. In order to obtain the soil with fine particle content between the two areas, the Zhengzhou soil and Jinan soil were mixed in a ratio of 3:7, and the fine particle content of the mixed silt was 63%. As can be seen from the specification, the soil whose particle size is less than 0.075 mm and the mass of the fine-grained group is greater than or equal to 50% of the total mass in the sample is called fine-grained soil (GBJ 145-90), so the three kinds of soil are all fine-grained soil. The liquid limit, plastic limit and plastic index of the three kinds of soil were measured by the combined liquid plastic limit instrument, as shown in Table 1. The three kinds of fine-grained soil were further subdivided according to the plastic diagram (liquid limit 17 mm), and all of them were low-liquid limit silt soil.

Figure 1. The location of the test soil.

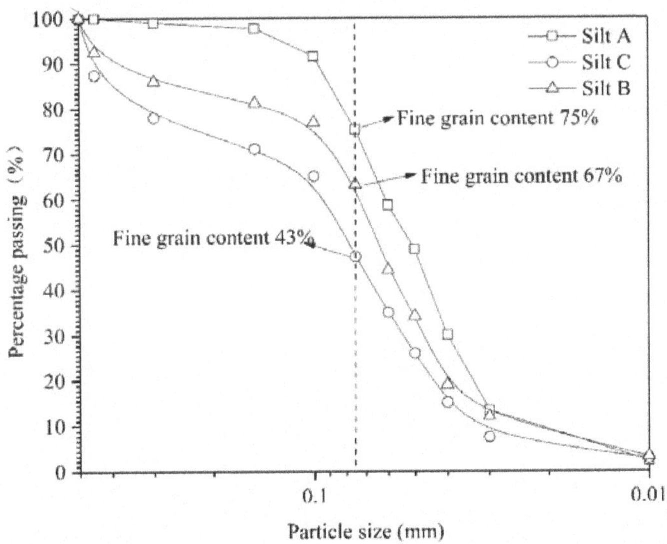

Figure 2. Grading curve of soil particles.

Table 1. Physical characteristics of soil.

Soil Type \ Physical Properties	Liquid Limit (%)	Plastic Limit (%)	The Plastic Index
Silt A	30.5	14.2	16.3
Silt B	24.8	10.9	13.9
Silt C	24.5	11.3	13.2

Two porosity ratios (0.75 and 0.9) were considered, and a plexiglass mold with an inner diameter of 30 mm and a length of 60 mm was used. All samples were prepared by static compaction. For each type of silt sample, control the weight of the soil to achieve the desired two porosity ratios. Six types of samples were prepared by considering three kinds of fine particle contents and two kinds of pore ratios. For the convenience of description, Silt A, Silt B and Silt C are used to represent the three silts according to their fine particle contents from high to low, respectively. The pore ratios of 0.75 and 0.9 are used to represent I and II, respectively.

2.3. MICP Treatment

A two-phase injection scheme with peristaltic pump was used at 25 ± 2 °C [19]. From top to bottom, bacterial suspensions of 30 mL were injected into the sample at 1.0 mL/min, and OD_{600} of the effluent was tested after injecting. Then, 150 mL cementation solutions of the urea and $Ca(CH_3COO)_2$ with 1 M equalmolar concentrations at 1.5 mL/min were pumped. After a curing time of 8 h, cementation treatment was repeated once for each individual sample. In order to study the effect of the injection amount of cement on the physical and chemical properties of silt, the injection amount of cement was increased to 300 mL per round as a control when other factors remained unchanged. $Ca(CH_3COO)_2$ was adopted here following by Zhang et al. [20], who considered $Ca(CH_3COO)_2$ is a better source of calcium compared with $CaCl_2$ and $Ca(NO_3)_2$. The grouting device is shown in Figure 3. After the end of grouting, the treated samples were static and dried, and then measured for various physical and chemical properties. Each test was carried out in three groups of parallel tests. In order to better evaluate the influence of different factors on the physical and chemical properties of silt, additional blank control samples were prepared.

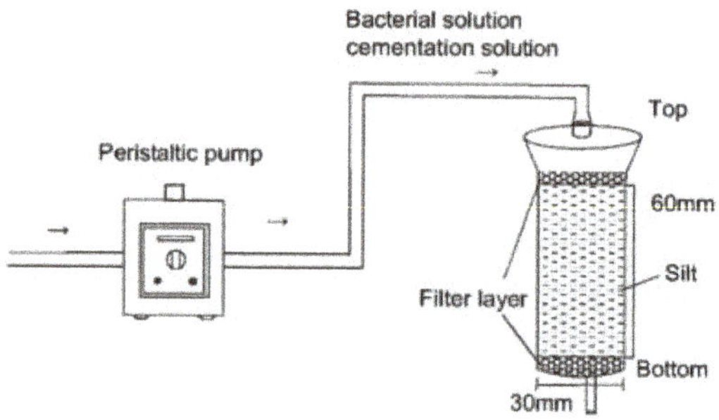

Figure 3. Grouting device.

2.4. Tests and Methods

Samples cured by microorganisms were immersed in deionized water for 48 h to eliminate the soluble substances on the surface, and then the biological treated and control samples were dried at 60 °C for 48 h. The corresponding dry mass was denoted as M_d. Unconfined compressive strength tests were carried out for all the samples at a loading rate of 1 mm/min. After the unconfined compressive strength tests, all fractions of each biological treated sample were collected and mixed with excessive 2 M HCl, and then the final residues were collected by filter paper and were dried at 105 °C for 24 h. The dry mass of the soil was subsequently measured and denoted as M_{silt}. Hence, the mass of calcium carbonate M_{Ca} was calculated through the result of the sample before and after treatment [12,21]:

$$M_{Ca} = M_d - M_{silt} \tag{1}$$

The calcium carbonate content, C_m, was defined as the ratio of the dry mass of calcium carbonate crystals and mixed sands, which can be expressed as:

$$C_m\ (\%) = M_{ca}/M_{silt} \times 100\% \tag{2}$$

The enzyme activity of the bacterial suspension was measured as U_{ini} before grouting, and the enzyme activity of the effluent bacterial suspension was measured as U_{eff} after the bacterial suspension was injected into each sample. The change in bacterial activity can be assessed by the difference in urease activity between the injected bacterial suspension (U_{ini}) and the effluent suspension (U_{eff}) and determined by the following equation [19]:

$$\text{Bacterial activity retention (\%)} = (1 - U_{eff}/U_{ini}) \times 100\% \quad (3)$$

It can be seen that the more retained bacterial cells and urease released in the samples, the higher the retention of bacterial activity.

Mineralogical composition was examined by X-ray diffraction (XRD) test, and microstructure was examined by the mercury intrusion porosimeter (MIP) tests and scanning electron microscope (SEM) tests. For this MIP and SEM test, trimmed small pieces were all from the central of each sample.

3. Results and Discussion

3.1. Generated Calcium Carbonate

As can be seen from Figures 4 and 5, after curing by microorganism, the content of calcium carbonate in silt is between 9% and 20%, and the generated calcium carbonate will attach to the surface of silt particles or fill in the pores between particles, cementing the loose silt particles into columns with a certain strength. This was confirmed by Montoya et al. [22], it is believed that precipitated calcium carbonate influences soil behavior by bonding particles together and densifying the soil by pore space precipitation, which decreases its void ratio. When the injection amount of cementing fluid was increased to 300 mL each round, the production amount of $CaCO_3$ was still significantly increased by controlling other factors unchanged. This indicated that the production amount of $CaCO_3$ could be increased by appropriately increasing the injection amount of cementing fluid. However, the content of calcium carbonate does not increase linearly with the amount of cement injection. On the one hand, the urease activity of the bacterial fluid is limited, which cannot promote the hydrolysis of more urea to produce CO_3^{2-} and Ca^{2+} binding. On the other hand, with the injection of bacterial liquid and cementing liquid, $CaCO_3$ is constantly generated in the pore between particles, which makes the silt structure more compact, and it is difficult to inject subsequent bacterial liquid and cementing liquid.

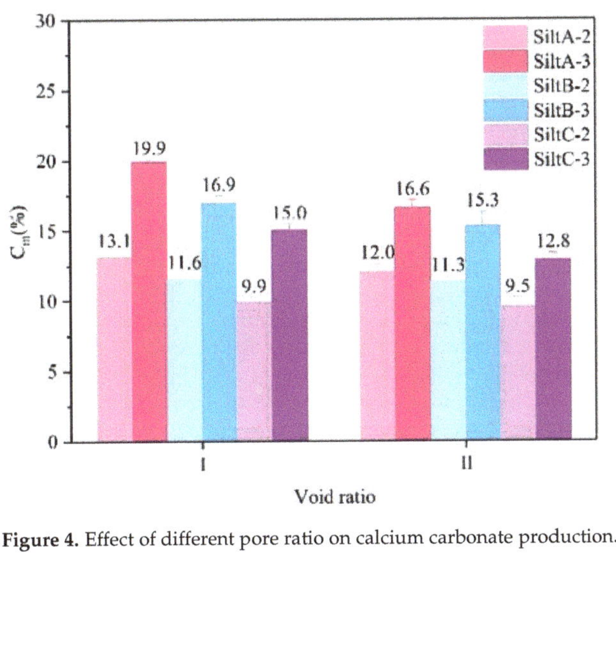

Figure 4. Effect of different pore ratio on calcium carbonate production.

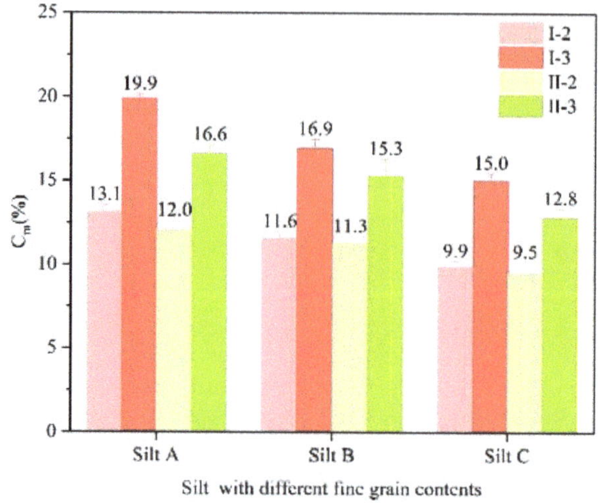

Figure 5. Effect of different fine particle content on calcium carbonate production.

As can be seen from Figure 4, when the content of fine particles is the same, more $CaCO_3$ is generated in silt with a pore ratio of 0.75. This is because the silt sample with a pore ratio of 0.9 has a loose structure and a large proportion of large pores. In the process of grouting, bacterial liquid and cementing liquid are more likely to flow out under the action of peristaltic pump, which is not conducive to the generation of $CaCO_3$. As can be seen from Figure 5, when the porosity ratio is the same, $CaCO_3$ production increases with the increase in fine particles content, fine particles will increase the adhesion rate of bacterial fluid, and $CaCO_3$ is more easily generated on the surface of particles. It can be seen that after curing by microorganism, the maximum $CaCO_3$ production amount is produced in silt A with pore ratio of 0.75 and fine particle content of 75%. Md Touhidul Islam et al. [23] believed that this increase in calcium carbonate precipitation with the increase in clay content indicates that the activity of soil bacteria present in the clay portion of the soil was essential for precipitating calcite.

3.2. SEM/XRD

Calcium carbonate generated from biological treated samples Silt A, Silt B and Silt C was analyzed by XRD, as shown in Figure 6a–c, respectively. Three polymorphs types of calcium carbonate, including calcite, vaterite and aragonite, were identified in three samples by comparison with standard cards. In sample A, the calcium precipitation consisted of 81.7% vaterite, 17.5 aragonite, and 0.8% calcite (Figure 6a). In sample B, the calcium precipitation consisted of 73.8 vaterite, 20.6% aragonite, and 5.6% calcite (Figure 6b) In sample C, the calcium precipitation consists of 60.3% vaterite, 32.3 aragonite, and 7.4% calcite (Figure 6c). This indicates that the calcium polymorph types of the three samples are mainly composed of vaterite and aragonite, and the calcium precipitation polymorph type composition in the samples is not affected by the content of fine particles. Calcite, vaterite and aragonite are the three polymorph types of calcium carbonate, and many researchers have found that calcite is the main calcium polymorph type in the process [24]. There may be two reasons for this change. First, spherical arponite is the primary mineral formed at high levels of hydrolysis (above 18 mM Urea h^{-1}) [25]. In addition, calcium sources influence the polymorph type of calcium [20]. Calcite may be formed when $CaCl_2$ is used, while vaterite and aragonite precipitate easily when $Ca(CH_3COO)_2$ is used. This result is consistent with the research results of using $Ca(CH_3COO)_2$ as calcium source, and the main calcium polymorph types are vaterite and aragonite [20].

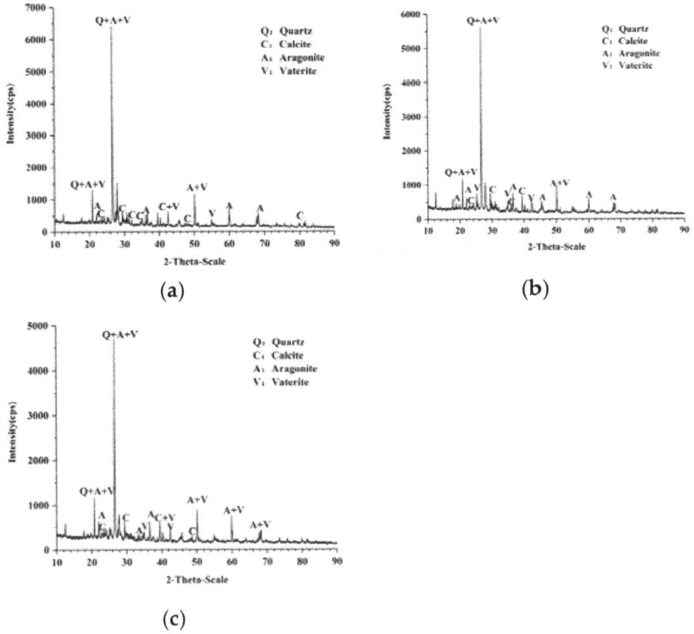

Figure 6. X-ray diffraction results of different silts. (**a**) X-Ray diffraction results of Silt A; (**b**) X-Ray diffraction results of Silt B; (**c**) X-Ray diffraction results of Silt C.

At room temperature, calcite is the most stable crystalline phase, while needle-like aragonite and spheroidal aragonite are metastable crystalline phases, which can be transformed into stable crystalline calcite under certain conditions. Different from calcite and needle-like aragonite, spheroidal aragonite has more excellent water solubility, spherical and porous structure and high specific surface product. Few literatures have explored the relationship between the crystal phase of $CaCO_3$ and experimental conditions in the presence of ammonium salts in a gas–liquid system. Studies have shown that under the condition of pH = 11.1, spheraragonite and calcite are generated at the same time, but with the decrease in pH value, the composition of calcite decreases, until pH = 7.9 only spheraragonite crystal phase.

SEM images of three different fine particle contents of silty soils treated with calcium carbonate precipitation induced by microorganisms were obtained to examine the morphology of samples. After curing by microorganism, obvious mineral precipitation traces can be observed on the surface of silt particles and between pores (Figure 7a–c). At the same time, two different shapes of mineral polymorphs (spherical vaterite and acicular cluster aragonite) can be identified (Figure 7b,c), which are consistent with the corresponding mineral composition revealed by XRD tests. Compared with samples B and C, the calcium polymorph type of sample A is mainly spherical vaterite, and more calcium carbonate polymorphs are deposited on its surface with more uniform size (Figure 7a). With the decrease in fine particle content in silt, the content of acicular cluster aragonite in samples B and C increases (Figure 7b,c). Chu et al. [26] believed that the biological blocking and biocementation of sand particles caused by calcareous precipitation enhanced the strength of sand particles, in other words, a more obvious effect led to a more compact structure.

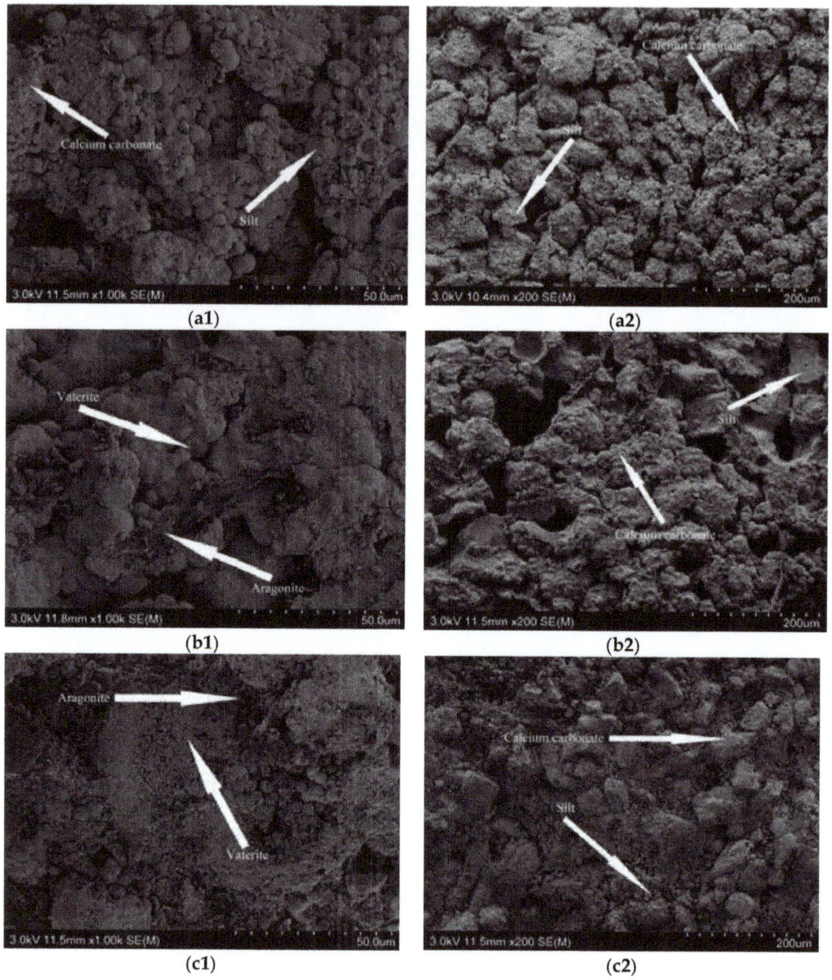

Figure 7. SEM images of different silts. (**a1**,**a2**) SEM image of silt A; (**b1**,**b2**) SEM image of silt B; (**c1**,**c2**) SEM image of silt C.

3.3. *Ability to Retain Bacteria*

As can be seen from Figure 8, when the pore ratio was 0.9, the capacity of residual bacteria decreased by about 10% compared with that of the same type when the pore ratio was 0.75. When the pore ratio was 0.75, the retention capacity of bacteria was between 60% and 80%, which indicated that more than half of the bacteria remained in the silt during the first injection of bacterial liquid. The higher the retention capacity of bacteria, the more Bacillus pasteurii octahedrin used for catalyzing urea hydrolysis. In the process of grouting, the bacteria liquid is usually injected first and then the cement liquid is injected. For the controls, similar to the findings of Harkes et al. [19], more than 70% of bacterial activity could not be retained. The retention capacity of bacteria is related to the bacterial suspension OD_{600} that flows out during the first round of grouting, and has nothing to do with the injection amount of cement liquid. Therefore, there is little difference between OD_{600} when the injection amount of cement liquid is 150 mL and 300 mL. As can be seen from Figure 9, in the case of the same pore ratio, the capacity of silt retention bacteria increases with the increase in fine particle content, which is also related to the adsorption of fine particles. It can be seen from the figure that silt A with pore ratio of 0.75 and fine particle content of 75% has the highest bacteria retention capacity, which is 77%.

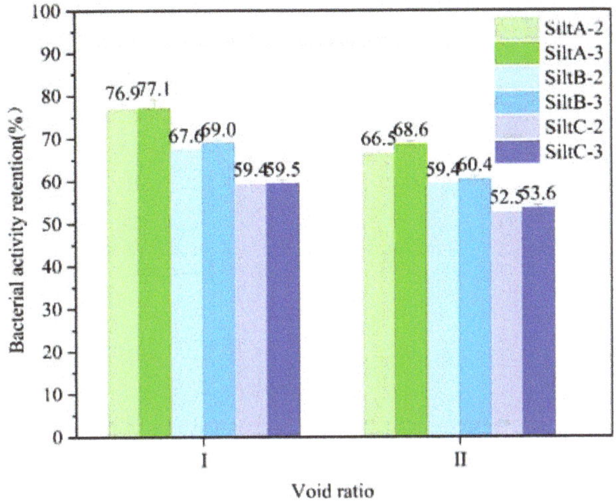

Figure 8. Effect of different pore ratios on the ability to retain bacteria.

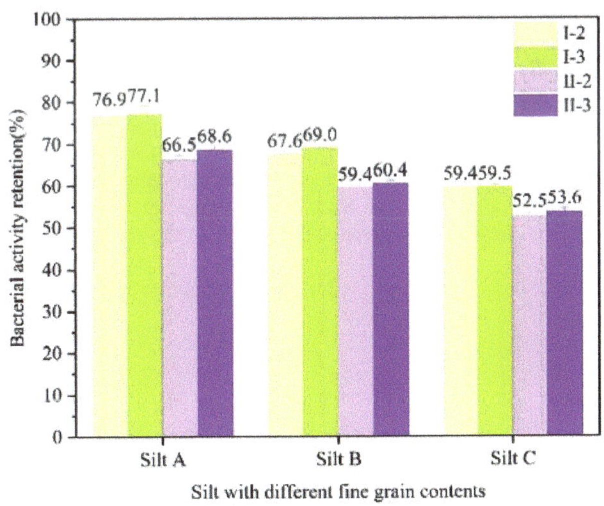

Figure 9. Effect of different fine particle content on the ability to retain bacteria.

3.4. *Relationship between the Ability of Retaining Bacteria and $CaCO_3$*

As can be seen from Figure 10, after curing by microorganism, the $CaCO_3$ content of silt increased with the improvement of silt's ability to retain bacteria, while the ability of silt to retain bacteria increased with the decrease in porosity ratio and with the increase in fine particle content. In particular, the retention ability of silt A with pore ratio of 0.9 and fine particle content of 75% was better than that of silt B with pore ratio of 0.75 and fine particle content of 47%, indicating that the retention ability of silt bacteria was jointly affected by the pore ratio and the fine particle content. When the pore size was large, the fine particle content began to play a decisive role. On the whole, when all kinds of silt have similar retention bacteria capacity, the percentage of $CaCO_3$ is higher when the injection amount of cement solution is 300 mL per round.

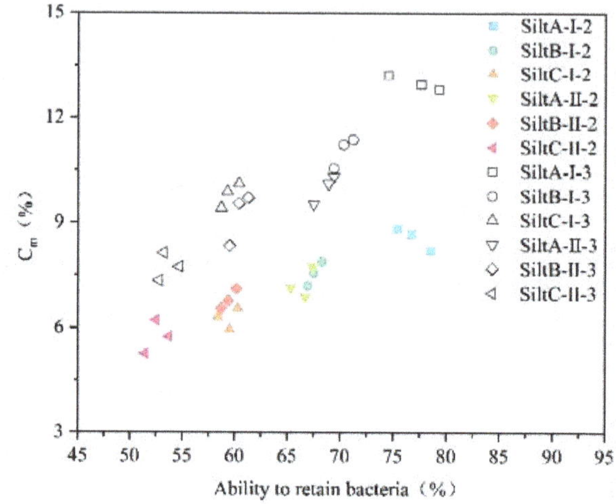

Figure 10. Relationship between retention bacteria capacity and $CaCO_3$.

3.5. Comparison of Uniaxial Compressive Strength

As can be seen from Figures 11 and 12, the unconfined compressive strength of silt after reinforcement is significantly improved compared with that before treatment. Gao et al. [17] reported that the shear strength of the biocemented samples was higher than that of untreated samples at all relative density levels. The more treatments, the higher the shear strength, and the biocemented samples showed greater initial stiffness than untreated samples. In the case of the same amount of bacterial fluid injection, the effect of cement fluid injection on silt strength is not linear. With the injection of cementing fluid, the pores between particles are filled with calcium carbonate, which leads to the decrease in the conversion rate of calcium carbonate, so the strength of silt will not be doubled.

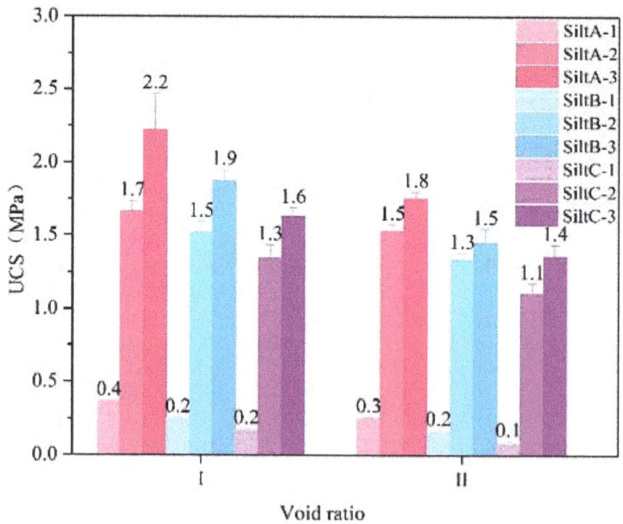

Figure 11. Influence of different pore ratios on unconfined compressive strength.

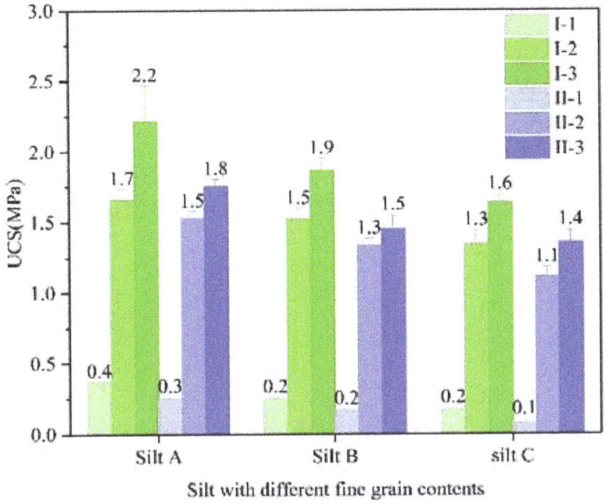

Figure 12. Effects of different fine particle content on unconfined compressive strength.

As can be seen from Figure 12, unconfined compressive strength increases with the increase in fine particle content under the condition of the same porosity ratio, and this rule can be reflected in both untreated silt and strengthened silt. When silt is untreated, the fine particles provide cohesive force for the particles to bond and thus have a certain strength. When the silt is treated through biochemical techniques, the fine particles will make more bacterial liquid and cementation liquid remain in the silt, and generate calcium carbonate on the particle surface and in the pores to cement the soil. Martinez and DeJong [27] reports that when silica sand is used, the degradation of cemented sand particles at the micro scale is controlled by calcite bonds. Therefore, it can be expected that denser soils with higher particle contact numbers will have larger increases in unconfined compressive strength than loose soils.

As can be seen from Figure 11, in the case of the same fine particle content, the unconfined compressive strength of silt with a pore ratio of 0.75 is slightly higher than that of silt with a pore ratio of 0.9 of the same type. When the pore ratio is 0.75, the pore volume of silt sample is small and large. When the microbe induced calcium carbonate precipitation technique was applied to the sample, the number of bacteria in the pores of the samples increased, which increased the nucleation site of calcium carbonate when the cementing solution was injected, and the content of calcium carbonate increased correspondingly and the strength increased. When the pore ratio is 0.9, the pore volume of the silt sample is too large, and the bacterial liquid and the cementing liquid are not easily retained in the sample during the grouting process, resulting in the reduction in calcium carbonate production and strength. By comparing the amount of calcium carbonate in sand and residual soil samples, Soon et al. [13] found that soil particle size had a significant effect on the amount of calcium carbonate. The calcium carbonate content of the treated sand samples is generally higher than that of the residual soil samples, but the improvement effect of microbial-induced calcium carbonate precipitation on the shear strength of the residual soil is better than that of the sand, and the increase in the shear strength of the residual soil increases with the increase in density.

3.6. Relationship between $CaCO_3$ and Unconfined Compressive Strength

As can be seen from Figure 13, unconfined compressive strength of the sample is significantly correlated with $CaCO_3$ content. With the increase in $CaCO_3$ content, the distribution of $CaCO_3$ in the sample is more uniform. When the content of $CaCO_3$ is accumulated to a certain extent, the strength of the sample will be significantly improved. Chu et al. [28] found that there was a strong linear correlation between unconfined compressive strength of solidified sand column and calcite deposition, and the corresponding linear empirical formula was obtained.

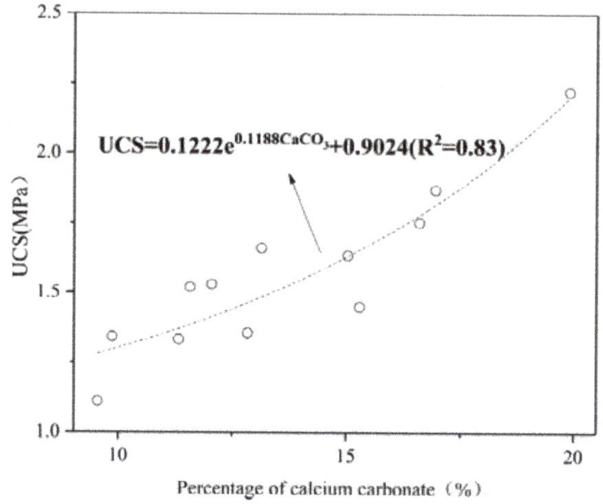

Figure 13. Relationship between the content of calcium carbonate and unconfined compressive strength.

3.7. Comparison of Permeability Coefficient

As can be seen from Figures 14 and 15, after curing by microorganism, the permeability coefficient of Silt A with pore ratio of 0.75 and fine particle content of 75% is 7.08×10^{-8} cm/s, while that of the same type of untreated silt is 1.18×10^{-4} cm/s, which decreases by 4 orders of magnitude. This indicates that calcium carbonate precipitation induced by microorganisms technology can significantly improve the permeability coefficient of silt, making it nearly impervious to water. It can be seen from the figure that, when the injection amount of single round cementing fluid is 150 mL, the permeability coefficient of silt has decreased significantly compared with that of untreated silt, and when the injection amount of cementing fluid continues to increase, the permeability coefficient can still decrease by 1~2 orders of magnitude.

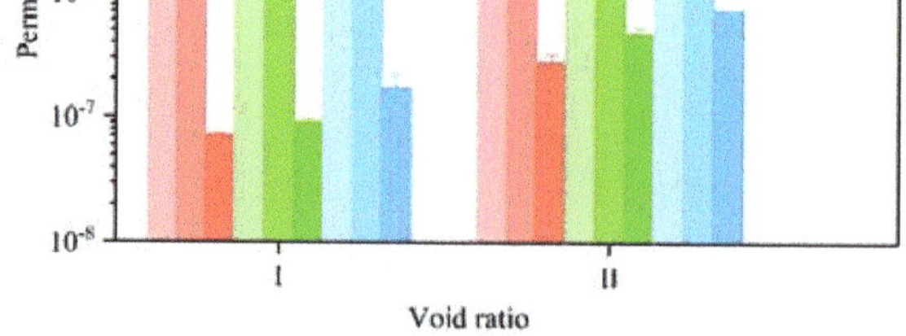

Figure 14. Influence of different pore ratio on permeability coefficient.

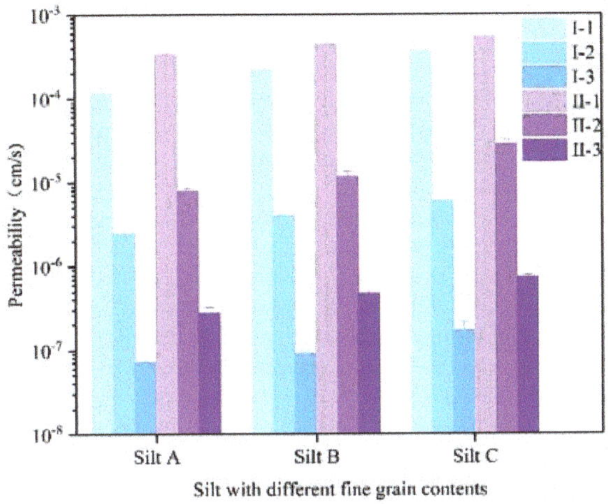

Figure 15. Effect of different fine particle content on permeability coefficient.

As can be seen from Figure 14, when the porosity ratio was 0.75 and the cement injection amount was 150 mL, the permeability coefficient of silt with different fine particle contents decreased by 2 orders of magnitude compared with that of untreated silt. When the pore ratio is 0.9, the permeability coefficient of silt with different fine particle content does not decrease significantly because of the large and abundant pores in the samples. On the contrary, the permeability coefficient of silt with fine particle content of 63% and 47% only decreases by one order of magnitude compared with that without treatment, which is similar to the production amount of $CaCO_3$. All are caused by the large porosity in the sample. Al Qabany et al. [29] conducted permeability tests on samples with different relative densities, and the results showed that with the increase in calcium carbonate precipitation amount, the permeability of both loose and compact samples decreased.

4. Conclusions

In this paper, based on calcium carbonate precipitation induced by microorganisms technology, considering the influence of three different factors, such as fine particle content, porosity ratio and cement injection amount, on the curing effect of silt, a series of tests, such as $CaCO_3$, unconfined compressive strength, permeability, XRD, SEM and so on, show that the macroscopic properties and microstructure of the strengthened silt samples have been improved to varying degrees.

1. After curing by microorganism, the $CaCO_3$ content of silt increased with the decrease in pore ratio and the increase in fine particle content. Therefore, the $CaCO_3$ content of Silt A with pore ratio of 0.75 and fine particle content of 75% was the highest. There is no linear relationship between the content of $CaCO_3$ in silt and the injection amount of cementing liquid. When the injection amount of cementing liquid is 150 mL in a single round, the $CaCO_3$ content is 13%, and when the injection amount is 300 mL, the $CaCO_3$ content is 19%.
2. By XRD analysis, there are three crystal types of calcium carbonate generated in the biological treatment samples, namely aragonite, vaterite and calcite, among which vaterite and aragonite are the main crystal types. The SEM images of three kinds of biological treatment samples of silt with different fine particle contents were obtained, and the crystal shapes were mainly spherical and acicular clusters, which were consistent with the XRD results. With the decrease in fine particle content, the crystal shape changes from spherical uniform distribution to spherical and acicular cluster distribution.

3. In the process of calcium carbonate precipitation induced by microorganisms, the capacity of residual bacteria in silt under different conditions is all over 50%. The higher the capacity of residual bacteria is, the higher the content of $CaCO_3$ in silt after curing. The retention capacity of silt A was affected by silt porosity ratio and fine particle content, and had nothing to do with the injection amount of cementing fluid. The retention capacity of silt A with pore ratio of 0.75 and fine particle content of 75% was the highest, which was 77%.
4. Untreated silt has a certain strength due to its cohesion. After curing by microorganism, the unconfined compressive strength of silt can reach more than 1 MPa. When the content of fine particles is the same, the unconfined compressive strength of silt with a pore ratio of 0.75 is higher than that of silt with a pore ratio of 0.9 because of its small pore volume, large number of pores and strong retention bacteria capacity. There is an obvious correlation between unconfined compressive strength and $CaCO_3$ content of silt. When $CaCO_3$ content accumulates to a certain extent, the strength of silt will be significantly improved.
5. The permeability coefficient of untreated silt is in the range of 1.18×10^{-4}~5.37×10^{-4} cm/s. After curing by microorganism, the permeability coefficient of silt is significantly improved. In particular, when the single injection amount of cementing fluid is 300 mL, the permeability coefficient of Silt A with pore ratio of 0.75 and fine particle content of 75% is 7.08×10^{-8} cm/s, and the silt is close to the impermeable state.

Author Contributions: Conceptualization, Z.X. and X.H.; Data curation, Q.W.; Formal analysis, Q.W. and Z.X.; Funding acquisition, Y.Z., J.Z. and Q.T.; Investigation, Y.Z.; Methodology, Y.Z. and Q.W.; Project administration, Q.T.; Resources, Y.Z.; Software, Q.W., M.Y. and X.C.; Supervision, Y.Z., X.H. and J.Z.; Validation, Q.T.; Writing—original draft preparation, Y.Z.; Writing—review and editing, Q.W. All authors have read and agreed to the published version of the manuscript.

Funding: The research was supported by National Natural Science Foundation of China (No.51409102), Postdoctoral Science Foundation of China (No. 2018M640683), Postdoctoral Research Grant of Henan (No. 001801006), Henan Science Foundation (No. 19A560003; 202102310011), Key Research Projects of Henan Higher Education Institutions (No. 19A320045); National Nature Science Foundation of China (52078317); Natural Science Foundation of Jiangsu Province for Excellent Young Scholars (BK20211597); Bureau of Housing and Urban-Rural Development of Suzhou (2021–2025) and Youth talent teacher of Henan (No. 2018GGJS078).

Institutional Review Board Statement: Not applicable.

Informed Consent Statement: Not applicable.

Data Availability Statement: Not applicable.

Acknowledgments: The Map of China quoted in this paper comes from the Standard Map Service and is produced under the supervision of the Ministry of Natural Resources. The map number is GS(2016)1569.

Conflicts of Interest: The authors declare no conflict of interest.

References

1. Tang, Q.; Gu, F.; Zhang, Y.; Zhang, Y.Q.; Mo, J.L. Impact of biological clogging on the barrier performance of landfill liners. *J. Environ. Manag.* **2018**, *222*, 44–53. [CrossRef] [PubMed]
2. Liu, P.; Zhang, Y.; Tang, Q.; Shi, S. Bioremediation of metal-contaminated soils by microbially-induced carbonate precipitation and its effects on ecotoxicity and long-term stability. *Biochem. Eng. J.* **2021**, *166*, 107856. [CrossRef]
3. Huang, Y.-C.; Chen, J.; Tian, A.-R.; Wu, H.-L.; Zhang, Y.-Q.; Tang, Q. Mechanical properties of fiber and cement reinforced heavy metal-contaminated soils as roadbed filling. *J. Central South. Univ.* **2020**, *27*, 2003–2016. [CrossRef]
4. Cui, M.-J.; Zheng, J.-J.; Zhang, R.-J.; Lai, H.-J.; Zhang, J. Influence of cementation level on the strength behaviour of bio-cemented sand. *Acta Geotech.* **2017**, *12*, 971–986. [CrossRef]
5. DeJong, J.T.; Soga, K.; Banwart, S.; Whalley, R.; Ginn, T.R.; Nelson, D.C.; Mortensen, B.M.; Martinez, B.C.; Barkouki, T. Soil engineering in vivo: Harnessing natural biogeochemical systems for sustainable, multi-functional engineering solutions. *J. R. Soc. Interface* **2010**, *8*, 1–15. [CrossRef]

6. Zhang, Y.; Tang, Q.; Shi, P.; Katsumi, T. Influence of bio-clogging on permeability characteristics of soil. *J. Geotext. Geomembr.* **2021**, *49*, 707–721. [CrossRef]
7. Pan, X.; Chu, J.; Yang, Y.; Cheng, L. A new biogrouting method for fine to coarse sand. *Acta Geotech.* **2020**, *15*, 1–16. [CrossRef]
8. Zhao, J.; Tong, H.; Shan, Y.; Yuan, J.; Peng, Q.; Liang, J. Effects of Different Types of Fibers on the Physical and Mechanical Properties of MICP-Treated Calcareous Sand. *Materials* **2021**, *14*, 268. [CrossRef]
9. Fang, X.; Yang, Y.; Chen, Z.; Liu, H.; Xiao, Y.; Shen, C. Influence of Fiber Content and Length on Engineering Properties of MICP-Treated Coral Sand. *Geomicrobiol. J.* **2020**, *37*, 582–594. [CrossRef]
10. DeJong, J.T.; Mortensen, B.M.; Martinez, B.C.; Nelson, D.C. Bio-mediated soil improvement. *Ecol. Eng.* **2010**, *36*, 197–210. [CrossRef]
11. Mortensen, B.M.; Haber, M.J.; DeJong, J.T.; Caslake, L.F.; Nelson, D.C. Effects of environmental factors on microbial induced calcium carbonate precipitation. *J. Appl. Microbiol.* **2011**, *111*, 338–349. [CrossRef]
12. Soon, N.W.; Lee, L.M.; Khun, T.C.; Ling, H.S. Improvements in engineering properties of soils through microbial-induced calcite precipitation. *J. Civ. Eng.* **2013**, *17*, 718–728. [CrossRef]
13. Soon, N.W.; Lee, L.M.; Khun, T.C.; Ling, H.S. Factors Affecting Improvement in Engineering Properties of Residual Soil through Microbial-Induced Calcite Precipitation. *J. Geotech. Geoenviron. Eng.* **2014**, *140*, 04014006. [CrossRef]
14. Zamani, A.; Montoya, B.M.; Gabr, M.A. Investigating challenges of in situ delivery of microbialinduced calcium carbonate precipitation (MICP) in fine-grainsands and silty sand. *J. Can. Geotech. J.* **2019**, *56*, 1889–1900. [CrossRef]
15. Jiang, N.J.; Soga, K.; Kuo, M. Microbially induced carbonate precipitation (MICP) for seepage-induced internal erosion control in sand-clay mixtures. *J. Geotech. Geoenviron. Eng.* **2016**, *143*, 04016100. [CrossRef]
16. Li, M.M.; Fang, C.L.; Kawasaki, S.; Achal, V. Fly ash incorporated with biocement to improve strength of expansive soil. *J. Sci. Rep.* **2018**, *8*, 2565. [CrossRef] [PubMed]
17. Gao, Y.F.; Hang, L.; He, J.; Chu, J. Mechanical behaviour of biocemented sands at various treatment levels and relative densities. *J. Acta. Geotech.* **2019**, *14*, 697–707. [CrossRef]
18. Whiffin, V.S.; van Paassen, L.; Harkes, M.P. Microbial Carbonate Precipitation as a Soil Improvement Technique. *Geomicrobiol. J.* **2007**, *24*, 417–423. [CrossRef]
19. Harkes, M.P.; van Paassen, L.A.; Booster, J.L.; Whiffin, V.S.; van Loosdrecht, M.C.M. Fixation and distribution of bacterial activity in sand to induce carbonate precipitation for ground reinforcement. *Ecol. Eng.* **2010**, *36*, 112–117. [CrossRef]
20. Zhang, Y.; Guo, H.; Cheng, X. Role of calcium sources in the strength and microstructure of microbial mortar. *Constr. Build. Mater.* **2015**, *77*, 160–167. [CrossRef]
21. Yasuhara, H.; Neupane, D.; Hayashi, K.; Okamura, M. Experiments and predictions of physical properties of sand cemented by enzymatically-induced carbonate precipitation. *Soils Found.* **2012**, *52*, 539–549. [CrossRef]
22. Montoya, B.; DeJong, J.T. Stress-Strain Behavior of Sands Cemented by Microbially Induced Calcite Precipitation. *J. Geotech. Geoenviron. Eng.* **2015**, *141*, 04015019. [CrossRef]
23. Islam, M.T.; Chittoori, B.C.S.; Burbank, M. Evaluating the applicability of biostimulated calcium carbonate precipitation to stabilize clayey soils. *J. Mater. Civ. Eng.* **2020**, *32*, 04019369. [CrossRef]
24. Cheng, L.; Cord-Ruwisch, R.; Shahin, M.A. Cementation of sand soil by microbially induced calcite precipitation at various degrees of saturation. *J. Can. Geotech.* **2013**, *50*, 81–90. [CrossRef]
25. Van Paassen, L.A. Biogrout, Ground Improvement by Microbial Induced Carbonate Precipitation. Ph.D. Thesis, Delft University of Technology, Delft, The Netherlands, 2009.
26. Chu, J.; Stabnikov, V.; Ivanov, V. Microbially induced calcium carbonate precipitation on surface or in the bulk of soil. *J. Geomicrobiol.* **2012**, *29*, 544–549. [CrossRef]
27. Martinez, B.C.; DeJong, J.T. Bio-mediated soil improvement: Load transfer mechanisms at the micro-and macro-scales. Proceedings of conference on advances in ground improvement: Research to practice in the United States and China. *J. ASCE Geotech. Spec. Publ.* **2009**, *188*, 242–251.
28. Chu, J.; Ivanov, V.; Naeimi, M.; Stabnikov, V.; Liu, H.-L. Optimization of calcium-based bioclogging and biocementation of sand. *Acta Geotech.* **2014**, *9*, 277–285. [CrossRef]
29. Qabany, A.; Soga, K. Effect of chemical treatment used in MICP on engineering properties of cemented soils. *Géotechnique* **2013**, *63*, 331–339. [CrossRef]

applied sciences

MDPI

Article

Effects of Hydroxypropyl Methylcellulose (HPMC) on the Reinforcement of Sand by Microbial-Induced Calcium Carbonate Precipitation (MICP)

Wanyi Zhu [1,2], Mengnan Yuan [1], Fanmin He [3], Yang Zhao [1,*], Zhiyang Xiao [1], Qian Wang [1], Fanyou Meng [1] and Qiang Tang [4,*]

1 College of Geosciences and Engineering, North China University of Water Resources and Electric Power, Zhengzhou 450046, China; 15981919387@163.com (W.Z.); ymengnan1223@163.com (M.Y.); 201401519@stu.ncwu.edu.cn (Z.X.); wq13276901507@163.com (Q.W.); 18838917536@163.com (F.M.)
2 Chengdu Xingcheng Capital Management Co., Ltd., Chengdu 610041, China
3 Chengdu Surveying Geotechnical Research Insitiute Co., Ltd. of MCC, Chengdu 610023, China; simba163@163.com
4 School of Rail Transportation, Soochow University, Suzhou 215131, China
* Correspondence: china.zhaoyang@hotmail.com (Y.Z.); tangqiang@suda.edu.cn (Q.T.)

Abstract: Microbial-induced calcium carbonate precipitation (MICP) is a new technology used for reinforcing soils through microbial mineralization. However, the existing MICP treatment technology is more suiTable for deeper soils due to its high permeability. In this study, HPMC, a cohesive material combined with *Sporosarcina pasteurii*-induced calcium carbonate precipitation was used to improve the surface layer of the soil. It was also tested in different contents of bacterial solutions and cementation solutions and with a different number of MICP treatments, and was analyzed and discussed by measuring the surface layer strength, calcium carbonate content generated in the crust, rainfall erosion resistance, wind erosion resistance, and ammonia retention rate of the specimens. The microstructure of the samples was investigated by XRD and SEM. It was shown that the addition of HPMC and increasing the number of MICP treatments were effective in the increase in the calcium carbonate content in the surface crust, but had no effect on the total amount of induced calcium carbonate. The combination of HPMC addition and MICP technology can effectively reduce rainwater scour loss and wind erosion loss, and increase its rainfall erosion resistance and wind erosion resistance. Rainfall losses were reduced by up to 30% in specimens treated with HPMC for MICP technology. When the wind speed was 12 m/s, the maximum mass loss rate of the specimens treated with HPMC for MICP was only 0.828%. The addition of HPMC can effectively improve the ammonia absorption rate and reduce the release of ammonia in the process of MICP technology, which is of great significance for environmental protection. The microstructure showed that the addition of HPMC and the increase in the number of treatments using MICP technology can make the surface structure of the specimens more compact, and the calcium carbonate can more effectively fill the pores and cement the soil particles, while the addition of HPMC may not change the calcium carbonate crystal type.

Keywords: MICP; calcium carbonate; HPMC; surface strength; ammonia retention rate; wind erosion resistance

Citation: Zhu, W.; Yuan, M.; He, F.; Zhao, Y.; Xiao, Z.; Wang, Q.; Meng, F.; Tang, Q. Effects of Hydroxypropyl Methylcellulose (HPMC) on the Reinforcement of Sand by Microbial-Induced Calcium Carbonate Precipitation (MICP). *Appl. Sci.* **2022**, *12*, 5360. https://doi.org/10.3390/app12115360

Academic Editors: Jaroslaw Rybak, Gabriele Chiaro, Małgorzata Jastrzębska and Krystyna Kazimierowicz-Frankowska

Received: 5 April 2022
Accepted: 24 May 2022
Published: 25 May 2022

1. Introduction

Soil surface treatment can improve soil erosion resistance, strength, and stability, which is of great significance in soil–water conservation and dust prevention. It is practical to improve the soil surface using physical methods (such as covering dust-proof net) and using chemical substances (such as cement, lime, fly ash, gypsum, surfactant, polymer, asphalt products, waterborne polyurethane, etc.) [1]. In recent years, the application of MICP in soil improvement has attracted extensive attention [2]. MICP technology mainly uses bacteria to produce urease to the hydrolyze urea and generate $CaCO_3$ to achieve cementation and reinforcement. Although MICP technology presents greater implementation complexity due to the use of living organisms, and has a higher cost compared with traditional technologies [1], it has broad application prospects in soil improvement, concrete crack repair, erosion prevention, cultural relic restoration, and heavy metal pollution control [3–7] due to its environmental friendliness and sustainability [8,9].

MICP can effectively cement soil particles together and improve the performance of soil. Cheng et al. [10] used surface filtration technology to place bacteria and cementation solution. Under high permeability, bacteria and cementation solutions could be used alternately many times to fix bacteria in a column of more than 1 m, and the homogenization of the improved sand column reached a reasonable degree. Van Paassen et al. [11] used MICP technology to carry out a reinforcement test on a 100 m^3 large sand foundation, which effectively improved the bearing capacity of the sand foundation. Moreover, soil surface could also be reinforced by MICP. Stabnikov et al. [12] used a mixture of calcium salts, urea, and bacterial suspension to make a crust on the surface of the sandy soil, which reduced the permeability of the sand layer from 10^{-4} m/s to 16×10^{-7} m/s (14 mm/d). Xiao et al. [13] used a surface spraying method and an MICP technique to surface treat a series of model slopes prepared by different grades of sand, which effectively reduced the rainfall erosion of sandy slopes. Liu et al. [14] applied MICP treatment to five groups of soil slope model samples, and showed that MICP treatment is an effective method to control the surface erosion of clayey soil by analyzing the soil erosion pattern and soil loss rate. Zomorodian et al. [15] studied the application of the MICP technique in different grades of silica sand and carbonate rocks, determined the optimum reagent concentration, and investigated the variation pattern of shear strength and resistance to wind erosion after a number of different treatments. The wind tunnel tests showed that the treated crustal sands could remain stable at a wind speed of 20 m/s.

The desired result of surface reinforcement should be a significant increase in soil strength within the shallow layer [16]. However, the existing MICP treatment solutions, such as grouting or surface penetration, may be difficult to achieve this desired treatment due to their higher permeability, which is more suiTable for deeper treatment. Due to the large permeability of the soil to be treated, existing studies generally use micro-dose injection of MICP solution in order to make the MICP treatment solution and to more precisely reinforce the shallow soil. In addition, in order to achieve the desired treatment effect, the MICP solution is generally infiltrated several times. Salifu et al. [17] injected the bacterial solution and the cementing solution into the soil 18 treatments. In the experiments of Shanahan and Montoya et al. [18], they also made multiple infiltrations to reach the target due to the large permeability.

Materials such as hydrogels (xcan, ganlan) or polyvinyl alcohol (PVA) can reach such shallow depths of cementation, whereby the crust is formed by cementing the shallow layers to improve the soil surface [19]. Hamdan et al. [20] used hydrogel-assisted enzyme-induced carbonate precipitation (EICP), which showed that hydrogels retained water for a longer time, reduced the infiltration of solution in the soil, prolonged the reaction time, improved the efficiency of precipitation, and enhanced the formation of the crust. Sun et al. [21] utilized a biopolymer-assisted EICP approach to improve soil water retention and reduce permeability by retaining carbonate around soil particles. Based on MICP technology, Wang et al. [22] used polyvinyl alcohol (PVA) solution instead of water to prepare cementitious solution. The viscous polymer solution immobilized bacteria and cementitious medium on the surface, promoted calcium carbonate precipitation, formed uniform soil crust on the surface area, and reduced the corrosion ability of the soil.

Hydroxypropyl methylcellulose (HPMC) is abundant, cheap, and biocompatible [23]. A gel can be formed by heating the solution. HPMC, because of its good water retention performance, is widely used in cement mortar as water retention agent and retarder, to improve the hardening strength, save the amount of cement, and then achieve the purpose of cost-saving. Because of its viscosity, HPMC is combined with MICP technology to react in the soil surface better, forming a hard crust on the surface and enhancing the surface strength.

As mentioned above, MICP reinforcement of surficial soil is usually not ideal due to the high permeability of the soil; therefore, it has great potential for research on effectively stabilizing the surface layer of soil via MICP technology. In this study, a series of laboratory experiments were conducted to investigate the feasibility of using MICP and HPMC in the surficial treatment of soil for the purpose of improving the properties of the soil surface. The surficial treatment and ammonia adsorption were achieved by utilizing MICP and HPMC on the soil surface. The macro-scale characteristics of the treated specimens were studied, including the surface strength of sand, the content of $CaCO_3$, and the resistance to wind and rain erosion. Scanning electron microscopy (SEM) and X-ray diffraction (XRD) were used to further analyze the microstructure and formation mechanism under different treatments.

2. Materials and Methods

2.1. Biological Treatment Process

The microorganism used in this study was *Sporosarcina pasteurii*, a kind of bacteria that can produce urease [24], purchased from China General Microbiological Culture Collection Centre (CMCC1.3687). In this study, optical density was measured by a spectrophotometer at a wavelength of 600 nm (OD600), and the urease activity of bacteria was determined at 25 °C by the conductivity method [25].

Under aerobic conditions at 32 °C, first, the bacteria were pre-cultured for about 8 h in a medium containing peptone (5 g/L), beef extract (3 g/L), and urea (20 g/L). Then, the bacteria were cultured in a medium containing yeast powder (20 g/L) and ammonium sulfate (10 g/L) at 5% inoculum for about 24 h. Both media used in the study had a pH of 8 [26]. The urease activity of the bacterial solution used in the study was controlled at 22 ± 2 mM urea/min, and the OD600 was 3 ± 0.5.

2.2. Properties of the Soil and HPMC

The gradation curve of the tested soil is shown in Figure 1, wherein D_{10}, D_{30}, and D_{60} were 0.02 mm, 0.18 mm, and 0.62 mm, respectively. In addition, the coefficient of uniformity (C_u) and the coefficient of gradation (C_c) of the soil sample were 3.10 and 2.61, respectively. According to ASTM D2487, the tested soil used in this study was silty sand, which was sterilized at 160 °C for 24 h.

In this study, HPMC was used as the gelling material, which was purchased from Tianjin Kemeiou Chemical Reagent Co., Ltd. (Tianjin, China). HPMC is a fibrous white powder at room temperature. HPMC has relatively stable properties, superb thickening, cohesiveness, and good compatibility with water. Moreover, HPMC has good film-forming properties and is safe, non-toxic, and non-polluting. The basic physical parameters of HPMC are shown in Table 1. The viscosity of HPMC used in this study was 25 mpa.s.

Table 1. Basic physical parameters of HPMC.

Name	Density	Specific Gravity	Viscosity (mpa.s)	Surface Tension (dyn/cm)
HPMC	0.5	1.3	5~200,000	42~56

2.3. Sample Preparation and MICP Procedure

In this study, cuboid boxes made of transparent plastic were used as molds, whose length, width, and height were 100 mm, 70 mm, and 40 mm, respectively. To evenly pour the sand into the mold, a funnel was used [27], and the height was flush with the mold. A number of uniform 1 mm holes were drilled at the bottom of each mold to allow the gas to be discharged in the pores during grouting and the gas generated during MICP.

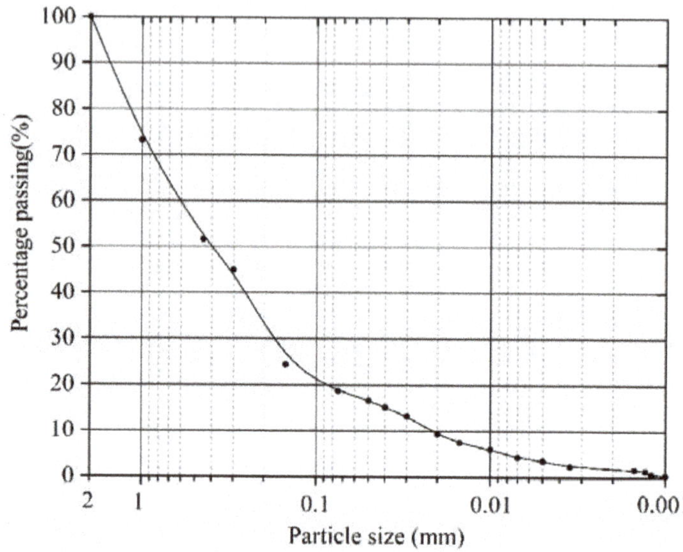

Figure 1. Particle gradation curve.

MICP treatment was performed in the laboratory at 25 ± 2 °C. The cementation solution is a mixed solution of urea (1.5 M) and $CaCl_2$ (1.0 M). The ratio of bacterial solution to cementation solution was 1:4. There were four gradients in total, and each gradient was sprayed once, twice, and three times for surface MICP treatment, with an interval of 12 h. The bacterial solution and the cementation solution were mixed in a ratio of 1:4, and the mixed solution was sprayed evenly on the surface layer of the specimen using a spray bottle. Four solution volume gradients were set in the test, and the ratio of spraying was controlled when spraying the mixed solution so that the same volume of solution was sprayed under the same gradient. The mixed solutions were sprayed at 0.7, 1.1, 1.4, and 1.7 mL/cm^2 under the four gradients. Among them, for the HPMC combined with MICP-treated specimens, HPMC was dissolved in a mixture of bacterial solution and cementation solution at room temperature of 25 °C before spraying at the same ratio as the MICP-treated specimens.

2.4. Tests and Methods

The content of $CaCO_3$ was determined using acid digestion [28]. It is worth noting that the determination of the $CaCO_3$ content was not carried out on the whole sample, but separated the crust formed after MICP treatment from the lower soil to measure the content of calcium carbonate in each part.

Using an ultraviolet–visible spectrophotometer, the concentration of ammonium ion was tested using the Nessler method [26]. Under a constant temperature of 25 °C, the MICP-treated sample was placed in a conical flask with distilled water and oscillated at a rate of 120 r/min for 12 h. Then, 50 mL of supernatant was directly taken, and the absorbance was measured. The ammonia concentration in the supernatant can be calculated as follows:

$$\rho_N = \frac{A_s - A_b - a}{b \times V} \tag{1}$$

where ρ_N is the mass concentration of ammonia in the supernatant (mg/L); A_s is the absorbance of water specimen; A_b is the absorbance of blank control; a is the intercept of calibration curve; b is the calibrate the slope of the curve; V is the volume of the specimen solution.

The concept of the ammonia retention rate was introduced in the study. Ammonia retention rate N was defined as the ratio of ammonia amount after MICP complete reaction to ammonia amount before MICP. The ammonia retention rate N was calculated using the following formula:

$$N = \frac{M_2}{M_1} \quad (2)$$

where M_2 is the mass of ammonia in the specimen after reaction; M_1 is the mass of ammonia in the sample before the reaction.

The soil particles washed out by rainfall were collected by simulating the scouring of the surface layer of the MICP cured specimens. Three rainfall intensities of 0.8, 1.5, and 2.5 mm/min were set for this test, corresponding to light, medium, and heavy rainfall, respectively. The slope of the test specimen was adjusted to 30°, and the continuous scouring was carried out for 10 min under each rainfall intensity. The mass loss before and after the simulated rainfall test was calculated. The apparatus used in the trial is shown in Figure 2.

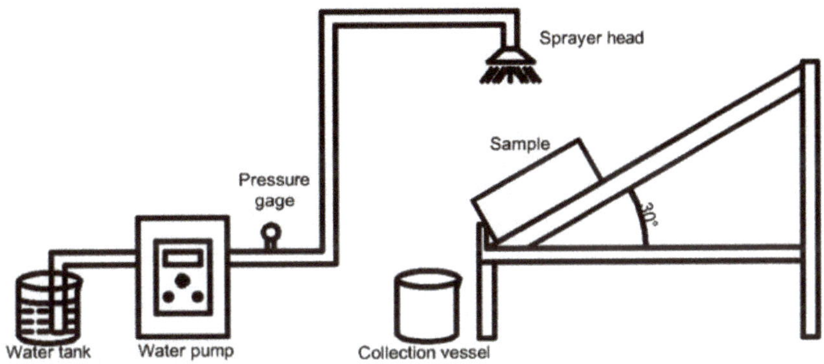

Figure 2. Rainfall device diagram.

Using the apparatus in Figure 3, wind erosion resistance tests were performed on the MICP-treated and MICP combined with HPMC-treated specimens at wind speeds of 4 and 12 m/s, respectively. The mass loss of the specimens before and after the wind erosion tests was measured, and the mass loss rate of the specimens was calculated for different amounts of MICP solution and at different times of treatment.

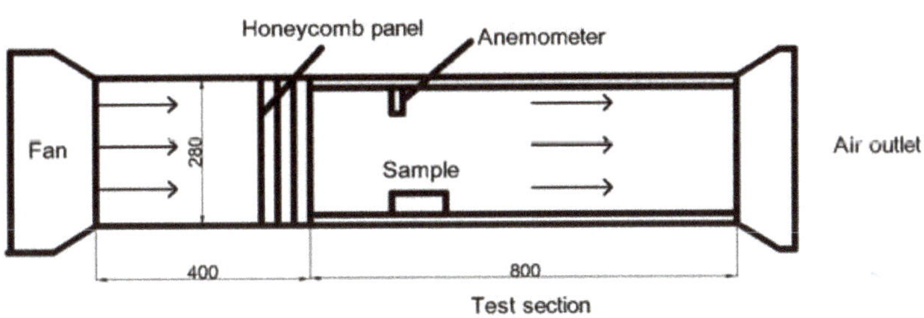

Figure 3. Diagram of wind erosion resistance device.

There are four gradients and three processes for MICP grouting. The four gradients are I: 10 mL bacterial solution + 40 mL cementation solution; II: 15 mL bacterial solution + 60 mL cementation solution; III: 20 mL bacterial solution + 80 mL cementation solution; IV: 24 mL bacterial solution + 96 mL cementation solution. The three processes were divided into one time, two times, and three times to spray uniformly on the surface of the specimen, with an interval of 12 h. The solution with HPMC composition was sprayed one time, two times, and three times as G1, G2, and G3, respectively. The viscosity of the solution was 25 mpa.s by adding HPMC. Similarly, the experiment was carried out according to the above scheme without the addition of HPMC and was compared with the specimen with the addition of HPMC for MICP treatment. Among them, the solution without HPMC component was sprayed once, twice, and three times, which were respectively marked as G4, G5, and G6. The samples after MICP treatment were dried for 48 h in an oven at 60 °C. Under the same process and the same amount of solution, eight samples were prepared, respectively, with and without HPMC. The surface strength of 2 samples was measured by a pocket penetrometer, and the samples collected after the determination were tested for ammonia mass concentration. Two samples were tested for calcium carbonate content in the upper crust and lower crust, two specimens were tested for rain erosion resistance under simulated rainfall conditions, and two specimens were tested for wind erosion resistance in a wind tunnel instrument.

3. Results and Discussion

3.1. Calcium Carbonate Content

The acid digestion method was used to measure the content of calcium carbonate in specimens treated by MICP. The content of calcium carbonate generated by each experimental group is shown in Figure 4, and the percentage of calcium carbonate production in the 3–7 mm surficial layer in the total amount of calcium carbonate generated is shown in Figure 5. The results showed that the precipitation of calcium carbonate at the thinner skin layer was effectively improved by the addition of HPMC, and the highest percentage of calcium carbonate to total calcium carbonate in the hard shell reached 78.05%. However, as for the total $CaCO_3$ content in the samples, the total $CaCO_3$ mass of the samples with and without HPMC is roughly the same after MICP treatment. This phenomenon illustrated that HPMC did not improve the total content of calcium carbonate, but could enhance the viscosity of reactants and increase the time of liquid infiltration so that more calcium carbonate precipitation was generated at the thinner surficial layer. At the same reaction amount, the highest calcium carbonate produced in the surficial layer with the addition of HPMC for MICP treatment could reach 1.82 times higher than that with only MICP treatment.

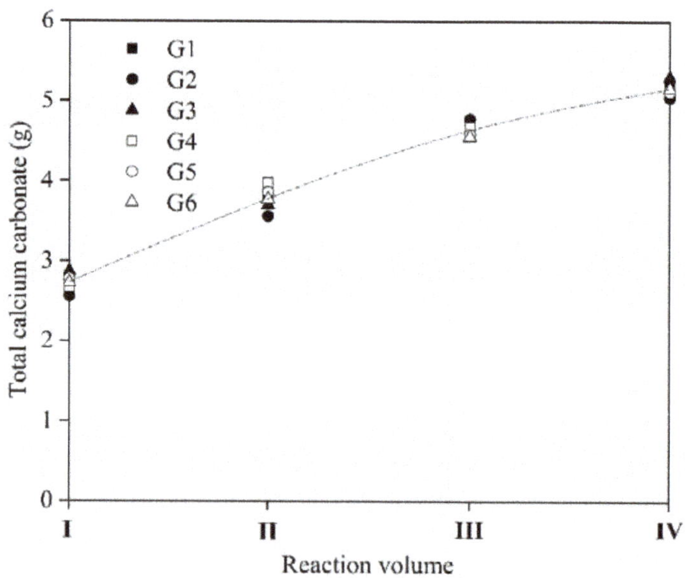

Figure 4. Total calcium carbonate content.

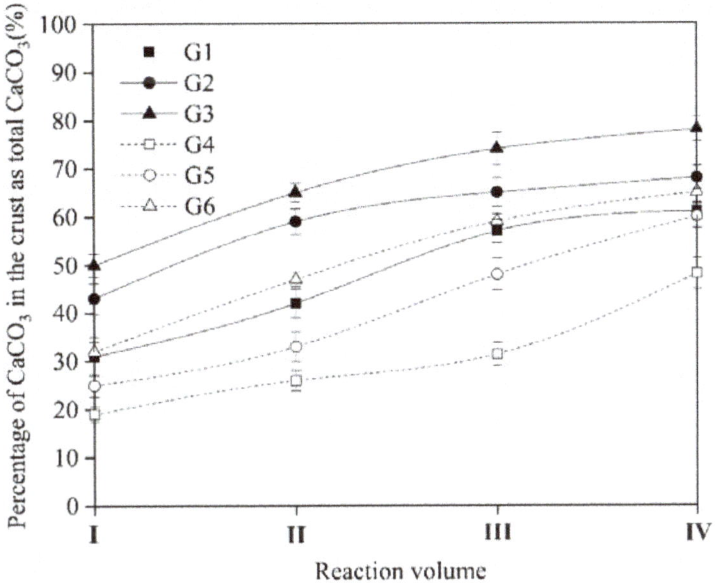

Figure 5. Percentage of $CaCO_3$ in the crust as total $CaCO_3$.

In addition, it can be seen from Figure 5 that under the same reaction amount, with the increase in MICP treatment times, the percentage of $CaCO_3$ content in the surface hard shell increases. This trend is consistent with experimental results of Cheng et al. [29]. They found the average calcium carbonate content could almost double from 1.9% to 3.7%, as the MICP treatment cycles increase from 3 to 7. The possible reason is that with a rise in the number of MICP treatments, the first cycle MICP treatment could generate $CaCO_3$ on the surface layer and effectively block the pores of the sample, and more bacteria and cementation solution could be retained in the surface layer for the following treatment cycles.

3.2. Surface Strength Test

The penetration test is one of the strength test methods (e.g., triaxial test and shear test) used to evaluate soil strength. The pocket penetrometer was used by many scholars to assess the strength of soil surface layers [27,30]. Followed by them, the surface strength was evaluated with a pocket penetrometer. The average thickness of the treated hard crust layer of the specimens in this study was 3–7 mm, and the strength results are shown in Figure 6. The surface layer strength of the specimens could be improved using HPMC, a viscous material, and the highest surface layer strength of the specimens with HPMC added was 1.61 times the surface layer strength of the specimens without HPMC added. It can be seen from Figure 6 that the surface layer strength gradually increased with an increase in the reaction amount, but for the slight reaction amount, the surface layer strength of the specimens could be improved quickly by increasing the amount of reactants, but for the reactants larger than III, the surface layer strength of the specimens increased more slowly by increasing the bacterial solution and the cementation solution. The calcium carbonate generated by MICP treatment glued the soil particles together and improved the consolidation of the soil sample. When there were less reactants, less calcium carbonate was generated, and as the reactants increased, the calcium carbonate content also gradually increased, filling the surface pores of the specimen more to bond the soil particles. In the case of sufficient reactants, the generated calcium carbonate adhered more to the surface and did not play a role in increasing the strength of the bonded soil particles (as shown in Figure 7a,b). In Figure 6, it can also be seen that the strength of the surface layer of the specimen increased with the number of MICP treatments at the same reaction volume, a trend similar to that of the $CaCO_3$ content (Figure 5). Furthermore, it is interesting to compare the strength data obtained from different strength test (e.g., penetration test, triaxial test, and shear test) in future MICP research.

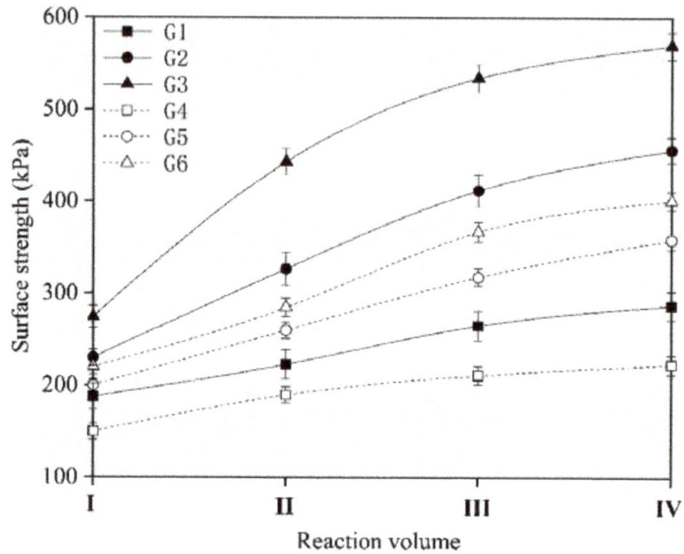

Figure 6. Surface strength of the specimens.

(a) (b)

Figure 7. Specimen surface features: (**a**) with HPMC; (**b**) without HPMC.

3.3. Ammonia Retention Rate of Specimens

In addition to the precipitation of $CaCO_3$ during the MICP reaction, ammonia was also released, which has a negative impact on air pollution, so it is important to determine the quality of ammonia. The concentration of ammonium (NH_4^+) was measured and the ammonium retention rate was calculated based on the concentration of ammonium (NH_4^+) (Figure 8). It can be seen that by simply conducting MICP without the addition of HPMC component, the specimen could retain less ammonia, due to the fact that the soil also has a certain adsorption capacity. The maximum amount of ammonia adsorbed was only 44%, while when this component of HPMC was added to it, the amount of ammonia adsorbed reached up to 87%. Furthermore, as the number of treatments increased, the ammonia retention rate gradually decreased, as shown in Figure 8. When the reactants were sprayed several times for MICP treatment, among them, when the first spray was made, a thinner crust was formed in the surface reaction, so that more reactions of the reactants could be exposed to air in the subsequent treatment, and the ammonia retention rate of the specimens could certainly be reduced. The concentration of ammonium (NH_4^+) in the specimens after MICP treatment was measured by Cheng et al. [29], and the concentration of ammonium (NH_4^+) in the specimens after 7 treatments was 14 g/L in their study. They also found that the concentration of ammonium (NH_4^+) in the specimens increased with the number of MICP treatments.

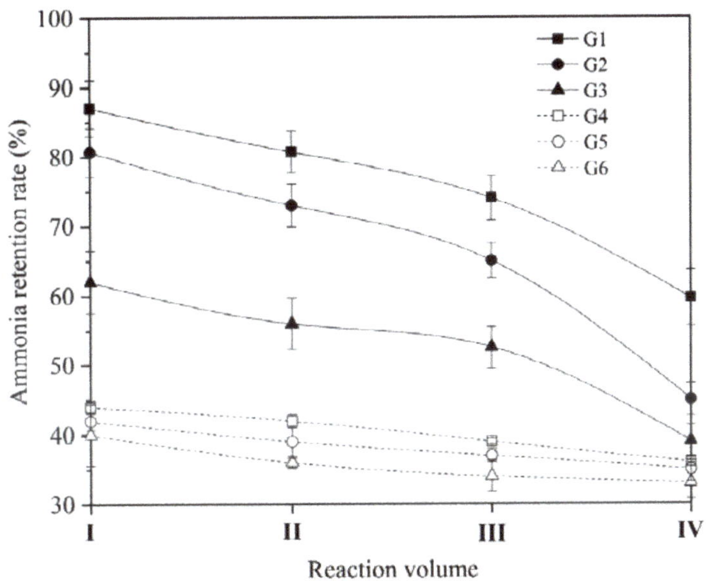

Figure 8. Ammonia retention rate.

3.4. Simulated Rainfall Test

By simulating three rainfall levels, i.e., large, medium, and light, the masses of the specimens lost by rainfall under these conditions were collected separately. The rainfall losses for each test group are shown in Figures 9 and 10. The loss of specimens treated with HPMC for MICP (Figure 9) was less than that of specimens treated without HPMC for MICP (Figure 10) under the three rainfall levels. In particular, under heavy rainfall levels, the rainfall loss with HPMC and three MICP treatments was about 30% less than that without HPMC under the same conditions. By adding HPMC as a component, the viscosity of the reactants was increased so that more $CaCO_3$ was produced in the surface layer of the specimens, and because of its excellent solubility in cold water, the combined effect was to improve the water resistance of the surface layer. Moreover, it can be seen from the Figure that the rain erosion resistance of the surface layer of the specimen could be effectively improved after several treatments, and more calcium carbonate was generated in the surface layer after several treatments (Figure 5), making the surface layer of the specimen more compact and steady in the structure and enhancing its rain erosion resistance.

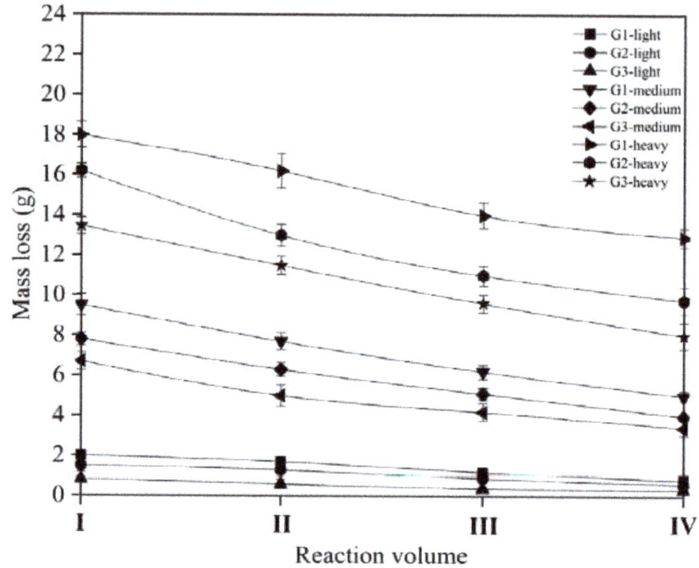

Figure 9. Mass loss under rainfall conditions when HPMC was added.

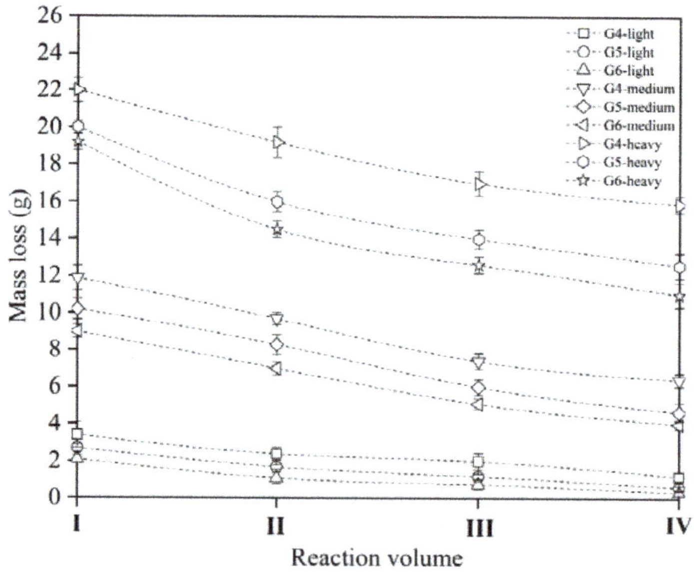

Figure 10. Mass loss under rainfall conditions when HPMC was not added.

3.5. Wind Erosion Test

Wind erosion resistance tests were performed on MICP-treated and MICP combined with HPMC-treated specimens at wind speeds of 4 m/s and 12 m/s, respectively. The mass loss of the specimens before and after the wind erosion resistance test was measured, and the mass loss rate of the specimens was calculated for different amounts of MICP solution and different times of treatments, and the results are shown in Figures 11 and 12. The mass-loss rate of the specimens treated with HPMC for MICP (Figure 11) was lower than that of the specimens treated with MICP without HPMC (Figure 12) at both wind speeds.

When the wind speed was 12 m/s, the maximum mass loss rate of the specimens treated with HPMC for MICP was 0.828%, while the maximum mass loss rate of the specimens treated with MICP was 0.1487%. Devrani et al. [31] found that the mass loss of the MICP-treated specimens was 1.02% at a wind speed of approximately 12 m/s when the urea concentration was 1 M. The addition of HPMC, a viscous substance, increased the viscosity of the MICP solution, which allowed more calcium carbonate to be produced in the surface layer of the specimens and improved the resistance of the surface layer to wind erosion. It can also be seen from Figures 10 and 11 that the mass loss rate of the specimen surface layer gradually decreased with an increase in the number of MICP treatment, a trend consistent with the results of Wang et al. [32].

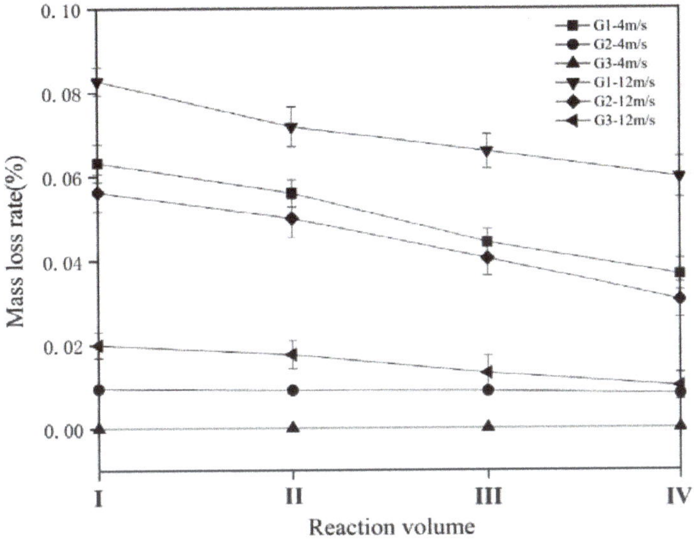

Figure 11. Mass loss rate when adding HPMC.

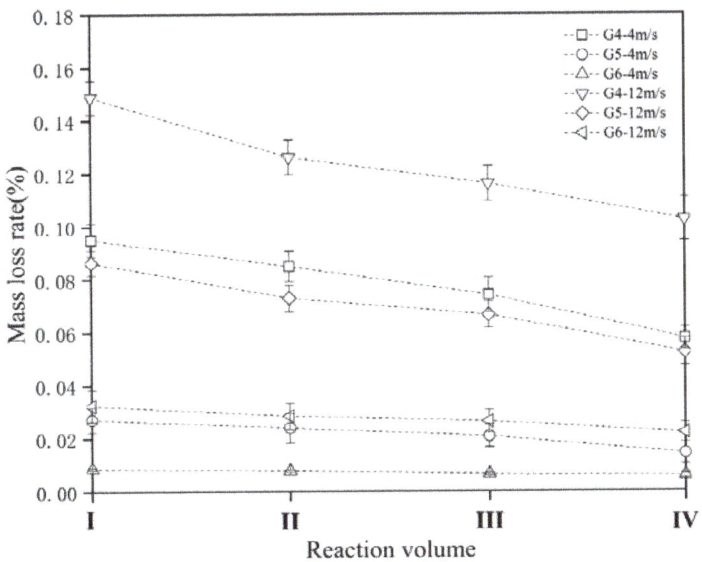

Figure 12. Mass loss rate when HPMC was not added.

3.6. SEM/XRD

To investigate whether the addition of HPMC changed the type of $CaCO_3$ crystals produced by the MICP treatment, the MICP reaction was first carried out in sterile glassware A with the addition of bacterial solution and cementation solution. Then, the same amounts of bacterial solution, cementation solution, and HPMC (viscosity of 25 mpa.s) were added in sterile glassware B for the reaction. The products in glassware A and glassware B were washed, filtered, and dried to obtain calcium carbonate. The calcium carbonate in Glass Dish A and Glass Dish B were also analyzed by XRD, as shown in Figure 13. Three types of calcium carbonate crystals were identified in these two samples, including calcite, sphalerite, and aragonite. In sample A, the $CaCO_3$ precipitate consisted of 91.6% calcite, 6.3% aragonite, and 2.1% vaterite (Figure 13). Sample B was identified as consisting of 89% calcite, 9.6% aragonite, and 1.4% vaterite (Figure 13). This indicates that the primary $CaCO_3$ crystal type of both samples was calcite. HPMC did not change the predominant crystal type of $CaCO_3$ generated by the MICP treatment. In this study, $CaCl_2$ was used as the calcium source, and the primary crystal type of $CaCO_3$ obtained from MICP treatment was calcite, which is also consistent with the conclusion obtained from the study by Zhang et al. [33]. Moreover, the addition of HPMC slightly increased the aragonite content, and it is interesting to investigate if, in time, the addition of HMPC can increase the formation of aragonite and evaluate aragonite's behavior in long-term conditions in future MICP studies.

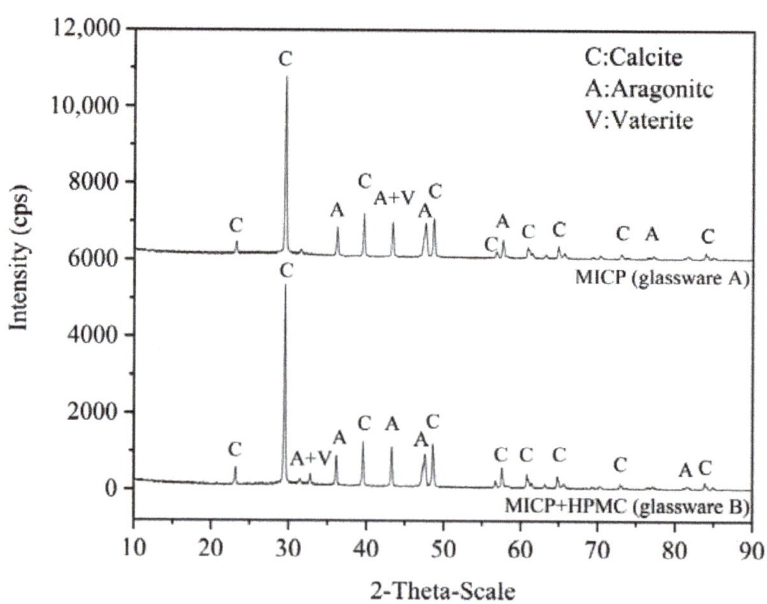

Figure 13. X-ray diffraction results.

SEM tests were performed on the MICP-treated specimens in G3 (reactant III) and G6 (reactant III), and the SEM images clearly showed the structural differences between the two specimens with and without the addition of HPMC for the MICP technique. Compared with the specimen without HPMC (Figure 14), the surface layer of the specimen treated with HPMC was more compact, and the soil particles and the pores between the soil particles were filled with more induced calcium carbonate crystals (Figure 15). HPMC increased the viscosity of the reactants and helped to induce more calcium carbonate precipitation in the surface layer of the specimens, thus improving the surface strength of the specimens.

Figure 14. SEM image without HPMC addition.

Figure 15. SEM image when adding HPMC.

4. Conclusions

This study investigated the application of MICP technology in soil surface cementation curing. The viscosity of the reaction solution was increased by adding HPMC, a cohesive material, and a series of tests, including mechanical properties, rainfall erosion resistance, wind erosion resistance, ammonia retention, XRD, and SEM, were performed on each test group using four reactant dosage gradients and three solution injection rounds. The results of the study showed that the MICP technique incorporating HPMC is effective for soil surface improvement.

1. The content of $CaCO_3$ in the curst of the sample induced by MICP treatment increased with an increase in the number of treatments, and the addition of HPMC can effectively improve the content of $CaCO_3$ in the curst of the sample. MICP treatment with HPMC can produce up to 1.82 times as much calcium carbonate in the crust as MICP treatment alone, but has no effect on the total calcium carbonate. In addition, the incorporation of HPMC did not significantly change the formation of primary crystal type calcite.

2. The rainfall erosion resistance of the soil surface layer was enhanced by the addition of HPMC combined with MICP technology. Rainfall losses were reduced by up to 30% in specimens treated with this method.
3. HPMC combined with MICP technology enhanced the wind erosion resistance of the surface layer. When the wind speed was 12 m/s, the maximum mass loss rate of the specimens treated with HPMC for MICP was only 0.828%.
4. The addition of HPMC and the increase in treatment cycles can effectively improve the surface strength of the sample. By adding HPMC and multiple treatments, more calcium carbonate precipitated in the surface layer of the sample. SEM images showed that more $CaCO_3$ filled the pores between soil particles and cohesive soil particles, forming a dense structure. The surface strength of the sample treated with HPMC was 1.61 times that of those without HPMC.
5. The ammonia retention rate of MICP treated with HPMC was significantly higher than that without HPMC. The ammonia retention rate of adding HPMC was 1.98 times that of not adding HPMC, which could effectively reduce the release of ammonia gas in the process of MICP, which is of great significance to environmental protection.

Author Contributions: Conceptualization, F.H., Y.Z. and Z.X.; data curation, M.Y.; formal analysis, M.Y., F.H., Y.Z. and Z.X.; funding acquisition, W.Z., F.H., F.M. and Q.T.; investigation, W.Z.; methodology, W.Z. and M.Y.; project administration, W.Z. and F.H.; resources, W.Z. and F.H.; software, M.Y., Y.Z., Z.X., Q.W. and F.M.; supervision, W.Z., F.H. and Y.Z.; validation, W.Z., F.H. and Y.Z.; writing—original draft, W.Z. and M.Y.; writing—review and editing, M.Y. All authors have read and agreed to the published version of the manuscript.

Funding: The research was supported by the Postdoctoral Science Foundation of China (No. 2018M640683), the Postdoctoral Research Grant of Henan (No. 001801006), the National Natural Science Foundation of China (52078317), the Natural Science Foundation of Jiangsu Province for Excellent Young Scholars (BK20211597), the Bureau of Housing and Urban–Rural Development of Suzhou (2021-25; 2021ZD02; 2021ZD30), the Bureau of Geology and Mineral Exploration of Jiangsu (2021KY06), and the China Tiesiju Civil Engineering Group (2021-19).

Institutional Review Board Statement: Not applicable.

Informed Consent Statement: Not applicable.

Data Availability Statement: Not applicable.

Conflicts of Interest: The authors declare no conflict of interest.

References

1. Naveed, M.; Duan, J.; Uddin, S. Application of microbially induced calcium carbonate precipitation with urea hydrolysis to improve the mechanical properties of soil. *Ecol. Eng.* **2020**, *153*, 105885. [CrossRef]
2. Martinez, B.C.; DeJong, J.T.; Ginn, T.R.; Montoya, B.M.; Barkouki, T.H.; Hunt, C.; Tanyu, B.; Major, D. Experimental optimization of microbial-induced carbonate precipitation for soil improvement. *J. Geotech. Geoenviron. Eng.* **2013**, *139*, 587–598. [CrossRef]
3. Cheng, L.; Shahin, M. Urease active bio-slurry: A novel soil improvement approach based on microbially induced calcite precipitation. *Can. Geotech.* **2016**, *53*, 1376–1385. [CrossRef]
4. Choi, S.G.; Wang, K.; Wen, Z.; Chu, J. Mortar crack repair using microbial induced calcite precipitation method. *Cem. Concr. Comp.* **2017**, *83*, 209–221. [CrossRef]
5. Comoss, E.J.; Kelly, D.A.; Leslie, H.Z. Innovative erosion control involving the beneficial use of dredge material, indigenous vegetation and landscaping along the Lake Erie shoreline. *Ecol. Eng.* **2002**, *19*, 203–210. [CrossRef]
6. Tiano, P.; Biagiotti, L.; Mastromei, G. Bacterial bio-mediated calcite precipitation for monumental stones conservation: Methods of evaluation. *J. Microbiol. Methods* **1999**, *36*, 139–145. [CrossRef]
7. Li, M.; Cheng, X.H.; Guo, H.X. Heavy metal removal by biomineralization of urease producing bacteri-a isolated from soil. *Int. Biodeterior. Biodegrad.* **2013**, *76*, 81–85. [CrossRef]
8. Sel, I.; Ozhan, H.B.; Cibik, R. Bacteria-induced cementation process in loose sand medium. *Mar. Georesources Geotechnol.* **2015**, *33*, 403–407. [CrossRef]
9. Brossi, D.L.; Julia, M. Soil-derived microbial consortia enriched with different plant biomass reveal distinct players acting in lignocellulose degradation. *Microb. Ecol.* **2016**, *71*, 616–627. [CrossRef]

10. Cheng, L.; Cord-Ruwisch, R. In situ soil cementation with ureolytic bacteria by surface percolation. *Ecol. Eng.* **2012**, *42*, 64–72. [CrossRef]
11. Van Paassen, L.A.; Ghose, R.; van der Linden, T.J.M. Quantifying biomediated ground improvement by ureolysis: Large-scale biogrout experiment. *J. Geotech. Geoenviron. Eng.* **2010**, *136*, 1721–1728. [CrossRef]
12. Stabnikov, V.; Naeimi, M.; Ivanov, V. Formation of water-impermeable crust on sand surface using biocement. *Cem. Concr. Res.* **2011**, *41*, 1143–1149. [CrossRef]
13. Xiao, Y.; Ma, G.; Wu, H.; Lu, H.; Zaman, M. Rainfall-induced erosion of biocemented graded slopes. *Int. J. Geomech.* **2022**, *22*, 04021256. [CrossRef]
14. Liu, B.; Xie, Y.H.; Tang, C.S. Bio-mediated method for improving surface erosion resistance of clayey soils. *Eng. Geol.* **2021**, *293*, 106295. [CrossRef]
15. Zomorodiana, S.M.A.; Hamideh, G.; Brendan, C.O. Stabilisation of crustal sand layer using biocementation technique for wind erosion control. *Aeolian Res.* **2019**, *40*, 34–41. [CrossRef]
16. Wang, X.; Tao, J. Polymer-modified microbially induced carbonate precipitation for one-shot targeted and localized soil improvement. *Acta Geotech.* **2018**, *14*, 657–671. [CrossRef]
17. Salifu, E.; MacLachlan, E.; Iyer, K.R.; Knapp, C.W.; Tarantino, A. Application of microbially induced calcite precipitation in erosion mitigation and stabilisation of sandy soil foreshore slopes: A preliminary investigation. *Eng. Geol.* **2016**, *201*, 96–105. [CrossRef]
18. Shanahan, C.; Montoya, B.M. Erosion reduction of coastal sands using microbial induced calcite precipitation. In Proceedings of the Geo-Chicago 2016, Chicago, IL, USA, 14–18 August 2016.
19. Wang, J.Y.; Snoeck, D.; Vlierberghe, S.V. Application of hydrogel encapsulated carbonate precipitating bacteria for approaching a realistic self-healing in concrete. *Constr. Build. Mater.* **2014**, *68*, 110–119. [CrossRef]
20. Hamdan, N.Z.; Zhao, M.; Mujica, E.; Kavazanjian, E., Jr.; He, X. Hydrogel-assisted enzyme-induced carbonate mineral precipitation. *J. Mater. Civ. Eng.* **2016**, *28*, 04016089. [CrossRef]
21. Sun, X.; Miao, L.; Wang, H. Enhanced rainfall erosion durability of enzymatically induced carbonate precipitation for dust control. *Sci. Total Environ.* **2021**, *791*, 148369. [CrossRef]
22. Wang, X.; Tao, J.; Bao, R. Surficial Soil Stabilization against Water-Induced Erosion Using Polymer-Modified Microbially Induced Carbonate Precipitation. *J. Mater. Civ. Eng.* **2018**, *30*, 4018267. [CrossRef]
23. Ty, A.; Ren, H.A.; Uo, A. Biomimetic mineralization of the carbonates regulated by using hydroxypropyl methylcellulose macromolecules as organic templates. *J. Cryst. Growth* **2019**, *508*, 72–81.
24. Li, M.D.; Wen, K.J.; Li, Y.; Zhu, L.P. Impact of oxygen availability on microbially induced calcite precipitation (MICP) treatment. *Geomicrobiol. J.* **2017**, *35*, 15–22. [CrossRef]
25. Whiffin, V.S.; van Paassen, L.; Harkes, M.P. Microbial Carbonate Precipitation as a Soil Improvement Technique. *Geomicrobiol. J.* **2007**, *24*, 417–423. [CrossRef]
26. Cheng, L.; Cord-Ruwisch, R.; Shahin, M.A. Cementation of Sand Soil by Microbially Induced Calcite Precipitation at Various Degrees of Saturation. *Can. Geotech. J.* **2013**, *50*, 81–90. [CrossRef]
27. Almajed, A.; Lemboye, K.; Arab, M.G. Mitigating wind erosion of sand using biopolymer-assisted EICP technique. *Soils Found.* **2020**, *60*, 356–371. [CrossRef]
28. Choi, S.G.; Park, S.S.; Wu, S.; Chu, J. Methods for calcium carbonate content measurement of biocemented soils. *J. Mater. Civ. Eng.* **2017**, *29*, 06017015. [CrossRef]
29. Cheng, Y.; Tang, C.; Pan, X.H.; Liu, B.; Xie, Y.H.; Cheng, Q.; Shi, B. Application of microbial induced carbonate precipitation for loess surface erosion control. *Eng. Geol.* **2021**, *294*, 106387. [CrossRef]
30. Chen, R. Bio Stabilization for Geopolymer Enhancement and Mine Tailings Dust Control. Ph.D. Thesis, The University of Arizona, Tucson, AZ, USA, 2014.
31. Devrani, R.; Dubey, A.A.; Ravi, K.; Sahoo, L. Applications of bio-cementation and bio-polymerization for aeolian erosion control. *J. Arid. Environ.* **2021**, *187*, 104433. [CrossRef]
32. Wang, Z.Y.; Zhang, N.; Lin, F.; Ding, J.H.; Yang, H.M. Experimental Study on Wind Erosion Resistance and Strength of Sands Treated with Microbial-Induced Calcium Carbonate Precipitation. *Adv. Mater. Sci. Eng.* **2018**, *2018*, 3463298. [CrossRef]
33. Zhang, Y.; Guo, H.X.; Cheng, X.H. Role of calcium sources in the strength and microstructure of microbial mortar. *Constr. Build. Mater.* **2015**, *77*, 160–167. [CrossRef]

applied sciences

MDPI

Article

Geotechnical Evaluation of Diesel Contaminated Clayey Soil

Christian E. Hernández-Mendoza [1,*], Pamela García Ramírez [2] and Omar Chávez Alegría [2]

1 CONACYT—Laboratory of Environmental Geotechnics, Facultad de Ingeniería, Universidad Autónoma de Querétaro, Circuito Universitario S/N, Querétaro 76010, Mexico
2 Facultad de Ingeniería, Universidad Autónoma de Querétaro, Circuito Universitario S/N, Querétaro 76010, Mexico; carmenchis3@gmail.com (P.G.R.); omar.chavez@uaq.mx (O.C.A.)
* Correspondence: cehernandezme@conacyt.mx

Abstract: Soil contamination by different hydrocarbons has rapidly expanded worldwide, surpassing the self-purification capacity of soils and increasing the number of contaminated sites. Although much effort has been devoted to study the effects of diesel contamination on the geotechnical properties of soil, there is still limited available information about it. Moreover, there is no available information about the maximum diesel retention that soil can have and its effect on the geotechnical behavior of the soil. Thus, in this paper, we determined the maximum diesel retention by an unsaturated clayey soil and evaluated the impact of diesel contamination on its geotechnical properties. The results showed that the soil could only retain 12.6% of the added diesel and the excess was expulsed. At such a diesel concentration, the saturation rate of the soil was lower than 80%. Diesel contamination increased the plasticity and the internal friction angle of the soil, while its cohesion was considerably decreased. It should be noted that the matric suction of contaminated soil was lower than the one obtained for natural soil. However, its osmotic suction was considerably higher. This indicates that osmotic suction must be considered to evaluate the shear strength of contaminated soils.

Keywords: geotechnical properties; diesel; contaminated soil; clay; environmental geotechnics

Citation: Hernández-Mendoza, C.E.; García Ramírez, P.; Chávez Alegría, O. Geotechnical Evaluation of Diesel Contaminated Clayey Soil. *Appl. Sci.* **2021**, *11*, 6451. https://doi.org/10.3390/app11146451

Academic Editor: Daniel Dias

Received: 10 June 2021
Accepted: 9 July 2021
Published: 13 July 2021

1. Introduction

Hydrocarbons are the main energy source used to satisfy our energy needs, but during their life cycle, they can produce many environmental impacts. Hydrocarbon leaks/spills, which can occur from its exploitation to use, cause environmental pollution, where, commonly, the first receptor is the soil.

Soil contamination by hydrocarbons can be originated by accidental spills during its transportation and disposal, leakage of storage tanks and pipes, and/or by clandestine fuel intakes. The unsaturated (aerated) zone of the soil, that is, the area in which the infrastructure's foundations are frequently laid, is the first one to be contaminated by hydrocarbon leaks and/or spills.

It is also possible that hydrocarbons permeate through the soil up to the groundwater level zone, contaminating the saturated zone of soil. Therefore, hydrocarbon leaks and spills can affect the deep foundations located on it as well as it can contaminate groundwater. In this case, the hydrocarbons can be transported by the groundwater to the surrounding areas contaminating the soil and may affect the foundations located in such areas.

The permanence of hydrocarbons in the unsaturated zone depends on different factors, such as the solubility, volatility, toxicity, biodegradation rate, and sorption of the hydrocarbons; the nature and permeability of the soil or soils forming the unsaturated zone; and the natural or man-made liquid flow (water and/or contaminants).

Unfortunately, soil contamination by different hydrocarbons has rapidly expanded worldwide, surpassing, in many cases, the self-purification capacity of soils. As a result, the number of hydrocarbon-contaminated sites has increased in both developed and developing countries, latter being the more affected ones.

Soil is the most frequent source of risk to construction projects [1] due to not only the geotechnical or geological aspects but also the contaminants present in it. Nevertheless, there is still limited available information about the impact of hydrocarbons contamination on the geotechnical properties of the soils. Some of the risks associated with soil contamination are loss of foundation and structural integrity, damage to construction materials, and loss of facility [1]. Independent from the time the soil gets polluted—before or after the infrastructure construction—the changes caused in its engineering properties can place both the integrity of the structures and its users at risk.

It is known that soil contamination by hydrocarbons affects its engineering properties. However, such an effect depends on many variables, such as the type of soil and the physical and chemical properties of the spilled/leaked hydrocarbon [2–4].

Published papers have been focused on the effect of diesel contamination on the geotechnical properties of soil by varying the contaminant concentration from 0% to 20% (weight basis) [5–9]. However, none of them has been focused on the quantity of diesel that soil is capable to retain. This can have important repercussions since the exceeding hydrocarbons can be subjected to evaporation or transportation depending on the type of hydrocarbon and the local environmental conditions. Nevertheless, the retained hydrocarbons can have a major influence on the geotechnical behavior of the contaminated soil.

Izdebska-Mucha and Trzcinski [10] reported that after contaminating a clay sample with diesel, clay particles showed a change in their microstructure, resulting in a denser structure due to change in the pore size distribution. Ijimdiya [4] studied the effects of used motor oil on the properties of lateritic soil and found a reduction in the amount of fine-grained size particles of the soil as the concentration of the contaminant was increased from 0% to 8%.

Son et al. [11] estimated the effect of diesel contamination on the electrical properties of unsaturated soils. These authors reported that contaminated soils with a water content of 5% had higher electrical resistivities and decreased permittivity compared to uncontaminated soils. In contrast, contaminated soils having a water content of 15% showed smaller resistivities and higher permittivity compared to non-contaminated soils.

It has been reported that soil contamination affects its consistency limits. Salimnezhad et al. [12] reported that the plastic limit of a crude-oil-contaminated soil increased as the contaminant concentration increased, but the plastic index decreased as the contaminant concentration increased. Yazdi and Teshnizi [7] found that the liquid limit of a gasoline-contaminated soil increased, reaching maximum value at a contaminant concentration of 6%. Although a further increase of gasoline concentration reduced the liquid limit of the soil, it was higher than the one obtained for the natural soil.

Salimnezhad et al. [12] reported that swelling pressure of natural soil was higher than that of contaminated soil. However, for 12% oil content, the contaminated soil had a higher free swell percentage than the natural soil.

Other researchers have studied the effects of diesel [5], oil [13], gasoline [6], and crude oil [12] on the shear resistance of different clay soils, finding that soils' cohesions tended to decrease as the contaminant concentration increased. However, while in some cases, the internal friction angle was reported to increase as the contaminant concentration increased [5,6], in other cases, a reduction of its value was reported as the contaminant concentration increased [7,12,13]. These results may mean that clay minerals exhibit different behaviors according to the hydrocarbon used to contaminate the soil.

Yazdi and Teshnizi [7] showed that the settlement and ductility of a gasoline-contaminated soil increased as the contaminant concentration increased. They also found that the soil failure mechanisms changed depending on the contaminant concentration in the soil.

Taheri et al. [9] studied the effects of soil contamination in a soil that was simultaneously contaminated with lead and diesel. These authors found that the increase in diesel concentration reduced the hydraulic conductivity of the contaminated soil. Such decrease was because diesel occupied the pore spaces and limited the water flow through the soil samples.

Salimnezhad et al. [12] found that the settlement of the contaminated soil increased as the crude oil concentration increased because the contaminant reduced the specific surface area of the soil, decreasing the water absorption of the particles of the soil facilitating the water drainage.

Although many efforts had been carried out to study the effects of diesel contamination on the geotechnical properties of soils, most of the studies had focused on saturated soil conditions. As a result, the available information about the effect of diesel contamination on unsaturated soils is scarce. Moreover, to the widest knowledge of the authors, there is no available information about the maximum diesel retention that natural soil can have and its effect on the geotechnical behavior of the soil.

Thus, the objective of this study was to determine the maximum diesel retention by an unsaturated clayey soil, and to evaluate the impact of diesel contamination on its geotechnical properties. The test performed in both natural and contaminated soil conditions showed that although the soil could not retain more diesel, its saturation rate did not exceed 80%. Furthermore, the changes in the geotechnical properties of the contaminated soil promoted an increase in its shear strength resistance. Such difference was due to an increase in the osmotic suction in the contaminated soil.

2. Materials and Methods

2.1. Experimental Procedure

The experiment was performed in two phases: natural soil characterization and soil contamination and characterization. In the first phase, the extraction of disturbed natural (uncontaminated) soil samples from a place in the north of Querétaro City, Mexico, was done. An undisturbed cubic soil sample was extracted to evaluate the natural soil's water content and volumetric unit weight. All soil samples were stored in a controlled temperature chamber (20 ± 1 °C) until its use for geotechnical characterization.

All the disturbed soil samples were mixed into the controlled temperature chamber to obtain a uniform mass of soil to perform the remaining geotechnical tests. This soil mass was divided into subsamples, poured into hermetically sealed containers, and stored in the same chamber. The soil sample quantity that is required for each test could now be taken with ease, avoiding opening and manipulating the whole soil sample every time a quantity of soil was needed for analysis.

The tests were performed according to the methods described in the Test Methods section. The unit weight of the natural soil was $\gamma_m = 15.10\ \text{kN/m}^3 \pm 0.10\ \text{kN/m}^3$, and its water content was $w = 36.94\% \pm 0.19\%$; this corresponded to an average saturation rate of $s_r = 70.82\% \pm 0.18\%$. After finished the natural soil characterization, the soil contamination process was started.

In the second phase of this work, the remaining soil sample was contaminated with diesel for its geotechnical characterization. Table 1 shows the characteristics of the commercially available diesel used to contaminate the soil. To define the quantity of diesel needed for soil contamination, a 50-g natural soil sample was placed in a sealed glass container. Then, this soil subsample was contaminated with diesel, at an initial concentration of 4% by weight, and stored for seven days in the temperature-controlled room.

During this curing period, the contaminant could distribute throughout the soil and react with the pore fluid to achieve ionic stability [2]. After the curing period, the sample was visually inspected to verify if the soil could retain the added diesel. This process was repeated by increasing the diesel concentration (i.e., 4%, 8%, 12%, and 14%) until the soil could not retain more diesel.

Table 1. Applied diesel fuel specifications [14].

Parameter	Value
Flash point (°C)	≥45
Density (kN/m^3)	8.16
Specific gravity	0.832
Kinematic viscosity at 40 °C (mm^2/s)	1.9–4.1
Cetane index	≥48
Sulfur content (ppm)	<500
Aromatics content (%, in volume)	≤30
Water and sediments (%, in volume)	≤0.05

When the soil could not retain more diesel, it was transferred to a smaller glass container to allow the soil to drain the exceeding contaminant. Then, a contaminated soil subsample was taken from the top of the container to determine the retained diesel concentration by a gravimetric method [15]. The retained diesel concentration was 12.62% ± 0.27%. Therefore, such a concentration of diesel was used to contaminate the natural soil to be used during this second phase.

Once the diesel concentration to contaminate the soil was defined, the following procedure was applied. First, the remaining natural soil was air-dried to reduce its water content. This was done because in the previous experiment, a considerable increase in the liquids content of the contaminated soil was noticed (see Section 3.1). Then, the soil was placed in a container to mix it manually with diesel to simulate the long-term contamination effect [16].

Next, the contaminated soil was poured into hermetically sealed containers and stored in the temperature-controlled room for six weeks. After this curing period, the contaminated soil was tested as indicated in the Test Methods section.

The selected contamination procedure is an alternative way to reduce the sample preparation time [16]. It also allows the contaminant to bring about a long-term change in the initial soil structure [16]. Although a curing time of seven days is enough to allow the soil to react with the contaminant [2], the curing period was extended to six weeks to enhance the long-term contamination effect.

2.2. Test Methods

The evaluated properties of the soil are summarized in Table 2 and were executed as indicated in the corresponding standards. The "water content" of the contaminated soil was obtained by applying the same standard used for the natural soil. However, the term "water content" was transformed to "liquids content" to consider the difference in specific gravity between diesel and water [17].

To determine the modified free swell index (MFSI) for both soil conditions, 10 g of oven-dried (105 ± 1 °C) soil passing a No. 40 (425 μm) sieve were poured into a 100-mL graduated cylinder filled with distilled water. The soil was manually mixed with the distilled water and left undisturbed for the soil to swell/settle and achieve an equilibrium state of volume. The MFSI was calculated as the ratio of the equilibrium sediment volume to the dry weight of the oven-dried soil.

The liquid limit was evaluated by the fall cone method as indicated in the standard procedure adopted. Distilled water was gradually added to the soil—both natural and contaminated—to obtain a semiliquid consistency. The soil and the added water were thoroughly mixed until a homogeneous mass was obtained. This mass was poured into a sealed container. It was stored in the controlled-temperature chamber for 24 h to allow the added water to permeate through the soil before the cone penetrometer tested the sample. For both soil conditions, the liquid limit was evaluated by considering the equivalent water content of the samples.

The direct shear test was executed in a 100 mm × 100 mm shear box at a shear velocity of 1 mm/min. Normal stresses applied were 50, 100, and 200 kPa. Matric and total suction of the soils were evaluated by the filter paper method, using the calibration curves for wetting the filter paper, Whatman No. 42.

Table 2. Evaluated properties of both natural and diesel contaminated soils.

Property	Method Applied
Water content [1]	ASTM D2216 [18]
Unit weight	ASTM D7263 [19]
Soil mineralogy	X-ray diffraction analysis
Chemical elements analysis	Total reflection X-ray fluorescence spectroscopy
Specific gravity of soil solids	ASTM D854 [20]
pH	NOM-021-RECNAT-2000 [21]
Particle-size distribution of soil	ASTM D6913 [22], ASTM D1140 [23]
Liquid limit and plastic limit	BS EN ISO 17892-12 [24]
Soil classification	ASTM D2487 [25]
Modified free swell index	Sridharan and Rao [26]
Direct shear	ASTM D3080 [27]
Soil suction	ASTM D5298 [28]
Soil texture	NOM-021-RECNAT-2000 [21]
Electrical conductivity	NOM-021-RECNAT-2000 [21] (by the saturated paste extract, soil to water ratio of 1:2.5)
Organic content	AASHTO T 267 (by loss-on-ignition) [29]
Cation exchange capacity	NOM-021-RECNAT-2000 (Kjendahl method) [21]

[1] As a technical precision, in the case of the contaminated soil, the term "water content" was replaced by the term "liquids content" [17].

The pH and the electrical conductivity (EC) of the soils were measured with a pH/EC/DO/ISE multiparameter (Orion Versa Star Pro, Thermo Scientific, Waltham, MA, USA) with automatic temperature compensator, equipped with a glass electrode for pH measurements and an EC sensor. Before each natural and contaminated soil sample set's pH measurement, the equipment was calibrated with the traceable to the National Institute for Standards and Technology (NIST) buffer solutions of pH 4, pH 7, and pH 10 obtained from Thermo Fisher. The equipment calibration was repeated until a correlation coefficient of the calibration curve of the multiparameter was at least of 0.98. For the EC measurements the equipment was calibrated with the traceable to NIST standard solutions of 100 μS/cm and 1413 μS/cm obtained from Thermo Fisher (Austin, TX, USA).

2.2.1. Chemical Elements Analysis of the Soil

The elemental chemical characterization of soils was performed with a total reflection X-ray fluorescence spectroscope (S2 PICOFOX TXRF, Bruker Nano), equipped with a molybdenum tube and silicon drift detector. The S2 PICOFOX is a portable benchtop spectrometer for multi-element analysis of liquids, suspensions, and solids.

Before analysis of the samples, a gain correction was performed with a mono-element standard sample (1 μg As). Further, the spectroscopic resolution (1 μg Mn) and sensibility analysis (1 ng Ni) of the spectroscope were performed according to the manufacturer instructions.

The standard solutions of H_3AsO_4 (1000 mg/L); SiO_2 (10,000 mg/L), $Ni(NO_3)_2$ (1000 mg/L); $Fe(NO_3)_3$ (1000 mg/L); $Al(NO_3)_3$ (10,000 mg/L); $Ca(NO_3)_2$ (10,000 mg/L), $NaNO_3$ (10,000 mg/L), KNO_3 (10,000 mg/L), $(NH_4)_2TiF_6$ (1000 mg/L), $Mg(NO_3)_2$ (10,000 mg/L), $Mn(NO_3)_2$ (10,000 mg/L), and $Zn(NO_3)_2$ (10,000 mg/L) were obtained from Merck, with a quality for Inductively Coupled Plasma (ICP) test, traceable to the Standard Reference Material (SRM) from NIST. The necessary dilutions have been made from the stock solutions in order to achieve the linearity of the standard curve.

The soil samples for elemental analysis were prepared as follows: 20 mg of soil was ground in an agate mortar and then passed through a No. 230 (63 μm) sieve. Then, the powder was suspended in 1 mL of dispersant solution (1% Triton X-100 in distilled water).

The soil suspension was spiked with 10 μL of gallium (1000 mg/L, ICP standard traceable to SRM from NIST, Sigma Aldrich, St. Louis, MO, USA) as internal standard.

The suspension was then sonicated for 15 min, after which it was homogenized in a vortex agitator. Then, 10 μL of a sample solution was transferred into a siliconized quartz glass sample carrier and dried at 50 °C on a hot plate. The prepared soil samples were analyzed with a live collection time of 1000 s. Spectra acquisition and data evaluation were done by using the software Spectra 7.8 (Bruker Nano GmbH, Berlin, Germany).

2.2.2. Cation Exchange Capacity of the Soil

The cation exchange capacity (CEC) of the soil was evaluated according to the Mexican standard NOM-021-RECNAT-2000 [21], using ammonium acetate (1N, pH 7, American Chemical Society grade, Meyer) as a saturating solution to replace the absorbed cations (Ca^{2+}, Mg^{2+}, Na^{+}, and K^{+}). Firstly, the soil was air-dried and sieved through the No. 10 (2 mm) sieve. Then, the exchange complex was saturated with the ammonium cation. After that, the ammonium cation was removed with a sodium chloride (NaCl, American Chemical Society grade, Merck) solution and the total capacity change was determined (Kjendahl method). The detection limit for the total CEC was 3.86 $cmol_{+}/kg$.

2.2.3. Soil Mineralogy

Soil samples were powdered in an agate mortar and sieved through a No. 200 sieve. The powders were characterized by X-ray diffraction (XRD Rigaku, MiniFlex) with Cu-Kα radiation (λ = 1.5406 Å). Diffraction intensity, in function of 2θ, was analyzed at the interval from 2° to 80°, with a 2θ step size of 0.02°, and a counting time of 0.06 s per point. Interplanar spaces (d) were evaluated by the Bragg's law equation:

$$n\lambda = 2d\ \sin(\theta) \tag{1}$$

where n is the class of diffraction peak ($n = 1$ for the first-order basal reflection), λ is the Cu-Kα wavelength of the incident X-ray wave, and θ is the diffraction angle.

3. Results and Discussion

3.1. Diesel Retention in the Contaminated Soil

At 4% diesel concentration, the natural soil readily absorbed the contaminant (Figure 1b). In contrast, at 14% diesel concentration, the soil could not retain the contaminant at all (Figure 1c). After analyzing the contaminated soil, it was found that the soil retained only 12.62% ± 0.27% of diesel.

Figure 1. Evolution of soil aspects as diesel concentration increased from: (**a**) natural soil, (**b**) 4% added diesel, and (**c**) 14% added diesel.

Clay fractions of soil can adsorb non-polar molecules—such as diesel—due to the van der Waals attraction forces [30]. Such attraction forces between particles rapidly decay with the surface-to-surface distance (varying inversely as the seventh power of distance between the plates). However, attractive forces are additive. So, van der Waals attraction forces can be calculated by adding the attractions among all the particles. This may result in a higher total force and a lower decay (varying inversely as the third power of distance between the plates) with the surface-to-surface distance [30].

Nonetheless, Anandarajah and Chen [31] found that van der Waals attraction forces get reduced when fluid other than water fills the medium between the attracted particles. Further, Chiou and Shoup [32] showed that organic compounds that are not potent wetting agents have lower sorption capacity on polar mineral surfaces [32]. Hence, this makes water a stronger opponent to displace non-ionic organic solutes from minerals in aqueous systems.

In the absence of water, clay particles can behave as conventional organic compound absorbers, its large surface area being the main contributor to its high absorptive capacity [33]. However, in the presence of water, the absorption of non-polar hydrocarbons can be suppressed since non-polar organic molecules are not solid competitors for adsorption sites on the clay surface compared to highly polar water [30].

Thus, in the present study, since the soil was contaminated at its natural water content, diesel adsorption by the soil minerals was limited to those places where soil absorption sites were available. This means that soil could only retain the required diesel to satisfy its hydration needs.

Diesel is a non-polar mixture of aromatic (including naphthalenes) and saturated (including isoparaffins and cycloparaffins) hydrocarbons obtained from petroleum. Diesel hydrocarbons have carbon numbers in the range of C9–C28 with high boiling points between 170 and 430 °C. As previously stated, the term "water content" was transformed to "liquids content" to consider the difference in specific gravity between diesel and water [17]. Thus, Equation (2) was used to consider the contribution of the liquid's specific gravity on the liquids content of the soil [17].

$$E_{wc} = l_c\left(\frac{c+1}{c+G_f}\right), \quad (2)$$

where E_{wc} is the equivalent water content, l_c corresponds to the liquid content, c is the water to diesel volume ratio in the pore's liquid, and G_f is the specific gravity of diesel. It must be mentioned that in Equation (2), for water that has not been mixed with any contaminant, $c = 0$.

Table 3 presents the impact on the E_{wc} and the saturation rate (s_r) of the soil due to its contamination. Diesel contamination increased the E_{wc} and the s_r of the contaminated soil by 18% and 12%, respectively, compared to the natural soil. Such an increase in both parameters influences the engineering behavior of the contaminated soil. It can even make it complicated to directly compare the results obtained by both soil conditions.

Table 3. Change in soil's saturation rate due to diesel contamination.

Parameter	Natural Soil	Contaminated Soil
Water content, w (%)	36.94 ± 0.19	----
Liquids content, l_c (%)	----	42.54 ± 1.21
Water to diesel volume ratio, c	0	6.94
Equivalent water content, E_{wc} (%)	36.94 ± 0.19	43.46 ± 1.21
Saturation rate, s_r (%)	70.82 ± 0.18	79.46 ± 1.03 [1]

Values in the table: mean value ± standard deviation (n = 3). [1] The s_r of the contaminated soil was calculated using E_{wc} instead of w.

Hence, we decided to air-dry the natural soil before its contamination and then add the diesel to achieve a concentration of 12.62%. Hence, the contaminated soil used for its characterization had a liquids content ($l_c = 34.39 \pm 0.22\%$), equivalent water content ($E_{wc} = 35.13 \pm 0.22\%$), and saturation rate ($s_r = 71.28 \pm 0.21\%$) closer to the same factors observed in the natural soil.

3.2. Influence of Diesel Contamination on the Soil's pH

Soil contamination with diesel had a significant impact on its pH. After soil contamination, the pH of the soil changed from moderately alkaline ($pH_{natural} = 7.95 \pm 0.14$) to neutral ($pH_{contaminated} = 6.98 \pm 0.03$). Nevertheless, this pH drop does not represent a limitation for biodegradation processes, since the pH for the biotransformation of the contaminant has been found to be in the range of 6 to 8 [34]. This tendency to decrease the pH as the soil gets contaminated agrees with the results obtained by Taheri et al. [9].

The pH reduction of soil after its contamination was due to the acidifying effect of the sulfur contained in diesel. It is known that the addition of elemental sulfur (S) to soil affects the oxidation rate and produces two hydrogen ions due to the microbial activity of soil (Equation (3)). Since pH is a measure of the hydrogen ion concentration ($pH = -\log_{10}[H^+]$) in soil solution, the increase in concentration resulted in a lower pH.

$$S + CO_2 + \frac{1}{2}O_2 + 2H_2O \rightarrow CH_2O + SO_4^{2-} + 2H^+ \tag{3}$$

Momeni et al. [35] studied the effect of acid and alkaline waters on the geotechnical properties of clay soil and found that the decrease of the pH value of the soil increased the liquid limit and the plastic index of the soil. Moreover, they indicated that the coefficient of permeability of the soil increased as the pH values of the soil shifted from the neutral condition to the acidic or alkaline conditions [35].

3.3. Influence of Diesel Contamination on the Soil's Texture

Diesel contamination of natural soil caused a change in its texture classification from silty clay to silty loam. This change, from a fine texture to a ridged one after soil contamination, was due to the agglomeration of soil particles as a response to soil acidification and the reduction of the dielectric constant of the pore fluid.

A decrease in pH promoted the exchange of calcium ions by hydrogen ions, which resulted in clay particle agglomeration/flocculation due to its surface charge alteration [36]. Furthermore, partial replacement of water contained in the pores of soil by the contaminant changed the dielectric constant value of the pore fluid. Hence, the attractive forces prevailed over repulsive ones, resulting in soil agglomeration/flocculation [37].

The van der Waals attraction force developed between two parallel plates can be calculated as shown in Equations (4) and (5), while the repulsion force can be obtained as indicated in Equations (6) and (7) [37].

$$Att = \frac{H}{6\pi d^3}, \tag{4}$$

where d is the distance between plates and H is the Hamaker constant obtained as follows [37]:

$$H = \frac{3}{4}kT\left(\frac{\varepsilon_1 - \varepsilon_2}{\varepsilon_1 + \varepsilon_2}\right)^2 + \frac{3hv_e}{16\sqrt{2}}\frac{\left(m_1^2 - m_2^2\right)^{3/2}}{\left(m_1^2 + m_2^2\right)^{3/2}}, \tag{5}$$

where k is the Boltzmann constant, T is the absolute temperature, ε_i denotes the dielectric constant of the medium, h is the constant of Planck, v_e corresponds to the main electronic adsorption frequency in ultraviolet light, and m_i is the reflective index of the medium in visible light.

$$Rep \propto \frac{1}{K_1\beta}, \tag{6}$$

where K_1 is a constant that depends on the valence of the ions and the ionic concentration and β can be obtained as follows [37]:

$$\beta = \frac{K_1}{\varepsilon k T K_3}, \tag{7}$$

with K_1 and K_3 being constants that depend on the ionic concentration.

It can be appreciated from Equations (4) and (6) that both attractive and repulsive forces depend on the dielectric constant value of the pore fluid. However, the variation of the dielectric constant value of the pore fluid has a higher effect over the repulsive forces [37]. Thus, a decrease on the value of the dielectric constant of the pore fluid reduced both attractive and repulsive forces, but the decrease in repulsive forces was bigger than the decrease in the attractive forces. Hence, attractive forces prevail over repulsive ones (the net effect was attractive) and gave as result the agglomeration of the soil particles.

3.4. Effect of Diesel Contamination on the Specific Gravity of Soil

The specific gravity of natural soil was reduced from 2.72 ± 0.02 to 2.60 ± 0.02 after soil contamination with diesel. The partial substitution of water (in the contaminated soil) explains this significant change ($p < 0.01$) since diesel has a density and specific gravity lower than water.

3.5. Impact of Diesel Contamination on the Soil's Particle Size Distribution

Grains' size content of both soils is presented in Table 4, and their granulometric curves are shown in Figure 2. Both natural and diesel contaminated soils corresponded to a fine-grained soil. After diesel soil contamination, a significative ($p < 0.01$) change in fine-grained particles sizes' content was observed. The silt-like sized granules increased by 37% while the clay-like sized granules decreased by 46%. This change was due to both the reduction of the dielectric constant and the acidification of the pore fluid of the soil, which caused that the clay-like sized particles agglomerated and increased their size reaching a silt-like size.

Table 4. Impact of diesel contamination in grain-like size distribution of soil.

Grain Size	Natural Soil	Contaminated Soil
Gravel (%)	0.25 ± 0.06	0.22 ± 0.03
Sand (%)	7.34 ± 0.49	6.43 ± 0.21
Silt (%)	52.03 ± 1.21	71.72 ± 1.44
Clay (%)	40.38 ± 2.30	21.63 ± 0.70

Values in the table: mean value ± standard deviation (n = 5).

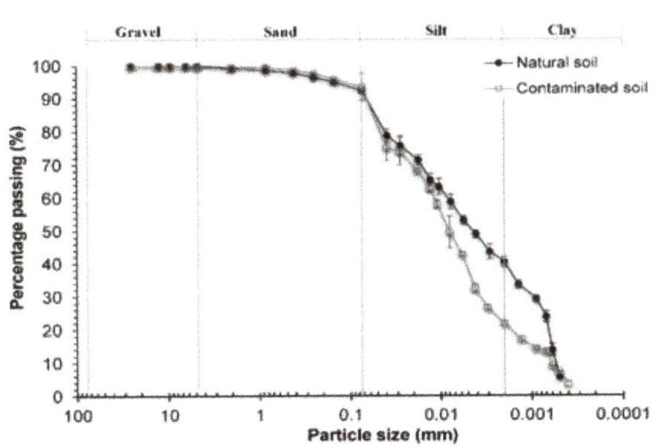

Figure 2. Impact of diesel contamination on particle size distribution of soil. Error bars denote standard deviations.

Lambe and Withman [38] indicated that the reduction of the dielectric constant and/or pH drop, as well as ion valence reduction and/or electrolyte concentration decrease, promotes soil flocculation. In low pH environments, the thickness of the double layer around clay particles can decrease [39]. The diffuse double layer theory indicates that the thickness of the double layer ($1/\chi$) can be calculated as indicated in Equation (8):

$$\frac{1}{\chi} = \sqrt{\frac{\varepsilon k T}{8\pi z^2 e^2 \eta}} \tag{8}$$

where ε corresponds to the dielectric constant of the pore fluid, k is the Boltzman constant, T is the absolute temperature, z is the valence of the cations, e is the electric charge, and η is the concentration in number of ions per milliliter.

When the soil was contaminated with diesel ($\varepsilon_{diesel} = 2.3$), it interacted with the water ($\varepsilon_{water} = 80.3$) contained in the pores of the soil and reduced the dielectric constant of the pore fluid. As a result, the thickness of the double layer was reduced and the soil particles began to join together due to the influence of the van der Waals forces. As the soil particles got closer to each other, the effect of the van der Waals attractive forces increased. This promoted the agglomeration of clay particles by forming flocs, which was appreciated as an increment in the silt-like size fraction content, with a correspondent decrement in the clay-like size fraction.

3.6. *Influence of Diesel Contamination on the Consistency Limits*

Table 5 provides the plastic characteristics of the soils. According to their Atterberg limits, both the natural and the contaminated soils are considered as very high and expansive ones. Despite the variations of the values of the Atterberg limits, both the natural and the contaminated soils were classified as silts of high compressibility. The natural soil was considered as a normal active clay, while contaminated soil was considered as a very active one. This increase in the activity of contaminated soil, in comparison with its activity at uncontaminated condition, was attributed to the physical and chemical effect of diesel on clay particles.

Table 5. Influence of diesel contamination in the consistency limits of soil.

Parameter	Natural Soil	Contaminated Soil
Liquid limit (%)	77.3 ± 0.4	83.9 ± 1.9
Plastic limit (%)	35.8 ± 2.2	37.3 ± 1.0
Plasticity index (%)	41.5 ± 2.5	46.7 ± 1.5
USCS classification	MH	MH
Average clay activity (%) *	1.03	2.14

Values in the table: mean value ± standard deviation ($n = 5$). * Calculated as: average plasticity index/average clay fraction.

After soil contamination, its liquid and plastic limits increased by 9% and 4%, respectively. It is known that the liquid behavior of soils is influenced by free water, which is not absorbed by the particles of soil and can freely move through soil pores. When diesel was added to the natural soil, it covered the surfaces of the clay particles, making the contact between the free water and the clay particles more difficult.

Since diesel is a non-polar liquid, it could not allow that the water added during the test, which created a link with clay surface, to provide the required condition for clay to flow. When water was added to the contaminated soil to evaluate its Atterberg limits, it gradually displaced the diesel adhered to clay surface, allowing the interaction between clay and water. Thus, more water was needed in the contaminated soil to achieve its plastic state and, thus, higher plasticity was obtained.

3.7. Effect of Diesel Contamination on the Modified Free Swell Index of the Soil

The MFSI test classified the natural soil as a high-swelling potential clay (2.33 ± 0.47 cm^3/g). However, this index classified contaminated soil as a very high-swelling potential soil (3.79 ± 0.77 cm^3/g). This finding agrees with those reported by Salimnezhad et al. [12], who observed an increase in the free swelling of oil-contaminated clayey soil as the contaminant concentration increased from 4% to 12%.

At low pH levels, acids attack the original minerals contained in soil and produce a modification in both the silica and alumina content, reducing the Si/Al ratio as well as the calcium content (see Table 6). Hence, the chemical composition of the soil changed when the pH of the pore fluid diminished, and due to the reduction of the calcium content there was an increase in its swell capacity.

Table 6. Impact of diesel contamination on the concentration of the chemical elements of the soil.

Element	Limit of Detection (%)	Concentration (%)	
		Natural Soil	Contaminated Soil
Si	0.033	50.94 ± 3.10	54.01 ± 1.97
Fe	0.001	15.97 ± 1.15	14.28 ± 2.20
Al	0.097	14.80 ± 0.77	17.06 ± 1.61
Ca	0.001	7.42 ± 0.16	5.40 ± 0.47
Na	0.71	5.04 ± 3.61	3.79 ± 0.88
K	0.002	2.17 ± 0.13	2.43 ± 0.43
Ti	0.001	2.13 ± 0.28	1.97 ± 0.23
Mg	0.165	0.96 ± 0.86	0.59 ± 0.14
Mn	0.001	0.29 ± 0.06	0.25 ± 0.07
Zn	0.001	0.28 ± 0.09	0.22 ± 0.04
Ratios			
Ca/Si		0.15 ± 0.01	0.10 ± 0.01
Si/Al		3.44 ± 0.04	3.18 ± 0.18
Mg/Al		0.06 ± 0.05	0.03 ± 0.01

Values in the table: mean value ± standard deviation (n = 3).

3.8. Consequence of Diesel Contamination on the Shear Strength of the Soil

Figure 3 shows the changes in shear stress resistance of both natural and contaminated soils. Cohesion and internal friction angle were obtained from the failure envelop plotted for each soil condition. The cohesion of the contaminated soil decreased by 20% (from 36.7 ± 0.2 kPa to 29.2 ± 0.1 kPa), and its internal friction angle increased by 30% (from 26.9° ± 0.2° to 34.8° ± 0.1°) after soil contamination.

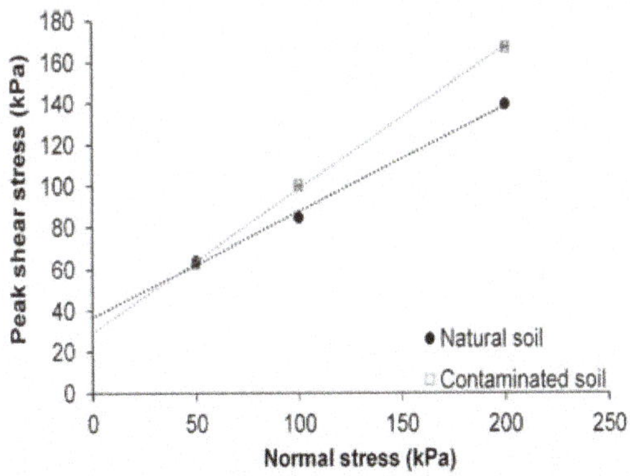

Figure 3. Effect of diesel contamination on the direct shear behavior of soil.

The cohesion of soil depends on the material type as well as the intermolecular bonds between clay particles and the absorbed water. As the natural soil particles got coated by diesel, clay particles agglomerated, formed silt-like size grains, and reduced the clay-like size grains content in the contaminated soil. Although both soils had nearly the same equivalent water content ($E_{wc-natural} = 36.94\% \pm 0.19\%$ vs. $E_{wc-contaminated} = 35.13\% \pm 0.22\%$) and saturation degree ($s_{r-natural} = 70.82 \pm 0.18$ vs. $s_{r-contaminated} = 71.28\% \pm 0.21\%$), the chemical characteristics of the diesel and soil water content reduction contributed to the decrease in the contaminated soil's cohesion.

The reduction of the soil's cohesion has been reported by other authors [5,12,13]. They attributed this decrease to reducing the specific surface area (SSA) of soil particles due to hydrocarbon deposition over the soil grains. The reduction of the SSA decreased the CEC of the soil and reduced the contact between the soil particles and water molecules [5,12,13]. Therefore, the cohesion of the soils was reduced since soil particles absorbed a lower quantity of water.

The agglomeration effect of diesel over the soil particles increased the friction angle of the contaminated soil. Thus, it can be inferred that under normal stresses between 100 and 200 kPa, contaminated soil will exhibit a higher shear resistance than natural soil. The quality of the soil is not only defined by its shear strength but also by its consolidation behavior.

So, before considering that the contaminated soil quality was enhanced due to its higher shear strength, it is necessary to consider its permeability and compressibility changes. As has already been reported, diesel deposition over soil particles reduced the soil's hydraulic conductivity [9] and SSA [12,13]. Hence, higher settlements may be observed compared to the ones obtained for the natural soil [7,12]. Furthermore, the slope stability of the trenches located over contaminated soils must be verified because of the decrease in its cohesion.

Peak shear stresses were plotted for each soil condition (Figure 3). For a normal stress of 50 kPa, natural and contaminated soils showed almost the same shear strength. However, for the 100 and 200 kPa normal stresses, the shear strength of contaminated soil was increased by 18% and 20%, respectively. According to Fredlund et al. [40], the shear strength of unsaturated soils (τ_f) can be evaluated in terms of two stress state variables, as shown in Equation (9):

$$\tau_f = c' + (\sigma_n - u_a)\ tan\phi^a + (u_a - u_w)\ tan\phi^b. \tag{9}$$

The first two terms of Equation (9) correspond to the Mohr–Coulomb criterion, c' is the effective cohesion, $(\sigma_n - u_a)$ denotes the net normal stress, u_a is the pore air pressure, and ϕ^a is the angle of frictional resistance related with normal stress. The third term of Equation (9) introduces an additional friction angle (ϕ^b) relative to matric suction ($\psi_m = u_a - u_w$) contribution to shear strength resistance, where u_w corresponds to the pore water pressure.

Total soil suction (ψ_t) considers the effect of the matric suction (ψ_m) and osmotic or solute suction (ψ_o), as expressed in Equation (10). Matric suction was defined as the difference between u_a and u_w, and is related to capillary effects. Osmotic suction arises from the presence of dissolved salts or solutes in the soil pore fluid.

$$\psi_t = \psi_m + \psi_o \tag{10}$$

The natural soil had a ψ_m of 4709 ± 1 kPa, while the contaminated soil had a ψ_m of 2077 ± 12 kPa. As the contaminated soil had a considerably lower matric suction than the natural soil, a decrease in its shear strength could be expected. This is because the tension force developed by the air–water interface or contractile skin on soil particles decreases as the matric suction decreases [41]. In contrast, an increase in shear strength of soil was observed after its contamination.

The ψ_t of natural and contaminated soils were 7521 ± 1 kPa and 18,484 ± 13 kPa, respectively. From Equation (5), the ψ_o can be obtained as the difference between ψ_t and ψ_m. As shown in Figure 4a, the ψ_m had a higher effect on the shear strength of natural soil, while the ψ_o had a major role in the shear strength of the contaminated soil (Figure 4b).

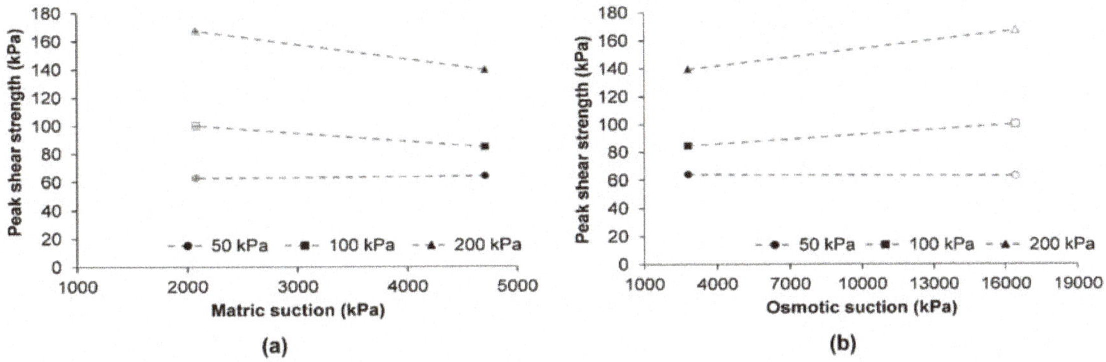

Figure 4. Influence of diesel contamination: (**a**) matric suction and (**b**) osmotic suction in the shear strength of soil. The symbols filled in black color correspond to natural soil while the symbols in grey color correspond to contaminated soil.

It is known that ψ_t influences the shear stress behavior of the unsaturated soils. However, in geotechnical engineering applications, the ψ_m component is more frequently considered than the contribution of ψ_o [41]. Nevertheless, in the case of contaminated soils, the impact of ψ_o on the shear strength of soil must be considered [42].

Therefore, as proposed by [42], and considering ψ_o as an independent stress-state variable, Equation (11) could be modified as:

$$\tau_f = c' + (\sigma_n - u_a)\ tan\phi^a + \psi_o\ tan\phi^c \quad (11)$$

where ϕ^c corresponds to the fiction angle relative to osmotic suction contribution to shear strength resistance.

Figure 5 shows the strength envelop, considering ψ_o as an independent stress-state variable, as well as the equation for shear strength behavior of the contaminated soil, according to the results obtained in this research, as a function of both normal stress and osmotic suction.

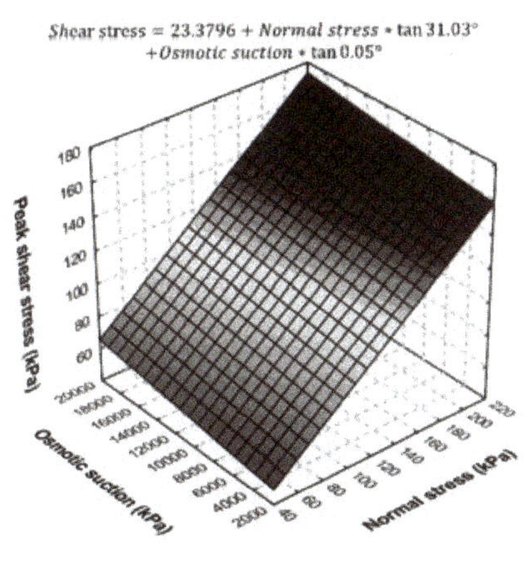

Figure 5. Contaminated soil shear strength model based on osmotic suction and normal stress.

3.9. Effect of Diesel Contamination on the Soil's Mineralogy

Figure 6 shows the X-ray diffractograms of both the natural and the contaminated soil. Both soil samples showed the presence of montmorillonite clay mineral [PDF 00-013-0135: montmorillonite-15A: $Ca_{0.2}(Al,Mg)_2Si_4O_{10}(OH)_2{\cdot}4H_2O$] and anorthite [PDF 00-018-1202: anorthite, sodian, intermediate: $(Ca,Na)(Si,Al)_4O_8$]. Montmorillonite belongs to the smectite group and is formed by two Si-based tetrahedral sheets. Anorthite is a plagioclase feldspar mineral that occurs in some igneous and metamorphic rocks.

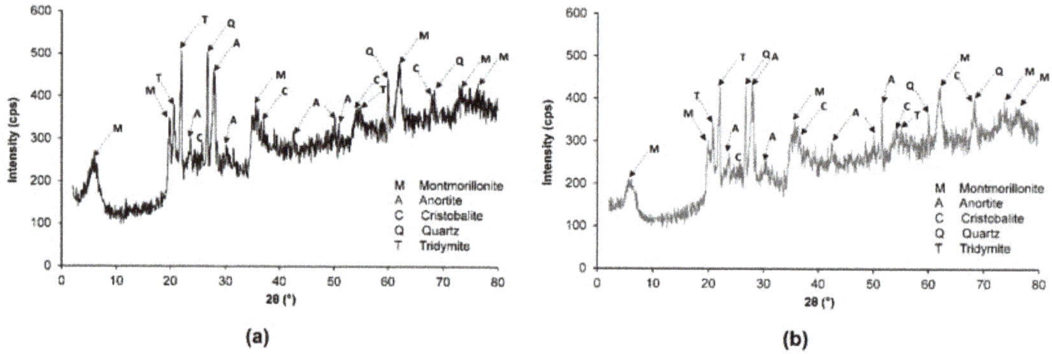

Figure 6. X-ray diffractogram of: (**a**) natural and (**b**) diesel-contaminated soils.

The presence of quartz, cristobalite, and tridymite was also observed. Silica or silicon dioxide (SiO_2) can exhibit both crystalline (quartz, tridymite, and cristobalite) and amorphous forms. Cristobalite and tridymite are silica mineral quartz polymorphs that exhibit the same chemical formula but have different crystal structures.

In most cases, cristobalite is the first crystallization product of the amorphous SiO_2 at high temperatures. In tridymite, the silica tetrahedra are packed in a two-layer structure, while in cristobalite, the silica tetrahedra are packed in a three-layer structure [43]. Cristobalite and tridymite have a more open structure than quartz, which allows the inclusion of other elements into its crystal structure [43].

A change was observed in the reflective angle of the montmorillonite in diesel-contaminated soil ($2\theta = 5.70°$) in comparison with natural soil ($2\theta = 5.88°$). This shift to the left of montmorillonite peak caused an increase in the lattice spacing of the montmorillonite mineral [44]. It implies that montmorillonite may undergo a swelling effect of the internal layer, from d_{001} = 15.018 Å (for natural soil) to d_{001} = 15.492 Å (for diesel-contaminated soil), due to its diesel contamination.

Soil contamination also decreased the peak area of the montmorillonite minerals, which was due to a reduction in the column length of the reflective minerals. This indicated a decrease in the montmorillonite minerals' concentration [44].

Although the peaks of quartz, cristobalite, and tridymite minerals did not show a shift in their reflective angle, their intensities were reduced after soil contamination. This was due to a coating effect on the mineral surface, caused by the peripheral adsorption of diesel on these minerals [44]. Such an effect resulted in an apparent decrease in the concentration of mineral crystals, since they were not able to reflect the incident X-rays as the natural soil could.

3.10. Impact of Diesel Contamination on the Chemical Elements of the Soil

The results of the analysis of the chemical elements of both natural and contaminated soils are presented in Table 6. For both soil conditions, the major elements detected were Si, Fe, Al, Ca, Na, and K, while minor elements were Mg, Mn, and Zn. In nature, inorganic exchange ions (e.g., Ca^{2+} and Na^+) frequently balance the clays' net negative charge [33].

Further, in clays, the hydrophilic nature of the mineral surfaces is due to the occurrence of Si–O groups and the hydration of those inorganic exchangeable ions [33].

Although the concentration of the elements detected varied from natural to contaminated condition, no significant changes were observed ($p > 0.09$), except for calcium ($p < 0.01$). The statistical analysis of this data set is presented in Appendix A. The reduction of calcium content after soil contamination was attributed to a chemical reaction with any of the diesel constituents; for example, the sulfur contained in the applied diesel. Due to the calcium concentration decrease in the contaminated soil matrix, both a pH and a cation exchange reduction can be expected [45].

3.11. *Effect of Diesel Contamination on Some Properties Related with Its Possible Remediation*

3.11.1. Electrical Conductivity of the Soil

The EC of soil has been used as a parameter related to the level of macronutrients and micronutrients available in the soil [46]. The EC of the natural soil decreased from 536 ± 86 μS/cm to 420 ± 38 μS/cm after its contamination. Due to this significant decrease ($p < 0.03$), it is possible that the available nutrients for both microorganisms and plants may be unbalanced, and an addition of external nutrients may be needed for soil bioremediation. The observed EC reduction has been previously reported [9], and it was attributed to the low dielectric constant of the diesel which reduced the dielectric constant of the soil's pore fluid [9].

3.11.2. Organic Matter Content in the Soil

The organic matter content of natural soil increased from 5.7% ± 0.5% to 11.6% ± 0.2% after its contamination. This increment was related to the presence of diesel in soil—no other sources of organic matter were added. Due to the organic nature of diesel, it is possible to infer that organic carbon concentration in contaminated soil also increased. Thus, the C:N:P ratio of contaminated soil has to be verified because if it is not in a suitable range, the efficiency of the available bioremediation and phytoremediation techniques may be reduced.

3.11.3. Soil's Cation Exchange Capacity

Soil's CEC is one of the factors that influence its pH buffer capacity. After soil contamination, its CEC was reduced by 50% in comparison with its CEC at its natural condition (from 75.94 to 38.41 $cmol_+/kg$). This indicates a drop in the pH buffering capacity of the contaminated soil that is reflected in its acidification because the higher the CEC capacity, the higher the pH buffering capacity [46].

Kaya and Fang [37] suggested that the decrease of the CEC was associated to the reduction of the dielectric constant of the pores' fluid, because the number of the exchangeable cations diminished when the dielectric constant of the pore fluid got reduced.

The CEC reduction indicates that contaminated soil decreased its capacity to retain and release cations, such as Ca^{2+}, Mg^{2+}, K^+, and Na^+, by electrostatic forces, in comparison with natural soil. So, a reduction in soil fertility can be expected since Ca^{2+}, Mg^{2+}, K^+, Mn^{2+}, and Fe^{2+} are considered as bioavailable metal cation species essential for green plant nutrition and soil microorganisms [47].

It is known that organic matter has a high CEC. Thus, an increase in the CEC of contaminated soil, due to the increment of its organic matter content, can be expected. However, a decreased CEC in contaminated soil was observed. When diesel was added to the natural soil, cation exchange between diesel and calcium cations (Ca^{2+}), contained in both clay minerals and organic matter, was promoted.

This process released Ca^{2+} in the soil's pores' solution that were used by clay particles to flocculate and, at the same time, to counteract the acidification effect caused by the added H^+. Thus, although the organic matter content increased in contaminated soil, it could not increase its CEC.

4. Conclusions

The geotechnical properties of unsaturated diesel-contaminated soil were evaluated and compared with the ones obtained for the soil under natural condition. Diesel concentration in the soil was gradually increased from 0% to 14%; however, the soil could only retain 12.62% ± 0.27% of the added diesel, which corresponded to a saturation rate of 79.46% ± 1.03%. This result indicated that the soil only retained the necessary diesel to fulfil its hydration demands.

Diesel contamination of soil reduced its calcium content and caused a decrease in the pH of the soil. Diesel promoted the agglomeration of fine-grain-sized particles of the soil and changed the particle size distribution of the contaminated soil.

The cohesion of contaminated soil decreased by 20%, while its internal friction angle increased by 30%. Although the matric suction of contaminated soil was considerably lower than the one obtained for natural soil; its osmotic suction was considerably higher. This contributed to an increase in the shear strength of the contaminated soil as compared to the natural soil. However, before considering that the contaminated soil quality improved due to its higher shear strength, the consolidation behavior of the contaminated soil should be assessed.

Author Contributions: Conceptualization, C.E.H.-M.; experiment set up, C.E.H.-M.; investigation, P.G.R. and C.E.H.-M.; data visualization, P.G.R. and C.E.H.-M.; writing–original draft, C.E.H.-M.; writing–review & editing, P.G.R., O.C.A. and C.E.H.-M.; project administration, C.E.H.-M.; funding acquisition, C.E.H.-M. and O.C.A. All authors have read and agreed to the published version of the manuscript.

Funding: This research was funded by the Consejo Nacional de Ciencia y Tecnología (CONACYT), grant number 270161, and by the Universidad Autónoma de Querétaro, grant number FIN-2020-03 (FONDEC-UAQ-2019).

Institutional Review Board Statement: Not applicable.

Informed Consent Statement: Not applicable.

Data Availability Statement: Data presented is contained within the article.

Conflicts of Interest: The authors declare no conflict of interest. The funders had no role in the design of the study; in the collection, analyses, or interpretation of data; in the writing of the manuscript, or in the decision to publish the results.

Appendix A

Statistical analysis of the concentration of the chemical elements of the soil reported on Table 4. The analysis was executed in Excel for Microsoft 365 with the "Data analysis" tool.

Table A1. Sodium.

Summary						
Groups	*Account*	*Sum*	*Average*	*Variance*		
Na—Natural soil	3	15.13	5.043333333	13.0270333		
Na—Contaminated soil	3	11.37	3.79	0.7761		
ANALYSIS OF VARIANCE						
Origin of variances	*Sum of squares*	*Degrees of freedom*	*Mean of squares*	*F*	*Probability*	*Critical value for F*
Between groups	2.356266667	1	2.356266667	0.3414104	0.590383	7.708647422
Within groups	27.60626667	4	6.901566667			
Total	29.96253333	5				

Table A2. Magnesium.

Summary						
Groups	*Account*	*Sum*	*Average*	*Variance*		
Mg—Natural soil	3	2.887	0.962333333	0.74157633		
Mg—Contaminated soil	3	1.768	0.589333333	0.01989033		
ANALYSIS OF VARIANCE						
Origin of variances	*Sum of squares*	*Degrees of freedom*	*Mean of squares*	*F*	*Probability*	*Critical value for F*
Between groups	0.2086935	1	0.2086935	0.54813562	0.500182374	7.708647422
Within groups	1.522933333	4	0.380733333			
Total	1.731626833	5				

Table A3. Aluminum.

Summary						
Groups	*Account*	*Sum*	*Average*	*Variance*		
Al—Natural soil	3	44.386	14.79533333	0.59737233		
Al—Contaminated soil	3	51.185	17.06166667	2.60753433		
ANALYSIS OF VARIANCE						
Origin of variances	*Sum of squares*	*Degrees of freedom*	*Mean of squares*	*F*	*Probability*	*Critical value for F*
Between groups	7.704400167	1	7.704400167	4.80787802	0.093411475	7.708647422
Within groups	6.409813333	4	1.602453333			
Total	14.1142135	5				

Table A4. Silica.

Summary						
Groups	*Account*	*Sum*	*Average*	*Variance*		
Si—Natural soil	3	152.826	50.942	9.635884		
Si—Contaminated soil	3	162.029	54.00966667	3.88626633		
ANALYSIS OF VARIANCE						
Origin of variances	*Sum of squares*	*Degrees of freedom*	*Mean of squares*	*F*	*Probability*	*Critical value for F*
Between groups	14.11586817	1	14.11586817	2.08781412	0.221990319	7.708647422
Within groups	27.04130067	4	6.761075167			
Total	41.16016883	5				

Table A5. Potassium.

Summary						
Groups	*Account*	*Sum*	*Average*	*Variance*		
K—Natural soil	3	6.521	2.173666667	0.01738433		
K—Contaminated soil	3	7.283	2.427666667	0.18305233		
ANALYSIS OF VARIANCE						
Origin of variances	*Sum of squares*	*Degrees of freedom*	*Mean of squares*	*F*	*Probability*	*Critical value for F*
Between groups	0.096774	1	0.096774	0.9656317	0.381407702	7.708647422
Within groups	0.400873333	4	0.100218333			
Total	0.497647333	5				

Table A6. Calcium.

Summary						
Groups	*Account*	*Sum*	*Average*	*Variance*		
Ca—Natural soil	3	22.247	7.415666667	0.02434433		
Ca—Contaminated soil	3	16.195	5.398333333	0.21904633		
ANALYSIS OF VARIANCE						
Origin of variances	*Sum of squares*	*Degrees of freedom*	*Mean of squares*	*F*	*Probability*	*Critical value for F*
Between groups	6.104450667	1	6.104450667	50.1617482	0.002097894	7.708647422
Within groups	0.486781333	4	0.121695333			
Total	6.591232	5				

Table A7. Titanium.

Summary						
Groups	*Account*	*Sum*	*Average*	*Variance*		
Ti—Natural soil	3	6.393	2.131	0.081003		
Ti—Contaminated soil	3	5.924	1.974666667	0.05164433		
ANALYSIS OF VARIANCE						
Origin of variances	*Sum of squares*	*Degrees of freedom*	*Mean of squares*	*F*	*Probability*	*Critical value for F*
Between groups	0.036660167	1	0.036660167	0.55274638	0.498494008	7.708647422
Within groups	0.265294667	4	0.066323667			
Total	0.301954833	5				

Table A8. Manganese.

Summary						
Groups	*Account*	*Sum*	*Average*	*Variance*		
Mn—Natural soil	3	0.878	0.292666667	0.00410233		
Mn—Contaminated soil	3	0.751	0.250333333	0.00423233		
ANALYSIS OF VARIANCE						
Origin of variances	*Sum of squares*	*Degrees of freedom*	*Mean of squares*	*F*	*Probability*	*Critical value for F*
Between groups	0.002688167	1	0.002688167	0.64505679	0.466896995	7.708647422
Within groups	0.016669333	4	0.004167333			
Total	0.0193575	5				

Table A9. Iron.

Summary						
Groups	*Account*	*Sum*	*Average*	*Variance*		
Fe—Natural soil	3	47.916	15.972	1.315753		
Fe—Contaminated soil	3	42.831	14.277	4.838479		
ANALYSIS OF VARIANCE						
Origin of variances	*Sum of squares*	*Degrees of freedom*	*Mean of squares*	*F*	*Probability*	*Critical value for F*
Between groups	4.3095375	1	4.3095375	1.40051188	0.302164965	7.708647422
Within groups	12.308464	4	3.077116			
Total	16.6180015	5				

Table A10. Zinc.

Summary						
Groups	*Account*	*Sum*	*Average*	*Variance*		
Zn—Natural soil	3	0.825	0.275	0.008757		
Zn—Contaminated soil	3	0.664	0.221333333	0.00189233		
ANALYSIS OF VARIANCE						
Origin of variances	*Sum of squares*	*Degrees of freedom*	*Mean of squares*	*F*	*Probability*	*Critical value for F*
Between groups	0.004320167	1	0.004320167	0.81134969	0.418651132	7.708647422
Within groups	0.021298667	4	0.005324667			
Total	0.025618833	5				

References

1. Johnson, S.T.; Jardine, F.M. A framework for assessing risk in contaminated land engineering. In *Contaminated Soil '95. Soil & Environment*; van den Brink, J.W., Bosman, R., Arendt, F., Eds.; Springer: Dordrecht, The Netherlands, 1995; Volume 5, pp. 1475–1486. [CrossRef]
2. Meegoda, N.J.; Ratnaweera, P. Compressibility of contaminated fine-grained soils. *Geotech. Test. J.* **1994**, *17*, 101–112. [CrossRef]
3. Rahman, Z.; Hamzah, U.; Taha, M.; Ithnain, N.S.; Ahmad, N. Influence of oil contamination on geotechnical properties of basaltic residual soil Zulfahmi. *Am. J. Appl. Sci.* **2010**, *7*, 954–961. [CrossRef]
4. Ijimdiya, T.S. The effect of oil contamination on the consolidation properties of lateritic soil. *Develop. Appl. Ocean. Eng.* **2013**, *2*, 53–59.
5. Safehian, H.; Rajabib, A.M.; Ghasemzadeh, H. Effect of diesel-contamination on geotechnical properties of illite soil. *Eng. Geol.* **2018**, *241*, 55–63. [CrossRef]
6. Heris, M.N.; Aghajani, S.; Hajialilue-Bonab, M.; Molamahmood, H.V. Effects of lead and gasoline contamination on geotechnical properties of clayey soils. *Soil Sediment. Contam.* **2020**, *29*, 340–354. [CrossRef]
7. Yazdi, A.; Teshnizi, E.S. Effects of contamination with gasoline on engineering properties of fine-grained silty soils with an emphasis on the duration of exposure. *SN Appl. Sci.* **2021**, *3*, 704. [CrossRef]
8. Correia, N.D.; Portelinha, F.H.M.; Mendes, I.S.; da Silva, J.W.B. Lime treatment of a diesel-contaminated coarse-grained soil for reuse in geotechnical applications. *Int. J. Geo-Eng.* **2020**, *11*, 8. [CrossRef]
9. Taheri, S.; Ebadi, T.; Maknoon, R.; Amiri, M. Predicting variations in the permeability and strength parameters of a sand-bentonite mixture (SBM) contaminated simultaneously with lead (II) and diesel. *Appl. Clay Sci.* **2018**, *157*, 102–110. [CrossRef]
10. Izdebska-Mucha, D.; Trzcinski, J. Effects of petroleum pollution on clay soil microstructure. *Geologija* **2008**, *50*, 69–75.
11. Son, Y.; Oh, M.; Lee, S. Influence of diesel fuel contamination on the electrical properties of unsaturated soil at a low frequency range of 100 Hz–10 MHz. *Environ. Geol.* **2009**, *58*, 1341–1348. [CrossRef]
12. Salimnezhad, A.; Soltani-Jigheh, H.; Soorki, A.A. Effects of oil contamination and bioremediation on geotechnical properties of highly plastic clayey soil. *J. Rock Mech. Geotech. Eng.* **2021**, *13*, 653–670. [CrossRef]
13. Kermani, M.; Ebadi, T. The effect of oil contamination on the geotechnical properties of fine-grained soils. *Soil Sediment. Contam.* **2012**, *21*, 655–671. [CrossRef]
14. DieselNet. Mexico: Diesel Fuel. *Fuel Regulations*. Available online: https://dieselnet.com/standards/mx/fuel.php#y2016 (accessed on 5 October 2020).
15. Villalobos, M.; Avila-Forcada, A.P.; Gutierrez-Ruiz, M.E. An Improved Gravimetric Method to Determine Total Petroleum Hydrocarbons in Contaminated Soils. *Water Air Soil Pollut.* **2008**, *194*, 151–161. [CrossRef]
16. Meegoda, N.J.; Rajapakse, R.A. Short-Term and Long-Term Permeabilities of Contaminated Clays. *J. Environ. Eng.* **1993**, *119*, 725–743. [CrossRef]
17. Meegoda, J.N.; Chen, B.; Gunasekera, S.D.; Pederson, P. Compaction Characteristics of Contaminated Soils-Reuse as a Road Base Material. Recycled Materials in Geotechnical Applications. In Proceedings of the American Society of Civil Engineers Annual Convention, Boston, MA, USA, 18–21 October 1998; Vipulanandan, C., Elton, D.J., Eds.; Geotechnical Special Publication 79. ASCE: New York, NY, USA, 1998; pp. 195–209.
18. ASTM D2216. *Test Methods for Laboratory Determination of Water (Moisture) Content of Soil and Rock by Mass*; ASTM International: West Conshohocken, PA, USA, 2019. [CrossRef]
19. ASTM D7263. *Standard Test Methods for Laboratory Determination of Density (Unit Weight) of Soil Specimens*; ASTM International: West Conshohocken, PA, USA, 2018. [CrossRef]
20. ASTM D854. *Standard Test Methods for Specific Gravity of Soil Solids by Water Pycnometer*; ASTM International: West Conshohocken, PA, USA, 2014. [CrossRef]

21. NOM-021-RECNAT-2000. Mexican Official Norm NOM-021-RECNAT-2000 That States the Specifications of Fertility, Salinity, and Classification of Soil. *Studies, Sampling, and Analysis.* Available online: http://www.ordenjuridico.gob.mx/Documentos/Federal/wo69255.pdf (accessed on 1 September 2019).
22. ASTM D6913. *Standard Test Methods for Particle-Size Distribution (Gradation) of Soils Using Sieve Analysis*; ASTM International: West Conshohocken, PA, USA, 2017. [CrossRef]
23. ASTM D1140. *Standard Test Methods for Determining the Amount of Material Finer than 75-μm (No. 200) Sieve in Soils by Washing*; ASTM International: West Conshohocken, PA, USA, 2017. [CrossRef]
24. BS EN ISO 17892-12. *Geotechnical Investigation and Testing. Laboratory testing of soil. Determination of Liquid and Plastic Limits*; British Standards Institution/International Organization for Standardization: London, UK, 2018.
25. ASTM D2487. *Standard Practice for Classification of Soils for Engineering Purposes (Unified Soil Classification System)*; ASTM International: West Conshohocken, PA, USA, 2017. [CrossRef]
26. Sridharan, A.; Rao, S.M.; Murthy, N.S. Free swell index of soils: A need for redefinition. *Indian Geotech. J.* **1985**, *15*, 94–99.
27. ASTM D3080. *Standard Test Method for Direct Shear Test of Soils Under Consolidated Drained Conditions*; ASTM International: West Conshohocken, PA, USA, 2011. [CrossRef]
28. ASTM D5298. *Standard Test Method for Measurement of Soil Potential (Suction) Using Filter Paper*; ASTM International: West Conshohocken, PA, USA, 2016. [CrossRef]
29. AASHTO T 267. *Standard Method of Test for Determination of Organic Content in Soils by Loss on Ignition*; American Association of State Highway and Transportation Officials: Washington DC, USA, 2018.
30. Yong, R.N.; Rao, S.M. Mechanistic evaluation of mitigation of petroleum hydrocarbon contamination by soil medium. *Can. Geotech. J.* **1991**, *28*, 84–91. [CrossRef]
31. Annandarajah, A.; Chen, J. Van der Waals attractive forces between clay particles in water and contaminants. *Soils Foundations.* **1997**, *37*, 27–37. [CrossRef]
32. Chiou, C.T.; Shoup, T.D. Soil sorption of organic vapors and effects of humidity on sorptive mechanism and capacity. *Environ. Sci. Technol.* **1985**, *19*, 1196–1200. [CrossRef]
33. Lee, J.-F.; Mortland, M.M.; Chiou, C.T.; Kile, D.E.; Boyd, S.A. Adsorption of benzene, toluene, and xylene by two tetramethylammonium-smectites having different charge densities. *Clays Clay Minerals.* **1990**, *38*, 113–120. [CrossRef]
34. Alexander, M.A. *Biodegradation and Bioremediation*, 2nd ed.; Academic Press: New York, NY, USA, 1999.
35. Moeni, M.; Bayat, M.; Ajalloeian, R. Laboratory investigation on the effects of pH induced changes on geotechnical characteristics of clay soil. *Geomech. Geoeng.* **2020**. [CrossRef]
36. Mitchell, J.K.; Soga, K. *Fundamentals of Soil Behavior*, 3rd ed.; John Wiley & Sons: New York, NY, USA, 2005.
37. Kaya, A.; Fang, H.-Y. The effects of organic fluids on physicochemical parameters of fine-grained soils. *Can. Geotech. J.* **2000**, *37*, 943–950. [CrossRef]
38. Lambe, T.W.; Whitman, R.V. *Soil Mechanics*; John Wiley & Sons: New York, NY, USA, 1969.
39. Wang, Q.; Tang, A.M.; Cui, Y.-J.; Delage, P.; Gatmiri, B. Experimental study on the swelling behaviour of bentonite/claystone mixture. *Eng. Geol.* **2012**, *124*, 59–66. [CrossRef]
40. Fredlund, D.G.; Morgenstern, N.R.; Widger, R.A. The shear strength of unsaturated soils. *Can. Geotech. J.* **1978**, *15*, 313–321. [CrossRef]
41. Fredlund, D.G.; Rahardjo, H.; Fredlund, M.D. *Unsaturated Soil Mechanics in Engineering Practice*; John Wiley & Sons: New York, NY, USA, 2012.
42. Zhang, Z.; Chen, Y.; Fang, J.; Guo, F. Study on shear behavior of kaolinite contaminated by heavy metal Cu (II). *Environ. Sci. Pollut. Res.* **2019**, *26*, 13906–13913. [CrossRef]
43. Gutiérrez-Castorena, M.C. Pedogenic siliceous features. In *Interpretation of Micromorphological Features of Soils and Regoliths*, 2nd ed.; Stoops, G., Marcelino, V., Mees, F., Eds.; Elsevier: Amsterdam, The Netherlands, 2018; pp. 127–155. [CrossRef]
44. Alazigha, D.P.; Indraratna, B.; Vinod, J.S.; Heitor, A. Mechanisms of stabilization of expansive soil with lignosulfonate admixture. *Transport. Geotech.* **2018**, *14*, 81–92. [CrossRef]
45. Frenkel, H.; Suarez, D.L. Hydrolysis and decomposition of calcium montmorillonite. *Soil Sci. Soc. Am. J.* **1997**, *41*, 887–891. [CrossRef]
46. Ujowundu, C.O.; Kalu, F.N.; Nwaoguikpe, R.N.; Kalu, O.I.; Ihejirika, C.E.; Nwosunjoku, E.C.; Okechukwu, R.I. Biochemical and physical characterization of diesel petroleum contaminated soil in southeastern Nigeria. *Res. J. Chem. Sci.* **2011**, *1*, 57–62.
47. Sposito, G. *The Chemistry of Soils*, 3rd ed.; Oxford University Press: New York, NY, USA, 2016.

applied sciences

MDPI

Article

Evaluating Biosedimentation for Strength Improvement in Acidic Soil

Ahmed Hassan Saad [1], Haslinda Nahazanan [1,*], Zainuddin Bin Md Yusoff [1], Muskhazli Mustafa [2], Mohamed Hamdy Elseknidy [3] and Angham Ali Mohammed [1]

[1] Department of Civil Engineering, Faculty of Engineering, Universiti Putra Malaysia, Serdang 43400, Malaysia; ahhassaad@gmail.com (A.H.S.); zmy@upm.edu.my (Z.B.M.Y.); nane.91912828@gmail.com (A.A.M.)
[2] Department of Biology, Faculty of Science, Universiti Putra Malaysia, Serdang 43400, Malaysia; muskhazli@upm.edu.my
[3] Department of Chemical and Environmental Engineering, Faculty of Engineering, Universiti Putra Malaysia, Serdang 43400, Malaysia; mhhse@outlook.com
* Correspondence: n_haslinda@upm.edu.my

Citation: Saad, A.H.; Nahazanan, H.; Yusoff, Z.B.M.; Mustafa, M.; Elseknidy, M.H.; Mohammed, A.A. Evaluating Biosedimentation for Strength Improvement in Acidic Soil. *Appl. Sci.* **2021**, *11*, 10817. https://doi.org/10.3390/app112210817

Academic Editor: Daniel Dias

Received: 29 September 2021
Accepted: 1 November 2021
Published: 16 November 2021

Publisher's Note: MDPI stays neutral with regard to jurisdictional claims in published maps and institutional affiliations.

Abstract: Marine clay soils are problematic soils in the construction industry when they are subjected to construction loads. When these soils are loaded, they lose their structure. This leads to the soil being unable to withstand loads of any magnitude without exhibiting significant, permanent deformations. In order to stabilize the marine soil, new methods for soil improvement were built upon biogrouting by incorporating physical, biological and chemical treatments into the soil. However, the biggest challenge of this method is the bacteria migration through the soil medium. To overcome this issue, the electrokinetic phenomenon can be utilized alongside biogrouting to prevent the bacteria migration. In this regard, the present study applied electrobiogrouting stabilization to investigate the improvement of acidic marine clay soil with a pH of 3.69. To accomplish this, two large-scale physical models with dimensions of 500 × 300 × 1200 mm were fabricated to examine the influence of two different treated distances between the inlet and outlet—450 mm (D45) and 600 mm (D60)—on the stability of the treated soil. It was observed that the shear strength of the treated soil improved significantly. The shear strength at the D45 treated distance increased from 3.65 kPa (untreated soil) to 28.14 kPa (treated soil). However, the strength increased by increasing the treated distance. In addition, compressibility and soil electrical conductivity were reduced significantly, and the Atterberg limits were significantly enhanced from OH to OL. The reasons for the enhancement of treated soil were the formation of $CaCO_3$, which filled the soil voids, and that the water content was reduced. To address issues with marine clay soil, this study aims to minimize the high cost of a special foundation system and the use of non-environmentally friendly materials such as calcium-based binders, aside from the reduction of deformations caused by loading. The findings of this study can be used for acidic soils and the improvement of soil's geotechnical behavior in general.

Keywords: biomineralization; *Bacillus pasteurii*; marine clay; MICP; soil improvement; acidic soil; soil treatment

1. Introduction

Marine clay soil is a type of soil that results from the weathering and erosion of limestone or dolomite, which contains a high content of magnesium, calcium and iron. This soil has high water content and soluble salt that makes it difficult for concrete to set and is porous. They regularly plunge to huge depths of more than 30 m, as shown in Figure 1 [1]. Due to its high-water content [2] and a high amount of organic materials that can exceed 10.0% of the total dry weight [3], it has low undrained shear strength. Additionally, marine clay soils are a thixotropic substance, meaning that when the water content increases, the thixotropy strength ratio increases [4]. Moreover, acidic clay soils with pH levels less than 4.0 show a considerable loss in shear strength [5]. As a result, significant settling occurs both before and after construction [2,6]. Thus, several treatment approaches using lime [7], basalt fibers [8] and biomineralization [9] have been emphasized and oriented toward the enhancement of "environmental soil treatment" that is ecologically friendly and resource-efficient over previous procedures [10].

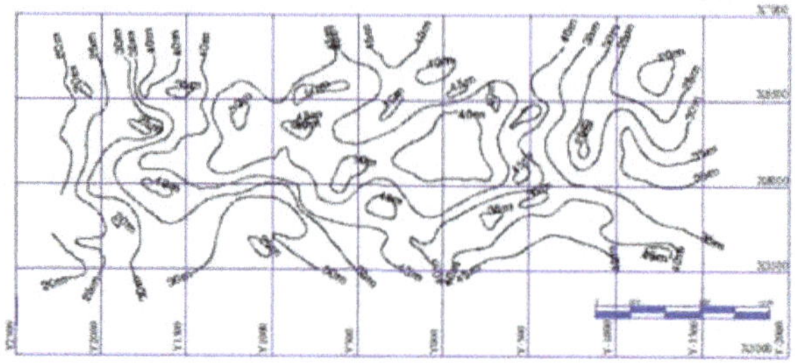

Figure 1. Typical isoline showing marine clay's total thickness [1].

In this respect, biomineralization treatment is one of the numerous environmentally friendly and cost-effective soil improvement strategies that may be used to reinforce weak soils without requiring their replacement or the deployment of a special foundation system [7,11]. Microbially induced carbonate precipitation (MICP) was studied to treat weak and loose soils, including a 7-m sand column [10], an upscaling sand treatment beneath footing [12] and a 15-cm clay specimen [13]. For MICP, species of the Bacillus family have the ability to precipitate calcium carbonate ($CaCO_3$) by the hydrolysis of urea ($CO(NH_2)_2$) into ammonium (NH_4^+) and carbonate (CO_3^{2-}) ions [14,15]. They result in an increase in the pH level of their cell contents due to the precipitation ability of calcium carbonate ($CaCO_3$) in the surrounding microenvironment [16]. The following are the reaction equations [17,18]:

$$CO(NH_2)_2 + H_2O \rightarrow 2NH_3 + CO_2 \tag{1}$$

Ammonia (NH_3) and dissolved inorganic carbon are generated in large amounts by the urease enzyme. Meanwhile, the pH rises, and the production of ammonia (NH_3) and dissolved inorganic carbon continues until the urea ($CO(NH_2)_2$) reduces dramatically:

$$2NH_3 + 2H_2O \rightarrow 2NH_4^+ + 2OH \tag{2}$$

$$2OH^- + CO_2 \rightleftharpoons HCO_3^- + OH^- \rightleftharpoons CO_3^- + H_2O \tag{3}$$

Alkalinity accumulates with the hydrolysis reaction within bacterial cells:

$$\text{Cell-Ca}^{2+-} + CO_3^- \rightarrow \text{Cell-CaCO}_3 \tag{4}$$

Sporosarcina pasteurii, also known as *Bacillus pasteurii* [18], is a bacillus family bacterium that produces urease enzymes intracellularly. *S. pasteurii* is an aerobic endospore-forming soil bacterium with a diameter of 0.5–1.2 μm and a length of 1.3–4.0 μm [19]. It can also survive in extremely alkaline (pH 10) circumstances, making it a relevant agent source for the MICP phenomena [20,21]. Sporosarcina pasteurii, on the other hand, has a peak activity in urease enzyme synthesis in pH ranges from 8 to 9 [22–24], while precipitation requires a pH of 9 [25]. Sporosarcina pasteurii also possesses a mechanical pressure resistance of 0.50 J/cm^3 [26]. In general, the precipitation rate of calcium carbonates ($CaCO_3$) during the MICP phenomenon is influenced by the bacterial type, cell concentrations, the pH value of the surrounding microenvironment, calcium concentration, urea concentration, temperature and humidity [27,28].

Soil treatment with the MICP phenomenon increased the strength by up to 570 kPa for sand soil [10] and 92 kPa for clay soil [29], with the carbonate concentration up to

60 kg/m^3 of soil specimen [10]. The period of treatment has a minor impact on clay treatment, as evidenced by the fact that the calcium carbonate ($CaCO_3$) content for 14 days of treatment was 6% greater than the calcium carbonate ($CaCO_3$) content for 7 days [29]. Laboratory investigations revealed that the calcium carbonate concentration improved the soil's hydraulic and mechanical properties, with a reduction in the permeability, void ratio, liquid limit, plastic limit and collapse index [30,31]. As a result, there was improved structural stability, performance and deformability [32].

Electrokinetic treatment efficiently resists the zeta potential phenomena which is caused by the negative charged surface of the clay and the low hydraulic conductivity [33], where the soil saturation level has a large impact on electrokinetic effectiveness [34,35]. In addition to that, it was used to improve surcharge preloading consolidation in the clay [36,37]. A pH gradient was created across the electrodes, and electrical conductivity was increased [37]. These benefits were employed to treat soils in conjunction with biomineralization [38]. Furthermore, electrokinetic treatment employing the MICP phenomenon increased the strength of soft clay by up to 10 times, with an associated calcium carbonate ($CaCO_3$) concentration of 10–16% of the soil specimen [39] as specialized techniques, and materials which requires substantial and considerable development [40,41].

In this regard, the present study aims to conduct electrokinetic (EK) stabilization of soft acidic soil using the MICP phenomenon on a large-scale specimen to enhance the physical properties of marine clay soil.

2. Materials and Methods

2.1. Bacterial Strain Preparation

In this study, *Bacillus pasteurii* bacteria (ATCC-6453) were utilized and harvested using NH4-YE (ATCC Medium 1376) medium, with the following ingredients produced for 1 L. In 200 mL of desalinated water, 10 g ammonium sulphate was dissolved, 20 g yeast extract was dissolved in 200 mL desalinated water, and 0.13 M Tris buffer ($C_4H_{11}NO_3$) was dissolved in 600 mL desalinated water. The three components were autoclaved individually for 20 min at 121 °C to sterilize them. After cooling, each component was mixed in an Erlenmeyer flask. The pH of the medium was determined to be 9.0, which is ideal for *Bacillus pasteurii* growth. The bacteria were then transferred to the broth in an incubator shaker set at 30 °C with 200 rpm shaking, and the Erlenmeyer flask was sealed for aerobic reduplication. The broth changed into a turbid dark yellow solution after 24–48 h, as illustrated in Figure 2, with the desired cell number for the bacterial concentration being 2.0×10^7 cells/mL.

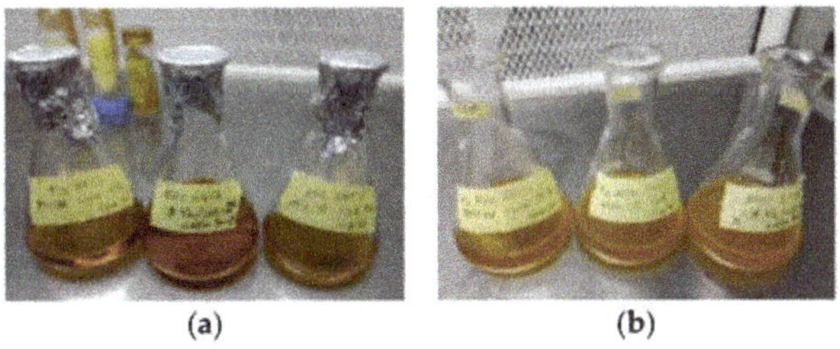

Figure 2. Bacterial broth (**a**) before culturing and (**b**) after 24–48 h in the incubator.

2.2. Soil Sampling

The soil type used in the study was chosen based on the area's history of severe geotechnical issues. Soil samples were obtained from the Malacca riverside at depths ranging from 2.5 to 4.5 m, with the sample site depicted in Figure 3. The groundwater level was 2 m deep from ground level following the river water level. The samples were kept in accordance with Part-1 1377 of the BSI instructions and recommendations [42], and the soil samples were described in accordance with Part-2 1377 of the BSI instructions [43] to avoid uncertainties caused by sample transfer. Table 1 displays the soil's engineering characteristics. A one-dimensional consolidation test revealed soil permeability of 1.88×10^{-7} cm/s [44]. With a pH of 3.69, the soil was very acidic. As indicated in Table 2, the samples were also analyzed for X-ray fluorescence (XRF) to determine the chemical composition of the soil.

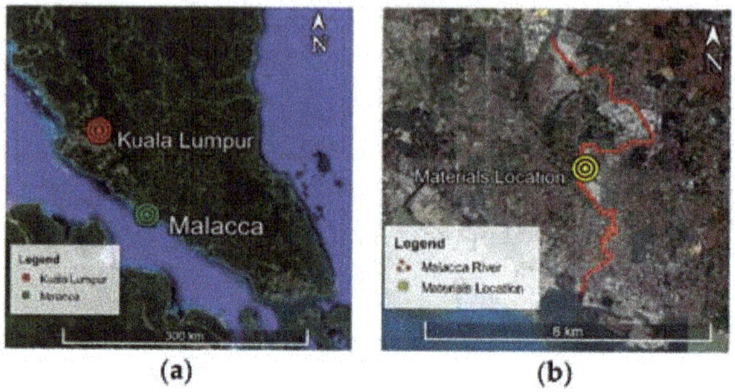

Figure 3. (**a**) Location of Malacca on the Peninsular Malaysia map and (**b**) location of the sampling site for marine clay soil at Malacca River in Malaysia.

Table 1. Engineering characteristics of the soil samples.

Properties (Unit)	Value
Specific gravity Gs	2.57
Liquid limit (%)	63.66
Plastic limit (%)	32.57
Plasticity index (%)	31.09
Hydraulic conductivity (cm/s)	1.88×10^{-7}
γ_d (kg/cm^3)	945
pH	3.69
Clay (%)	47.7
Mineral composition	Quartz, kaolinite and illite
Undrained shear strength (kPa)	3.65
Water content (%)	67
Organic content	2.15%

Table 2. Chemical composition of the soil samples.

Compound	SiO_2	Al_2O_3	Fe_2O_3	K_2O	SO_3	MgO
Percentage (%)	56.23	22.81	11.29	2.82	1.92	1.55

2.3. Biosedimentation Procedure

The physical model dimensions were set in terms of seepage velocity through the targeted soil sample for treatment and longitudinal elongation of the sample to investigate the treatment response with distance and to facilitate the samples' collection. The proposed electrode cell positions were designed to allow the treatment response to be evaluated along the flow direction as well as surrounding the electrode containers.

Figure 4 shows the dimensions of the used container. Acrylic glass was used to construct the set-up, which contained two specimens measuring 450 mm and 600 mm in length. In addition, two chambers of acrylic glass were attached within the container, allowing it to function as an electrolyte solution holder. Each chamber had a perforated wall with two fixed graphite bars on one side. The flow between the two compartments was appropriately allowed. Moreover, to regulate the pH value, a pH controller system was added in the cathode chamber and was adjusted using a Tris buffer.

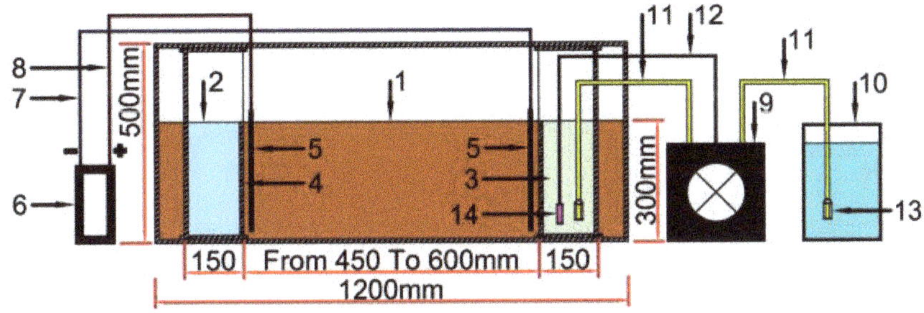

Figure 4. The physical model dimensions show the following: (**1**) soil sample, (**2**) anode ($CaCl_2$ solution), (**3**) cathode (urea and bacteria solution), (**4**) perforated wall of the cell, (**5**) graphite rod, (**6**) power supply, (**7**) wire of positive current, (**8**) wire of negative current, (**9**) pump and pH meter, (**10**) pH controller solution (tris buffer solution), (**11**) pump outflow pipe, (**12**) pH meter wire, (**13**) check valve and (**14**) pH meter sensor.

For a consistent fluid level inside the system, the surplus fluids created by the electro-osmotic flow were manually evacuated. As a result, the specimen did not produce the hydraulic gradient effect. Electro-osmotic flow usually occurs from the anode to the cathode chamber.

A power source with a direct current of 35–40 V was used in the system. A voltameter was also used to detect the change in voltage potential along the specimen. Vane blades were employed to estimate the shear strengths across the soil sample after 7 treatment days, in line with BSI1377-7-3 [45]. A thermometer was added to the cathode chamber to track temperature changes during the electrokinetic biosedimentation operation. In addition, the water content of each part was measured, and the $CaCO_3$ percent was determined using the EDX test.

As illustrated in Figure 3, to eliminate any ambiguities, two treatment lengths between electrode cells of 600 mm (D60) and 450 mm (D45) were investigated using two physical models under the same conditions for the EK effect, treatment period and solutions. As shown in Table 3, the soil samples were analyzed at the end of the treatment period to investigate any micromechanical changes. The experimental work design is shown in Figure 5.

Table 3. Test procedure summary.

No.	Code	Activates	Location	Duration
1	A1	Injection of $CaCO_3$ and $CO(NH_2)_2$	Anode chamber	1–7
2	A2	Cultured bacteria injection	Cathode chamber	1–7
3	A3	Electrical potential measurement	Across the specimens	1–7
4	A4	Vane shear test	Across the specimens	14
5	A5	Measurement of water content	Across the specimens	14
6	A6	Measurement of $CaCO_3$ percent	Across the specimens	14

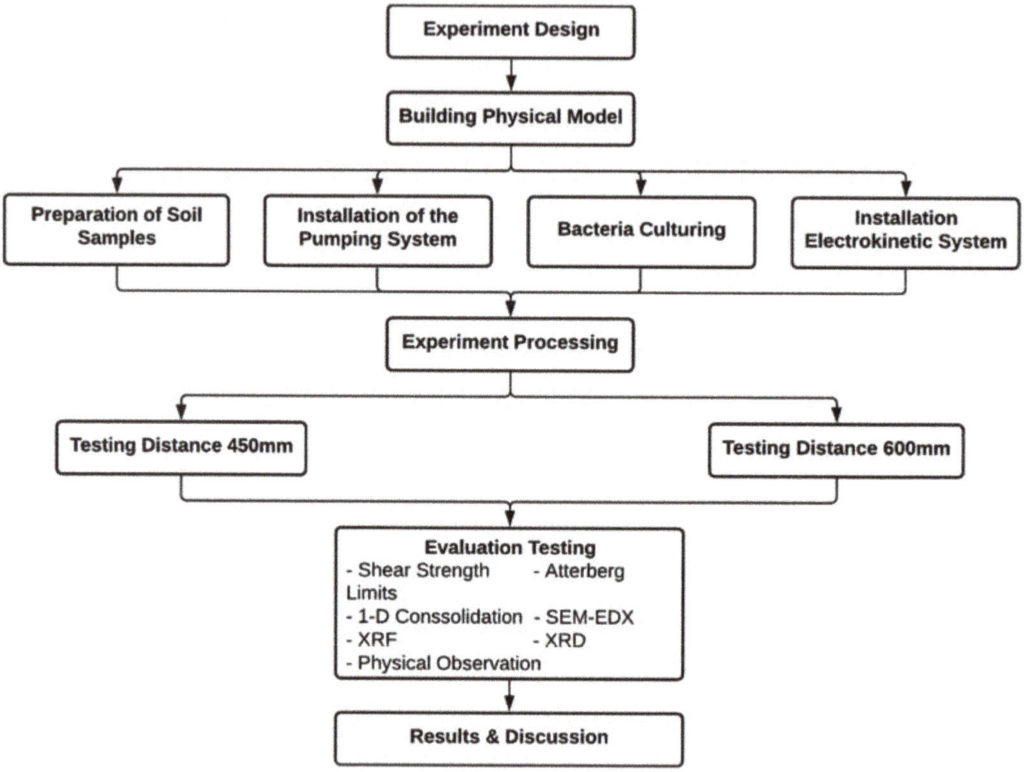

Figure 5. The experimental work's design.

3. Results and Discussions

3.1. Physical Observations

3.1.1. pH Value

It was found that the pH value in both electrode cells and along the soil specimen rose over time as the treatment progressed. According to earlier studies [46–50], the pH of the negative electrode, which had the solution combination of urea $CO(NH_2)_2$ and bacteria cells, rose slightly during the first 2 days of the treatment and then progressively declined within a pH value of slightly greater than 9.0. The pH values of the mediums inside the physical model rose after treatment, regardless of the length of the treated distances, as shown in Figure 6 for D45 (450 mm) and D60 (600 mm). The pH of the soil, on the other hand, was below 6.5, indicating that the bacterial activity inside the soil was restricted and not at its peak. Additionally, the pH along the specimen dropped as the distance extended by 150 mm, with the pH value for D45 being greater than that for D60. The high rate of NH_3 and NH_4OH generation by *Bacillus pasteurii* bacteria was the cause of the pH rise, as indicated in chemical Equations (1) and (2), which is consistent with van Paasen's [51] findings.

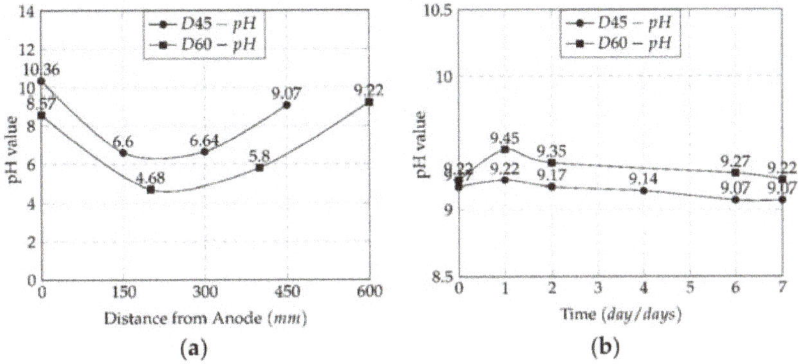

Figure 6. The pH values for D45 and D60 (**a**) along the specimen and (**b**) inside the cathode cell.

3.1.2. Precipitations

Regardless of the length of the treated distance, it was observed that the cementitious material calcium carbonate ($CaCO_3$) was significantly precipitated in the zone of the anode compared with the zone of the cathode, as shown in Figure 7. The precipitation occurred due to the chemical reaction between calcium carbonates ($CaCO_3$) and urea ($CO(NH_2)_2$) in the presence of the urease enzyme produced by *Bacillus pasteurii* bacteria, as shown in the series of chemical reactions from 1 to 4, which was infiltrated into the soil under the EK effect, consistent with the findings from prior research [27,30,52,53]. In addition, at the positive electrode cell, the perforated plate was clogged by calcium carbonate ($CaCO_3$) precipitation after treatment, as shown in Figure 8.

Figure 7. The precipitation of calcium carbonate ($CaCO_3$) at both (**a**) the anode and (**b**) the cathode for D60's test.

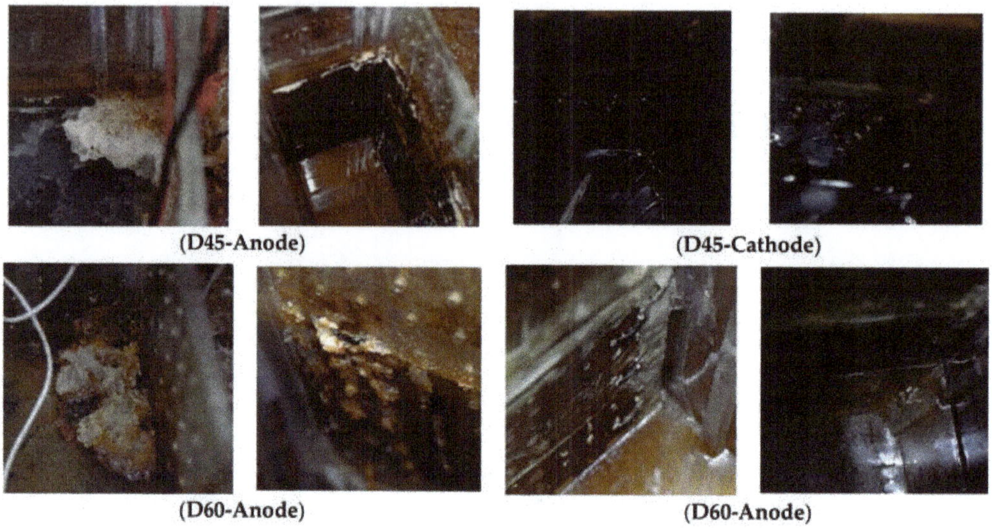

Figure 8. The clogging because of the anode's perforated wall calcium carbonate ($CaCO_3$) precipitation for the two treated distances at 450 mm (D45) and 600 mm (D60) in the tests.

3.2. *Macro-Mechanical Results*

3.2.1. Fines Activity

In general, the Atterberg limits were decreased considerably following treatment, with the liquid limit dropping by about 36% for both treatment distances D45 and D60 when compared with natural soil (N). In addition, as shown in Figure 9, the fines classification for D45 and D60 had changed from organic clay with high flexibility to that of moderate plasticity. The reasons for this were the buildup of $CaCO_3$ on the surface of the soil particles, as well as the water migration outside the soil specimen due to the electro-kinetic action, as evidenced by earlier research [37,38]. In Table 4, all of the data are tallied and compared to natural soil (N). Table 4 summarizes all of the findings and compares them to natural soil (N).

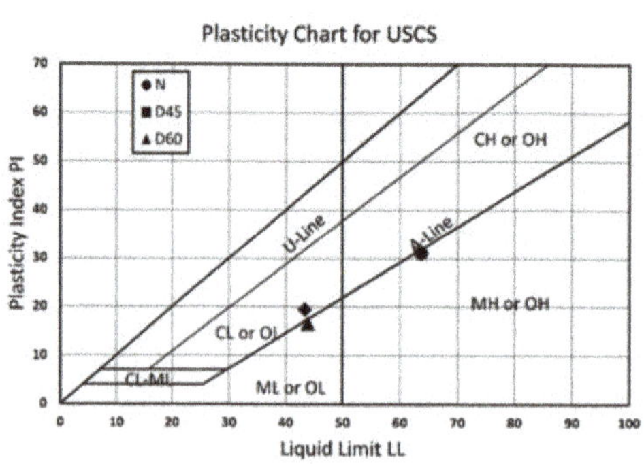

Figure 9. Atterberg limits and soil classification for untreated soil (N) and treated distances of 450 mm (D45) and 600 mm (D60).

Table 4. Atterberg limits and soil classification for untreated soil (N) and treated distances of 450 mm (D45) and 600 mm (D60).

Code (Unit)	w_c (%)	w_{LL} (%)	w_{PL} (%)	PI	LI	CI	USCS
N	67	63.66	32.57	31.1	1.151	−0.101	OH
D45	49.07	43.37	23.94	19.44	1.52	−0.24	OL
D60	48.02	43.88	27.27	16.62	1.89	−0.15	OL

3.2.2. Soil Compressibility

The 1-D consolidation test yielded a set of coefficients and indices, which are summarized in Tables 5 and 6. The consolidation characteristics were considerably improved after D45 treatment compared with D60 treatment. In general, compared with natural soil's compressibility, the soil compressibility decreased substantially following treatment (N). Furthermore, compared with natural soil's permeability, the permeability after treatment increased by eight times for both D45 and D60. This was due to the buildup of $CaCO_3$ on the soil surface and the dissipation of water outside the soil specimen during the MICP and EK processes, which was observed by Keykha et al. [39].

Table 5. Conducted and derived consolidation parameters for N, D45 and D60.

Code	C_C	CR	$C_{\alpha\epsilon}$	C_r
N	0.27995	0.10450	0.03351	0.01462
D45	0.24186	0.10555	0.03852	0.03282
D60	0.20796	0.09424	0.04121	0.02371

Table 6. Conducted and derived permeability and compression parameters for N, D45 and D60.

Code	C_V	k	m_v	a_v
(Unit)	(m^2/year)	(m/s)	(m^2/ton)	(m^2/ton)
N	2.55778	1.9×10^9	0.02916	0.07146
D45	8.10157	8.2×10^9	0.01110	0.01792
D60	3.62619	6.4×10^9	0.01027	0.02255

3.2.3. Specific Gravity and Density

The soil characteristics of the D45 and D60 soil specimens were considerably improved following treatment, as indicated in Table 7. The bulk density for D45 and D60 rose from 1560 kg/m^3 to 1710 kg/m^3 and 1670 kg/m^3, respectively. For both D45 and D60, the void ratio fell substantially. D60, on the other hand, maintained the same specific gravity of 2.51 as natural soil (N). Furthermore, the modulus of elasticity (Young's modulus) for both D45 and D60 was considerably increased, reaching up to two times that of natural soil (N). That was owing to the filling of soil voids by $CaCO_3$, as well as a reduction in the water content, both of which occurred as a result of the electrokinetic effect, as validated by research [37,38].

Table 7. Specific gravity, density, void ratio, modulus of elasticity and preconsolidation stress for N, D45 and D60 after treatment.

Code	G_S	γ_{wet}	γ_{dry}	e_o	E	P_c'
(Unit)		(t/m^2)	(t/m^2)		(kPa)	(kPa)
N	2.51	1.56	0.94	1.68	2130.64	42
D45	2.63	1.71	1.15	1.29	4848.07	320
D60	2.51	1.69	1.14	1.21	4670.79	380

3.2.4. Shear Strength

The shear strength of the treated soil improved as a result of this technique, which is thought to be related to calcium carbonate precipitation ($CaCO_3$). It was observed that the strength of the soil samples changed with the distance between the anode and the cathode, with maximum shear strengths of 28.14 and 14.50 kPa for D45 and D60, respectively, at the anode. Furthermore, as shown in Figure 10, the shear strength fell by as much as 36% for D45 and 23% for D60 from the maximum undrained shear strength as the treatment distance increased. This fluctuation in strength corresponded to the rate of calcium carbonate precipitation, with more $CaCO_3$ precipitation resulting in a higher shear strength. Both van Paassan et al. [53] and Keykha et al. [39] agreed that the void ratio was lowered owing to the buildup of $CaCO_3$ inside it and that the water content was considerably reduced by dissipation outside the voids of the soil specimen, which assisted in the shear strength's enhancement.

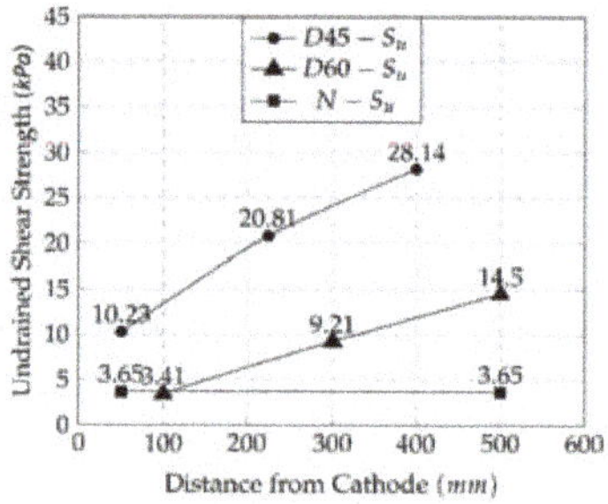

Figure 10. Shear strength after 7 days of biomineralization treatment under the EK method for D45 and D60.

3.3. *Micro-Mechanical Results*

3.3.1. XRD Results

The results of the X-ray diffraction (XRD) analysis showed no changes for the soil specimen mineralogy after the treatment for D45 compared to natural soil (N), and both were majorly formed by quartz, kaolinite and illite. Because the EK had little effect on the chemical composition and mineralogy of soil particles, it only changed the ions in the solution. Furthermore, the MICP solutions and their precipitation had a minor chemical interaction with the soil minerals. Therefore, it can be concluded that the biomineralization process under the electrokinetic method did not affect the recreation of soil minerals, as shown in Figure 11 with the peaks.

3.3.2. XRF Results

The results of X-ray powder fluorescence (XRF) showed that the amount of calcium carbonate ($CaCO_3$) rose significantly by 1.18% of the total weight of the sample, as shown in Table 8, indicating the effectiveness of the biomineralization process utilizing the electrokinetic technique and the movement of the materials inside the soil specimen. In addition, the calcium carbonate ($CaCO_3$) decreased as the distance between the anode and cathode increased from 1.18% to 0.83% for D45 and D60, respectively. Furthermore, for the fixed distance between the cathode and anode, the calcium carbonate ($CaCO_3$) percentage from the total weight decreased upon getting closer to the cathode, changing from 1.18% at 400 mm to 0.12% at 50 mm from the cathode cell for D45-1 and D45-3, respectively.

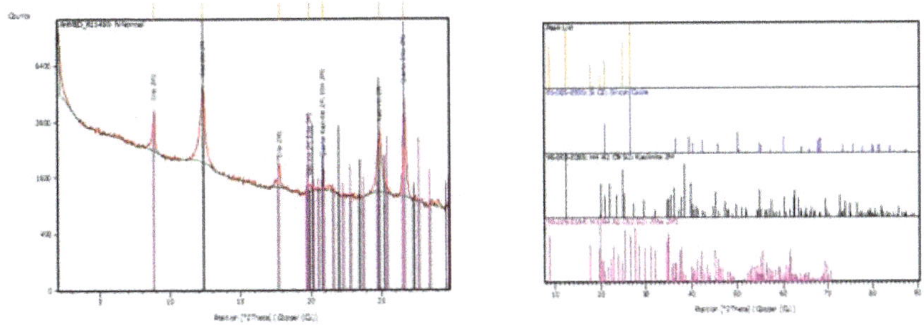

(N)

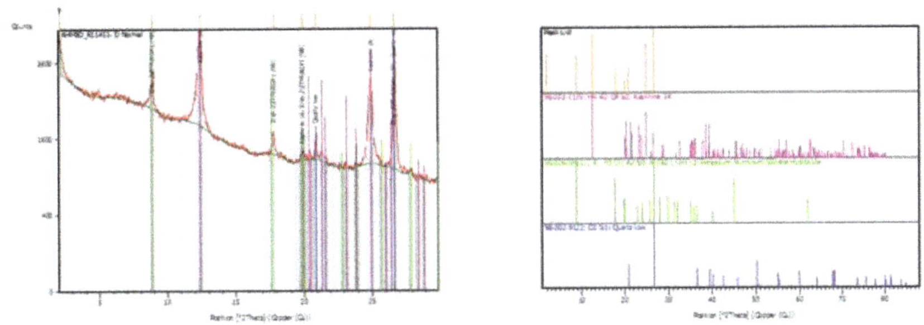

(D45)

Figure 11. XRD results and peaks for both untreated soil (N) and the treated distance of 450 mm (D45).

Table 8. Chemical composition of untreated soil (N) and treated distances of 450 mm (D45) and 600 mm (D60) after treatment.

Compound	(Unit)	Al_2O_3	SiO_2	$CaCO_3$	TiO_2	K_2O	SO_3	Cl	Others
N	(%)	27.88	63.46	-	-	-	1.6	-	7.06
D45-1	(%)	28.78	62.97	1.18	0.35	-	-	-	6.72
D45-2	(%)	31.08	58.05	0.71	-	-	2.47	-	7.69
D45-3	(%)	31.02	61.22	0.12	0.56	-	-	-	7.08
D60	(%)	29.14	58.24	0.83	-	-	-	-	11.79

3.3.3. SEM-EDX Results

The coating rate of the interfacial of clay particles rose when the SEM sample was collected closer to the anode cell, indicating that the precipitated quantity of calcium carbonate ($CaCO_3$) covered the interfacial of the clay particles. Following the coating rate, the edges of the clay layers sharpened and grew more interlocked. Figure 12 illustrates the calcium carbonate ($CaCO_3$) coating dispersion in the mapping picture for D60. Figure 13c, at a distance of 50 mm from the cathode cell for D45, shows that the soil particles got smaller as the amount of calcium carbonate ($CaCO_3$) precipitated decreased. Furthermore, as demonstrated in Figures 13a and 12d for D45 and D60, respectively, the quantity of calcium carbonate ($CaCO_3$) precipitated increased as the treatment distance between the cathode and anode cells decreased. To explain this, upon getting further away from the anode and the direction of ion migration within the soil specimen during the treatment process, the range of the EK effect decreased [37,38].

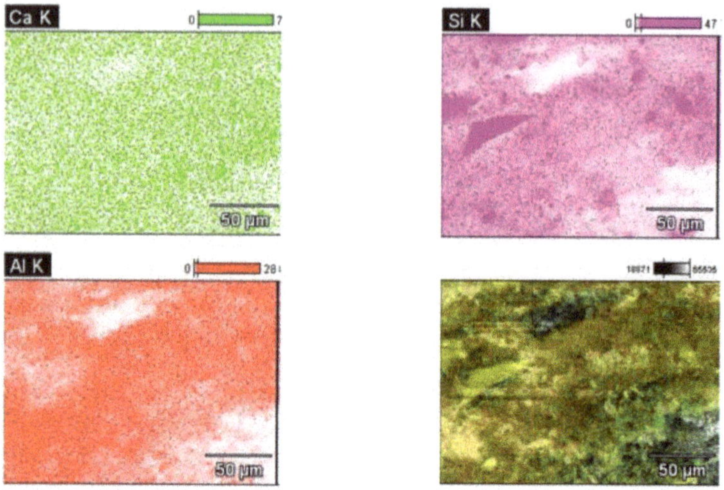

Figure 12. SEM-EDX mapping for a treated distance of 600 mm (D60), showing Ca, Si, Al and their overlapping mapping with the SEM image.

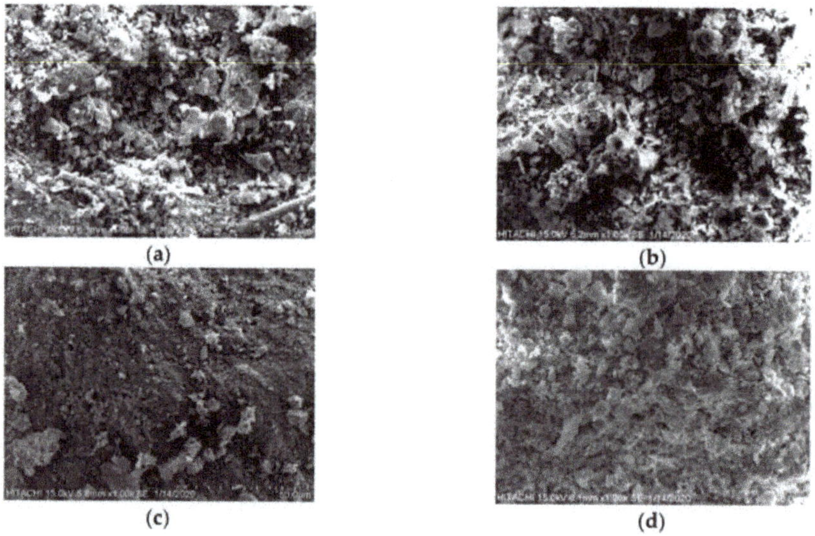

Figure 13. SEM-EDX results for a treated distance of 450 mm (D45) at (**a**) 400 mm, (**b**) 225 mm and (**c**) 50 mm, as well as for (**d**) a treated distance of 600 mm (D60) at 500 mm.

3.4. Chemical Observations

During the treatment, bubbles surrounding the graphite bars were seen to form shortly after the electric current was turned on. Following treatment, the graphite rods at the cathode were found to be considerably eroded, as illustrated in Figure 14. After erosion, the cross-section of graphite rods was 25% of the initial cross-section. The negative electrode liquid's color also changed to black, as seen in Figure 15. Furthermore, as the depth of the solution increased, the color of the solution became darker. This was due to the heavy weight of the chemical compounds produced from the chemical reaction with the graphite rod under the EK effect.

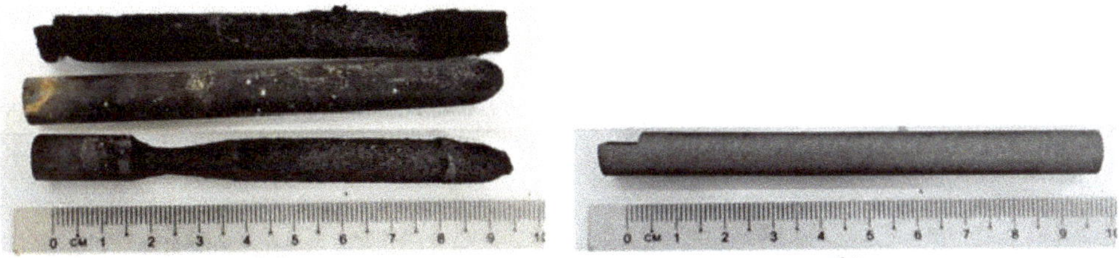

(a) (b)

Figure 14. Erosion of the graphite bars (**a**) after treatment compared with (**b**) the initial condition.

Figure 15. The negative electrode solution and the zone around the graphite bars are colored black after treatment.

4. Conclusions

Substantial construction projects impact the economy, society and the environment over their life cycles [54], where weak soil stability plays a key role in construction sustainability. As marine clay is a weak soil, it is normally found in coastal regions around the world, such as; seas, lakes, ocean shores and riverbanks. It usually extends to enormous depths underneath the earth's surface. It is also notorious for being involved in landslides and large settlements. Special foundation systems, such as piles or soil replacements, are usually costly solutions compared with the proposed structure cost itself and require high quality control to meet their purpose: balance the essentials to be generated among the economic and environmental aspects [54,55]. Conducting electrokinetic (EK) stabilization of soft acidic soil using the MICP phenomenon on a large-scale specimen is very important for improving soil environmentally. Nevertheless, no study has measured or assessed this previously mentioned experiment that encourages soil improvement adoption. This study aimed to cover this gap by highlighting the use of electrokinetic (EK) stabilization of soft acidic soil using the MICP for achieving soil project sustainability.

The feasibility of large-scale marine clay soil treatment was examined, where the following conclusions were drawn:

1. For the distance effect, as the distance increased by 150 mm between the two electrode cells, the shear strength decreased by 50%, where the maximum undrained shear strength was achieved by D45, which was 28.14 kPa, and 14.50 kPa was achieved for D60 at the anode cell.
2. Furthermore, the variation in shear strength after treatment along the soil specimen from the anode to the cathode was significant, where it decreased gradually until it reached the minimum values at the cathode. The minimum values were lowered by more than a half of the maximum values for both D45 and D60.
3. For the compressibility, both D45 and D60's compressibility indices were reduced significantly compared with natural soil (N), where for the fines activity, all Atterberg

limits were also significantly enhanced, and the soil changed from OH to OL for D45 and D60. Therefore, the treated soil could sustain loads with lesser deformations compared with non-treated soils.
4. In addition, the pH value increased gradually and remained higher than 9.0, which guaranteed bacterial activity during the treatment process. This was due to the high rate of bacterial activity in the first 48 h in the production of NH_3.
5. After treatment for both the D45 and D60 specimens, the perforated walls were clogged at both the anode and cathode but were more significant for the anode. This was a result of the maximum accumulation of $CaCO_3$ precipitation at the anode side compared with the cathode side by at least 10 times.
6. For soil electric conductivity, it was observed that the soil resistance increased gradually until the end of the treatment from 25–30 V to 55–60 V. This reflected the increase in soil density, as the electrical potential deference increased significantly after treatment.
7. XRF detected that the precipitation of calcium carbonate ($CaCO_3$) was achieved for both D45 and D60, with higher precipitation for D45. However, XRD detected that the soil mineralogy remained the same after EK treatment with biomineralization. The precipitation was 1.18% of the specimen's total weight for D45 and 0.83% for D60.
8. SEM-EDX showed soil texture enhancement, and mapping images showed the coating of particles by calcium carbonate ($CaCO_3$), which was the reason for void reduction and strength enhancement after treatment, and therefore the soil properties were enhanced in terms of compressibility.

Based on the results, biocementitiousness for acidic soil helped for shear strength improvement and soil compressibility, which led to a significant reduction in deformations caused by different loading processes. Accordingly, this method could help in strengthening weak clayey or acidic soils even without replacement. That aside, it would reduce the construction time of foundations and limit the costs of special foundation systems significantly. In addition, the method of treatment in this article helps with the reduction of soil permeability, which could help with the enhancement of water structure stability by increasing the hydraulic gradient underneath the water structures.

The effect of changing the urea injecting location along the soil specimen is to be studied in future research to demonstrate its effect on the treatment process and MICP rate. In addition, proposing new injection methods could be covered in prospective studies that would help increase the MICP treatment range and quality.

Author Contributions: Conceptualization, A.H.S. and H.N.; methodology, A.H.S., H.N. and M.M.; validation, A.H.S., H.N., Z.B.M.Y. and M.M.; formal analysis, A.H.S. and H.N.; investigation, A.H.S.; resources, H.N.; data curation, A.H.S.; writing—original draft preparation, A.H.S. and H.N.; writing—review and editing, A.H.S., H.N., Z.B.M.Y., M.M., M.H.E. and A.A.M.; supervision, H.N., Z.B.M.Y. and M.M; project administration, H.N.; funding acquisition, H.N. All authors have read and agreed to the published version of the manuscript.

Funding: This research was funded by the Putra Grant Scheme (Geran Putra): GP/2018/9635200, Universiti Putra Malaysia.

Institutional Review Board Statement: Not applicable.

Informed Consent Statement: Not applicable.

Data Availability Statement: All relevant data are within the manuscript.

Acknowledgments: The authors acknowledge the technical support by Mohd Razali and Azleen Ahmad in both the geotechnical laboratory and biology laboratory of Universiti Putra Malaysia.

Conflicts of Interest: The authors declare no conflict of interest.

References

1. Bo, M.W.; Arulrajah, A.; Sukmak, P.; Horpibulsuk, S. Mineralogy and geotechnical properties of Singapore marine clay at Changi. *Soils Found.* **2015**, *55*, 600–613. [CrossRef]
2. Indraratna, B. Interpretation of field and laboratory shear strength data of a soft marine clay. *Q. J. Eng. Geol. Hydrogeol.* **2007**, *30*, 197–203. [CrossRef]
3. Lu, T.; Bryant, W.R. Comparison of vane shear and fall cone strengths of soft marine clay. *Mar. Georesour. Geotechnol.* **1997**, *15*, 67–82. [CrossRef]
4. Yang, S.; Andersen, K.H. Thixotropy of Marine Clays. *Geotech. Test. J.* **2016**, *39*, 331–339. [CrossRef]
5. Gratchev, I.; Towhata, I. Stress-strain characteristics of two natural soils subjected to long-term acidic contamination. *Soils Found.* **2013**, *53*, 469–476. [CrossRef]
6. Wu, H.N.; Huang, R.Q.; Sun, W.J.; Shen, S.L.; Xu, Y.S.; Liu, Y.B.; Du, S.J. Leaking behavior of shield tunnels under the Huangpu River of Shanghai with induced hazards. *Nat. Hazards* **2014**, *70*, 1115–1132. [CrossRef]
7. Saadat, M.; Bayat, M. Prediction of the unconfined compressive strength of stabilised soil by Adaptive Neuro Fuzzy In-ference System (ANFIS) and Non-Linear Regression (NLR). *Geomech. Geoeng.* **2019**, *1–12*. [CrossRef]
8. Hadi Sahlabadi, S.; Bayat, M.; Mousivand, M.; Saadat, M. Freeze–Thaw Durability of Cement-Stabilized Soil Reinforced with Polypropylene/Basalt Fibers. *J. Mater. Civ. Eng.* **2021**, *33*, 04021232. [CrossRef]
9. Whiffin, V.S.; van Paassen, L.A.; Harkes, M.P. Microbial Carbonate Precipitation as a Soil Improvement Technique. *Geomicrobiol. J.* **2007**, *24*, 417–423. [CrossRef]
10. DeJong, J.T.; Soga, K.; Banwart, S.A.; Whalley, W.R.; Ginn, T.R.; Nelson, D.C.; Mortensen, B.M.; Martinez, B.C.; Barkouki, T. Soil engineering in vivo: Harnessing natural biogeochemical systems for sustainable, multi-functional engineering solutions. *J. R. Soc. Interface* **2011**, *8*, 1–15. [CrossRef]
11. Choi, S.G.; Chu, J.; Brown, R.C.; Wang, K.; Wen, Z. Sustainable Biocement Production via Microbially Induced Calcium Carbonate Precipitation: Use of Limestone and Acetic Acid Derived from Pyrolysis of Lignocellulosic Biomass. *ACS Sustain. Chem. Eng.* **2017**, *5*, 5183–5190. [CrossRef]
12. DeJong, J.T.; Mortensen, B.M.; Martinez, B.C.; Nelson, D.C. Bio-mediated soil improvement. *Ecol. Eng.* **2010**, *36*, 197–210. [CrossRef]
13. Keykha, H.; Huat, B.B.K.; Asadi, A. Electro-biogrouting stabilisation of soft soil. *Environ. Geotech.* **2015**, *2*, 292–300. [CrossRef]
14. Keykha, H.; Asadi, A.; Huat, B.B.K.; Kawasaki, S. Laboratory Conditions for Maximal Calcium Carbonate Precipitation Induced by Sporosarcina pasteurii and Sporosarcina aquimarina Bacteria. *Environ. Geotech.* **2018**, *6*, 1–20. [CrossRef]
15. DeJong, J.T.; Fritzges, M.B.; Nüsslein, K. Microbially Induced Cementation to Control Sand Response to Undrained Shear. *J. Geotech. Geoenviron. Eng.* **2006**, *132*, 1381–1392. [CrossRef]
16. Hammes, F.; Boon, N.; de Villiers, J.; Verstraete, W.; Siciliano, S.D. Strain-Specific Ureolytic Microbial Calcium Carbonate Precipitation. *Appl. Environ. Microbiol.* **2003**, *69*, 4901–4909. [CrossRef]
17. Achal, V.; Mukherjee, A.; Kumari, D.; Zhang, Q. Biomineralization for sustainable construction—A review of processes and applications. *EarthScience Rev.* **2015**, *148*, 1–17. [CrossRef]
18. Li, D.; Tian, K.L.; Zhang, H.L.; Wu, Y.Y.; Nie, K.Y.; Zhang, S.C. Experimental investigation of solidifying desert aeolian sand using microbially induced calcite precipitation. *Constr. Build. Mater.* **2018**, *172*, 251–262. [CrossRef]
19. Keykha, H.; Huat, B.B.K.; Asadi, A.; Zareian, M.; Kawasaki, S. Electrokinetic properties of pasteurii and aquimarina bacteria. *Environ. Geotech.* **2014**, *2*, 181–188. [CrossRef]
20. Dhami, N.K.; Reddy, M.S.; Mukherjee, A. Biomineralization of calcium carbonates and their engineered applications: A review. *Front. Microbiol.* **2013**, *4*, 314. [CrossRef]
21. Bhaduri, S.; Debnath, N.; Mitra, S.; Liu, Y.; Kumar, A. Microbiologically induced calcite precipitation mediated by sporosarcina pasteurii. *J. Vis. Exp.* **2016**, *2016*, e53253. [CrossRef]
22. Achal, V.; Mukherjee, A.; Basu, P.C.; Sudhakara Reddy, M. Strain improvement of Sporosarcina pasteurii for enhanced urease and calcite production. *J. Ind. Microbiol. Biotechnol.* **2009**, *36*, 981–988. [CrossRef] [PubMed]
23. Sarayu, K.; Iyer, N.R.; Murthy, A.R. Exploration on the biotechnological aspect of the ureolytic bacteria for the production of the cementitious materials—A review. *Appl. Biochem. Biotechnol.* **2014**, *172*, 2308–2323. [CrossRef] [PubMed]
24. Meier, A.; Kastner, A.; Harries, D.; Wierzbicka-Wieczorek, M.; Majzlan, J.; Büchel, G.; Kothe, E. Calcium carbonates: Induced biomineralization with controlled macromorphology. *Biogeosciences* **2017**, *14*, 4867–4878. [CrossRef]
25. Stocks-Fischer, S.; Galinat, J.K.; Bang, S.S. Microbiological precipitation of $CaCO_3$. *Soil Biol. Biochem.* **1999**, *31*, 1563–1571. [CrossRef]
26. Grabiec, A.M.; Starzyk, J.; Stefaniak, K.; Wierzbicki, J.; Zawal, D. On possibility of improvement of compacted silty soils using biodeposition method. *Constr. Build. Mater.* **2017**, *138*, 134–140. [CrossRef]
27. Khodadadi, H.; Bilsel, H. Application of microorganisms for improvement of liquefiable sand. In Proceedings of the 3rd international conference on new developments in soil mechanics and geotechnical engineering, Near East University, Nicosia, North Cyprus, 28–30 June 2012; pp. 857–863.
28. Peng, Y.; Ding, X.-M.; Xiao, Y.; Chu, J.; Deng, W.-T. Study of particle breakage behaviour of calcareous sand by dyeing tracking and particle image segmentation method. *Rock Soil Mech.* **2019**, *40*, 2663–2672. [CrossRef]

29. Keykha, H.; Asadi, A.; Zareian, M. Environmental Factors Affecting the Compressive Strength of Microbiologically Induced Calcite Precipitation-Treated Soil. *Geomicrobiol. J.* **2017**, *34*, 889–894. [CrossRef]
30. Umar, M.; Kassim, K.A.; Ping Chiet, K.T.; Chiet, K.T.P.; Ping Chiet, K.T.; Chiet, K.T.P. Biological process of soil im-provement in civil engineering: A review. *J. Rock Mech. Geotech. Eng.* **2016**, *8*, 767–774. [CrossRef]
31. Kaur, G.; Dhami, N.K.; Goyal, S.; Mukherjee, A.; Reddy, M.S. Utilization of carbon dioxide as an alternative to urea in biocementation. *Constr. Build. Mater.* **2016**, *123*, 527–533. [CrossRef]
32. Valencia González, Y.; Carvalho-Camapum, J.; Lara-Valencia, L.A. Influence of biomineralization on a profile of a tropical soil affected by erosive processes. *DYNA* **2015**, *82*, 221–229. [CrossRef]
33. Jamshidi-Zanjani, A.; Khodadadi, A.; Sheta, K.W. Review of Enhancement Techniques on Medical Images. *Int. J. Adv. Comput. Electron. Technol.* **2017**, *4*, 26–30. [CrossRef]
34. Virkutyte, J.; Sillanpää, M.; Latostenmaa, P. Electrokinetic soil remediation—Critical overview. *Sci. Total Environ.* **2002**, *289*, 97–121. [CrossRef]
35. Cameselle, C. Enhancement of Electro-Osmotic Flow during the Electrokinetic Treatment of a Contaminated Soil. *Electrochim. Acta* **2015**, *181*, 31–38. [CrossRef]
36. Saad, A.H.; Nahazanan, H.; Yusoff, Z.M.; Huat, B.K.; Mustafa, M. Properties of Biomineralization Process in Various Types of Soil and Their Limitations. *Int. J. Eng. Adv. Technol.* **2019**, *9*, 4261–4268.
37. Micic, S.; Shang, J.Q.; Lo, K.Y.; Lee, Y.N.; Lee, S.W. Electrokinetic strengthening of a marine sediment using intermittent current. *Can. Geotech. J.* **2001**, *38*, 287–302. [CrossRef]
38. Moghadam, M.J.; Moayedi, H.; Sadeghi, M.M.; Hajiannia, A. A review of combinations of electrokinetic applications. *Environ. Geochem. Health* **2016**, *38*, 1217–1227. [CrossRef]
39. Keykha, H.; Huat, B.B.K.K.; Asadi, A. Electrokinetic Stabilization of Soft Soil Using Carbonate-Producing Bacteria. *Geotech. Geol. Eng.* **2014**, *32*, 739–747. [CrossRef]
40. Švajlenka, J.; Kozlovská, M. Perception of User Criteria in the Context of Sustainability of Modern Methods of Construction Based on Wood. *Sustainability* **2018**, *10*, 116. [CrossRef]
41. Švajlenka, J.; Kozlovská, M.; Pošiváková, T. Analysis of Selected Building Constructions Used in Industrial Construction in Terms of Sustainability Benefits. *Sustainability* **2018**, *10*, 4394. [CrossRef]
42. British Standards Institution. *General Requirements and Sample Preparation (BS1377-1)*; British Standards Institution: London, UK, 1990.
43. British Standards Institution. *Classification Tests (BS1377-2)*; British Standards Institution: London, UK, 1990.
44. Kulhawy, F.H.; Mayne, P.W. *Manual on Estimating Soil Properties for Foundation Design: EL-6800*; Electric Power Research Institute: Washington, DC, USA, 1990; p. 299.
45. British Standards Institution. *BS:1377-7 Shear Strength Tests (Total Stress)*; Part 7; British Standards Institution: London, UK, 1990; pp. 1–52.
46. Achal, V.; Pan, X. Influence of Calcium Sources on Microbially Induced Calcium Carbonate Precipitation by Bacillus sp. CR2. *Appl. Biochem. Biotechnol.* **2014**, *173*, 307–317. [CrossRef] [PubMed]
47. Al-Thawadi, S.M.; Cord-Ruwisch, R. Calcium Carbonate Crystals Formation by Ureolytic Bacteria Isolated from Australian Soil and Sludge. *J. Adv. Sci. Eng. Res.* **2012**, *2*, 12–26.
48. Fujita, Y.; Taylor, J.L.; Gresham, T.L.T.T.; Delwiche, M.E.; Colwell, F.S.; Mcling, T.L.; Petzke, L.M.; Smith, R.W. Stimulation of microbial urea hydrolysis in groundwater to enhance calcite precipitation. *Environ. Sci. Technol.* **2008**, *42*, 3025–3032. [CrossRef]
49. Wang, L.; Wang, J.; Xu, Y.; Chen, P.; Yuan, J.; Qian, X. Novel surface treatment of concrete bricks using acid-resistance mineral precipitation. *Constr. Build. Mater.* **2018**, *162*, 265–271. [CrossRef]
50. Tobler, D.J.; Maclachlan, E.; Phoenix, V.R. Microbially mediated plugging of porous media and the impact of differing injection strategies. *Ecol. Eng.* **2012**, *42*, 270–278. [CrossRef]
51. Van Paassen, L.A. Biogrout: Ground Improvement by Microbially Induced Carbonate Precipitation. Ph.D. Thesis, Delft University of Technology, Delft, The Netherlands, 2009.
52. Lauchnor, E.G.; Schultz, L.N.; Bugni, S.; Mitchell, A.C.; Cunningham, A.B.; Gerlach, R. Bacterially Induced Calcium Carbonate Precipitation and Strontium Coprecipitation in a Porous Media Flow System. *Environ. Sci. Technol.* **2013**, *47*, 1557–1564. [CrossRef] [PubMed]
53. Al Qabany, A.; Soga, K. Effect of chemical treatment used in MICP on engineering properties of cemented soils. *Géotechnique* **2013**, *63*, 331–339. [CrossRef]
54. Emmanuel Oke, A.; Omoregie Aghimien, D.; Olusola Olatunji, S. Implementation of Value Management as an Economic Sustainability Tool for Building Construction in Nigeria. *Int. J. Manag. Value Supply Chain* **2015**, *6*, 55–64. [CrossRef]
55. Martens, M.L.; Carvalho, M.M. Key factors of sustainability in project management context: A survey exploring the project managers' perspective. *Int. J. Proj. Manag.* **2017**, *35*, 1084–1102. [CrossRef]

applied sciences

MDPI

Article

Strength and Compressibility of Ammonia-Soda Residue from the Solvay Sodium Plant

Jakub Zięba [1], Przemysław Rzepka [2,3] and Bartłomiej Szczepan Olek [1,*]

1 Department of Geotechnics and Strength of Materials, Cracow University of Technology, ul. Warszawska 24, 31-155 Kraków, Poland; jakub.zieba@pk.edu.pl
2 Institute for Chemistry and Bioengineering, ETH Zurich, 8093 Zurich, Switzerland; przemyslaw.rzepka@chem.ethz.ch
3 Laboratory for Catalysis and Sustainable Chemistry, Paul Scherrer Institute, 5232 Villigen, Switzerland
* Correspondence: bartlomiej.olek@pk.edu.pl; Tel.: +48-609068216

Abstract: This work presents the discussion of the results for an experimental study conducted to characterise the mechanical behaviour of ammonia-soda residue (ASR). The calcareous sludge is an alkaline waste formed during the production of soda ash and deposited at the area of the former Solvay Sodium Plant factory in Krakow, Poland. Isotropically consolidation drained (CID) triaxial tests and constant rate of strain (CRS) consolidation tests include the full saturation with water, completion of the consolidation, and the loading/strain rate choice. For this purpose, ASR undisturbed samples were collected from the ground and submitted to laboratory experiments. These samples show a distinct difference in the initial bulk density, the initial level of compaction, initial void ratio, and the natural water content. The CD triaxial tests were conducted under three different levels of confining pressure; in turn, CRS tests were run with two appropriate input strain rates. According to the physical state of ASR and the depth of sampling, two different evolutions of the critical state in the stress–strain space were observed. In the light of the assessed stress–strain–strength behaviour, key design engineering parameters of ASR were calculated.

Keywords: ammonia-soda residue; compressibility; constant rate of strain; drained; shear strength; triaxial test

Citation: Zięba, J.; Rzepka, P.; Olek, B.S. Strength and Compressibility of Ammonia-Soda Residue from the Solvay Sodium Plant. *Appl. Sci.* **2021**, *11*, 11305. https://doi.org/10.3390/app112311305

Academic Editors: Daniel Dias, Małgorzata Jastrzębska, Krystyna Kazimierowicz-Frankowska, Gabriele Chiaro and Jaroslaw Rybak

Received: 7 October 2021
Accepted: 26 November 2021
Published: 29 November 2021

1. Introduction

Ammonia-soda residue (ASR), also known as ammonia-soda white mud or ammonia-soda ash waste, is an alkaline calcareous sludge formed during the production of industrial soda ash (SA) (Solvay processes). ASR is a white-coloured fine-grained material commonly stored in landfills or settling ponds and may be potentially hazardous and negatively impacting natural and human systems. SA (sodium carbonate, Na_2CO_3) is an essential inorganic product that has several diversified applications in the chemical industry [1–3], metallurgy [4,5], textile [6,7], printing and dyeing [8], glass [9–11], medicine [12], and food industries [13,14]. SA is also exploited in geotechnics to stabilise the subgrade soft soil, improve the geotechnical properties of soil, and accelerate the hardening of soil-binder mixtures.

In 2020, the total global production of soda ash was estimated to be about 52 million metric tons [15], with a considerable amount of ASR as a by-product material from manufacturing. The ASR is usually slurry-deposited and exhibits a unique structure with high moisture content, high compressibility, and unusual geotechnical properties [16,17], which presented several challenges to the capping landfill design [18,19]. Another feature is the fact that the ASR consists of extremely fine particles with a large specific surface area that exhibits a strong adsorption capacity [20,21]. Coleman et al. [22] considered SA as an additive to soil-cement mixtures to determine its effect on soil-cement and soil-cement-fly ash strength properties. It was found that the beneficial effects of the addition of

0.5 wt.% sodium carbonate were most observed after short curing periods. Over a longer time, SA can be detrimental to mixtures containing low cement contents. However, the authors reported various effects of addition of SA on the strengths of the soil-cement and soil-cement-fly ash mixtures.

El-Rawi and Toma [23] reported that sodium hydroxide and carbonate were effectively used as chemical additives with Iraqi soils stabilised by cement to increase shear strength.

Sun et al. [24] studied the engineering properties of the solidified soil with 20% non-clinker incorporating soda residue curing agent. The strength properties of this sample were almost the same as that of the solidified soil with about 10% composite cement. Zha et al. [25] examined alkaline residue as a new binder to treat heavy metal-contaminated soil because of its strong adsorption capacity for heavy metal ions. The increase of the treated soil's unconfined compressive strength (UCS) was associated with the drop of the leached Pb^{2+} concentration and the rise of the alkaline residue content in the specimen. He et al. [26] investigated the shear strength properties of stabilised soft soil composed of ASR and ground granulated blast furnace slag and compared it to cement-stabilised soil in deep mixing and soil-cement columns.

Nazir et al. [27] studied the effect of SA on pH and the compaction characteristics of Malaysian laterite soil. Based on the experimental investigation, the addition of soda ash led to the increase of pH of the soil and improvement of the compaction properties. Han et al. [28] used calcium chloride and SA solutions to decrease the soil expansion potential. The reduction in the expansion parameters was primarily attributed to the strong short-term reactions between clay and stabilisers.

Bai et al. [29] considered the improvement of ASR properties by using liquid soda residue and liquid fly ash. Mixing ASR and fly ash resulted in the elevation of the water content, liquid limit, and plastic limit, and lowering of permeability and compressibility in the pretreated soil. Furthermore, potential benefits of the exploitation of SA for the preparation of ASR were assessed through laboratory and field tests. The field tests on ASR were also carried out by Ma et al. [30] to study the impact of gravel, sand, and lime on the mechanical properties of ASR. The authors indicated that the addition of fly ash contributes to the enhancement of the strength properties of ASR, and the lime, sand, and rubble have a significant effect on the subgrade bearing capacity of composite soils consisting ASR.

Wang et al. [31] showed the safety risks associated with the use of ASR. Their analysis supported previous findings on the use of ASR in the cement industry [32–36] and suggested that ASR without any pretreatment for reducing harmful pollutants and/or stabilisation [37] should not be considered for soil remediation or agricultural activities. Based on the findings mentioned above, this statement also seems to hold true for geotechnical works.

As discussed above, numerous examples of using SA and ASR in geotechnical applications have been attempted. Reclamation activities have been conducted in the "White seas" area since 2006 to develop the land for a John Paul II Centre, roads, and tunnel facilities. Ground investigations were carried out at the site by Politechnika Krakowska to evaluate the impact of the earthworks on the behaviour of the anthropogenic sediments. Thus, broad research is needed to assess the strength and compressibility of the ASR properly. The main motivation for undertaking this study was the lack of available strength and compression data for this type of anthropogenic material. The general objective of the performed research was to conduct an appropriate experimental testing programme and to derive necessary parameters that characterise ASR properties. In order to analyse future tunnel-soil-structure systems, a long-term stability problem was assumed. Due to hydrogeologic conditions at the site, the drainage of the ASR during consolidation and shearing was expected. Therefore, it was decided to perform a CD-type triaxial test. An inclination of the critical state line along the stress–strain behaviour was of prime interest in the analysis. The results of the mechanical parameters were referred to the mineral composition and microstructure characterisation provided.

2. Experimental Procedure

2.1. Materials

Material for the laboratory experiments was collected from the area of the Solvay Sodium Plant in Krakow. The subsoil contains Quaternary soils (sand and gravel interlayered by cohesive soils represented mainly by silt and mixtures of silt and clay), which lie directly on the Tertiary formations of the Miocene clays facies. Over the years, post-industrial waste has been deposited over the layers of natural soils.

The undisturbed samples of ASR were collected from a depth of ca. 7.5 and 16.5 m below the ground surface using a combined Shelby probe with the plastic insert. The high in situ water content in the analysed ASR up to 15 m below the sea level, i.e., around 200%, led to large deformations even at low pressures. When the standard probe is used for a sample extrusion, a significant consolidation can occur, and a substantial loss of water from the soil is usually observed. Therefore, the sample with an undisturbed structure was collected by combining the Shelby probe with a plastic insert. The probe with an exterior diameter of 80 mm was selected, which allowed for installing a plastic pipe with a diameter of 75 mm. The interior diameter of the probe was 71 mm. The thicker sampler wall, made of the steel and PE pipes, allowed adjustment of the cutting knife to avoid large deformations of the taken sample. A cylinder was attached to the probe to provide accurate sampling for the ASR with a natural structure. The vacuum was applied during the extraction of the probe to the borehole in the sampler. The Pagani static probe was used to push the probe to a specific depth of sampling.

The groundwater table (GWT) was found at 15 m below sea level. Figure 1 shows the results of the piezocone test (CPTU) at the site. As one can see, the geotechnical profile consists of two anthropogenic layers of ASR lined by stiff clay. From the trend of cone resistance (q_t) and pore water pressure measured behind the cone tip (u_2) (both parameters increase with depth), the deposit of ASR is considered normally consolidated. The natural water content of the ASR I and ASR II varies from 150% to 200% and from 38% to 56%, respectively.

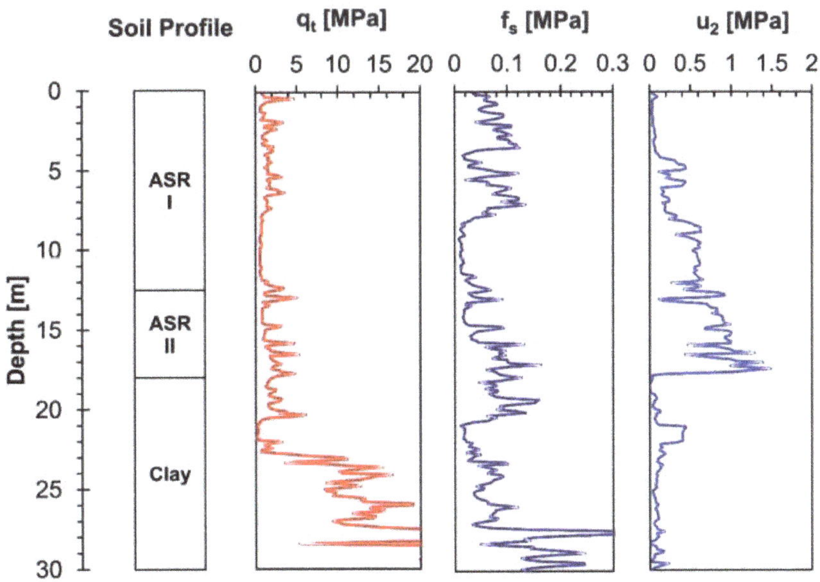

Figure 1. Soil profile with the piezocone test results.

2.2. Test Methods

Powder X-ray diffraction (XRD) and scanning electron microscopy (SEM) analyses were used to identify the mineral composition of waste material and to investigate the microstructure of ASR. The specimen was dried under 105 °C. The 1–2 cm grey inclusions inside the white cement were recognised and separated for the phase and elemental analyses. Two X-ray diffraction patterns on both samples were collected at Bragg–Brentano geometry by using an in-house PANalytical X'Pert PRO MPD diffractometer and exploitation of monochromatic X-ray radiation of $\lambda = 1.5406$Å. The data was recorded with step 0.08° in a range of 5–90°. The phase identification was performed using X'Pert High Score Plus.

The SEM images and elemental analyses were obtained using a JEOL JSM-7100F scanning electron microscope and an EDAX "Octane Pro" SDD Energy Dispersive X-Ray Spectrometer. Thin layers of both samples were spread on top of aluminium stumps covered by carbon tape. The SEM images and EDS spectra were taken at 15.0 keV with a working distance of 10.0 mm. For imaging, the samples were additionally coated by a 10 nm layer of carbon.

The physical properties of the ASR were determined as per the relevant International Organization for Standardization (ISO). The conventional grain size analysis was conducted in accordance with ISO 17892-4 standard [38], and the density of soil skeleton (G_s) was determined using the pycnometric method as per ISO 17892-3 [39]. The isotropically consolidated drained triaxial compression tests (CID) were carried out using the VJ-Tech triaxial compression apparatus. An appropriate testing procedure was adopted to gather high-quality data by performing full saturation with water, including increases of back-pressure, consolidation, and the choice of the loading rate.

The samples with dimensions: diameter 50 mm and height 90 mm were formed. The height was measured before each test. The samples placed in the cell were sealed with airtight membranes and double O-rings. The porous disks and the paper filters were placed at the bottom and the top of the sample, ensuring the free water flow. The saturation of the specimens was done by the ramp method. In this technique, the cell pressure and back-pressure at the specimen's faces are ramped, and the B-check is performed at regular intervals to examine whether the required Skempton's pore pressure coefficient B (B-parameter) has been reached. The B-parameter was calculated as the ratio of the increase in the excess pore water pressure to the applied stress increment. A mean total stress increment of 50 kPa was selected for all the B-checks. The ASR specimen was assumed to be saturated if the B-value was sufficiently high (the analysis criterion of $B > 0.98$ was expected).

To achieve the isotropic consolidation stage, the confining pressure on both end faces of the specimen was applied for one hour to establish target effective stresses between 50 and 400 kPa. For all the tests, the time needed to complete dissipation did not exceed 2 h.

The shearing and consolidation stages were carried out using one-way drainage conditions. The final loading rate was estimated to be ca. 50% of the maximum rate determined based on the consolidation results. Adopting such a rate limit results from the previous preliminary study [40].

The constant rate of strain (CRS) consolidation tests were performed in a Multi-Purpose Consolidation Cell (MPCC) produced by VJ-Tech. In the CRS test, the load increases continuously while a constant strain rate remains constant, so the continuous recording of stresses and strains can be made. The most important parameter that must be determined before starting the CRS test is the strain rate [41,42]. If the strain rate is too low, the pore pressure is very small or not generated at all during the test, leading to erroneous values of hydraulic conductivity (k_v) and coefficient of consolidation (c_v). On the other hand, if the strain rate is too high, the determined value of preconsolidation stress is significantly overestimated. Since the first use of the CRS test, many different methods have been developed to determine the appropriate strain rate. In this work, the method of Ozer et al. [43] was adopted as the reference strain rate, which is based on comparing the results of the CRS test with the results of the reference oedometer study. Thus, one reference oedometer test with unloading and reloading of the samples was performed. In this case, the following scheme was conducted: load–unload–repeated load: 1 $\rightarrow$ 6.25 $\rightarrow$ 12.5 $\rightarrow$ 25 $\rightarrow$ 50 $\rightarrow$ 100 $\rightarrow$ 200 $\rightarrow$ 400 $\leftarrow$ 200 $\leftarrow$ 100 $\leftarrow$ 50 $\leftarrow$ 25 $\leftarrow$ 12.5 $\rightarrow$ 25 $\rightarrow$ 50 $\rightarrow$ 100 $\rightarrow$ 200 $\rightarrow$ 400 $\rightarrow$ 800 $\rightarrow$ 1000 kPa.

3. Results and Discussion

3.1. Physical Properties

The grain size analysis showed that a silt fraction (2–63 μ mm) is a dominant component. In terms of consistency, the tested material revealed a soft-plastic state with liquidity index, LI = 0.64. Above the GWT, the ASR bulk density (ρ) ranged from 1.15 to 1.19 g/cm^3 (on average 1.17 g/cm^3), and below the GWT, from 1.58 to 1.68 g/cm^3 (on average 1.65 g/cm^3). The pycnometric method indicated the density of the soil skeleton ranged from 2.58 to 2.62 (on average 2.60). The physical properties and test conditions of the specimens are summarised in Table 1.

Table 1. Physical properties and test conditions for triaxial drained tests.

	Confining Pressure	G_s	n	e	S	Loading Rate
	[kPa]	[-]	[-]	[-]	[-]	[mm/min]
ASR I	75	2.61	0.84	5.32	0.98	0.01
	100	2.61	0.84	5.26	1.00	0.01
	200	2.61	0.83	5.00	0.98	0.008
	250	2.61	0.84	5.28	0.99	0.008
	400	2.61	0.85	5.51	0.99	0.008
	500	2.61	0.84	5.81	0.98	0.01
ASR II	75	2.60	0.55	1.24	1.00	0.01
	100	2.60	0.56	1.28	1.00	0.01
	200	2.60	0.58	1.36	1.00	0.01
	250	2.60	0.58	1.38	1.00	0.012
	400	2.60	0.58	1.40	1.00	0.015
	500	2.60	0.57	1.32	1.00	0.015

Note that: G_s—specific gravity, n—porosity, e—void ratio, S—degree of saturation.

3.2. XRD and Microstructural Analysis

The mineral composition of ASR was investigated by means of X-ray diffraction. The results are demonstrated in Figure 2. The phase analysis was based on the two diffractograms collected from the white cement, which is the major phase and the minor grey inclusions of 1–2 cm size. Both diffractograms show the distinct Bragg peaks, suggesting highly crystalline material. The phase analysis of the white cement revealed 57 wt% of calcite ($CaCO_3$), 31 wt% of portlandite ($Ca(OH)_2$), 7 wt% of brucite ($Mg(OH)_2$), and 5 wt% of barite ($BaSO_4$). The occurrence of barite results from Solvay production of barium carbonate, which was manufactured from natural barium sulfate. The grey inclusions were found as composed of mostly still calcite (83%) but associated with gypsum (8 wt%) and analcime (7 wt%). The formation of analcime rose from the highly alkaline environment provided by soda wastes disposal [19]. The euhedral crystals of portlandite ($Ca(OH)_2$) and gypsum ($CaSO_4 \cdot 2H_2O$) with well-outlined hexagonal and prismatic morphology, respectively, were also observed at SEM images (Figure 3). The crystals of both phases grew to several microns and were embedded in fine-crystal cement composed of mostly calcite ($CaCO_3$) (Figure 3).

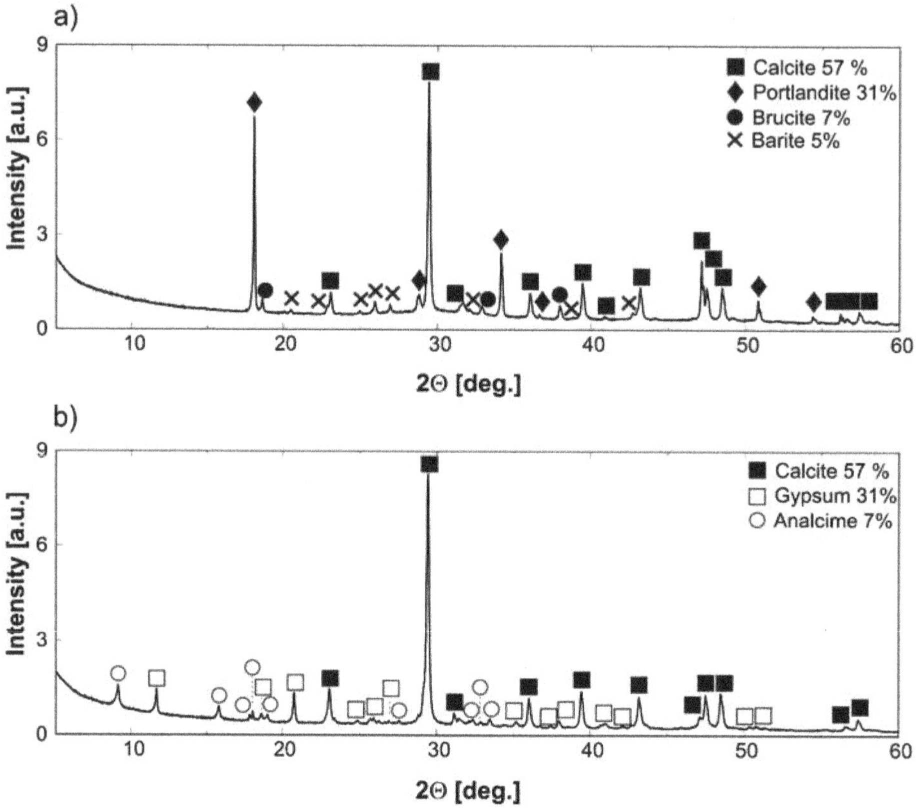

Figure 2. X-ray diffractograms and mineral compositions of two phases were found in the sample: the white cement (**a**) and the embedded grey inclusions (**b**).

Figure 3. SEM images collected from the white cement (**a**) and the grey inclusions (**b**). Large crystals of portlandite (**a**) and gypsum (**b**) are surrounded by the fine grains of calcite.

3.3. Drained Triaxial Tests

3.3.1. Stress–Strain Behaviour

The substantial differences between samples were the initial bulk density, initial level of compaction, apparent cementation, initial void ratio, and natural water content. Higher initial physical parameters are typical for samples collected from the first layer of ASR. The full stress–strain characterisation was conducted in the range from 0.2% to 35% (for samples collected from a depth of 8 m) and up to 30% (for samples from a depth of 16.5 m). Cauchy stress was used according to Equations (1) and (2) to calculate the mean effective stress (p') and deviator stress (q):

$$p' = \frac{\sigma_1' + 2\sigma_3'}{3} \quad (1)$$

$$q = \sigma_1' - \sigma_3' \quad (2)$$

where σ'_1 is the effective principal stress, and σ'_3 is the effective confining stress.

The deviator stress versus axial strain curves are shown in Figure 4. As one can see in Figure 4a, the samples from the smaller depth (ASR I) led to strain-hardening behaviour at high confining pressures.

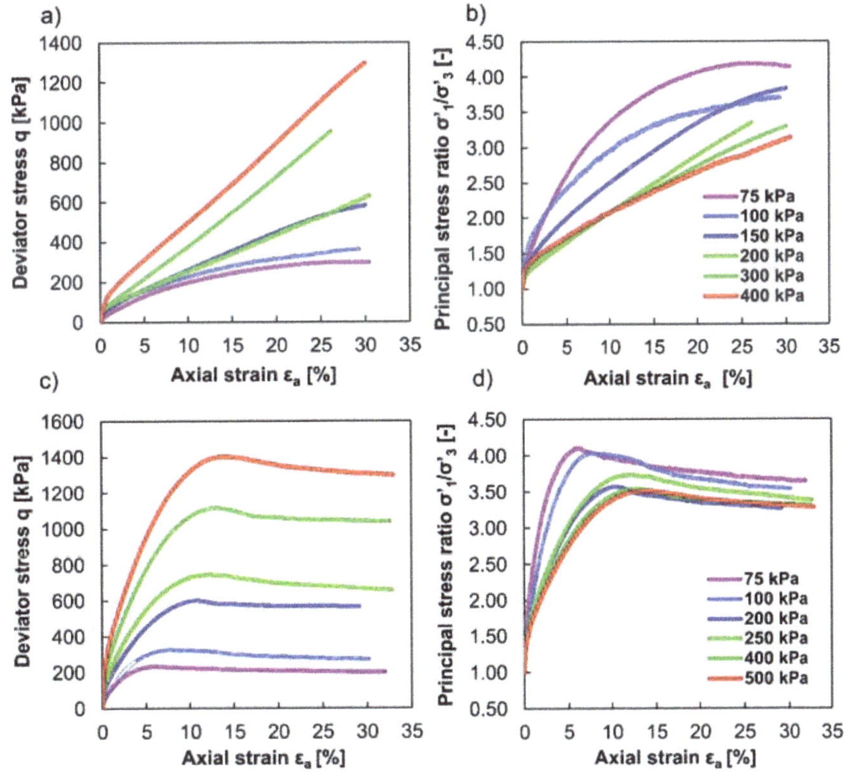

Figure 4. Drained triaxial test results: (**a**) axial strain ε_a vs. deviator stress q (ASR I), (**b**) axial strain ε_a vs. principal stress ratio σ'_1/σ'_3 (ASR I), (**c**) axial strain ε_a vs. deviator stress q (ASR II), and (**d**) axial strain ε_a vs. principal stress ratio σ'_1/σ'_3 (ASR II).

In this case, the deviator stress was gradually increased with the simultaneous increase of axial strain. No distinct failure or any tendency to rupture even up to 30% axial strain was observed. Such behaviour is typical for loose sands and is represented by a monotonically increasing curve (Figure 4a) with no clear drops and, therefore, no pronounced peaks (the failure points). The increasing strength of ASR I with the higher confining pressure probably comes from the cementation bonds among soil particles caused by the inclusions consisting of fine calcite. Moreover, the ASR may have greater stiffness due to the interlocking of particles and densification. When the confining pressure is high, the voids in the ASR closed quickly, and dislocations become blocked. It is also favoured by the considerable amount of water that drained from the sample during the consolidation stage, bringing the grains closer together and reducing porosity. At low confining pressure (i.e., $\sigma'_3 = 75$ kPa), the stress–strain curve showed a specific weak strain-softening. From the practical point of view, it is appropriate to consider the peak value of the deviator stress as the failure criterion. None of the specimens from layer I could be taken as failure as defined by the Mohr–Coulomb criterion. The maximum values of the deviator stress were identified after 25% of the axial strain was exceeded. Hence, as the criterion of the failure, 20% of the axial strain was accepted. Moreover, the high values of the ultimate axial deformation for the discussed case may be associated with the low consolidation of the specimen and the weak microstructure (no bonds) of the material [44].

Figure 4c shows deviator stress versus axial strain curves for the samples from the greater depth (ASR II). It can be noticed that the shear resistance of an ASR decreases with further shearing after the shear stress has reached its peak value. Hereafter, the slope of the deviator stress versus axial strain curve decreased to its residual value in an axial strain of about 30%. This phenomenon is referred to as strain-softening. Such behaviour is commonly associated with dense sands.

Notwithstanding, some studies have established that the observed strain-softening is not a material response but merely a consequence of nonhomogeneous deformations caused by the frictions between the specimen and the top and bottom loading platens or by the initial non-uniformity of the specimens [45–47]. Wanatowski and Chu [48] discussed two types of strain-softening behaviour under plane-strain conditions. For medium-dense sand, strain-softening occurred due to shear bands, mainly due to boundary constraints. The mentioned analysis also revealed that strain-softening is caused by the loading mode adopted to conduct the test. It is more pronounced in the tests with the deformation controlled (DC) loading mode when specimens are sheared using a constant rate of vertical displacement. In this way, strain-softening observed in the ASR samples from a greater depth is associated with the shear bands rather than the material's intrinsic behaviour. One should note that the traces of strain localisation were clearly visible at the end of the test in all specimens. However, it should be remembered that the strain localisation as a prelude to failure is initiated much earlier during the shearing. The strain localisation may emerge well before the peak shear stress, mainly due to heterogeneity in the specimen [49] or around the peak shear stress [50].

For the ASR I, the values of the principal stress ratio σ'_1/σ'_3 for the low confining pressures increased approximately linearly from unity to ~3.35 over the axial strain range of 5% to 30% (Figure 4b). In contrast, the values of the principal stress ratio σ'_1/σ'_3 increased non-linearly from unity to ~4.14 over the axial strain range of 0% to 30%, under higher confining pressures. The most pronounced non-linear specimen response was observed for the confining pressures of $\sigma'_3 = 300$ and 400 kPa. On the other hand, the values of the principal stress ratio σ'_1/σ'_3 calculated for ASR II increased non-linearly, reaching a peak and then decreasing slightly (Figure 4d).

As shown in Figure 4, ASR I and ASR II specimens exhibited strain hardening and strain-softening behaviour, respectively. The different strain behaviour was attributed to the fact that the ASR II specimens were denser than the ASR I specimens because the ASR II specimens were collected at a greater depth than the ASR I specimens.

3.3.2. Failure Modes

Figure 5 shows the typical failure modes for the ASR I and ASR II samples. There are no definite trends in the failure modes with respect to the confining pressure or type of ASR. The ASR I samples in the drained tests revealed, in general, the barrelling mode with the incomplete partially developed shear band or the single shear band pattern (Figure 5). For the ASR II samples, the barrelling mode with the "X" or "V" shear band pattern was observed (Figure 5). One can note that in most cases, the failure mode was a combination of the barrelling shape and shear plane. Wherein the shear band was not obvious, and barrelling was the predominant mode.

Figure 5. Typical failure modes of samples observed from the present study.

3.3.3. Stress Paths

The stress paths for two ASR are shown in Figure 6. The established failure and critical state lines are presented in the plots, defined by a line in the q-p′ plane, with a slope of M. The indices p and cs denote failure (peak strength) and the critical state, respectively. In particular, the critical state concept is the idea that soil and other granular materials, if continuously distorted until they flow as a frictional fluid, will come into a well-defined critical state [51].

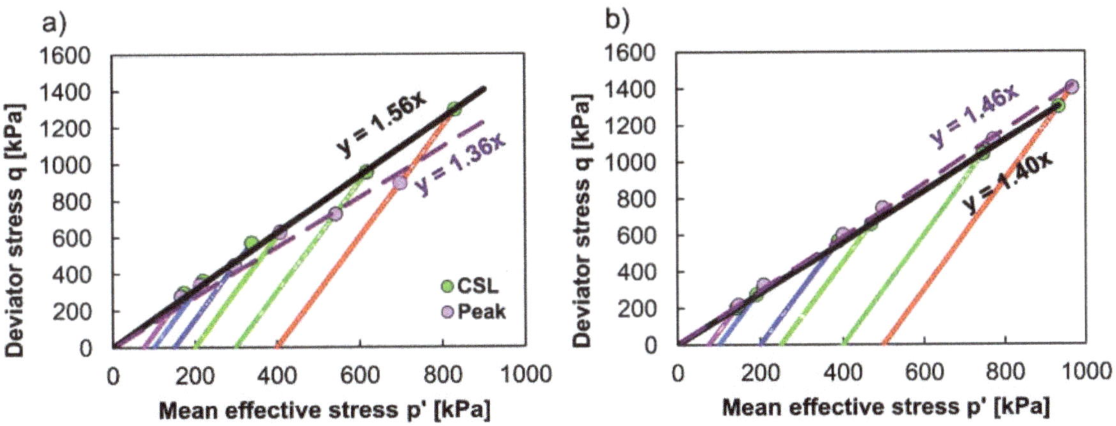

Figure 6. Effective stress paths in the drained triaxial compression tests with respect to the peak deviator stress and the critical state: (**a**) ASR I, (**b**) ASR II.

The slope of the critical state line (CSL) is commonly called the critical state frictional constant M or critical state ratio and should be treated as a function of the intermediate principal stress, expressed by the lode angle (θ) [52]. Parameter M defines the critical state stress ratio at failure conditions and is comparable to the critical friction angle (φ'_{cs}) for the Mohr–Coulomb failure line using the following expression:

$$M = \frac{q}{p'} = \frac{6sin\varphi'_{cs}}{3 - sin\varphi'_{cs}} \quad (3)$$

Note that the correlation between M and φ'_{cs} is not unique but depends on the stress conditions [53], and Equation (3) is valid only for the compression case (i.e., when $\sigma'_a > \sigma'_r$). Figure 6 shows the stress paths for two ASRs. For the ASR I, the peak points of the stress path in the drained condition were lower than for ASR II over the whole range of confining pressures. The slope of critical state lines in the q-p′ plane is M_{cs} = 1.56 and M_f = 1.36 for ASR I. In turn, the values of M_{cs} and M_f for ASR II are 1.40 and 1.46, respectively. Overall, it can be seen that the data points drawn in Figure 6 can accurately be fitted by a linear function (one for failure and one for the critical state). In this regard, the failure locus and the critical state locus are independent of the drainage conditions.

Figure 7 shows Mohr stress circles drawn for the peak deviator stress or for the axial strain of 25% as obtained from the results of CD tests on ASR I and II. Note that the solid lines indicate the critical state, and the dashed lines the failure condition corresponding to the peak shear strength. These lines are tangent to the Mohr stress circles over the entire range of applied confining pressures (Figure 7). The values of effective cohesion, effective friction angles at peak φ'_p, and critical friction angles φ'_{cs} for ASR I and ASR II (Figure 7) were comparable with the values for alluvial clayey silts [54] and many soft alluvial deltaic clays [55].

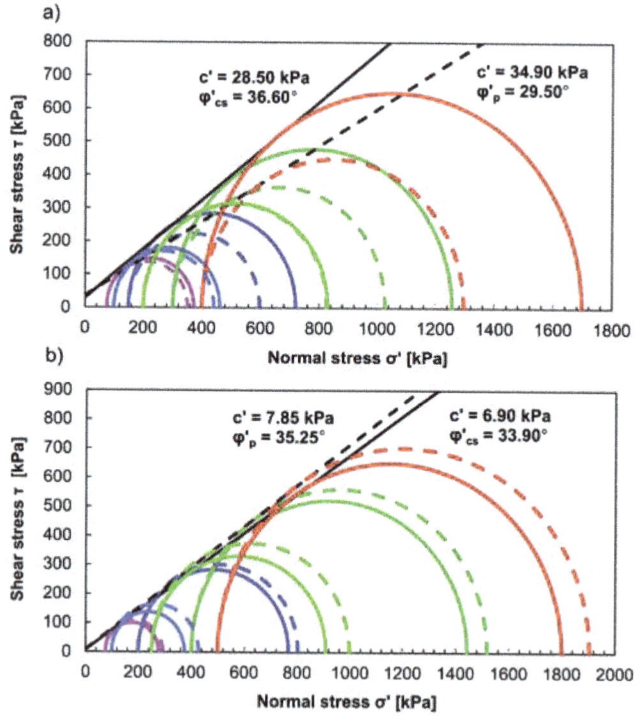

Figure 7. Mohr–Coulomb failure envelope based on the maximum deviator stress and the critical state from the drained triaxial tests: (**a**) ASR I, (**b**) ASR II.

Moreover, the values of the effective friction angles from the present study were consistent with the reported numbers by O'Kelly [56] for biosolids, sewage sludge, and water treatment residue materials. Considering the stability calculations for the intermediate and long term conditions, the derived values are far too high for the design purposes.

3.3.4. Stiffness

The small-strain stiffness can be assessed by the tangential shear modulus (G_{sec}) calculated using the relation between the deviator stress and deviator strain as follows:

$$G_{sec} = \frac{\delta q}{3\delta \varepsilon_s} \quad (4)$$

where δ_q is the increment in deviator stress and ε_s is the increment in deviator strain.

The degradation of G_{sec} with the deviator strain for different confining pressures is presented in Figure 8a. The stiffness of the ASR increases with the confining pressure. This trend is corroborated by the reported dependence of the normalised shear modulus and confining pressure [57,58]. The shape of the degradation curves varies slightly for the ASRs tested (Figure 8a). The typical S-curves were found in the most conducted tests. The samples of ASR I show the more gradual degradation of stiffness compared with the samples from ASR II. All tests, except the sample from ASR II under confining pressure of 500 kPa, indicate a threshold yield strain, which separates the plateau of the initial constant stiffness from the start of substantial stiffness degradation. Figure 8b illustrates the variation of stiffness with the deviator stress for the considered confining pressures. The stiffness of ASR reduced with the deviator stress.

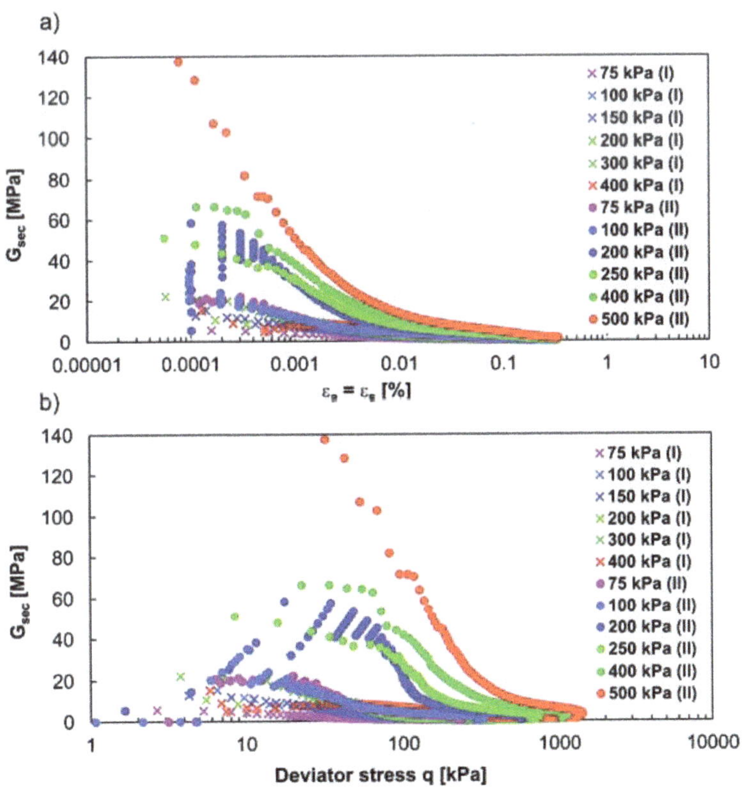

Figure 8. Small-strain stiffness of ASR: (**a**) deviator strain ε_s vs. tangential shear modulus G_{sec}, (**b**) deviator stress q vs. tangential shear modulus G_{sec}.

3.4. Constant Rate of Strain Consolidation Tests

During the CRS test, the vertical force (P), axial strain, and pore pressure (u_b) were measured continuously at the specified time intervals. The main problem with the interpretation of the CRS tests was a slight increase in the pore pressure at the sample bottom. The maximum increase in u_b was only 2-3 kPa, hampering the determination of the coefficient of consolidation c_v and the coefficient of permeability k. The lack of increase in u_b was also independent of the assumed strain rate in the test. Figure 9 shows the obtained axial deformations as a function of the effective stresses $\varepsilon_a = f(\sigma')$. The analysis of the results revealed the relationship between the stress–strain behaviour during compression and the strain rate. As can be seen from Figure 9, the overall compressibility of the samples decreased with the increasing strain rate.

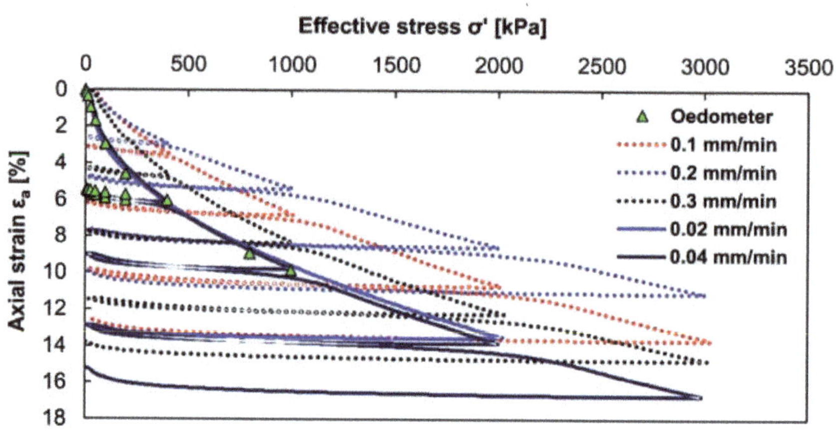

Figure 9. Compression curves for the different strain rates in CRS consolidation tests with superimposed data from the reference oedometer test.

The influence of the strain rate led to the proportional shift of the curve towards higher effective stresses (to the right). This effect was most pronounced at pressures around the yield stress. Using the graphic Casagrande method, the yield stress value of 155 kPa was determined based on the oedometer test. Figure 9 also shows superimposed data from the reference oedometer test. The shape of the primary compression curves for the different strain rates was similar. It was observed that the curves from the CRS study shifted to the right, showing higher yield stress than in the oedometer study.

Moreover, it can be observed that for a given strain, the higher strain rate leads to higher effective stress. The strain rate values selected for the CRS test were generally higher than the observed strain rate in the oedometer test, which caused the compressibility curves derived from the CRS test to lie above the compression curve from the oedometer test. The compressibility curve for the CRS test, obtained with strain rate of 0.04 mm/min, was the closest to the oedometer curve. Therefore, an attempt was made to run the test at a value lower than the strain rate of 0.04 mm/min to fit the results better. A new compression curve was obtained after taking the value of the strain rate of 0.005 mm/min and performing the test. However, the expected curve shift to the left did not take place. On the contrary, the curve shifted to the right and was higher than the curve obtained with the 0.04 mm/min strain rate. It should be remembered that the selection of very slow strain rates leads to the significantly extended test time, which may also cause the ageing effects and the gradual development of intermolecular bonds, which leads to an increase in preconsolidation pressure and a shift of the CRS compression curve to the right [59].

Additionally, in the course of the analysis, the compression index (C_c) was calculated, which was 0.16. This value indicates that ASR can be comparable with soils of a slight to low compressibility.

4. Conclusions

In the study presented herein, the strength and compressibility of ammonia-soda residue from the Solvay Sodium Plant in Krakow were evaluated using triaxial compression tests and the constant rate of strain consolidation tests. The collected material for the laboratory testing was also characterised in terms of the mineral composition and microstructure. The results presented in this work can be summarised as follows:

1. Consolidated drained triaxial tests on the ASR reveal two different evolutions of the critical state in the stress–strain space. The strain-hardening behaviour when the deviator stress was gradually increased with the increase of axial strain without manifesting any distinct failure was observed for the ASR I. On the other hand, the stress–strain response of ASR II indicated strain-softening behaviour associated with the creation of the shear bands in the specimen.

2. The shear strength of ASR I and ASR II evaluated from the present study is similar to the data reported for the alluvial and deltaic soft fine-grained soils. The shear strength of ASR I and ASR II tested in this study for the case of the critical state is quantified to be fi = 36.60 and fi = 33.90, respectively. Effective cohesion for ASR I and ASR II can be best characterised by $c' = 28.50$ kPa and $c' = 6.90$ kPa, respectively.

3. The determined strain rate in the CRS test significantly influences the stress generated in the ASR sample and, thus, the stress–strain response of the tested material. The optimal strain rate was successfully established using the results of the oedometric test. The results of the CRS test made it possible to obtain the compressibility index of the tested material. However, the slight increases in the pore pressure during the test hindered the determination of the c_v and k parameters.

4. The X-ray diffraction results indicated that the main mineral components of the ASR are calcite and portlandite. The analysis of the SEM images showed that the ASR is a porous material made up of crystalline particles with a wide range of shapes and sizes.

Author Contributions: Conceptualisation, B.S.O.; methodology, B.S.O. and P.R.; validation, J.Z. and B.S.O.; formal analysis, B.S.O. and P.R.; investigation, B.S.O.; resources, J.Z.; data curation, B.S.O.; writing—original draft preparation, B.S.O.; writing—review and editing, B.S.O.; visualisation, B.S.O.; supervision, B.S.O. All authors have read and agreed to the published version of the manuscript.

Funding: This research received no external funding.

Institutional Review Board Statement: Not applicable.

Informed Consent Statement: Not applicable.

Data Availability Statement: The datasets generated during and/or analysed during the current study are available from the corresponding author on reasonable request.

Conflicts of Interest: The authors declare no conflict of interest.

References

1. Rui, Y.; Liang, Y.; Wang, Y. Wet simulation flue gas desulfurisation with soda-ash dregs absorbent. *Environ. Sci. Technol.* **2006**, *29*, 21–25.
2. Şener, S. Use of solid wastes of the soda ash plant as an adsorbent for the removal of anionic dyes: Equilibrium and kinetic studies. *Chem. Eng. J.* **2008**, *138*, 207–214. [CrossRef]
3. Jiang, S.; Zhang, Y.; Li, Z. A new industrial process of $NaHCO_3$ and its crystallisation kinetics by using the common ion effect of Na_2CO_3. *Chem. Eng. J.* **2019**, *360*, 740–749. [CrossRef]
4. Pickles, C.A.; Togurib, J.M. Toguri Soda Ash Smelting of Lead Chloride. *Can. Metall. Q.* **1988**, *27*, 117–122. [CrossRef]
5. Pickles, C.A.; Togurib, J.M. The soda ash smelting of lead-acid battery residue. *Resour. Conserv. Recycl.* **1993**, *9*, 155–177. [CrossRef]
6. Allègre, C.; Moulin, P.; Maisseu, M.; Charbit, F. Treatment and reuse of reactive dyeing effluents. *J. Membr. Sci.* **2006**, *269*, 15–34. [CrossRef]
7. Ergas, S.J.; Therriault, B.M.; Reckhow, D.A. Evaluation of water reuse technologies for the textile industry. *J. Environ. Eng.* **2006**, 132. [CrossRef]
8. Chu, K.Y.; Provost, J.R. The dyeing and printing of silk fabrics. *Rev. Prog. Color. Relat. Top.* **1987**, *17*, 23–28. [CrossRef]
9. Roberts, P. Soda ash for the glass industry. *Glass Ceram.* **1965**, *22*, 276. [CrossRef]

10. Zeller, A.F. Soda Ash and the Glass Industry. In Proceedings of the 50th Conference on Glass Problems: Ceramic Engineering and Science Proceedings, Urbana, IL, USA, 7–8 November 1989; Wachtman, J.B., Ed.; American Ceramic Society: Westerville, OH, USA, 1990. [CrossRef]
11. Polkan, G.A.; Zavarina, S.V.; Gorokhovskii, V.A.; Pentko, V.L.; Bulanova, L.D.; Orlova, L.A.; Gorshkov, V.A. Soda Ash based on nepheline material used in the glass industry. *Glass Ceram.* **2003**, *60*, 208–212. [CrossRef]
12. Madeswaran, S.; Jayachandran, S. Sodium bicarbonate: A review and its uses in dentistry. *Indian J. Dent. Res.* **2018**, *29*, 672–677. [CrossRef]
13. Chazelas, E.; Deschasaux, M.; Srour, B.; Kesse-Guyot, E.; Julia, C.; Alles, B.; Druesne-Pecollo, N.; Galan, P.; Hercberg, S.; Latino-Martel, P.; et al. Food additives: Distribution and co-occurrence in 126,000 food products of the French market. *Sci. Rep.* **2020**, *10*, 3980. [CrossRef] [PubMed]
14. Wu, S.; Fitzpatrick, J.; Cronin, K.; Miao, S. Effect of sodium carbonate on the rehydration of milk protein isolate powder. *Food Hydrocoll.* **2020**, *99*, 105305. [CrossRef]
15. Statista. Available online: https://www.statista.com/statistics/587223/sodium-carbonate-production-worldwide (accessed on 9 August 2021).
16. Wotring, P.S.; Stadler, A.T. A field and laboratory investigation of ponded Solvay process residue behavior. In Proceedings of the Geo-Denver 2007: New Peaks in Geotechnics, Denver, CO, USA, 18–21 February 2007; American Society of Civil Engineers: Reston, VA, USA, 2007. [CrossRef]
17. Schmertmann, G.R.; Bachus, R.C. Geotechnical issues for closure of ammonia soda ash waste ponds. In Proceedings of the Conference from Research to Practice in Geotechnical Engineering, Geotechnical Special Publication 180, New Orleans, LA, USA, 9–12 March 2008.
18. Effler, S.W.; Matthews, D.A. Impacts of a soda ash facility on Onondaga Lake and the Seneca River, NY. Lake. *Reserv. Manag.* **2003**, *19*, 285–306. [CrossRef]
19. Kuang, S.P.; Zhang, C.J.; Jiang, Z.G.; Shi, Z.J. Review on comprehensive utilisation techniques of alkaline slag in soda ash factory. *China Res. Com. Util.* **2006**, *3*, 20–24. (In Chinese)
20. Zhang, G.; Li, X.; Li, Y.; Wu, T.; Sun, D.; Lu, F. Removal of anionic dyes from aqueous solution by leaching solutions of white mud. *Desalination* **2011**, *274*, 255–261. [CrossRef]
21. Yang, Y.B.; Zheng, M.R.; Guo, W.Y.; Wang, H.C. Experimental research on the durability of inorganic binder stabilised material with soda residue. *Key Eng. Mater.* **2014**, *629–630*, 189–194. [CrossRef]
22. Coleman, A.; O'Flaherty, M.M.; Davidson, D.T. Fly ash and sodium carbonate as additives to Soil-Cement Mixtures. In Proceedings of the 41st Annual Meeting of the Highway Research Board, Washington, DC, USA, 8–12 January 1962; pp. 108–123.
23. El-Rawi, N.M.; Toma, R.N. Investigation of the effect of some chemical additives on the strength of Iraqi soil-cement mixtures. *J. Build. Res. Cen.* **1987**, *6*, 1–12.
24. Sun, J.Y.; Gu, X. Engineering properties of the new non-clinker incorporating soda residue solidified soil. *J. Build. Mater.* **2014**, *17*, 1031–1035.
25. Zha, F.S.; Pan, D.D.; Xu, L.; Kang, B.; Yang, C.B.; Chu, C.F. Investigations on engineering properties of solidified/stabilised pb-contaminated soil based on alkaline residue. *Adv. Civ. Eng.* **2018**, *2018*, 8595419. [CrossRef]
26. He, J.; Wang, X. Shear strength of stabilised clay treated with soda residue and ground granulated blast furnace slag. *J. Mater. Civ. Eng.* **2019**, *31*, 06018029. [CrossRef]
27. Nazir, R.; Amiri, S.T.; Kassim, K.A. The effect of soda ash (Na_2Co_3) on pH and compaction characteristics of Malaysian laterite soil. Electron. *J. Geotech. Eng.* **2013**, *18*, 2963–2968.
28. Han, S.; Wang, B.; Gutierrez, M.; Shan, Y.; Zhang, Y. Laboratory study on improvement of expansive soil by chemically induced calcium carbonate precipitation. *Materials* **2021**, *14*, 3372. [CrossRef] [PubMed]
29. Bai, X.; Ma, J.; Liu, J.; Zhang, M.; Yan, N.; Wang, Y. Field experimental investigation on filling the soda residue soil with liquid soda residue and liquid fly ash. *Int. J. Damage Mech.* **2020**, 424. [CrossRef]
30. Ma, J.; Yan, N.; Zhang, M.; Liu, J.; Bai, X. Mechanical characteristics of soda residue soil incorporating different admixture: Reuse of soda residue. *Sustainability* **2020**, *12*, 5852. [CrossRef]
31. Wang, Q.; Li, J.; Yao, G.; Zhu, X.; Hu, S.; Qiu, J.; Chen, P.; Lyu, X. Characterisation of the mechanical properties and microcosmic mechanism of Portland cement prepared with soda residue. *Constr. Build. Mater.* **2020**, *241*, 7994. [CrossRef]
32. Shatov, A.A.; Dryamina, M.A.; Badertdinov, R.N. Potential utilisations of soda production wastes. *Energy Sustain. Dev.* **2004**, *12*, 565–571.
33. Kuznetsova, T.V.; Shatov, A.A.; Dryamina, M.A.; Badertdinov, R.N. Use of wastes from soda production to produce nonshrinking oil-well cement. *Russ. J. Appl. Chem.* **2005**, *78*, 698–701. [CrossRef]
34. Kesim, A.G. Properties of alinite cement produced by using soda sludge. *Adv. Cem. Res.* **2013**, *25*, 104–111. [CrossRef]
35. Yan, C.; Song, X.K.; Zhu, P.; Sun, H.Y.; Li, Y.P.; Zhang, J.F. Experimental study on strength characteristics of soda residue with high water content. *Chin. J. Geotech. Eng.* **2007**, *29*, 1683–1688.
36. Zhao, X.H.; Liu, C.Y.; Zuo, L.M.; Zhu, Q.; Ma, W.; Liu, Y. Preparation and characterisation of press-formed fly ash cement incorporating soda residue. *Mater. Lett.* **2020**, *259*, 126852. [CrossRef]

37. Vincevica-Gaile, Z.; Teppand, T.; Kriipsalu, M.; Krievans, M.; Jani, Y.; Klavins, M.; Hendroko Setyobudi, R.; Grinfelde, I.; Rudovica, V.; Tamm, T.; et al. Towards sustainable soil stabilisation in peatlands: Secondary raw materials as an alternative. *Sustainability* **2021**, *13*, 6726. [CrossRef]
38. ISO 17892-4: 2016. *Geotechnical Investigation and Testing—Laboratory Testing of Soil—Part 4: Determination of Particle Size Distribution*; International Standardization Organization: Geneva, Switzerland, 2016.
39. ISO 17892-3: 2015. *Geotechnical Investigation and Testing—Laboratory Testing of Soil—Part 3: Determination of Particle Density*; International Standardization Organization: Geneva, Switzerland, 2015.
40. Zięba, J. Mechanical behaviour of calcareous waste under consolidated drained triaxial compression testing in saturated conditions. In Proceedings of the 5th International Scientific Conference on Civil Engineering-Infrastructure-Mining, Kraków, Poland, 17–18 January 2019. [CrossRef]
41. Leroueil, S.; Kabbaj, M.; Tavenas, F.; Bouchard, R. Stress–strain–strain rate relation for the compressibility of sensitive natural clays. *Géotechnique* **1985**, *35*, 159–180. [CrossRef]
42. Mesri, G.; Feng, T.W. Constant rate of strain consolidation testing of soft clays and fibrous peats. *Can. Geotech. J.* **2019**, *56*, 1526–1533. [CrossRef]
43. Ozer, A.T.; Lawton, E.C.; Bartlett, S.F. New method to determine proper strain rate for constant rate-of-strain consolidation tests. *Can. Geotech. J.* **2012**, *49*, 18–26. [CrossRef]
44. Leroueil, S.; Locat, J.; Vaunat, J.; Picarelli, L.; Faure, R. Geotechnical characterisation of slope movements. In Proceedings of the 7th International Symposium on Landslides, Trondheim, Norway, 17–21 June 1996; pp. 53–74.
45. Drescher, A.; Vardoulakis, I. Geometric softening in triaxial tests on granular material. *Géotechnique* **1982**, *32*, 291–303. [CrossRef]
46. Chu, J.; Lo, S. On the measurement of critical state parameters of dense granular soils. *ASTM Geotech. Testing J.* **1993**, *16*, 27–35. [CrossRef]
47. Chu, J.; Lo, S.; Lee, I. Strain softening behavior of a granular soil in strain path testing. *J. Geotech. Eng. ASCE* **1992**, *118*, 191–208. [CrossRef]
48. Wanatowski, D.; Chu, J. Drained behaviour of Changi sand in triaxial and plane-strain compression. *Geomech. Geoengin.* **2007**, *2*, 29–39. [CrossRef]
49. Thakur, V. Strain Localisation in Sensitive Soft Clays. Ph.D. Thesis, Norwegian University of Science and Technology, Trondheim, Norway, 2007.
50. Cheng, Z.; Wang, J. Quantification of the strain field of sands based on X-ray micro-tomography: A comparison between a grid-based method and a mesh-based method. *Powder Technol.* **2019**, *244*, 314–334. [CrossRef]
51. Ghafghazi, M.; Shuttle, D.A. Accurate determination of the critical state friction angle from triaxial tests. In Proceedings of the 59th Canadian Geotechnical Conference, Vancouver, BC, Canada, 1–4 October 2006; Canadian Geotechnical Society: Vancouver, BC, Canada, 2006; pp. 278–284.
52. Sadrekarimi, A.; Olson, S. Critical state friction angle of sands. *Géotechnique* **2011**, *61*, 771–783. [CrossRef]
53. Roscoe, K.H.; Schofield, A.N.; Wroth, C.P. On the yielding of soils. *Géotechnique* **1958**, *8*, 22–53. [CrossRef]
54. Powell, J.; Lunne, T. Use of CPTU data in clays/fine grained soils. *Stud. Geotech. Mech.* **2005**, *27*, 29–65.
55. Ouyang, Z.; Mayne, P.W. Effective friction angle of clays and silts from piezocone penetration tests. *Can. Geotech. J.* **2018**, 55. [CrossRef]
56. O'Kelly, B.C. Geotechnics of municipal sludges and residues for land-filling. *Geotech. Res.* **2016**, *3*, 148–179. [CrossRef]
57. Ishibashi, I.; Zhang, X. Unified dynamic shear moduli and damping ratios of sand and clay. *Soils Found.* **1993**, *33*, 182–191. [CrossRef]
58. Darendeli, M.B. Development of a New Family of Normalised Modulus Reduction and Material Damping Curves. Ph.D. Thesis, University of Texas at Austin, Austin, TX, USA, 2001.
59. Qiao, Y.; Ferrari, A.; Laloui, L.; Ding, W. Nonstationary flow surface theory for modeling the viscoplastic behaviors of soils. *Comput. Geotech.* **2016**, *76*, 105–119. [CrossRef]

applied sciences

MDPI

Article

Modelling of Static Liquefaction of Partially Saturated Non-Cohesive Soils

Waldemar Świdziński and Marcin Smyczyński *

Department of Geomechanics, Institute of Hydro-Engineering of Polish Academy of Sciences, ul. Kościerska 7, 80-328 Gdańsk, Poland; w.swidzinski@ibwpan.gda.pl
* Correspondence: m.smyczynski@ibwpan.gda.pl; Tel.: +48-505-073-222

Abstract: Static soil liquefaction is widely known to be a serious danger to the stability of structures. The phenomena governing pore water generation, which leads to liquefaction in fully saturated soils, are already quite well described. However, much less is known of these phenomena occurring in partially saturated porous media, although this, too, is an important issue in geotechnics. This study presents the application of a semi-empirical model to predict the response of partially saturated soils under undrained conditions. The model proposed is based on an incremental equation describing the pre-failure undrained response of partially saturated non-cohesive soils during monotonic shearing in a standard triaxial test. Improved differential equations taking into account pore fluid compressibility were implemented together with empirical coefficients describing soil skeleton compressibility during the unloading phase. Model coefficients were determined in triaxial compression tests. The influence of the saturation level represented by Skempton's parameter B on the full spectrum of predicted stress paths was shown. For the analyzed saturation range, the maximum stress deviator normalized by initial mean effective stress varied from 0.38 to 1.67 for B values between 0.93 and 0.29, respectively. Model predictions were confronted with the results of triaxial tests for two types of non-cohesive soils (quartz medium sand and copper ore post-flotation industrial tailings). Good agreement between experimental data and theoretical predictions was achieved.

Keywords: incremental equations; non-cohesive soil; partial saturation; liquefaction

Citation: Świdziński, W.; Smyczyński, M. Modelling of Static Liquefaction of Partially Saturated Non-Cohesive Soils. *Appl. Sci.* **2022**, *12*, 2076. https://doi.org/10.3390/app12042076

Academic Editor: Cheng-Yu Ku

Received: 11 December 2021
Accepted: 10 February 2022
Published: 16 February 2022

1. Introduction

It is well known that monotonically loaded non-cohesive saturated soils under undrained conditions may be prone to liquefaction due to a static [1] or cyclic [2,3] load. It was originally thought that, in order for soil to liquefy, voids should be completely filled with water which corresponds to full saturation of the medium. This was due to the fact that any presence of air would significantly limit or prevent the process of pore pressure build-up. Such an approach was physically justified because the presence of air results in full or partial dissipation of the pore pressure excess and consequently scales it down by the Skempton's parameter B decreasing liquefaction potential [4]. However, observations in areas particularly vulnerable to earthquakes have shown that liquefaction can also occur in soils that are not fully saturated. The first studies proving that partially saturated soils may also undergo liquefaction concerned the response of the soil to cyclic loads [5,6]. There followed many others related to the same issue, e.g., [7–13].

Meanwhile, the large impact of saturation level on the liquefaction potential of monotonically loaded soils was proved by a series of laboratory tests [14–19].

In order to theoretically predict the behaviour of partially saturated non-cohesive media, the MODSOL model proposed by Bian and Shahrour [20] has been elaborated, though it is dedicated mainly to cyclic loads as well. An interesting approach to modelling the response of soil to monotonic loading is the elasto-viscoplastic model proposed by Kimoto et al. [21], which was derived for media characterized by relatively high suction values and low saturation levels. Another theoretical study on the behaviour of unsaturated soils can be found in [22–24], in which the authors also generally focus on suction pressure at low saturation levels. The implementation of constitutive laws governing the behaviour of soil in a partially saturated state in a SaniSand-Z model built by Chen et al. can be found in [25]. There are some other theoretical approaches to constitutive modeling of partially saturated soils, e.g., a deviatoric hardening model which can predict static liquefaction proposed by Lü et al. [26], solutions based on thermodynamic equations by Yang et al. [27,28] or a hypo-plastic model of sand developed by Zapata-Medina et al. [29].

According to [4,30,31], in partially saturated media, the location of the instability line tends to increase with a decrease in the saturation level, which causes a change in the moment of triggering of the liquefaction process.

The problem of liquefaction of partially saturated soils seems to be important in many areas, also, from an engineering point of view, e.g., wave-inducted seabed liquefaction [32,33], blast-induced densification in liquefiable sediments [34], submarine large-scale landslides triggered by gas dissociation [4], etc. Another interesting case relates to wet mine tailings storage facilities, where the zone of partial saturation could reach a dozen meters [35]. This is a consequence of the successive deposition of sediments in the form of soil–water mixtures by spigotting, which causes cyclic wetting and drying of stored waste mass [36]. As a result, relatively thick zones of partially saturated sediments appear above the full saturation line. In such a situation, the potential generation of pore pressure within this zone could, in an extreme case, lead to the liquefaction of tailing deposits, significantly reducing the stability of the surrounding dams. This is the reason why post-flotation cooper tailings have been selected as one of the investigated soils in this analysis.

In order to examine the influence of partial saturation on the undrained shear strength of soil, an extensive research programme has been carried out in recent years at the Institute of Hydro-Engineering of the Polish Academy of Sciences. The programme was aimed both at the empirical recognition and analysis of this phenomenon under laboratory conditions [37] as well as at its theoretical description [38].

First tests (Series A, [37]) were conducted on reconstructed specimens of post-flotation tailings (OZM50) from TSF Żelazny Most, the largest industrial copper tailings facility in Europe. Next, a complementary series of tests, the results of which are presented in this paper, was performed on the same soil (Series B), supplemented by tests carried out on medium quartz Skarpa sand (Series C). All tests were conducted under triaxial conditions.

The first series of tests provided a basis for theoretical modelling of liquefaction phenomena of monotonically loaded partially saturated media [38]. The initial theoretical approach reproduced test results fairly well in the qualitative sense but quantitative agreement was insufficient, showing that the model needed some corrections. The present paper describes and discusses the results of the development and improvement of the theoretical model, reproducing the response of partially saturated non-cohesive soil subjected to monotonic shearing under undrained conditions in a standard triaxial test. In addition, verification of the model by laboratory tests on two different types of non-cohesive soils is also presented. The experimental results shown in the paper provide knowledge about unsaturated soil liquefaction mechanisms and constitute a basis for the theoretical description and prediction of partially saturated soil responses in triaxial conditions.

2. Generation of Pore Pressure in Partially Saturated Soils

The phenomenon of soil liquefaction is closely related to the volume changes which dry soil is subjected to during shearing (see the critical state theory [39–44]). In general, the behaviour of non-cohesive soil during shearing depends on its initial state, which determines its contractive or dilative response and/or contractive–dilative intermediate states [39]. Contractive and dilative states are separated by the critical state line (CSL) [39], which is also identified as a steady-state line (SSL) [45] at which sheared soil reaches its critical state [46] (Figure 1a). As a measure of the intensity of soil response (contractive or dilative behaviour), a state parameter Ψ [40] is usually used, which is defined as a change in void ratio that must occur for the medium to reach a critical state (steady state) at a constant mean effective stress.

Differences between undrained shear stress paths corresponding to each initial state (schematically shown in Figure 1b) are caused by changes in pore pressure build-up and subsequent redistribution between mean effective and total stresses.

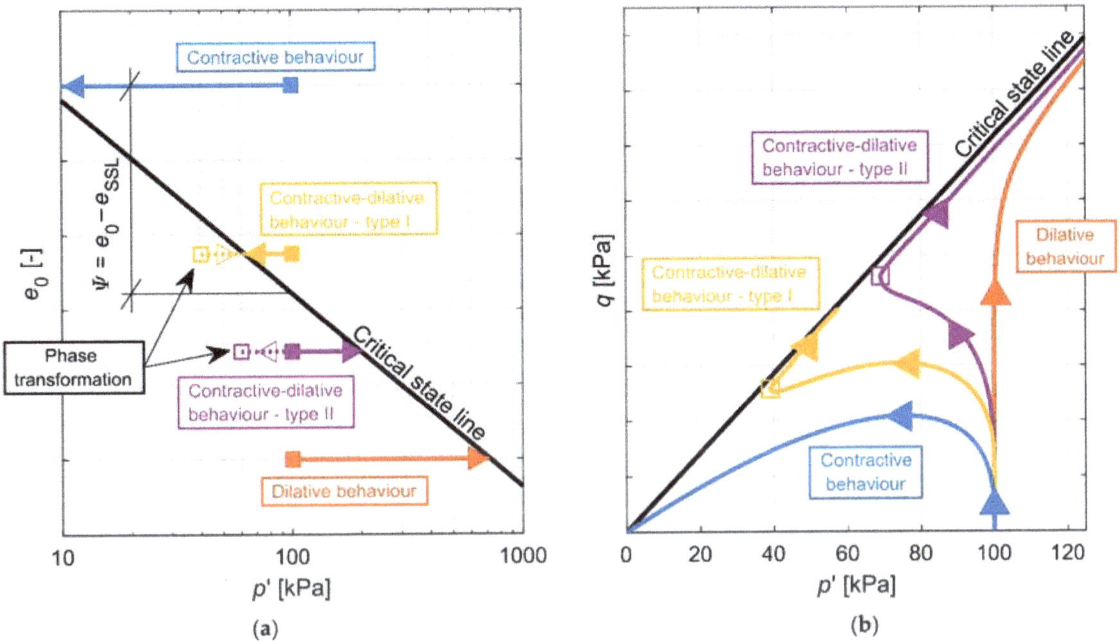

Figure 1. (**a**) Critical state line with definition of state parameter. (**b**) Typical responses of saturated soil during shearing under undrained conditions.

For initially contractive soils, this results in a decrease in the mean effective stress and subsequently a reduction of their shear strength. This reduction occurs after crossing an instability line [4,47], where the stress deviator reaches its maximum value. In the extreme case, effective stress vanishes, the soil is unable to transfer the shear stress and liquefies [48], which can be formally written as:

$$p' = p - u = 0. \quad (1)$$

The degree to which voids are filled with water may change between full saturation, represented by degree of saturation $S_r = 1$, and a completely dry medium (no water), $S_r = 0$. In between there occur intermediate states identified with unsaturated soils, which also include partial saturation. Partial saturation should be distinguished from an unsaturated state in which the gaseous phase of different levels is present, partial saturation being characterised by the lack of a continuous gas phase and the presence of only isolated air bubbles surrounded by water [18]. The model presented in this paper refers to partially saturated soils in which the saturation degree is usually above approximately 80% (Figure 2).

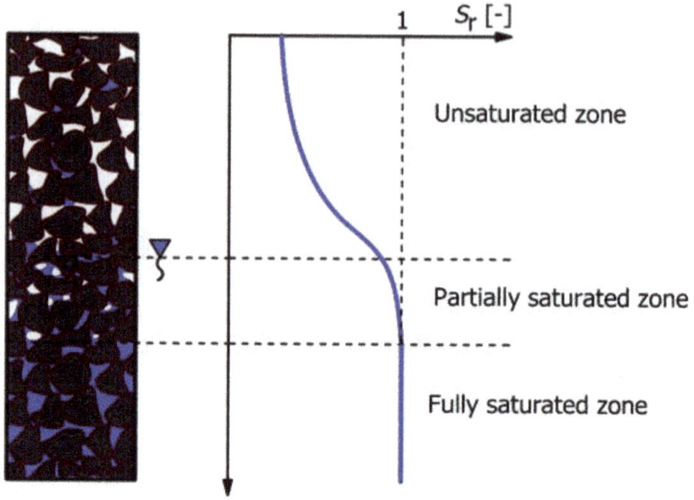

Figure 2. Saturation distribution in a soil medium.

Furthermore, in non-cohesive media with a degree of saturation corresponding to a partial saturation state, suction pressure is negligibly small or does not occur at all [20,49,50]. In Figure 3, water retention curves for some selected non-cohesive soils, i.e., Hostun sand [20] and Singapore fine sand [51], as well as suction pressures for tested soils (determined in Sand Apparatus for the range of high saturation degree) together with their grain size distribution curves, are presented. It can be seen that for a degree of saturation above 0.8 we are dealing with very small suction pressure values of the order of 1 or 2 kPa and therefore the application of Terzaghi's law given by Equation (1) can be justified [52].

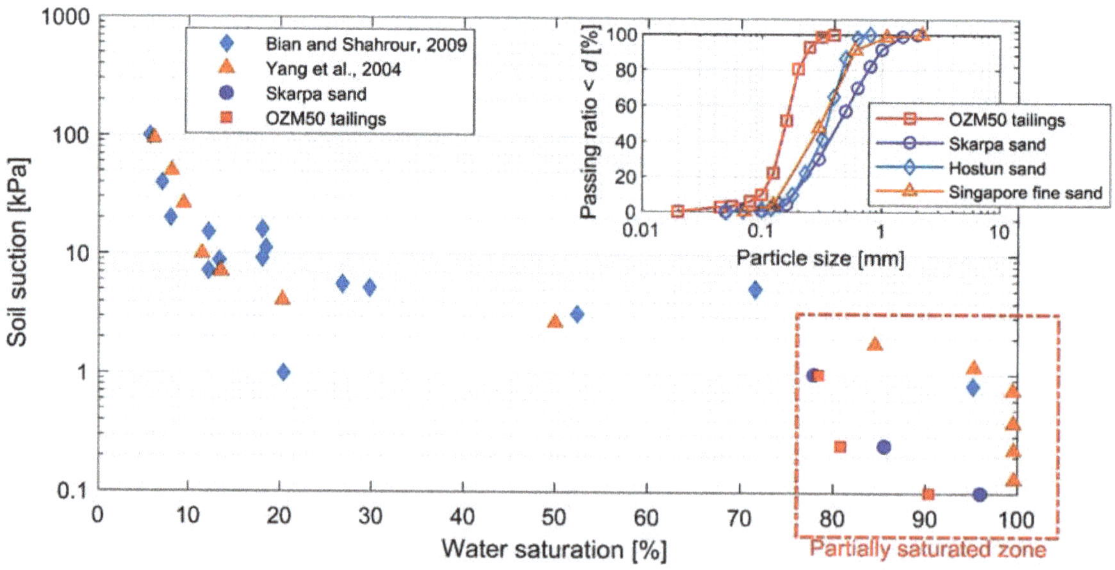

Figure 3. Suction pressures for tested soils and water retention curve for: Hostun sand [20] and Singapore fine sand [51].

Soils in a state of partial saturation are characterized by values of Skempton's parameter B [53] in a range from $B = 1$, which is equivalent to full saturation, to $B \approx 0^+$, corresponding to the lower boundary of a partially saturated state.

Isotropically loaded saturated soil has a tendency to decrease in volume, but compaction under undrained conditions is prevented by bonds imposed by hardly compressible pore fluid, which leads to the generation of excess pore pressure, Δu. Skempton's parameter B is defined as a portion of generated pore pressure, u, caused by corresponding isotropic stress. In an isotropic compression test in a triaxial apparatus, it is equivalent to the ratio of generated excess pore pressure to the change in cell pressure, $\Delta\sigma_3$:

$$B = \frac{\Delta u}{\Delta \sigma_3}. \tag{2}$$

Skempton's parameter B can also be described as a function of the compressibility of the pore fluid, κ_f, and the soil skeleton, $\kappa_s^{p'}$ [53]:

$$B = \frac{1}{1 + \frac{n\kappa_f}{\kappa_s^{p'}}} = f\left(\kappa_s^{p'}, \kappa_f\right), \tag{3}$$

where n is soil porosity.

2.1. Materials and Methods

To examine the influence of saturation level on the liquefaction potential of soils, tests in a triaxial compression apparatus were carried out.

The first series of tests, described in [37] (Series A), and subsequent additional tests (Series B) were performed on anthropogenic post-flotation tailings, denoted as OZM50, which are sharp-edged fine sand grains with a rough surface (Figure 4a). The third series (Series C) was carried out on a medium sand called Skarpa, which has rounded quarzitic grains with a smooth surface (Figure 4b) [54].

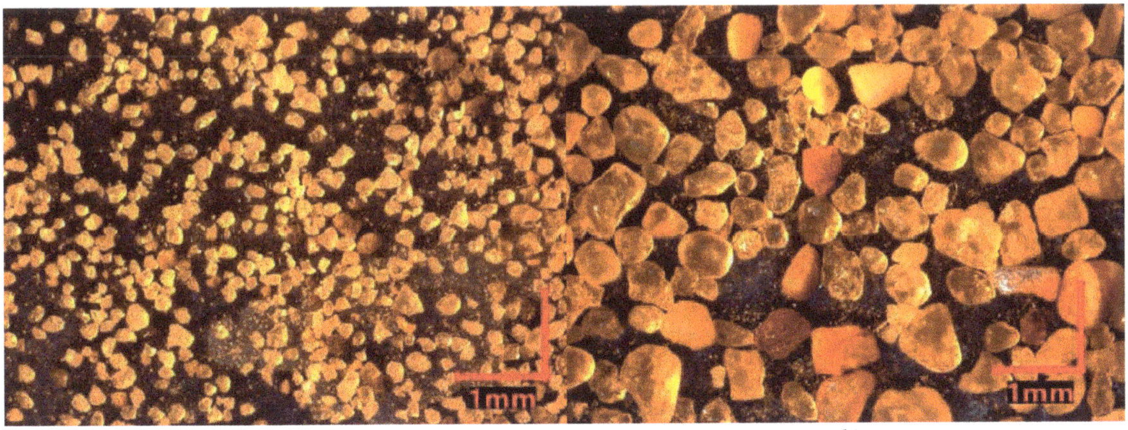

(a) (b)

Figure 4. Microphotographs of (**a**) OZM50 tailings and (**b**) Skarpa sand grains.

The basic index properties and strength parameters of both soils are presented in Table 1; grain size distribution curves are shown in Figure 5.

Table 1. Basic physical characteristics and mechanical properties of the soils tested.

Soil Type	ρ_s (g/cm^3)	d_{50} (mm)	d<0.075 mm (%)	e_{min} (–)	e_{max} (–)	ϕ'_{loose} (°)
Skarpa	2.650	0.420	0.25	0.432	0.677	34.8
OZM50	2.675	0.157	5.8	0.643	1.026	33.0

ρ_s is specific gravity, d_{50} is mean grain diameter, e_{min} and e_{max} are, respectively, minimum and maximum void ratios and ϕ'_{loose} is an internal friction angle for soils in loose state.

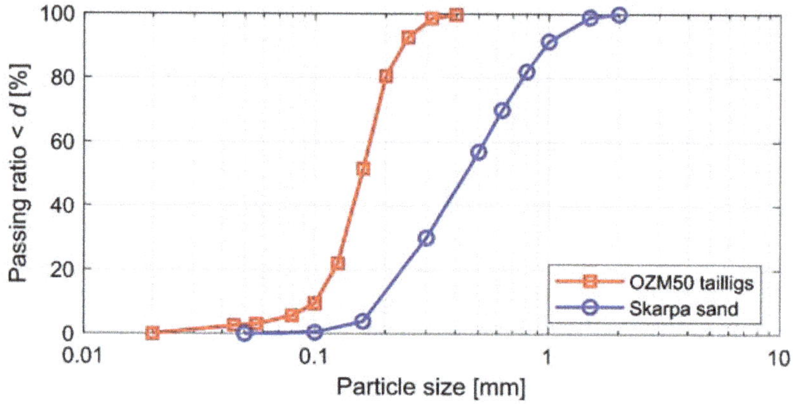

Figure 5. Particle size distribution curves of tested soils.

In order to identify the initial state of particular samples of soils tested, critical state lines for both soils were independently determined based on standard triaxial tests (Figure 6): in drained and undrained tests for Skarpa sand [55] and undrained tests for OZM50 tailings [56].

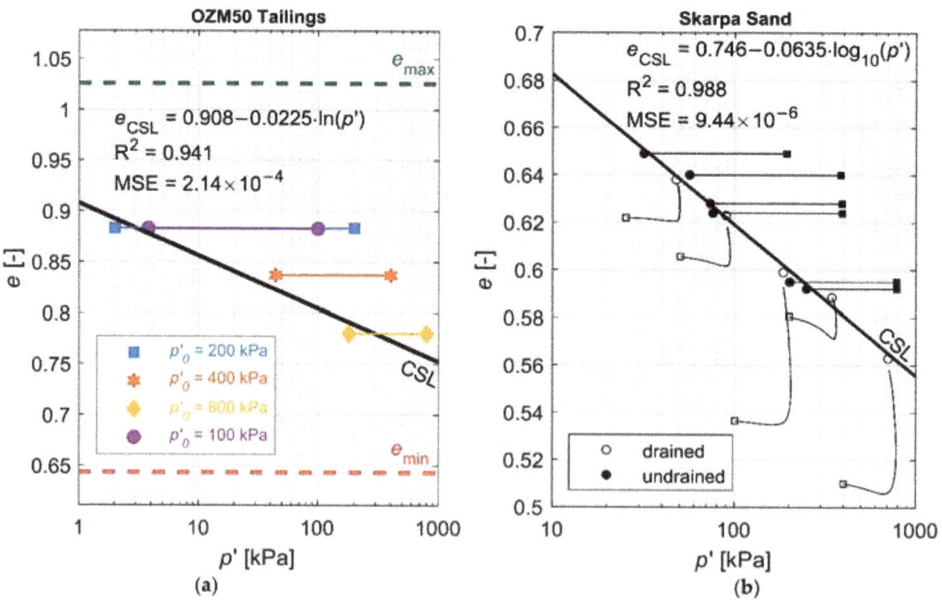

Figure 6. Critical state lines: (**a**) OZM50 tailings [56], (**b**) Skarpa sand, after [55].

The primary aim of the laboratory investigations was to empirically isolate the effect of saturation, expressed by the measured value of Skempton's parameter B on the response of soils during shearing in triaxial conditions. Therefore, in all tests performed, only the saturation level of samples was being changed in a controlled manner, whereas other parameters representing initial state conditions, for both soil types, were kept constant. Special attention was focused on the careful reconstitution of sandy samples in the triaxial apparatus, particularly by controlling void ratio from the very beginning of the specimens' installation as one of two parameters defining the initial state of non-cohesive soil. In order to fully control the value of the void ratio just prior to shearing, its changes during the first phases of the test (saturation, back pressure ramp, consolidation stages) were measured by local gauges making use of the Hall effect. A typical example of such changes is shown in Figure 7, representing the results obtained for test b1.

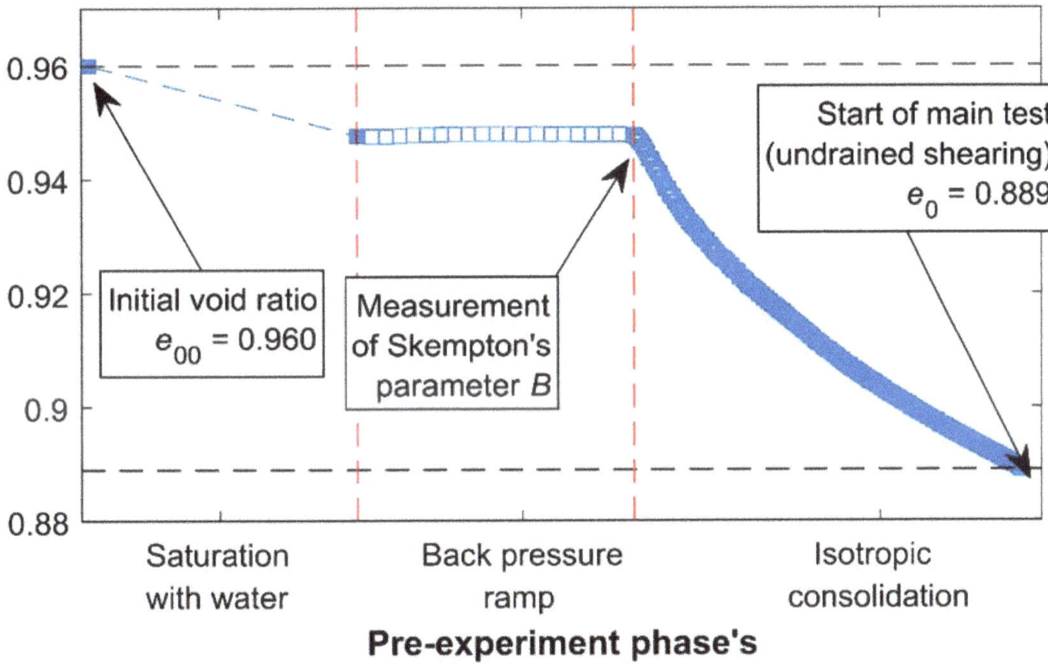

Figure 7. Tracking changes in void ratio in the b1 test.

All tested samples were in an initially contractive state. The samples were first subjected to isotropic consolidation by applying mean effective stress, $p'_0 = 400$ kPa and $p'_0 = 200$ kPa, for OZM50 tailings and Skarpa sand, respectively. All the tests were carried out in a strain-controlled mode at a constant strain rate, $\dot{\varepsilon_1} \cong 10\%/\text{h}$, and a constant cell pressure, $\sigma_3 = \text{const}$.

The saturation level of the soil was determined by Skempton's parameter B, measured during the test. In order to obtain assumed values of the parameter, back pressure was gradually increased in a controlled manner, which resulted in a decrease in the volume of air bubbles contained in soil voids, causing a reduction in pore air compressibility and thus a change in the compressibility of pore fluid as a mixture.

In total, thirteen experiments were carried out, covering a broad spectrum of saturation states represented by values of Skempton's parameter B between 0.29 and 0.93. In all cases, B was measured under a mean effective stress of 20 kPa after sample saturation and back pressure ramp but before consolidation. Such a level of mean effective stress was applied in each test for the sake of proper preparation of soil specimens to achieve the assumed initial void ratios as well as to install local gauges to control sample deformation prior to shearing. Table 2 contains the basic initial parameters of each experiment.

Table 2. Initial parameters of the tests.

Sample	B (–)	u_b (kPa)	p'_0 (kPa)	I_D (–)	Ψ (–)	e_{00} (–)	e_0 (–)
			OZM50 tailings—series A [37]				
a1	0.92	300	400	0.36	0.12	0.978	0.887
a2	0.93	200	400	0.41	0.10	0.943	0.873
a3	0.84	100	400	0.43	0.09	0.920	0.864
a4	0.55	100	400	0.49	0.07	0.915	0.839
a5	0.74	180	400	0.49	0.07	0.916	0.836
a6	0.39	68	400	0.46	0.08	0.917	0.846
a7	0.29	34	400	0.36	0.12	0.920	0.890
			OZM50 tailings—series B				
b1	0.88	499	400	0.36	0.12	0.960	0.889
b2	0.36	51	400	0.26	0.15	0.986	0.927
			Skarpa sand—series C				
c1	0.93	251	200	0.15	0.038	0.660	0.638
c2	0.78	396	200	0.18	0.031	0.653	0.631
c3	0.64	203	200	0.12	0.046	0.652	0.646
c4	0.34	48	200	0.13	0.044	0.665	0.644

u_b is a back pressure, p'_0 is mean effective stress at the start of shearing, I_D is a relative density, Ψ is the state parameter, e_{00} end e_0 are void ratios corresponding to an initial mean stress of 20 kPa and p'_0, respectively.

The location of the critical state lines for each soil in relation to initial void ratios just after sample formation and before applying vertical load are shown in Figure 8.

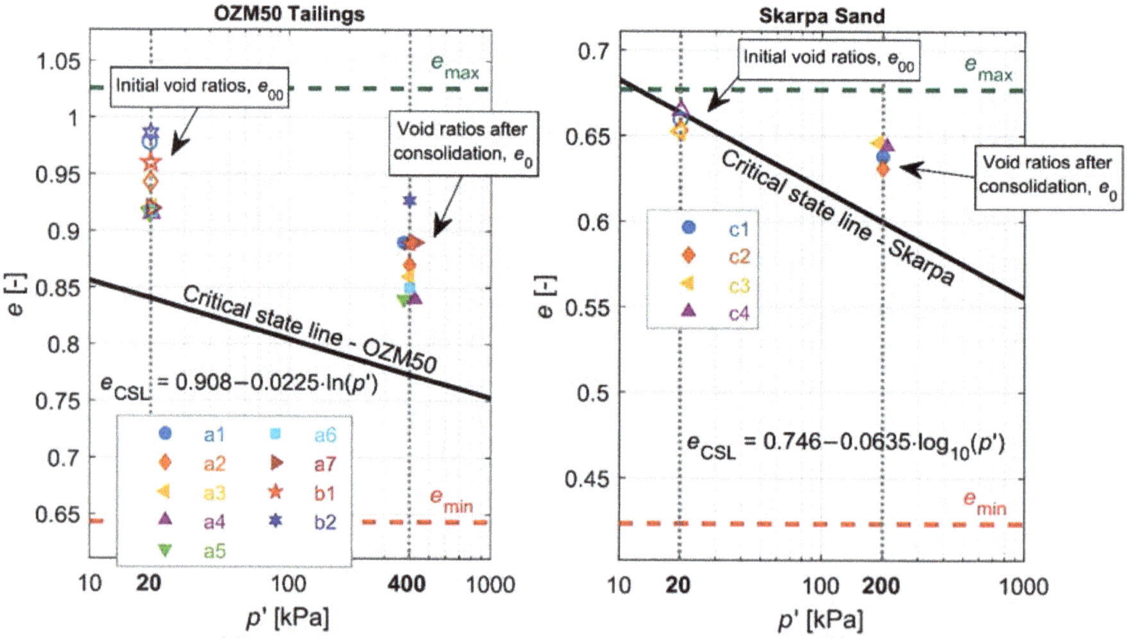

(a) (b)

Figure 8. Initial states of soil samples before testing.

2.2. Experimental Results

Basic results of the laboratory investigations are shown in Figures 9–14. A detailed description and deeper analysis of the results from the first experimental programme (Series A) can be found in the study [37]. Detailed laboratory data for all the experiments analysed in this paper are enclosed in the online Supplementary Materials (Repository S1: Experimental Data).

Changes in pore pressure, normalized by initial mean effective stress in relation to vertical strain ε_1 are shown in Figure 9 for Skarpa sand and in Figure 10 for OZM50 tailings, respectively. It can be seen that generated excess pore pressure increased with saturation and led to the liquefaction of samples with the highest saturation level.

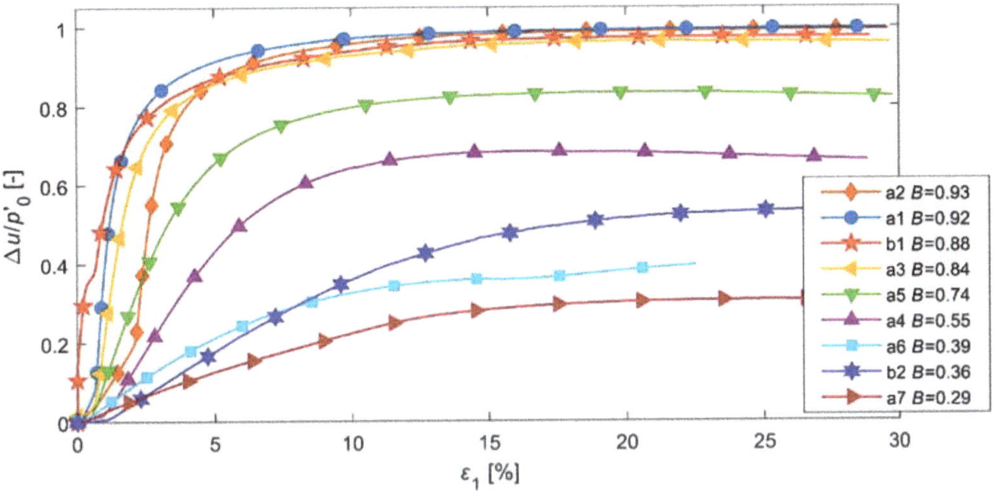

Figure 9. Response of OZM50 tailings at different levels of saturation: normalized excess pore pressure vs. vertical strain.

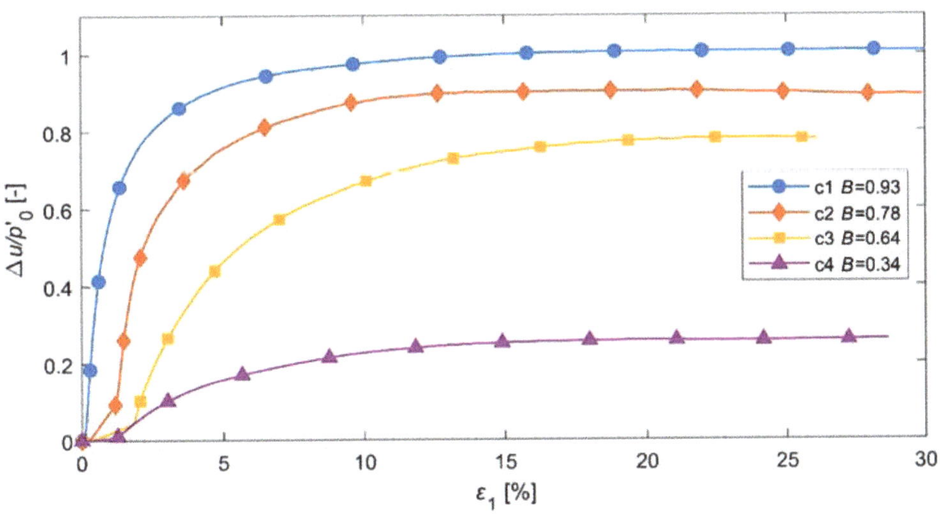

Figure 10. Response of Skarpa sand at different levels of saturation: normalized excess pore pressure vs. vertical strain.

In turn, changes of corresponding deviatoric stress in relation to vertical strain, ε_1, are presented in Figures 11 and 12 for Skarpa sand and for OZM50 tailings, respectively. It can be seen that lower saturation results in higher maximum and residual stress deviators.

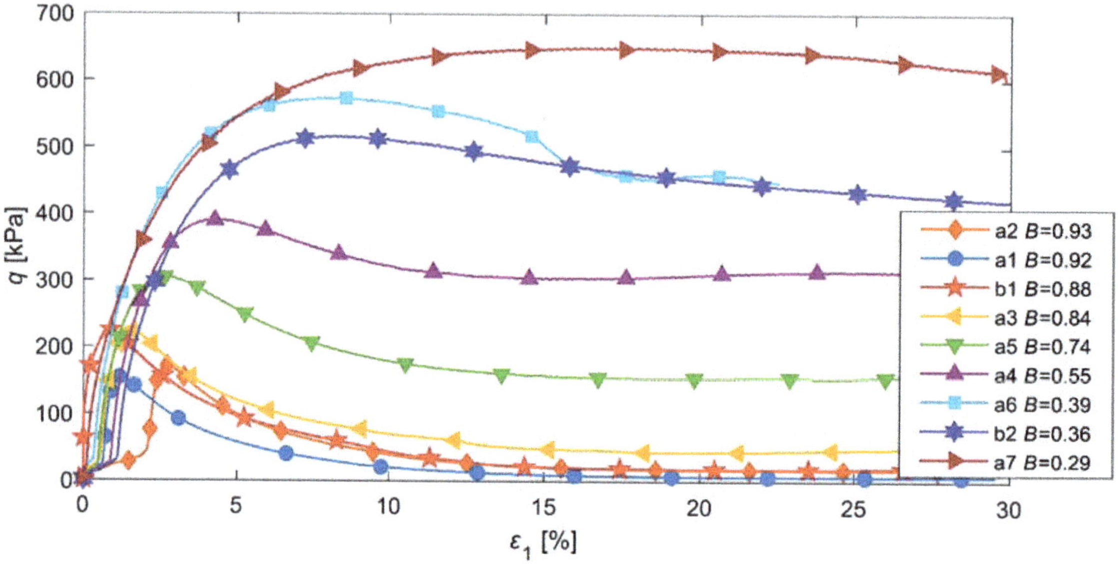

Figure 11. Stress–strain response of OZM50 tailings at different levels of saturation: stress deviator vs. axial strain.

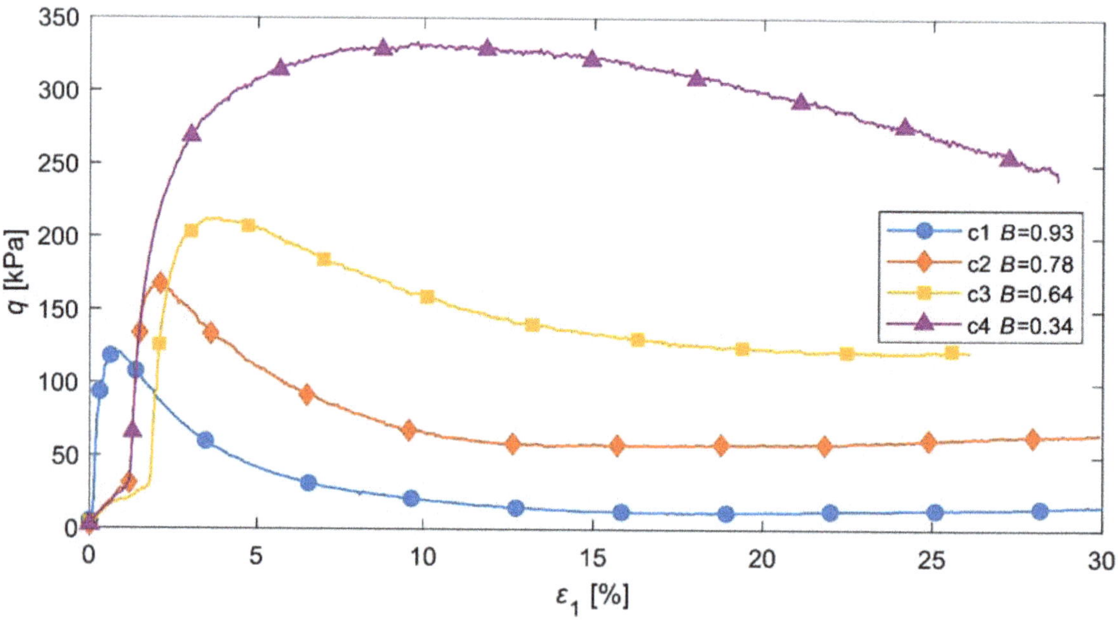

Figure 12. Stress–strain response of Skarpa sand at different levels of saturation: stress deviator vs. axial strain.

The stress paths corresponding to triaxial tests are collected in Figures 13 and 14. As a further confirmation, a clear decrease in the maximum stress deviator can be observed along with an increase in the saturation of the medium for the entire spectrum of saturation levels represented by Skempton's parameter B. When Skempton's parameter B increases, the response of contractive soil changes from one characteristic of dry soil (dashed line) to a typical response of a fully saturated medium.

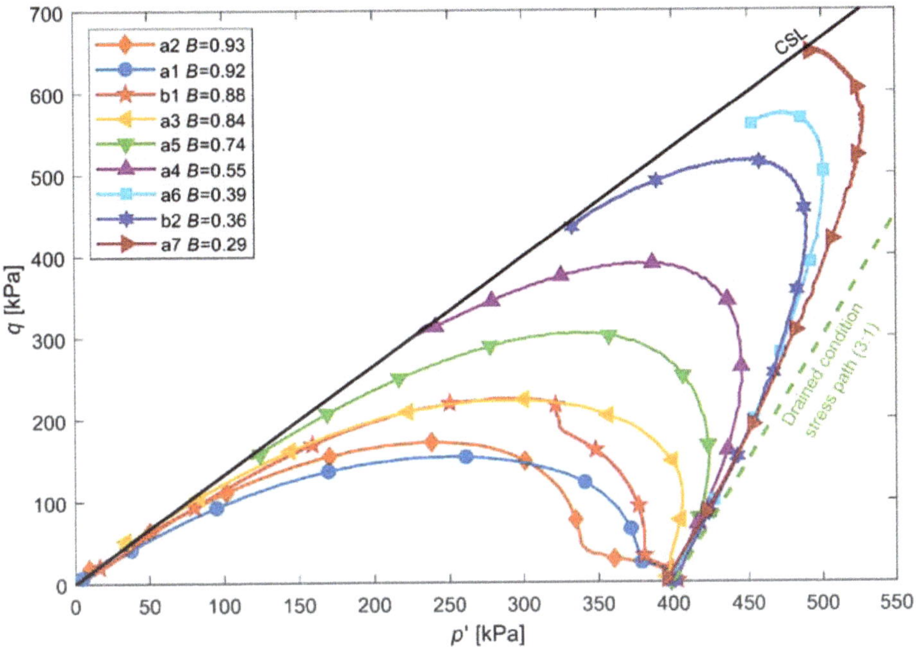

Figure 13. Stress paths corresponding to the behaviour of OZM50 tailings at different levels of saturation.

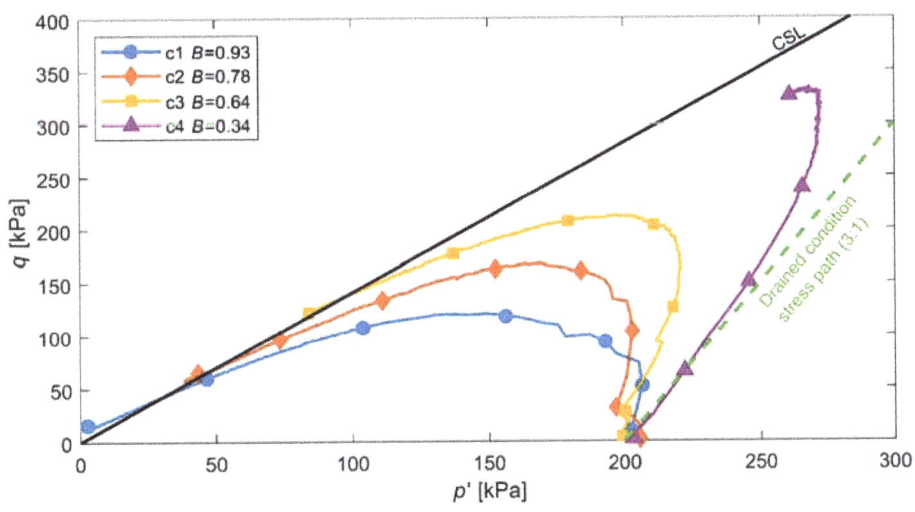

Figure 14. Stress paths corresponding to the behaviour of Skarpa sand at different levels of saturation.

In the presented research results, two stress paths corresponding to tests a6 and c4 differ from the rest. In both cases, failure of the sample took place before the boundary surface characteristic of each soil was reached. In these tests, unlike the others, clear shear surface was observed, typical for the deformation process in a highly compacted non-cohesive soil specimen.

3. Theoretical Model

3.1. Governing Equations

Based on the experimental results (Series A), a theoretical model was proposed to describe the behaviour of partially saturated non-cohesive soil in an initially contractive state subjected to axisymmetric monotonic loading under undrained conditions [38]. The core element of the model is a differential equation the integration of which allows the reproduction of any stress path within the stress space, including the generation of pore pressure in relation to the value of Skempton's parameter B characterizing the saturation level. A significant role in the proposed equation is played by two compressibility coefficients, the first corresponding to the compressibility of fluid treated as a mixture of incompressible water and compressible gas, the second to the compressibility of the soil skeleton. In general, these coefficients, which also depend on the type of loading (isotropic, deviatoric), are functions of soil state (void ratio and stress state expressed by mean effective stress and the stress deviator). The functions should be determined experimentally for each type of non-cohesive soil on the basis of isotropic consolidation and pure shearing tests under triaxial conditions.

The presented solutions are based on an idea of a semi-empirical incremental model for non-cohesive soils proposed originally by [57], which makes it possible to predict the liquefaction of fully saturated sandy soils, whereas the idea of developing a model for partially saturated soils was presented originally by Świdziński in an internal report [58] issued in 2015 and then published in the form of a research paper [38]. In general, the model can predict the full spectrum of soil behaviors shown in Figure 1 (from contractive to dilative ones); however, in this paper we focused on analyzing and theoretically reproducing the liquefaction phenomenon, therefore only the contractive state is analyzed. The basic assumptions of the model proposed are detailed below.

Volumetric strains of soil skeleton, ε_v, under triaxial compression conditions are dependent on mean effective stress and the stress deviator, q.

$$\varepsilon_v = f(p', q), \tag{4}$$

or alternatively

$$\varepsilon_v = f(p', \eta), \tag{5}$$

where η is the stress ratio expressed by the following equation:

$$\eta = \frac{q}{p'}. \tag{6}$$

The stress ratio, η, is a useful and convenient variable that makes it possible to define precisely the loading and unloading process in the stress space.

Functions describing strain–stress dependence (Equations (4) and (5)) take different forms for loading and unloading. Consequently, with a change in load direction, respective parameters have to be adopted.

Stress invariants $p\prime$ or η and the strain invariant ε_v are linked by functions of soil skeleton compressibility [38], where: $\kappa_s^{p'}$ corresponds to isotropic compression (η = const.), denoted as $\kappa_s^{p',l}$ under loading conditions and $\kappa_s^{p',un}$ for unloading, and κ_s^{η} corresponds to pure shearing ($p\prime$ = const.), denoted as $\kappa_s^{\eta,l}$, $\kappa_s^{\eta,un}$ under loading and unloading conditions, respectively, which may be formally written as:

$$\frac{\partial \varepsilon_v}{\partial p'} = \kappa_s^{p',l} \quad \partial p' > 0 \quad \text{or} \quad \frac{\partial \varepsilon_v}{\partial p'} = \kappa_s^{p',un} \quad \partial p' < 0, \tag{7}$$

$$\frac{\partial \varepsilon_v}{\partial \eta} = \kappa_s^{\eta,l} \quad \partial \eta > 0 \quad \text{or} \quad \frac{\partial \varepsilon_v}{\partial \eta} = \kappa_s^{\eta,un} \quad \partial \eta < 0 \tag{8}$$

Applying Equation (6), the change in the volumetric strain of the soil skeleton, ε_v, can be formally presented in the following form:

$$\mathrm{d}\varepsilon_v = \left(\frac{\partial \varepsilon_v}{\partial p'}\right)\mathrm{d}p' + \left(\frac{\partial \varepsilon_v}{\partial \eta}\right)\mathrm{d}\eta. \tag{9}$$

Equation (9) is similar in its form to two incremental equations proposed by Sawicki [59] describing the development of volumetric and deviatoric deformations in dry sand through the superposition of the effects of stress tensor invariants: deviator and isotropic stress. Partial derivatives in Equation (9) can be replaced by corresponding functions (recall Equations (7) and (8)) and formally rewritten as:

$$\mathrm{d}\varepsilon_v = \kappa_s^{p'}\mathrm{d}p' + \kappa_s^{\eta}\mathrm{d}\eta. \tag{10}$$

In the present case of partially saturated soils, it is assumed that the condition regarding the equality of potential changes in the volume of the soil skeleton structure and voids remains valid:

$$\Delta V_p = \Delta V_s, \tag{11}$$

where V_p is the volume of pores and V_s is the total volume of the soil skeleton structure.

Using a definition of soil porosity in the form:

$$V_p = nV_s, \tag{12}$$

the corresponding increments of pore fluid and soil skeleton volumetric strains $\mathrm{d}\varepsilon_v^f$, $\mathrm{d}\varepsilon_v^s$, respectively, are

$$\mathrm{d}\varepsilon_v^f = -\frac{\Delta V_p}{V_p} = -n\frac{\Delta V_s}{V_s}, \tag{13}$$

$$\mathrm{d}\varepsilon_v^s = -\frac{\Delta V_s}{V_s}. \tag{14}$$

Combining Equations (11)–(13), we obtain:

$$\mathrm{d}\varepsilon_v = \mathrm{d}\varepsilon_v^s = n\mathrm{d}\varepsilon_v^f. \tag{15}$$

According to the definition of pore fluid compressibility κ_f, we can write:

$$\mathrm{d}\varepsilon_v^f = \kappa_f\mathrm{d}u. \tag{16}$$

A general model equation describing changes in pore pressure caused by triaxial monotonic loading under undrained conditions is obtained by substituting Formulas (15) and (16) into Equation (10):

$$n\kappa_f\mathrm{d}u = \kappa_s^{p'}\mathrm{d}p' + \kappa_s^{\eta}\mathrm{d}\eta. \tag{17}$$

Equation (17) contains three variables and cannot be solved by numerical methods. However, in the case of a standard triaxial compression stress path, when a constant cell pressure is maintained, we can easily eliminate one of the unknowns [38].

Stress deviator and mean total stress are defined by total principal stresses σ_1 and σ_3 as follows:

$$q = \sigma_1 - \sigma_3, \tag{18}$$

$$p = \frac{1}{3}(\sigma_1 + 2\sigma_3), \tag{19}$$

and in the present case:

$$\sigma_3 = \text{const.} \rightarrow \mathrm{d}\sigma_3 = 0 \tag{20}$$

Thus, the corresponding increments of stress invariants can be written as:

$$\mathrm{d}q = \mathrm{d}\sigma_1, \tag{21}$$

$$\mathrm{d}p = \frac{1}{3}\mathrm{d}\sigma_1, \tag{22}$$

and finally, we obtain:

$$\mathrm{d}p = \frac{1}{3}\mathrm{d}q. \tag{23}$$

After transforming and differentiating Equation (6), we obtain:

$$\mathrm{d}q = \mathrm{d}(\eta \cdot p') = \mathrm{d}\eta p' + \mathrm{d}p'\eta. \tag{24}$$

On the other hand, the transformation and differentiation of Terzaghi's fundamental equation linking effective and total stresses (Equation (1)) leads to the relationship:

$$\mathrm{d}u = \mathrm{d}p - \mathrm{d}p'. \tag{25}$$

Next, after substituting Equations (23) and (24) into Equation (25), we obtain the dependence of the pore pressure increment on just two unknowns from Equation (17):

$$\mathrm{d}u = \frac{1}{3}(\mathrm{d}\eta p' + \mathrm{d}p'\eta) - \mathrm{d}p'. \tag{26}$$

Finally, substituting Equation (26) to Equation (17) and rearranging it, we obtain the following expression:

$$\frac{\mathrm{d}p'}{\mathrm{d}\eta} = \frac{n\kappa_f p' - 3\kappa_s^{\eta}}{3\left(\kappa_s^{p'} + n\kappa_f\right) - n\kappa_f\eta}. \tag{27}$$

Integration of Equation (27) makes it possible to reproduce a stress path in the (p',η), stress space from which one can easily go to the (p',q) stress space through Relationship (6). In the same manner, we can also solve Equation (17) for an arbitrary stress path defined by any σ_3-to-σ_1 stress ratio.

The originally proposed theoretical model predictions [38], confronted with experimental data, yielded a very good qualitative agreement of the corresponding stress paths for the same B values, but a much worse quantitative fit, especially for low values of Skempton's parameter B, for which the model strongly underestimated the results of the experimental studies.

In order to improve the model and thus obtain better quantitative predictions for the whole range of parameter B, some modifications to the compressibility functions were made.

3.2. Pore Fluid Compressibility Function, κ_f

The volumetric compressibility of pore fluid, which is a mixture of water and air contained in voids, is defined as [60]:

$$\kappa_f = (1 - S_r)\kappa_a + S_r\kappa_w, \tag{28}$$

where κ_a and κ_w are, respectively, the compressibilities of air and water.

Pore air can be treated as a perfect gas, so, assuming no temperature changes, its compressibility can be described by the formula:

$$\kappa_a = \frac{1}{u_c}, \tag{29}$$

where $u_c = u + u_{\text{atm}}$ is total pressure.

Water is very poorly deformable; its compressibility under atmospheric pressure is $\kappa_w = 0.45 \times 10^{-9}\ \text{Pa}^{-1}$ and does not show significant changes with its increase, so it can be treated as a material constant.

In laboratory tests, Skempton's parameter B was used to determine the saturation level of the soil because of its easy measurement during triaxial compression tests, in which a direct determination of saturation is extremely difficult. Since parameter B is not a physical measure nor is it useful in numerical calculations, Relationship (30) was used to determine the initial degree of saturation S_r^0:

$$S_r^0 = \frac{\frac{(1-B)}{nB}\kappa_s^{p'} - \kappa_a}{\kappa_w - \kappa_a}. \tag{30}$$

Formula (30) was obtained by substituting Equation (3) into (28), whereas many authors use the inverse relationship in their studies, e.g., [61–63]. In the above form, the formula makes it possible to determine the saturation level on the basis of Skempton's parameter B, porosity and the isotropic compressibilities of soil skeleton, pore water and pore air. Please note that the effect of solubility has a significant influence on the compressibility of the water–air mixture [49] described by Equation (28). The approach presented to determine the degree of saturation based on an experimentally determined Skempton's parameter B also takes into account the change in pore fluid compressibility due to dissolved air.

In the course of a change in pore pressure, the volume of pores, V_p, also changes, which consequently changes the degree of saturation. Based on the definition of the degree of saturation: $S_r = V_w/V_p$ and using Equations (13) and (16) for the pore voids and the analogical set for pore water, current soil saturation can be expressed in terms of the compressibility of pore fluid components according to the following equation:

$$S_r = S_r^0 \frac{(1 - \kappa_w \Delta u)}{\left(1 - \kappa_f \Delta u\right)}. \tag{31}$$

In a similar manner, by combining the definitions of soil porosity and compressibility of the pore fluid and keeping in mind the constant volume of the soil skeleton we can also easily determine changes in porosity caused by pore fluid compression during the test, which can be described as:

$$n = \frac{\left(1 - \kappa_f \Delta u\right)}{\left(1 - \kappa_f \Delta u\right) + \frac{1-n_0}{n_0}}. \tag{32}$$

3.3. Soil Skeleton Compressibility Functions

In addition to soil porosity and pore fluid compressibility, the basic parameters of the model, there are two functions of soil skeleton compressibility, i.e., the isotropic function, $\kappa_s^{p'}$, and the deviatoric function, κ_s^{η}, whose coefficients should be determined in triaxial tests for soils characterised by similar state parameters.

The function of deviatoric compressibility is determined in pure shearing tests under triaxial conditions during which mean effective stress is kept constant. For dry post-flotation copper tailings OZM50 in contractive state, volumetric strains as a function of stress ratio can be approximated by the following function (Figure 15):

$$\varepsilon_{\mathrm{v}} = D_1 \exp[D_2(\eta - \eta_{\mathrm{CM}})], \tag{33}$$

where η_{CM} corresponds to the Coulomb–Mohr yield surface.

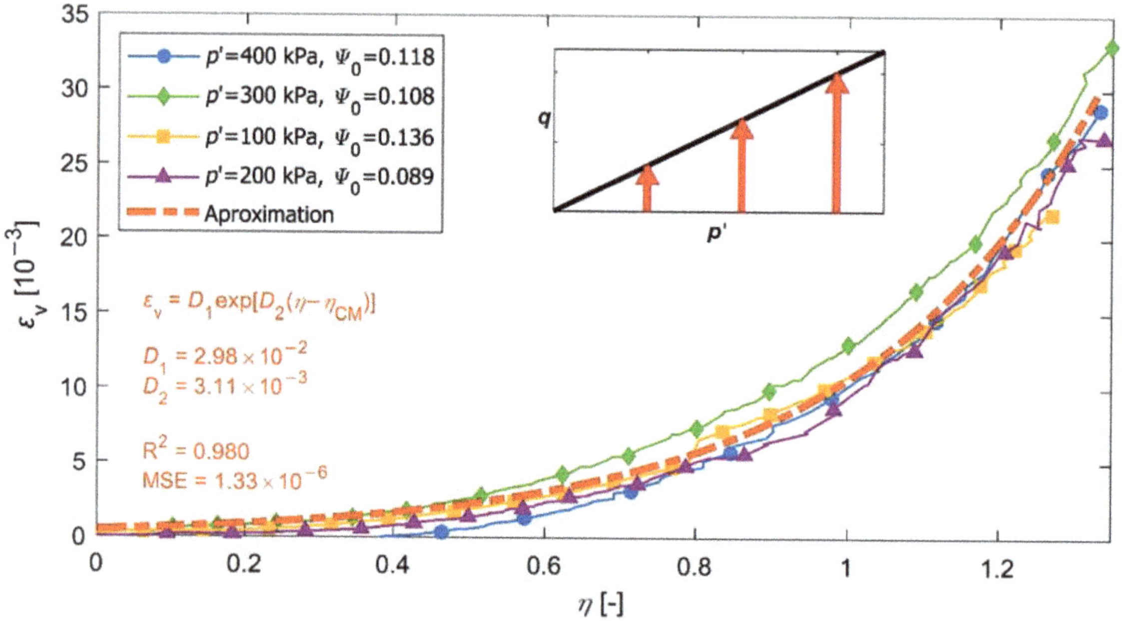

Figure 15. Volumetric strains versus stress ratio during pure shearing under triaxial conditions for contractive OZM50 tailings.

The coefficients in Function (33), estimated by the least squares method for OZM50 tailings, are $D_1 = 2.98 \times 10^{-2}$ and $D_2 = 3.11 \times 10^{-3}$.

According to Equations (8) and (33), the deviatoric compressibility function can take the following form:

$$\kappa_{\mathrm{s}}^{\eta,l} = \frac{\partial \varepsilon_{\mathrm{v}}}{\partial \eta} = D_1 D_2 \exp[D_2(\eta - \eta_{\mathrm{CM}})]. \tag{34}$$

In turn, the isotropic compressibility function can be determined in isotropic consolidation tests. Figure 16 shows volumetric strains that developed during isotropic compression applied to a specimen of the same post-flotation tailings OZM50. In this case, they can be nicely approximated by the logarithmic function:

$$\varepsilon_v = A_1 \ln(1 + A_2 p'), \tag{35}$$

whose coefficients estimated by the method of least squares are $A_1 = 2.97 \times 10^{-2}$ and $A_2 = 6.7 \times 10^{-3}$.

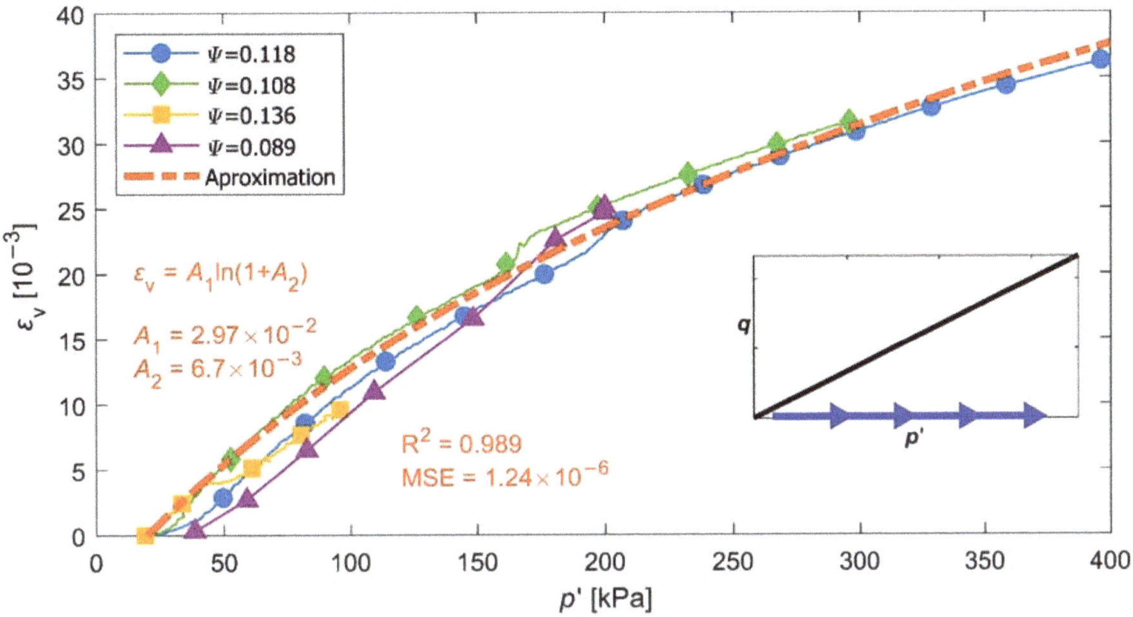

Figure 16. Volumetric strains developed during isotropic consolidation in OZM50 tailings.

The isotropic compressibility function is a derivative of the volumetric strains function, this time calculated with respect to mean effective stress:

$$\kappa_s^{p',l} = \frac{\partial \varepsilon_v}{\partial p'} = \frac{A_1 A_2}{1 + A_2 p'}. \tag{36}$$

Under undrained conditions, the contractive response of sheared soil manifests itself in a reduction of mean effective stress as a result of pore pressure build-up. In the primary model, the reduction of mean effective stress which corresponds to unloading was not formally taken into account but it has quite a significant influence on its predictions. Therefore, it was necessary to differentiate between isotropic compressibility for loading and unloading and to introduce into the model the isotropic compressibility under unloading, $\kappa_s^{p',un}$, which should also be determined by laboratory tests.

The results of an isotropic unloading test on contractive OZM50 tailings are shown in Figure 17. In order to approximate volumetric strains, which decrease during isotropic unloading, an exponential function was adopted:

$$\varepsilon_v = A_1^{un} p'^{A_2^{un}} + \text{const}. \tag{37}$$

The coefficients estimated by the method of least squares are $A_1^{un} = -2.23 \times 10^{-2}$ and $A_2^{un} = -1.92 \times 10^{-1}$.

Then the function of isotropic compressibility under unloading takes the following form:

$$\kappa_s^{p',un} = \frac{\partial \varepsilon_v}{\partial p'} = A_1^{un} A_2^{un} p'^{(A_2^{un}-1)}. \tag{38}$$

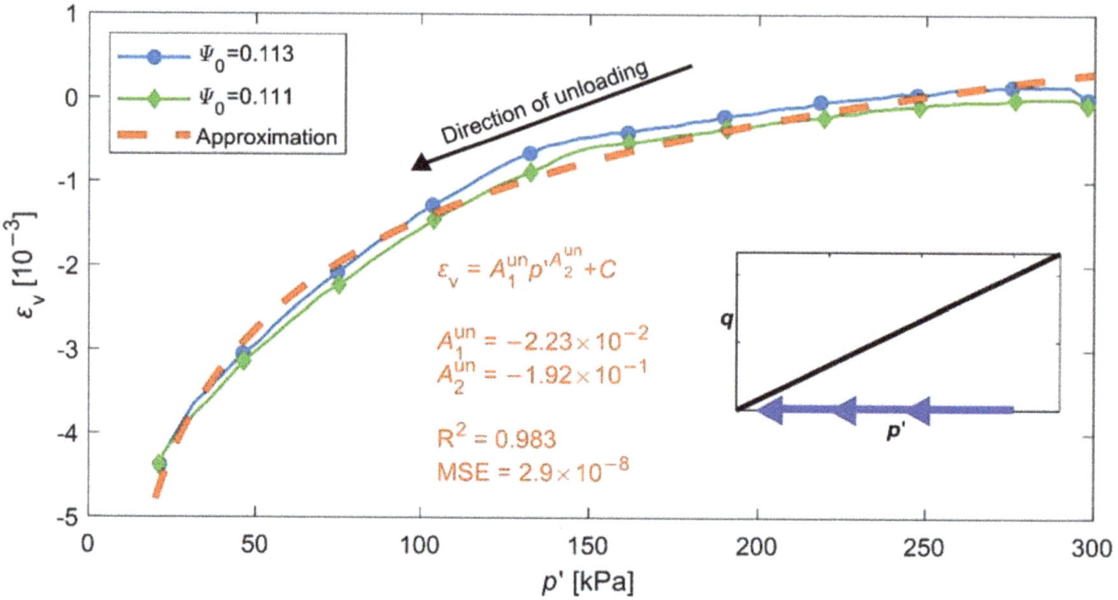

Figure 17. Volumetric strains developing during isotropic unloading in OZM50 tailings.

3.4. Predicted Stress Paths

Before stress path integration, the following initial conditions have been defined: the degree of saturation calculated from the value of Skempton's parameter B (Equation (30)), the value of the stress ratio, η (in case of isotropic consolidation $\eta_0 = 0$), porosity, mean effective stress and back pressure, with corresponding compressibilites of soil skeleton and pore fluid. Equation (27) can be solved numerically, assuming an incremental step of the stress ratio, $\Delta\eta$. For all cases presented in the paper the increment of $\Delta\eta = 0.001$ has been used, ensuring the stability of the solution. At the beginning of a single step increment, the mean effective stress corresponding to the current stress ratio η is calculated by using Equation (27). Afterwards, the stress deviator and excess pore pressure are calculated by employing Equations (24) and (26). At the end of a given step, compressibilities, porosity and degree of saturation are updated.

Integration of Equation (27) for different values of the B parameter enables the coverage of stress space (p',q) by corresponding stress paths. Under partial saturation conditions (see Figure 2), stress paths are located between responses characteristic of a fully saturated soil and a completely dry soil. Figures 18 and 19 show the distribution of stress paths predicted by the model for different degrees of saturation and the corresponding Skempton's B parameters for OZM50 tailings and Skarpa sand, respectively.

Predicted changes of degree of saturation (Equation (31)) and porosity (Equation (32)) corresponding to the model simulations shown in Figures 18 and 19 were presented in Figures 20 and 21, respectively.

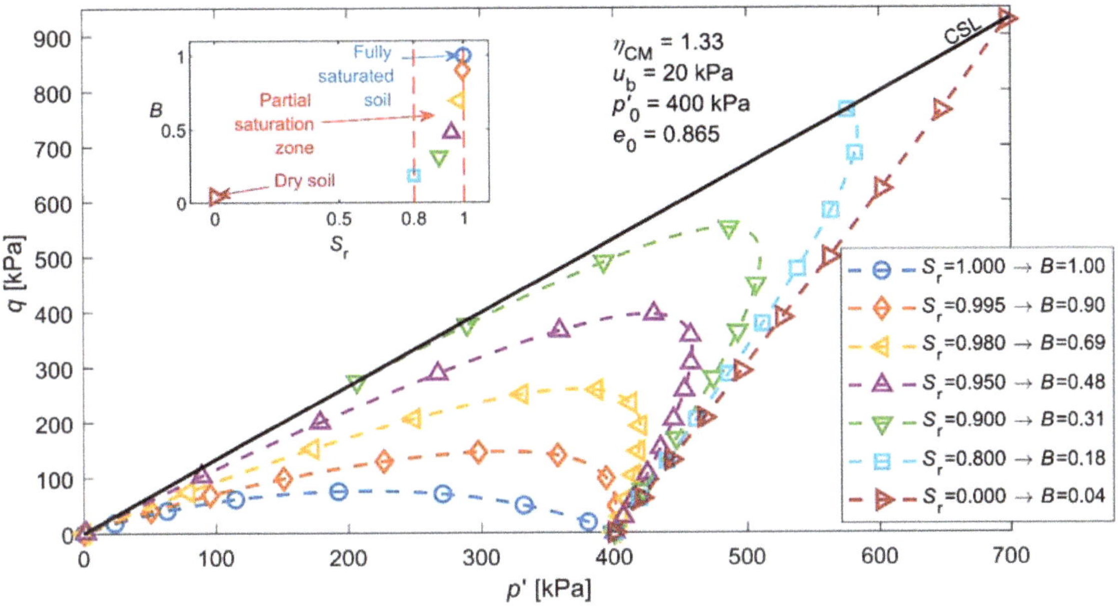

Figure 18. Theoretical prediction of stress paths for different saturations levels for OZM50 tailings.

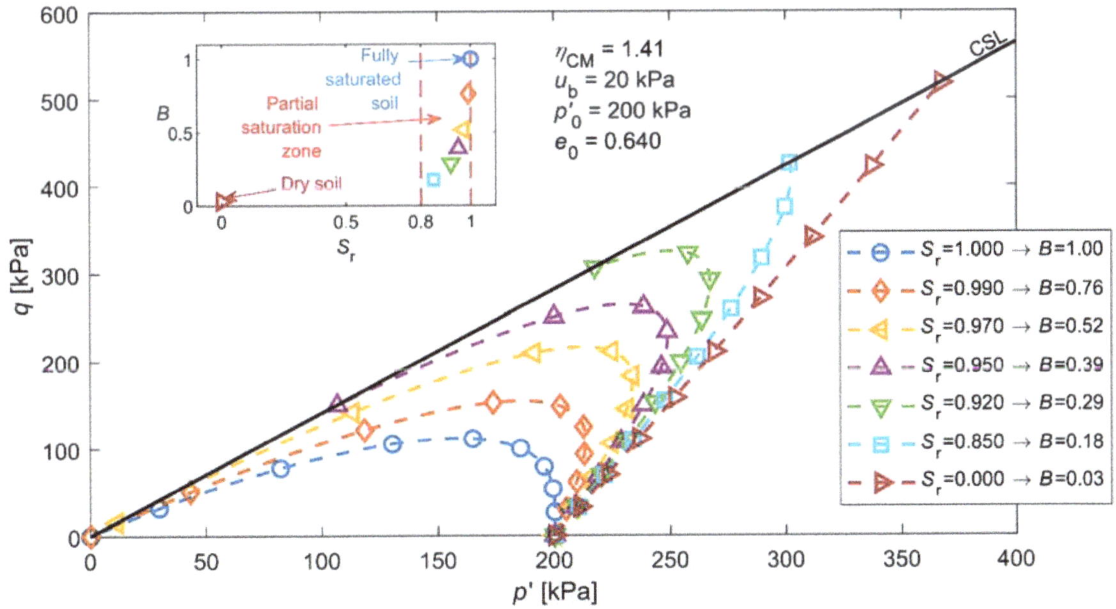

Figure 19. Theoretical prediction of stress paths for different saturations levels for Skarpa sand.

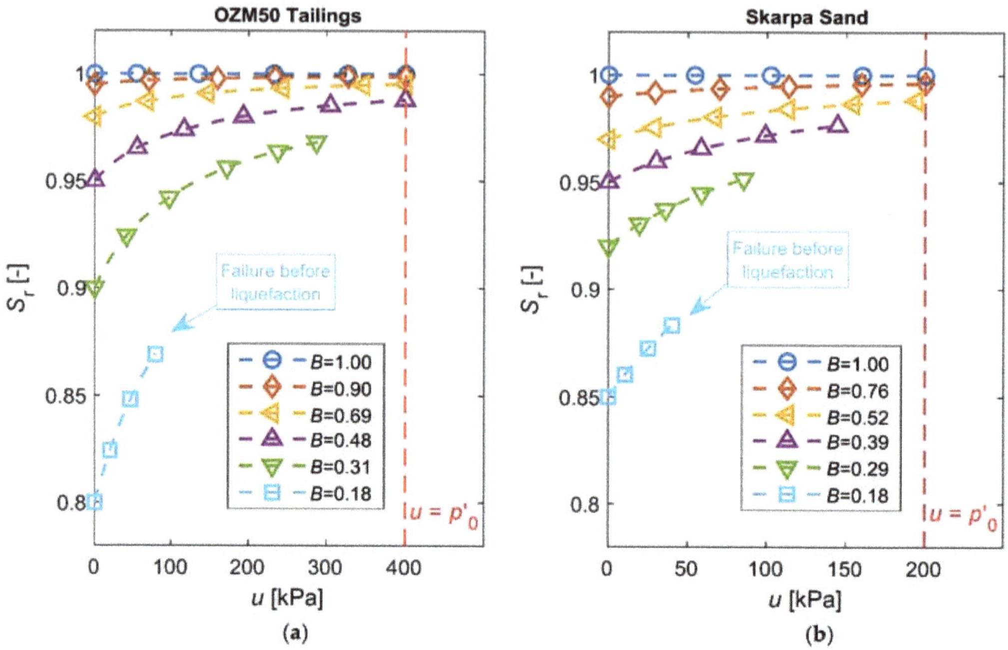

Figure 20. Model prediction of changes in degree of saturation during shearing presented in: (**a**) Figure 18 and (**b**) Figure 19.

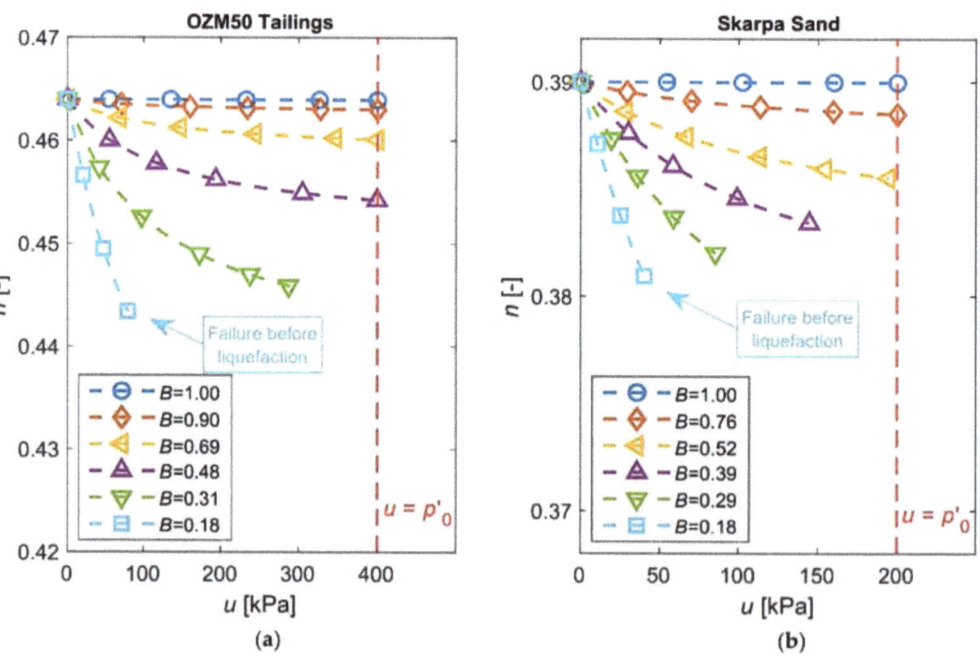

Figure 21. Model prediction of changes in porosity corresponding to simulations presented in: (**a**) Figure 18 and (**b**) Figure 19.

4. Verification of the Model

In order to verify the correctness of the model, its predictions were confronted with the results of laboratory tests (presented in Section 2.2).

The functions of soil skeleton compressibility were determined in the laboratory (for OZM50 tailings, see Section 3.3; for Skarpa sand, see studies [57,64]). All functions used together with their coefficients are collected in Table 3.

Table 3. Functions of soil skeleton compressibility and their coefficients.

Soil	Isotropic Load	Isotropic Unload	Deviatoric Load
Skarpa sand	$\kappa_s^{p',l} = \frac{A}{2\sqrt{p'}}$ $A = 9.33 \times 10^{-4}$	$\kappa_s^{p',\mathrm{un}} = \frac{A^{\mathrm{un}}}{2\sqrt{p'}}$ $A^{\mathrm{un}} = 4.59 \times 10^{-4}$	$\kappa_s^{\eta,l} = 4D\eta^3$ $D = 2.97 \times 10^{-3}$
OZM50 tailings	$\kappa_s^{p',l} = \frac{A_1 A_2}{1+A_2 p'}$ $A_1 = 2.97 \times 10^{-2}$ $A_2 = 6.7 \times 10^{-3}$	$\kappa_s^{p',un} = A_1^{\mathrm{un}} A_2^{\mathrm{un}} p'^{(A_2^{\mathrm{un}}-1)}$ $A_1^{\mathrm{un}} = -2.23 \times 10^{-2}$ $A_2^{\mathrm{un}} = -1.92 \times 10^{-1}$	$\kappa_s^{\eta,l} = D_1 D_2 \exp[D_2(\eta - \eta_{\mathrm{CM}})]$ $D_1 = 2.98 \times 10^{-2}$ $D_2 = 3.11$

The computations were made using the standard relationship between porosity, n, and void ratio, e, and Equations (28), (30) and (31), which relate pore fluid compressibility to Skempton's parameter B and pore pressure. Datasets with initial conditions for each test are given in Table 2.

Equation (27) was numerically integrated with the stress ratio increment $\Delta\eta = 0.001$. Predictions of stress paths obtained in this way could be easily transferred from the (p',η) coordinate system to the typical stress space (p',q) using Equation (6). The results of computations and laboratory tests are summarized in Figures 22 and 23.

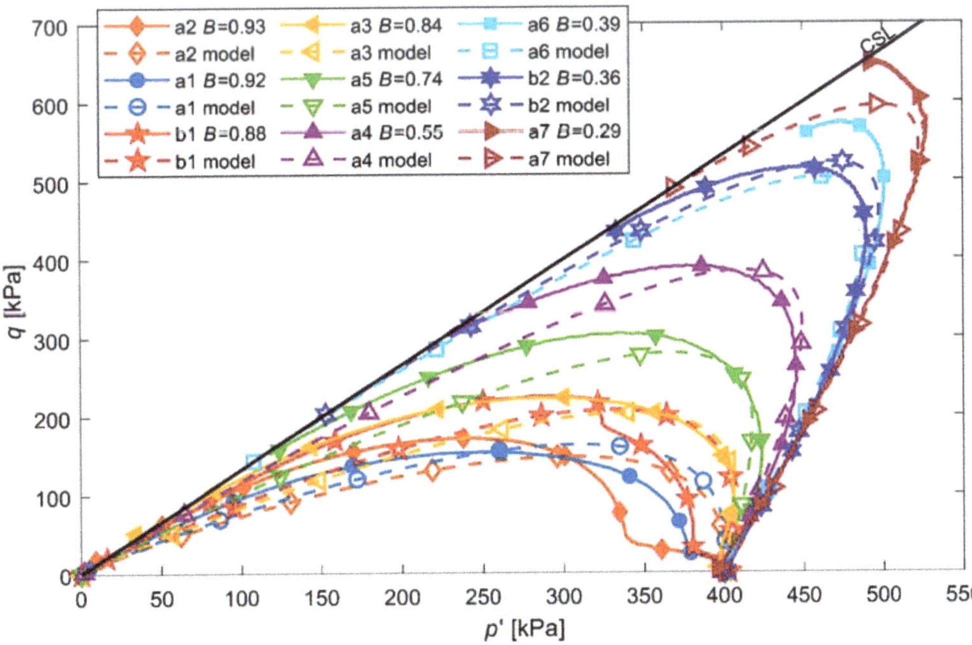

Figure 22. Theoretical predictions of triaxial compression tests of partially saturated OZM50 tailings vs. experimental results.

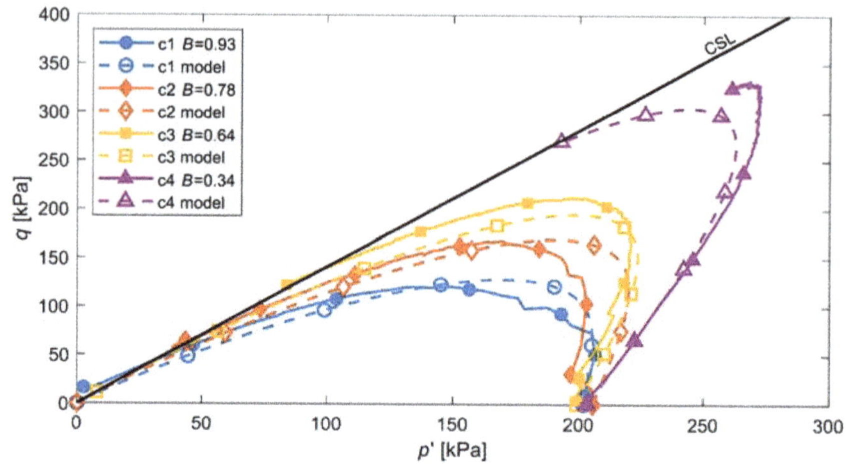

Figure 23. Theoretical predictions of triaxial compression tests of partially saturated Skarpa sand vs. experimental results.

For both types of soil, the model predicts stress paths quite well, both qualitatively and quantitatively. To quantify the prediction accuracy of the model, the maximum deviators obtained in the laboratory tests were compared with those obtained theoretically. For this purpose, the percentage measure of error, ζ, was applied.

$$\zeta = \left(\frac{q_{\max}^{\text{experiment}}}{q_{\max}^{\text{model}}} - 1 \right) \cdot 100\% \tag{39}$$

The results presented in Figure 24 show that for all cases the error was less than 16.1% and the average error was 7.7%, which in the domain of soil mechanics is quite satisfactory, although in the majority of tests analysed the model somewhat overestimated the test results.

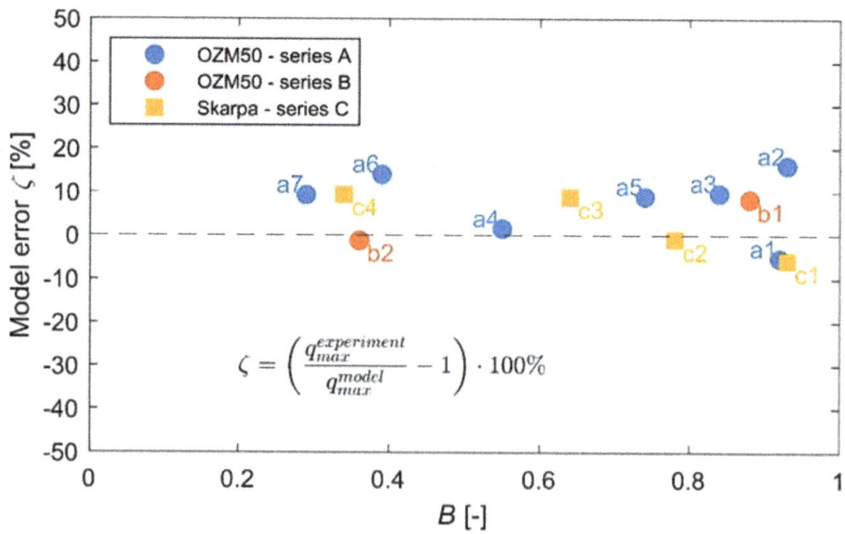

Figure 24. Prediction errors for maximum deviators.

5. Discussion and Conclusions

The results of laboratory tests presented in this paper have confirmed that the liquefaction phenomenon may also occur in partially saturated non-cohesive soils in a contractive state subjected to monotonic loading and that the potential for liquefaction decreases as saturation decreases. Furthermore, our investigations have shown that, under conditions of partial saturation characterized by positive values of Skempton's parameter B corresponding to the degree of saturation over 80% or higher, the behaviour of non-cohesive soil changes smoothly from that typical of a dry medium (or a medium under drained conditions) to that characteristic of a fully saturated medium.

Both experimental data and model predictions (Figure 25) show the decrease of undrained shear strength with the increasing saturation level expressed by Skempton's parameter B. For the analyzed saturation range, the maximum stress deviator normalized by initial mean effective stress varies from 0.38 to 1.67 for B values between 0.93 and 0.29, respectively. Furthermore, linear dependency can be observed across the whole spectrum of partially saturated states.

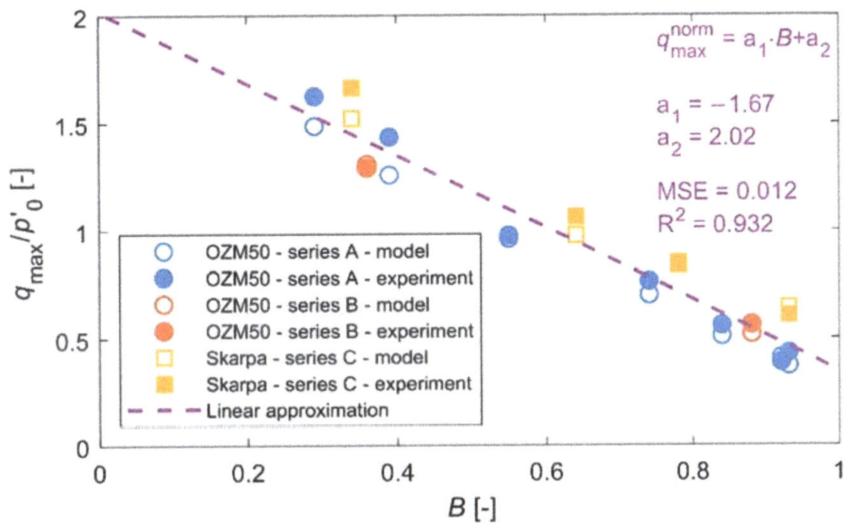

Figure 25. Linear changes of normalized maximum stress deviator with Skempton's parameter B.

In order to reproduce the test results obtained, a modified theoretical model describing the process of excess pore pressure generation in partially saturated soils has been developed. The modification of the original model consisted in extending the formula for calculating the compressibility of pore fluid using soil skeleton compressibility functions based on dedicated laboratory tests and incorporating soil response under isotropic unloading to reflect the behaviour of soil during mean effective stress reduction due to an increase of pore water pressure. The proposed model reproduces the results of triaxial tests much better than the original one.

The model has been verified for two non-cohesive soils with different granulation characteristics, showing acceptable qualitative and quantitative agreement between predicted values and experimental results.

The research reported here offers a semi-empirical incremental approach to modelling the phenomenon of liquefaction or a significant weakening of partially saturated non-cohesive soils under undrained conditions.

Supplementary Materials: The following are available online at https://www.mdpi.com/article/10.3390/app12042076/s1, Repository S1: Experimental data.

Author Contributions: Conceptualization, W.Ś.; methodology, W.Ś. and M.S.; software, M.S.; validation, M.S. and W.Ś.; formal analysis, W.Ś. and M.S.; investigation, M.S.; resources, W.Ś. and M.S.; data curation, M.S.; writing—original draft preparation, W.Ś. and M.S.; writing—review and editing, W.Ś. and M.S.; visualization, M.S.; supervision, W.Ś.; project administration, W.Ś.; funding acquisition, W.Ś. All authors have read and agreed to the published version of the manuscript.

Funding: This research received no external funding.

Institutional Review Board Statement: Not applicable.

Informed Consent Statement: Not applicable.

Data Availability Statement: The data presented in this study are available in Supplementary Materials: "Repository S1: Experimental data" and in external repository at https://doi.org/10.5281/zenodo.5774184.

Conflicts of Interest: The authors declare no conflict of interest.

References

1. Fourie, A.B.; Tshabalala, L. Initiation of static liquefaction and the role of K0 consolidation. *Can. Geotech. J.* **2005**, *42*, 892–906. [CrossRef]
2. Mele, L.; Tian, J.T.; Lirer, S.; Flora, A.; Koseki, J. Liquefaction resistance of unsaturated sands: Experimental evidence and theoretical interpretation. *Géotechnique* **2019**, *69*, 541–553. [CrossRef]
3. Mele, L.; Flora, A. On the prediction of liquefaction resistance of unsaturated sands. *Soil Dyn. Earthq. Eng.* **2019**, *125*, 105689. [CrossRef]
4. Marinelli, F.; Buscarnera, G. Instability criteria for quasi-saturated viscous soils. *Int. J. Numer. Anal. Methods Géoméch.* **2018**, *42*, 379–400. [CrossRef]
5. Sherif, M.; Ashibashi, I.; Tsuchiya, C. Saturation effect on initial soil liquefaction. *J. Geotech. Eng. Div.* **1977**, *103*, 914–917. [CrossRef]
6. Yoshimi, Y.; Tanaka, K.; Tokimatsu, K. Liquefaction Resistance of a Partially Saturated Sand. *Soils Found.* **1989**, *29*, 157–162. [CrossRef]
7. Arab, A.; Shahrour, I.; Lancelot, L. A laboratory study of liquefaction of partially saturated sand. *J. Iber. Geol.* **2011**, *37*, 29–36.
8. Ishihara, K.; Tsuchiya, H.; Huang, Y.; Kamada, K. Recent studies on liquefaction resistance of sand-effect of saturation. In Proceedings of the 4th International Conferences on Recent Advances in Geotechnical Earthquake Engineering and Soil Dynamics, San Diego, CA, USA, 26–31 March 2001.
9. Okamura, M.; Soga, Y. Effects of Pore Fluid Compressibility on Liquefaction Resistance of Partially Saturated Sand. *Soils Found.* **2006**, *46*, 695–700. [CrossRef]
10. Tsukamoto, Y.; Kawabe, S.; Matsumoto, J.; Hagiwara, S. Cyclic resistance of two unsaturated silty sands against soil liquefaction. *Soil Found.* **2014**, *54*, 1094–1103. [CrossRef]
11. Unno, T.; Kazama, M.; Uzuoka, R.; Sento, N. Liquefaction of Unsaturated Sand Considering the Pore Air Pressure and Volume Compressibility of the Soil Particle Skeleton. *Soils Found.* **2008**, *48*, 87–99. [CrossRef]
12. Wang, H.; Koseki, J.; Sato, T.; Chiaro, G.; Tian, J.T. Effect of saturation on liquefaction resistance of iron ore fines and two sandy soils. *Soils Found.* **2016**, *56*, 732–744. [CrossRef]
13. Xia, H.; Hu, T. Effects of Saturation and Back Pressure on Sand Liquefaction. *J. Geotech. Eng.* **1991**, *117*, 1347–1362. [CrossRef]
14. Arab, A.; Belkhatir, M.; Sadek, M. Saturation effect on behaviour of sandy soil under monotonic and cyclic loading: A laboratory investigation. *Geotech. Geol. Eng.* **2016**, *34*, 347–358. [CrossRef]
15. Della, N.; Arab, A. Laboratory investigation on the saturation and initial structure effects on the undrained behavior of granular soil under static loading. *Acta Polytech. Hung.* **2010**, *7*, 143–160.
16. He, J.; Chu, J.; Liu, H. Undrained shear strength of desaturated loose sand under monotonic shearing. *Soils Found.* **2014**, *54*, 910–916. [CrossRef]
17. Ishihara, K.; Tsukamoto, Y.; Kamada, K. Undrained behaviour of near-saturated sand in cyclic and monotonic loading. In *Cyclic Behaviour of Soils and Liquefaction Phenomena*, 1st ed.; Triantafyllidis, T., Ed.; CRC Press: New York, NY, USA, 2004; pp. 27–39.
18. Kamata, T.; Tsukamoto, Y.; Ishihara, K. Undrained shear strength of partially saturated sand in triaxial tests. *Bull. N. Z. Soc. Earthq. Eng.* **2009**, *42*, 57–62. [CrossRef]
19. Maleki, M.; Bayat, M. Experimental evaluation of mechanical behavior of unsaturated silty sand under constant water content condition. *Eng. Geol.* **2012**, *141–142*, 45–56. [CrossRef]
20. Bian, H.; Shahrour, I. Numerical model for unsaturated sandy soils under cyclic loading: Application to liquefaction. *Soil Dyn. Earthq. Eng.* **2009**, *29*, 237–244. [CrossRef]

21. Kimoto, S.; Ishikawa, R.; Akaki, T. Behavior of unsaturated sandy soil during triaxial compression tests under fully undrained conditions and its modelling. In Proceedings of the 19th International Conference on Soil Mechanics and Geotechnical Engineering, Seoul, Korea, 17–22 September 2017; pp. 1187–1190.
22. Duan, X.; Zeng, L.; Sun, X. Generalized stress framework for unsaturated soil: Demonstration and discussion. *Acta Geotech.* **2018**, *14*, 1459–1481. [CrossRef]
23. Gens, A.; Sánchez, M.; Sheng, D. On constitutive modelling of unsaturated soils. *Acta Geotech.* **2006**, *1*, 137–147. [CrossRef]
24. Ng, C.W.W.; Zhou, C.; Chiu, C.F. Constitutive modelling of state-dependent behaviour of unsaturated soils: An overview. *Acta Geotech.* **2020**, *15*, 2705–2725. [CrossRef]
25. Chen, C.; Wang, Y.; Zhang, X.; Kong, L.; Xu, G. Numerical modelling of gassy sand behaviour under monotonic loading. *Acta Geotech.* **2021**, 1–14. [CrossRef]
26. Lü, X.; Huang, M.; Andrade, J.E. Modeling the static liquefaction of unsaturated sand containing gas bubbles. *Soils Found.* **2018**, *58*, 122–133. [CrossRef]
27. Yang, G.; Bai, B. A thermodynamic model to simulate the thermomechanical behaviour of fine-grained gassy soil. *Bull. Eng. Geol. Environ.* **2020**, *79*, 2325–2339. [CrossRef]
28. Yang, G.; Bai, B.; Liu, Y.; Chen, P. Constitutive modelling for undrained shear behaviour of gassy sand considering energy dissipation at the mesoscopic scale. *Ocean. Eng.* **2021**, *219*, 108307.
29. Zapata-Medina, D.; Vergara, C.Y.; Vega-Posada, C.A.; Arboleda-Monsalve, L.G. On the use of Fredlund gas–fluid compressibility relationship to model medium-dense gassy sand behavior. *Can. Geotech. J.* **2019**, *56*, 1070–1079. [CrossRef]
30. Mihalache, C.; Buscarnera, G. Controllability Criteria for Soils Saturated by a Compressible Fluid. *J. Eng. Mech.* **2016**, *142*, 4016076. [CrossRef]
31. Shi, Z.; Tong, S.; Huang, M. Evaluation of instability of quasi saturated sand with entrapped gas due to decreasing total confining stress. *Eng. Geol.* **2021**, *293*, 106296. [CrossRef]
32. Zhou, M.-Z.; Jeng, D.-S.; Qi, W.-G. A new model for wave-induced instantaneous liquefaction in a non-cohesive seabed with dynamic permeability. *Ocean Eng.* **2020**, *213*, 107597. [CrossRef]
33. Hsu, J.R.C.; Jeng, D.S.; Lee, C.P. Oscillatory soil response and liquefaction in an unsaturated layered seabed. *Int. J. Numer. Anal. Methods Géoméch.* **1995**, *19*, 825–849. [CrossRef]
34. Gallant, A.P.; Finno, R.J. Measurement of Gas Released during Blast Densification. *Geotech. Test. J.* **2017**, *40*, 1011–1025. [CrossRef]
35. Gudehus, G. Mechanisms of partly flooded loose sand deposits. *Acta Geotech.* **2016**, *11*, 505–517. [CrossRef]
36. Kido, R.; Higo, Y.; Takamura, F.; Morishita, R.; Khaddour, G.; Salager, S. Morphological transitions for pore water and pore air during drying and wetting processes in partially saturated sand. *Acta Geotech.* **2020**, *15*, 1745–1761. [CrossRef]
37. Świdziński, W.; Mierczyński, J.; Mikos, A. Response of partially saturated non-cohesive soils. *Arch. Hydro-Eng. Environ. Mech.* **2017**, *64*, 187–207. [CrossRef]
38. Świdziński, W.; Mierczyński, J.; Smyczyński, M. Modelling of the response of partially saturated non-cohesive soil subjected to undrained loading. *Arch. Hydro-Eng. Environ. Mech.* **2018**, *65*, 11–29. [CrossRef]
39. Jefferies, M.; Been, K. *Soil Liquefaction: A Critical State Approach*, 2nd ed.; CRC Press: New York, NY, USA, 2015.
40. Been, K.; Jefferies, M.G. A state parameter for sands. *Géotechnique* **1985**, *35*, 99–112. [CrossRef]
41. Lashkari, A. A simple critical state interface model and its application in prediction of shaft resistance of non-displacement piles in sand. *Comput. Geotech.* **2017**, *88*, 95–110. [CrossRef]
42. Gao, Z.; Zhao, J. Constitutive Modeling of Anisotropic Sand Behavior in Monotonic and Cyclic Loading. *J. Eng. Mech.* **2015**, *141*, 4015017. [CrossRef]
43. Nguyen, H.B.K.; Rahman, M.; Fourie, A. The critical state behaviour of granular material in triaxial and direct simple shear condition: A DEM approach. *Comput. Geotech.* **2021**, *138*, 104325. [CrossRef]
44. Wood, M. *Soil Behaviour and Critical State Soil Mechanics*; Cambridge University Press: Cambridge, UK, 1991.
45. Poulos, S.J. The Steady State of Deformation. *J. Geotech. Geoenvironmental Eng.* **1981**, *107*, 501–516. [CrossRef]
46. Woo, S.I.; Salgado, R. Bounding surface modeling of sand with consideration of fabric and its evolution during monotonic shearing. *Int. J. Solids Struct.* **2015**, *63*, 277–288. [CrossRef]
47. Nguyen, H.; Rahman, M.; Fourie, A. How particle shape affects the critical state and instability triggering of granular material: Results from a DEM study. *Geotéchnique* **2021**, *71*, 749–764. [CrossRef]
48. Castro, G.; Poulos, S.J. Factors Affecting Liquefaction and Cyclic Mobility. *J. Geotech. Eng. Div.* **1977**, *103*, 501–516. [CrossRef]
49. Fredlund, D.G.; Rahardjo, H.; Fredlund, M.D. *Unsaturated Soil Mechanics in Engineering Practice*; John Wiley & Sons: New York, NY, USA, 2012.
50. Moscariello, M.; Cuomo, S.; Salager, S. Capillary collapse of loose pyroclastic unsaturated sands characterized at grain scale. *Acta Geotech.* **2017**, *13*, 117–133. [CrossRef]
51. Yang, H.; Rahardjo, H.; Leong, E.; Fredlund, D.G. Factors affecting drying and wetting soil-water characteristic curves of sandy soils. *Can. Geotech. J.* **2004**, *41*, 908–920. [CrossRef]
52. Finno, R.J.; Zhang, Y.; Buscarnera, G. Experimental Validation of Terzaghi's Effective Stress Principle for Gassy Sand. *J. Geotech. Geoenvironmental Eng.* **2017**, *143*, 4017092. [CrossRef]
53. Skempton, A.W. The Pore-Pressure Coefficients A and B. *Géotechnique* **1954**, *4*, 143–147. [CrossRef]

54. Smyczyński, M. Reakcje nawodnionych gruntów niespoistych o niepełnym nasyceniu w warunkach bez odpływu wody z porów. Responses of Partially Saturated Non-Cohesive Soils under Undrained Conditions. Ph.D. Thesis, Institute of Hydro-Engineering of Polish Academy of Sciences, Gdańsk, Poland, 2019. (In Polish).
55. Świdziński, W.; Mierczyński, J. Instability line as a basic characteristic of non-cohesive soils. *Arch. Hydro-Eng. Environ. Mech.* **2005**, *52*, 59–85.
56. Świdziński, W. *Instability of Post-Flotation Tailing Ponds due to Static Liquefaction;* Internal Report Falling into Commercial Restrictions; Institute of Hydro-Engineering of Polish Academy of Sciences: Gdańsk, Poland, 2011. (In Polish)
57. Sawicki, A.; Świdziński, W. Stress-Strain Relations for Dry and Saturated Sands. Part I: Incremental Model. *J. Theor. Appl. Mech.* **2010**, *48*, 309–328.
58. Świdziński, W. *Teoretyczny opis Reakcji Nawodnionego Piasku Poddanego Ścinaniu w Warunkach Częściowego Nasycenia. Theoretical Description of the Reaction of Sand Exposed on Sheared in Partial Saturation Conditions;* Institute of Hydro-Engineering of Polish Academy of Sciences: Gdańsk, Poland, 2015. (In Polish)
59. Sawicki, A. 3D and 2D formulations of incremental stress-strain relations for granular soils. *Arch. Hydro-Eng. Environ. Mech.* **2008**, *55*, 45–53.
60. Verruijt, A. Elastic storage of aquifers. In *Flow through Porous Media;* De Wiest, R.J.M., Ed.; Academic Press: New York, NY, USA, 1969; pp. 331–376.
61. He, J.; Chu, J. Undrained Responses of Microbially Desaturated Sand under Monotonic Loading. *J. Geotech. Geoenvironmental Eng.* **2014**, *140*, 4014003. [CrossRef]
62. Lade, P.V.; Hermamdez, S.B. Membrane Penetration Effects in Undrained Tests. *J. Geotech. Eng. Div.* **1977**, *103*, 109–125. [CrossRef]
63. Yang, J. Liquefaction resistance of sand in relation to P-wave velocity. *Géotechnique* **2002**, *52*, 295–298. [CrossRef]
64. Sławińska-Budzich, J.; Mierczyński, J. Deformations and stability of granular soils: Classical triaxial tests and numerical results from an incremental model. *Studia Geotech. Et Mech.* **2020**, *42*, 137–150. [CrossRef]

applied sciences

MDPI

Article

Measurement-While-Drilling Based Estimation of Dynamic Penetrometer Values Using Decision Trees and Random Forests

Eduardo Martínez García [1,2], Marcos García Alberti [1,*] and Antonio Alfonso Arcos Álvarez [3]

1 Departamento de Ingeniería Civil: Construcción, E.T.S de Ingenieros de Caminos, Canales y Puertos, Universidad Politécnica de Madrid, 28040 Madrid, Spain; emartinez@menard.es
2 Menard España, 28001 Madrid, Spain
3 Departamento de Ingeniería y Morfología del Terreno, E.T.S de Ingenieros de Caminos, Canales y Puertos, Universidad Politécnica de Madrid, 28040 Madrid, Spain; antonio.arcos@upm.es
* Correspondence: marcos.garcia@upm.es; Tel.: +34-91-0674121

Abstract: Machine learning is a branch of artificial intelligence (AI) that consists of the application of various algorithms to obtain information from large data sets. These algorithms are especially useful to solve nonlinear problems that appear frequently in some engineering fields. Geotechnical engineering presents situations with complex relationships of multiple variables, making it an ideal field for the application of machine learning techniques. Thus, these techniques have already been applied with a certain degree of success to determine such things as soil parameters, admissible load, settlement, or slope stability. Moreover, dynamic penetrometers are a very common type of test in geotechnical studies, and, in many cases, they are used to design the foundation solution. In addition, its continuous nature allows us to know the variations of the terrain profile. The objective of this study was to correlate the drilling parameters of deep foundation machinery (Measurement-While-Drilling, MWD) with the number of blows of the dynamic penetrometer test. Therefore, the drilling logs could be equated with said tests, providing information that can be easily interpreted by a geotechnical engineer and that would allow the validation of the design hypotheses. Decision trees and random forest algorithms have been used for this purpose. The ability of these algorithms to replicate the complex relationships between drilling parameters and terrain characteristics has allowed obtaining a reliable reproduction of the penetrometric profile of the traversed soil.

Keywords: machine learning; decision trees; random forests; penetrometer; MWD; rigid inclusions

Citation: García, E.M.; Alberti, M.G.; Arcos Álvarez, A.A. Measurement-While-Drilling Based Estimation of Dynamic Penetrometer Values Using Decision Trees and Random Forests. *Appl. Sci.* **2022**, *12*, 4565. https://doi.org/10.3390/app12094565

Academic Editors: Małgorzata Jastrzębska, Krystyna Kazimierowicz-Frankowska, Gabriele Chiaro and Jaroslaw Rybak

Received: 9 March 2022
Accepted: 27 April 2022
Published: 30 April 2022

1. Introduction

Machine learning (ML) is a field in great health. As a concept, it is not something new (the first decision tree algorithm dates from 1963 [1]), but the current capacity to acquire and handle a huge amount of data has exponentially increased its popularity as an analysis tool. ML is a branch of artificial intelligence that consists of the application, through computer programs, of a series of algorithms to obtain information from large data sets. The term learning is used given that these systems are capable of improving and adapting to the information supplied to them.

Perhaps some of the best-known applications are those related to advertising and economics. For example, based on our browsing data, the algorithms can predict our preferences and offer us personalized advertising [2]. ML has also been widely used in medicine, where algorithms work as a medical diagnosis. As a mode of example, Scikit-learn, a machine learning library for Python, comes preloaded with small data sets on diabetes, exercise, and breast cancer [3]. In areas such as the industrial sector, the main aim of ML has been to predict when breakdowns will occur [4]. Mining, a field with problems similar to those of civil and geotechnical engineering, has also applied machine learning techniques [5].

In civil engineering, there have already been important contributions to the monitoring of dams [6], excavations [7], and traffic prediction [8]. However, there is still a lack of studies considering the large amount of data and ML in certain fields such as geotechnics.

Geotechnical engineering is a field whose problems present complex, non-linear relationships among its variables. These problems have been solved using empirical formulations that approximate the solution with more or less accuracy. ML algorithms have shown to be extremely efficient to analyze non-linear problems. Therefore, they are an ideal tool to be applied to geotechnics. One of the obvious applications is to predict the behavior of the soil from parameters obtained from laboratory tests. This approach achieves higher degrees of accuracy than the traditional analytical formulas [9]. However, the application of ML has not only been limited to geotechnical parameters but also to computing load-bearing capacity, settlement, liquefaction and slope stability [10–12].

Modern geotechnical machinery and equipment has abundant instrumentation that allows the attainment of a multitude of drilling parameters such as the drilling speed or the applied torque. This constant collection of data is usually named Measurement-While-Drilling (MWD). This record supplies an interesting source of information for the application of machine learning techniques.

On another note, continuous dynamic penetration tests consist of the introduction of a penetration element into the ground by striking a hammer with a defined weight. The result of the test is the number of blows required to advance a certain length. This type of test has become one of the most attractive alternatives giving several advantages:

- They provide data along the entire length of the borehole.
- They are very common given their simplicity and economy, it being possible to carry out several of them at each site.
- They can be correlated with the Standard Penetration Test (SPT).

The data obtained from MWD is in some way related to the characteristics of the terrain. These data are recorded continuously throughout the perforation and should be able to be correlated with the blows of the dynamic penetrometer. However, this relationship is unknown, complex, and non-linear.

In this study, a significant amount of data obtained from the MWD system from deep foundation and ground improvement machinery is analyzed. Figure 1 shows the plan views of the four sites studied. The circles represent the perforations made; the crosses are the penetrometers available. The aim was to correlate said penetration tests with the values obtained from MWD using ML techniques, specifically decision trees and random forests.

The significance of this research lies in the fact that, at the time of writing, it is the first time that this approach has been applied to rigid inclusions execution data.

The input variables have been the drilling speed (m/h), rotation speed (rpm), rotation torque (t·m) and thrust (t). The objective variable has been the blow values of dynamic penetrometers.

The results obtained show that the algorithms have been able to relate, with a certain degree of precision, the drilling parameters with the blows of the penetrometers. This information allows an increase in the level of security, an adaption of the execution to the real conditions of the ground and detection of possible anomalies, among other things.

The structure of the paper is the following:

- The related works are described.
- MWD in pile driving and soil improvement machinery is presented.
- The decision trees and random forests algorithms are summarized.
- The methodology followed is explained.
- The analyzed sites are described.
- The results obtained are shown and discussed.
- The conclusions are drawn.

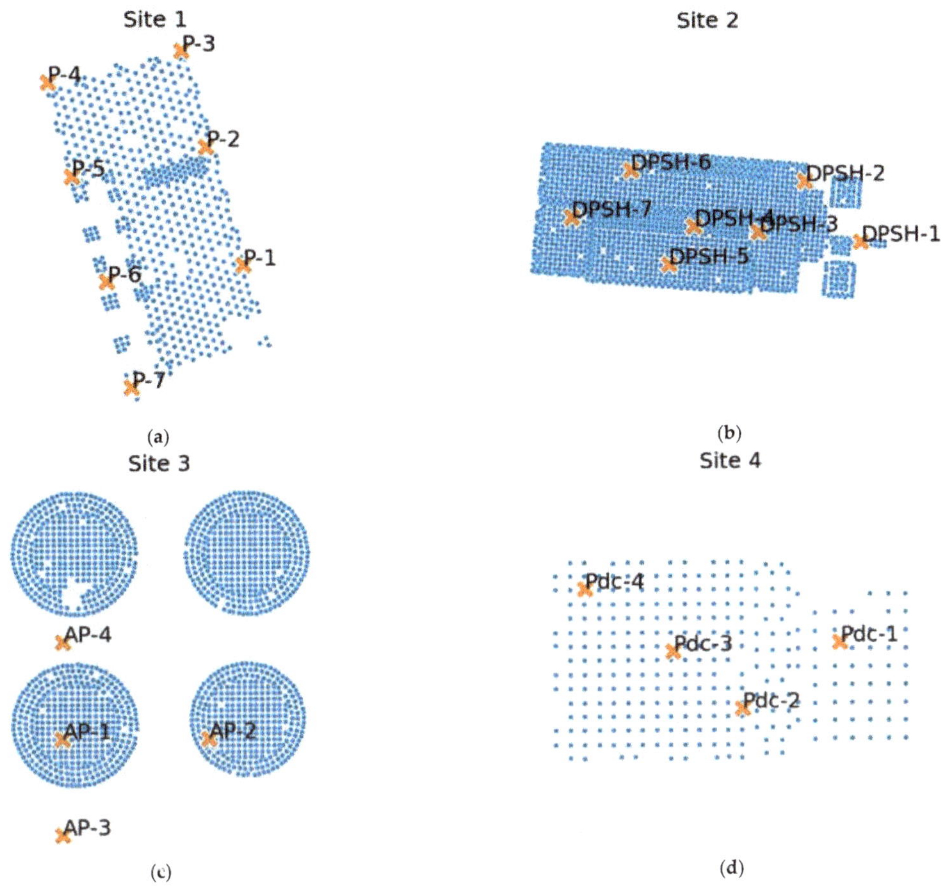

Figure 1. Distribution of columns (circles) and penetrometers (crosses) available in the analyzed sites. (**a**) Site 1. (**b**) Site 2. (**c**) Site 3. (**d**) Site 4.

2. Related Works

Geotechnical engineering hosts problems with many uncertainties in which the true relationships among the variables are unknown. This makes it an interesting field for the application of ML algorithms. In this way, several authors have used this approach [10], especially through the application of neural networks (NN).

In the field of layer characterization, Shuku et al. [13] proposed a new method for estimating trends and layer boundaries in depth-dependent soil data based on lasso. This new method (SBLasso) was applied to synthetic data, as well as an actual CPT sounding taken at Texas A&M University and it provided stratification consistent with existing methods.

Zhao and Wang [14] proposed an interpolation first method to characterize a multilayer soil property profile when measurements within each layer are sparse and limited. This method interpolates the multilayer soil property profile using measurements from all layers together as input to a Bayesian supervised machine learning first. Then, the interpolated multilayer soil property profile is stratified using an unsupervised machine learning method, e.g., the modified k-means clustering method.

Focusing on the use of MWD, Rai et al. [15] reviewed MWD techniques in the extractive industry, and therefore focused on rock drilling. Kadkhodaie-Ilkhchi et al. [16] conducted a comparative study of three machine learning techniques using MWD data from drilling for explosives at an iron mine in Australia. Other studies related to the recognition of rocks using MWD and ML have also been published with such references [17–20].

In the field of soils and civil engineering, Goh [21,22] estimated the load-bearing capacity of driven piles using NN in non-cohesive soils from data collected by Flaate [23] for wooden, precast concrete and steel piles. The input parameters were hammer weight, hammer drop height, hammer type, pile length, pile weight, pile modulus of elasticity, pile cross-sectional area, and pile set. The target value was the load-bearing capacity of the pile. The model was able to obtain high correlation coefficients between the input parameters and the load-bearing capacity for both the train set and the test set. Furthermore, NN were found to perform better than classical formulas when predicting load-bearing capacity.

Lee and Lee [24] also used NN to calculate the load-bearing capacity of piles. Five input variables were used: penetration depth ratio, the average SPT strike along the axis of the pile, the average SPT strike near the tip of the pile, the penetration rate for each strike and the hammer energy. The results obtained were compared with the Meyerhof formula [25]. The values obtained through NN correlated better with the measured values than those obtained through the Meyerhof equation.

Although the input parameters did not come from MWD, Teh et al. [26], proposed an NN model to estimate the load-bearing capacity of piles from dynamic pressure wave data. The data came from 37 precast piles from 21 different sites. The objective parameters were the values of the soil parameters predicted by the CAPWAP model [27]. This approach to predicting parameter values is similar to what is considered in this article.

In another use of NN, Diaz et al. [28] predicted the rate of penetration during wellbore drilling using data obtained as the drilling progresses within a given well.

Pal and Deswal [29] used a Gaussian regression process (GP) to predict the load-bearing capacity of the pile. Part of the input data was the same as that used by Goh. The performance of the proposed model was compared with support vector machines (SVM) and empirical relationships, obtaining a better result.

Galende-Hernández et al. [30] carried out a study estimating the value of the RMR from the characterization of the excavation face of a tunnel, using MWD and expert knowledge in the execution of a tunnel using explosives. The results obtained showed a good correlation. Such work posed a similar philosophy to that of this article since it sought to estimate the value of a soil parameter (the RMR) from the drilling parameters.

Although several examples of the use of MWD and ML data have been collected, the use of data obtained directly from machinery is relatively scarce. In this proposal, data from construction sites was taken directly from the MWD system and used to estimate the values of the dynamic penetrometer. The objective was to reach results for each perforation to be equivalent to having carried out a penetration test. Although data from rigid inclusions with ground displacement has been used, the methodology is general enough to be applicable to other works.

3. MWD in Pile Driving and Soil Improvement Machinery

The current pile machinery has a series of measuring instruments that allow the collection of performance data. The main information obtained during rotary drilling usually consists of the following [31]: the length of the drilling, drilling speed (V_A), rotation speed (V_R) and the hydraulic pressures of the rotary and thrust motors. These data are obtained with a high frequency, for example, every 10 cm of perforation.

3.1. Direct and Indirect Parameters

The parameters obtained from MWD can be divided into two groups depending on the process of attainment: direct parameters and indirect parameters.

Directly obtained parameters are those in which it is not necessary to apply any type of correlation to the values obtained from the sensors. Depth, rotational speed, and drilling speed can be considered direct parameters as they are obtained directly or as a function of time. Perhaps the most relevant of this type of parameter is the drilling speed as it is very indicative of the hardness of the terrain.

However, other parameters such as thrust force (P_O) and rotation torque (C_R) are obtained through mechanical correlations from the hydraulic pressure of the motors. These values (although they can be provided by the manufacturer) depend on a multitude of factors: the efficiency of the motor, the configuration of the machinery at any given time, the wear of the components, and so forth. Consequently, it is exceedingly difficult to know the true value of these parameters. These possible sources of error must be considered when analyzing the results obtained when using parameters that do not come from direct measurements. To this type of error attributed to the correlations, the errors and tolerances of the measuring devices must also be added.

3.2. Compound Indices and Parameters

Drilling parameters are related to soil resistance. However, by taking these parameters separately, the relationship is not clear. For this reason, various authors have developed compound parameters that combine individual parameters into energy expressions or empirical indices seeking to show the strength of the ground from the drilling data.

Most of these parameters or compound indices follow the same basic structure. In this way, the hardness of the ground is directly proportional to the rotation torque and the applied load and inversely proportional to the area of the drilling and drilling speed. These relationships tend to soften the profile, giving them greater physical meaning and making their interpretation easier. The most common of these indices are listed in Table 1 [32].

Table 1. Most used compound parameters.

Name	Formula	Units	Ref.
Penetration resistance	$R_p = (t)_{dZ=0.2m}$	s/0.2 m	Möller et al., 2004 [33]
Somerton index	$S_d \approx P_E\left(\frac{V_R}{V_A}\right)^{\frac{1}{2}} \approx \frac{P_O}{\sqrt{V_A}}$	kPa	Somerton 1959 [34]
Drilling specific energy	$SDE = \frac{P_O}{S_0} + \frac{2\pi}{S_0} \cdot \frac{V_R x C_R}{V_A}$	kJ/m^3	Teale 1965 [35]
Specific energy	$E_S = C_R \cdot \frac{V_R}{V_A}$	N·m/m	Pfister 1985 [36]
Normalized energy	$E_N = \frac{\alpha P_o V_A + \beta C_R 2\pi V_R + \gamma P_M f}{V_A}$	N·m/m	Nishi et al., 1998 [37]
Alteration index	$I_a = 1 + k_0\left(\frac{P_O}{P_{max}} - k_1\frac{V_A}{V_{max}}\right)$	-	Pfister 1985 [36]
Entropy of S	$L(z) = \sum_{Z_0}^{Z} \lvert S(z+dz) - S(z) \rvert$	-	Duchamp 1988 [38]

Of these parameters, penetration resistance, Somerton's index, and drilling specific energy are intended for research in hard rock and soils or for recording parameters during on-site testing. Their application to soft soils, based on MWD data, can be difficult to interpret.

The main obstacle with these formulas is that they assume a form of the relationship between the parameters and the strength that, although logical, does not necessarily correspond to reality. Therefore, these formulas give us a qualitative idea of the resistance to perforation and are useful when the differences in behavior within the soil are clear, but they do not supply precise figures.

4. Decision Trees and Random Forests

4.1. Decision Trees

Decision trees (DT) are a popular algorithm for classification or regression capable of providing reliable and easily interpretable results.

Although there are a variety of DT algorithms, they all have a similar structure. To be concise, the algorithm consists of dividing the data set into successive subsets following established rules. This process can be represented graphically in structures that resemble a tree (Figure 2).

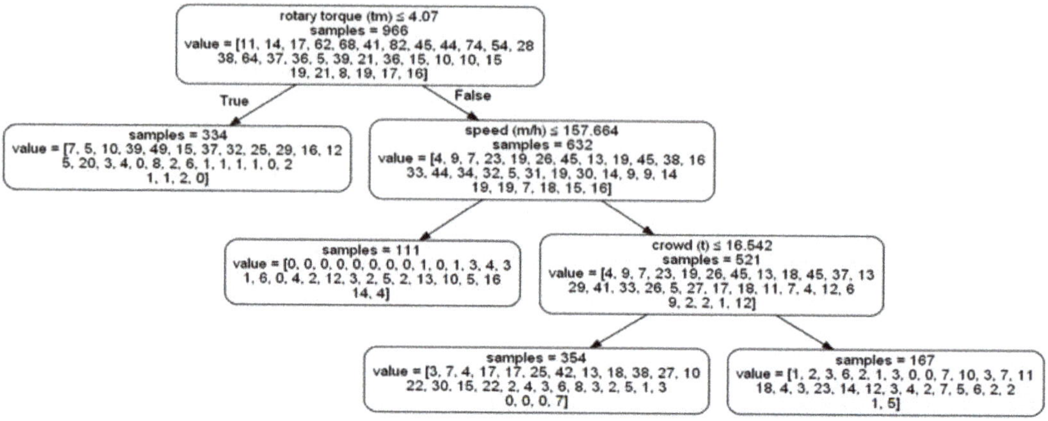

Figure 2. Decision tree example for site 1, limited to 4 leaf nodes.

The success of DTs is explained by several factors that make them quite useful in practice [39]:

- DTs are non-parametric. They do not assume the way in which the variables are related, and they are able to model complex interactions.
- They handle heterogeneous data (numbers, categories, or a mix of both).
- In their own operation, DTs implement the choice of the most important variables, making them robust (up to a point) against irrelevant variables.
- They are equally robust against outliers and sorting errors.
- They are easy to interpret, even for users without previous knowledge of statistics.

Building a decision tree consists of three parts: division rules, stop criteria and assignment rules.

4.2. Division Rules

A division at a node can be understood as a question whose answer divides the set at the node into two or more subsets. These questions or division rules vary depending on the type of decision tree chosen. Once these rules have been applied to each of the variables in the set, they are ordered according to the information they provide, establishing the most useful division as a node.

The application of these criteria defines the shape of the tree.

4.3. Stop Criteria

Overfitting refers to the situation where the model fits the training data too much, leading to larger errors in the test data. To avoid this problem stopping criteria can be set. This set of rules can be regulated to achieve trees that are neither too short (too general) nor too long (overfitting problem).

The tree will stop its development by itself in two cases:

- When the objective values are homogeneous in all samples of the node.
- When the input values are constant in the node.

In the previous cases, it is not possible to gain purity in the nodes by continuing the divisions, so the algorithm stops. These nodes that are not divided are called leaf nodes. In addition to these situations, other criteria can be added. The most common approaches are:

- Set a node as a leaf node if it contains less than a minimum number of samples.

- Set a maximum node depth. In this way, the number of divisions that the algorithm performs is limited.
- Establish a node as a leaf node when the impurity decrease is less than a certain threshold.
- Establish a node as a leaf node if it is not possible to establish dependent nodes with a minimum number of samples.

All these criteria must be defined by the user looking for the most adequate balance. Such balance can be difficult to achieve, which is why specially dedicated models are commonly used at the expense of a greater computational load.

These stopping criteria can be understood as a method of pruning the tree. Specifically, they are pre-pruning methods that are carried out during the growth of the tree.

Another method would be to develop the entire tree and perform the pruning eliminating those nodes that provide the worst results in a different data set than that used to generate the tree. This method usually provides better results than pre-pruning.

4.4. Assignment Rules

The assignment rules refer to the criteria followed to assign a value to a leaf node. In this way, in regression problems, it would be assigned to the value that provides the smallest mean square error. In a classification problem, the value would be that of the most probable category.

4.5. Implementation of DT in Scikit-Learn

For the application of DT and RF, the Scikit-learn library has been used. Scikit-learn [40] is a Python module that integrates a wide range of machine learning algorithms. It is a package designed for non-experts, focused on the ease of use, documentation, and consistency of the application programming interface (API).

The learning is carried out using the fit method together with the train values (in the case of supervised learning, these values would be X_train and y_train tables for input variables and target variables to predict, respectively). During the definition of the model, the hyperparameters that control the algorithm are introduced. The following pseudocode shows an example of DT training where the minimum number of samples to divide the node is five.

```
# Import the required module
from sklearn.tree import DecisionTreeClassifier,
# Definition of the model and hyper-parameters
clf = DecisionTreeClassifier(min_samples_split=5),
# Learning the model from the data
clf.fit(X_train, y_train),
```

The predictions are made with the prediction interface once the algorithm is trained, as shown in the next pseudocode.

```
# Make predictions from new data.
y_pred = clf.predict(X_test),
```

In addition, the prediction interface implements methods to check the confidence level of the prediction, as well as a score function to evaluate the performance of the model.

Lastly, the transformation interface allows the making of changes to the data, such as adjusting it to a normal distribution, which may be necessary when executing some algorithms.

As already mentioned, the selection of the best hyperparameters of a model can be a complex task. For this, scikit-learn implements two meta-estimators, GridSearchCV and RandomizedSearchCV. Both take as data an algorithm, those hyperparameters that must be used and a series of values to search through. GridSearchCV performs all possible combinations, while RandomizedSearchCV performs a fixed number of iterations. These meta-estimators perform a cross-validation method (CV). To do this, the data set is divided again into a train set and a test set. For each combination of hyperparameters and each generated train/test pair, GridSearchCV and RandomizedSearchCV adjust their estimator on the train set and evaluate their performance on the test set. In the end, the best-performing model is retained as the best set of hyper-parameters.

Within the multiple decision tree algorithms, scikit-learn implements an optimized version of Classification and Regression Trees (CART [41]), although it does not yet support non-numeric input variables.

4.6. Hyper-Parameters of Decision Trees in Scikit-Learn

Table 2 shows the hyper-parameters of the implementation in scikit-learn of the classification decision trees.

Table 2. Hyper-parameters for classification decision trees in scikit-learn 0.21.

Parameter	Default Value	Description
criterion	'gini'	Function to assess the quality of the division. The criteria supported are 'gini' for Gini impurity and 'entropy' for information gain.
splitter	'best'	Strategy used for choosing the division. The supported strategies are 'best' and 'random'.
max_depth	None	Maximum depth of the tree. If the value is 'None', the nodes will continue until all are pure or until all leaves contain less than min_samples_split samples.
min_samples_split	2	Minimum number of samples required to divide a node.
min_samples_leaf	1	Minimum number of samples required in a leaf node. A node will only be split if the result has at least min_samples_leaf samples.
min_weight_fraction_leaf	0.0	Minimum weight of a leaf node. Samples have equal weight when sample_weight is not provided.
max_features	None	The number of characteristics to consider when looking for the best division.
random_state	None	Parameter that controls the randomness of the model.
max_leaf_nodes	None	Maximum number of leaf nodes.
min_impurity_decrease	0.0	A node will divide if that division induces a decrease in impurity greater than or equal to this value.
min_impurity_split	0	A node will split if its impurity is above this value.
class_weight	None	Weights associated with the classes. If no value is provided, all classes will be assumed to have a weight of one.

4.7. Implementation of RF in Scikit-Learn

Unlike the original publication [42], the scikit-learn implementation combines classifiers by averaging their probabilistic prediction, instead of letting each classifier vote for a single class.

When building an RF in scikit-learn you can control the same parameters as for a DT in addition to specific RF hyperparameters. That is, the characteristics of the individual trees that make up the forest can be controlled. However, the main objective of the calibration of these parameters is to avoid overfitting the tree, which is partly avoided by the very way of constructing the RF. In this way, we can allow fully developed trees to form, at the cost of a higher computational cost. There are therefore two parameters to adjust: n_estimators and max_features.

n_estimators is the number of trees in the forest. In principle, the more the merrier. However, from a certain number of trees, there is no improvement in the prediction, and it must be considered that each tree increases the calculation time.

max_features is the size of the random subsets of features to consider when splitting a node. The lower this value, the greater the reduction in variance, but also the greater the increase in bias. In practice, as stated in the scikit-learn documentation, values of max_features = None (always considering all characteristics instead of a random subset) work well for regression problems, and max_features = "sqrt" (using a subset random size sqrt (n_features)) for classification tasks (where n_features is the number of features in the data).

Since we are treating the problem as classification, we will use max_features = "sqrt".

To define the optimal number of trees, an iterative calculation can be carried out to check the increase in the quality of the prediction. In this way, it can be verified at which point the increase in the number of trees no longer compensates.

5. Methodology

The main objective of this study was to predict the values of the dynamic penetrometer based on the drilling parameters. This test was chosen because it is abundant and easy to interpret, which makes the results obtained more useful. In fact, these tests are frequently used for the design of deep foundations and soil treatments.

The first step was to collect the data analyzed in an easy-to-use format from the four sites. This translated into transforming the data from the MWD files into a depth-based table format and adding necessary data such as the positions of each column.

Once the data had been adapted, it was necessary to assign a blow value to each row of data according to the closest penetrometer. In this sense, it was essential to pay attention to the execution levels of both the penetrometers and the columns.

The data were treated by eliminating anomalous values based on knowledge of the characteristics of the soil and the machinery used. The resulting values were divided into a train set and a test set, using the first set to train the decision tree and random forests algorithms. Once trained, the performance of the algorithms was checked with the test set (Figure 3).

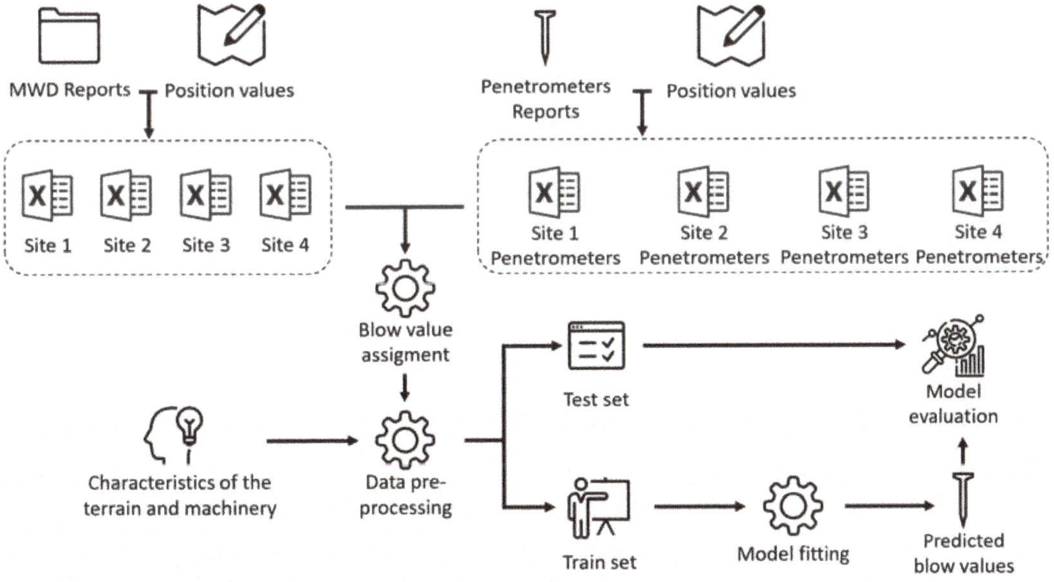

Figure 3. Procedure applied in the analysis.

6. Summary of the Analyzed Sites

Four different sites have been studied (Figure 1) in which the soil improvement method controlled modulus columns have been used [43]. The execution procedure has been by means of a displacement helix. This method moves the drilled soil to the sides without extracting it.

It must be considered that this drilling method presents a different behavior than a cutting tool, mobilizing the lateral resistance of the ground traversed.

These columns were executed using a drilling rig similar to an Enteco E6050.

6.1. Site 1

This site consisted of treating the ground under the slab and footings of an industrial building, making a total of 665 rigid inclusions of 300 and 360 mm in diameter.

During the geotechnical campaign, a total of 7 Dynamic Probing Super Heavy (DPSH) penetrometers were carried out, among other works.

The surface layer of the soil is formed by an anthropic fill of poured silty sands with an interpreted thickness that varies between 1.60 and 2.20 m. Under these fillings, a layer of vegetal soil of little thickness appears (between 0.4 and 1.0 m), also composed of silty sands. This is followed by an alluvial level of fine sand with silt of very loose compacity and a thickness of 1.60 to 2.20 m. The last layer investigated consists of the altered granite of the area with increasing compactness in depth.

The material is, therefore, of a homogeneous nature and mainly sandy, due to the water table being close to the surface. Figure 4 shows the DPSH tests carried out on site 1. The points represent the number of blows required to advance 20 cm depending on the depth.

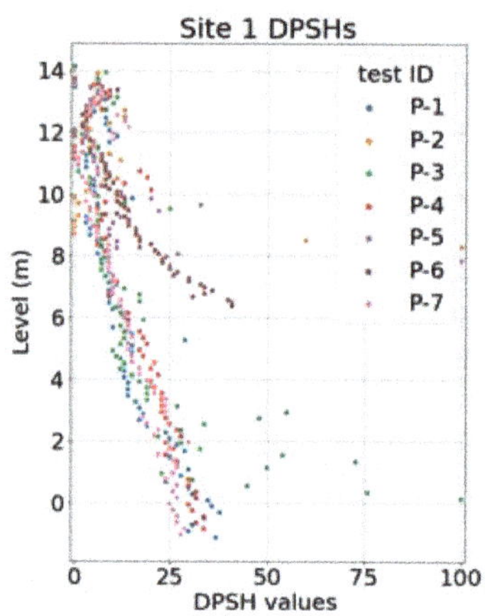

Figure 4. DPSH values for site 1.

6.2. Site 2

Site 2 also consisted of treating the ground under an industrial warehouse using 360 mm diameter inclusions.

In this case, a total of 7 DPSH penetrometers are available.

According to the geotechnical information, the first level is formed by anthropic fillings with an average thickness of 6 m made up of sands and clayey sands (SC). Below this level, alluvial materials appear arranged in layers of diverse nature, forming an interlayer from clays to sands, although with some lateral continuity. This level can reach a depth of about 30 m. The last level detected coincides with the Pliocene substrate in the area, made up of dense, highly compact sands. This level is at such a depth that it is not reached by the soil treatment.

The water table has been detected at very different levels depending on the time of the measurements. It has varied between 2 and 8 m deep.

Therefore, unlike site 1, this site presents a much more heterogeneous geology, with materials of different nature and a variable position of the water table. Figure 5 shows the DPSH tests carried out on site 2.

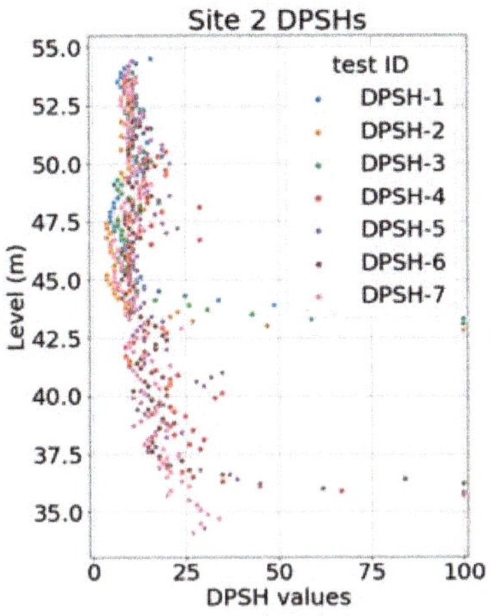

Figure 5. DPSH values for site 2.

6.3. Site 3

Site 3 consisted of soil treatment using 360 mm diameter controlled modulus columns for the construction of hydrocarbon tanks on a port landfill.

In this work, the information on four penetrometers is available. However, these were made before knowing the final position of the tanks, which finally did not coincide with that of all penetrometers. Due to this circumstance, only two of them are relevant (see lower left of Figure 1).

The port landfill consisted of material from dredging used to fill various areas reclaimed from the sea. The composition is variable but mainly granular, formed by sands with traces of fines and gravel. The thickness of this fill is between four and eight meters. Below this first level is the natural terrain composed of quaternary sands that increase in compactness with depth.

The water table is very superficial and will depend mainly on the level of the sea.

In this case, the terrain is again mainly sandy and reasonably homogeneous.

DPSH tests for site 3 are shown in Figure 6.

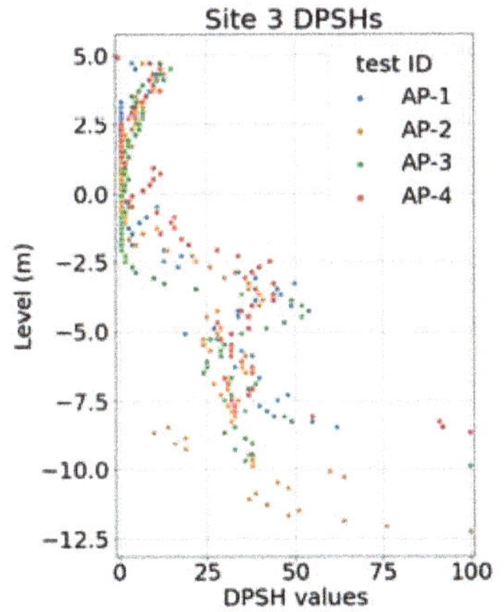

Figure 6. DPSH values for site 3.

6.4. Site 4

This works consisted of treating the ground under an industrial warehouse and exterior pavements with inclusions of 300 and 360 mm in diameter.

The number of penetrometers available for this work was four.

In this case, the soil is made up of an anthropic fill of sand and silt of low plasticity with a thickness of about 7 m. Below this level, there is a terrace deposit formed by silts and sandy clays and silty sands of reddish colors, fine grain and low plasticity.

The water table is about 6 m deep on average.

Unlike the previous cases, the penetrometric tests carried out in this site were Borro type instead of DPSH.

In the case of site 4, the Borro test values are shown in Figure 7.

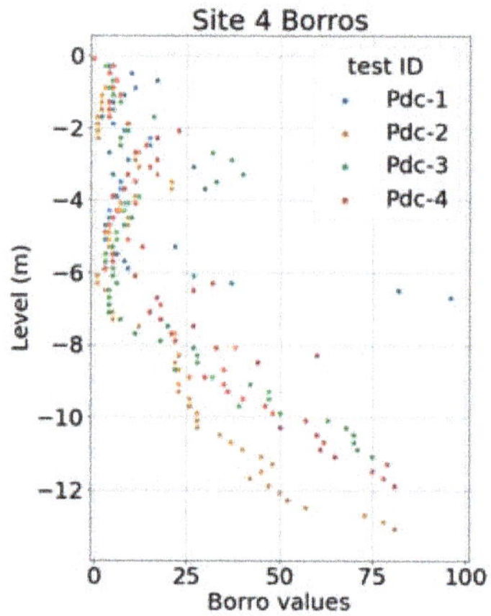

Figure 7. Borro values for site 4.

6.5. Available Data

The parameters were recorded with Menard's Emparex data acquisition system which provides data approximately every 8 cm of drilling. Said values, ignoring those of general information about the site, are shown in Table 3.

Table 3. MWD available data.

Parameter	Name of the Column	Depth from Start of Drilling	Drilling Speed	Rotation Speed	Rotary Torque Pressure	Crowd Pressure	Torque	Crowd
Unit	-	m	m/h	rpm	bar	bar	t·m	t

It should be noted that torque and crowd parameters are calculated from rotary torque pressure and crowd pressure, respectively.

In order to narrow down the analysis carried out, a single diameter has been considered for each work (360 mm in sites 1, 2 and 3 and 300 mm in site 4, according to the most common diameter in each site) and the nature of the terrain has not been taken into account as a variable (this is mainly granular in the sites analyzed). However, in the future, it will be necessary to add at least both parameters to the analysis. In this way, the number of initially available data is shown in Table 4.

Table 4. Data initially available.

Site	Columns	Data Rows
Site 1	159	20,916
Site 2	2096	269,921
Site 3	1504	111,428
Site 4	287	27,797

The input variables for the algorithms are drilling speed, rotation speed, torque, and crowd. These variables are the same used for the compound parameters of Table 1. The target variable is the number of blows to advance 20 cm of the penetrometers.

6.6. Initial Treatment of the Data

It has been verified that there is no missing data. That data without associated penetrometer value (in the case of the closest penetrometer being short or has started at a lower level than the level of execution of the inclusions) has been deleted. In the type of problem of this research, missing data should not be an issue. Nevertheless, for an in-depth discussion about that topic, the work of Duy-Tai Dinh et al. [44] can be studied.

The penetrometer data is highly affected by specific high values that do not reflect the real resistance of the ground and are therefore not translated into the resistance to perforation. In order to attenuate this effect, the stroke values at each point have been transformed to the mean of the point itself and the values above and below it. Stroke values greater than 30 have also been removed.

It has also been verified that the values obtained in the registry are in line with what is expected depending on the machinery used. In this type of machine, the relationship between the torque pressure and the applied torque depends on the rotational speed of the tool. Figure 8 represents the measured torque pressure and the rotation torque calculated by the parameter recording system for the case of site 1. The result is several lines with different slopes depending on the rotation speed. Abnormal values associated with low rotational speeds are observed, such as negative or very high torque values. It is also verified in the graph on the right that these values are associated with shallow depths. That is, these values are associated with the beginning of drilling and are not representative.

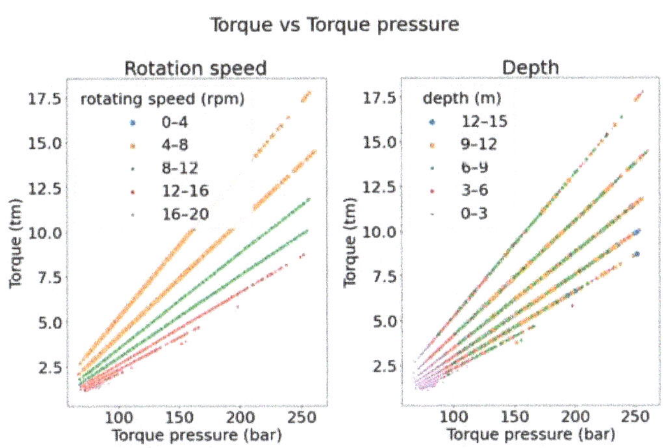

Figure 8. Rotation torque vs torque pressure measured on site 1. To the left depending on the rota–tion speed and to the right depending on the depth.

In this way, these values, which do not correspond to the expected normal behavior, are not considered when carrying out the study. The corrected graph is shown in Figure 9.

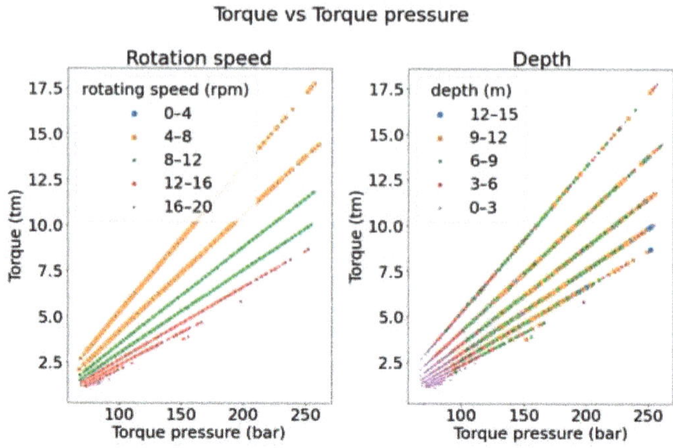

Figure 9. Rotation torque vs torque pressure on site 1 reviewed. To the left depending on the speed of rotation and to the right depending on the depth.

7. Results

The algorithms have been trained to predict the blow values of the DPSH penetrometer. These values are discrete in nature and the problem can be treated as both a regression or a classification. In this case, it was decided to use the classification algorithms with 31 classes (from 0 to 30 blow values). However, since the target variable values are numbers, regression metrics can also be used.

To evaluate the performance of the predictions, the accuracy (Equation (1)), Pearson's correlation coefficient (Equation (2)), MAE, MSE and R^2 were used.

$$accuracy\ (y, \hat{y}) = \frac{1}{n_{samples}} \sum_{i=0}^{n_{samples}-1} I(\hat{y}_i = y_i) \tag{1}$$

$$\rho_{X,Y} = \frac{\sigma_{XY}}{\sigma_X \cdot \sigma_Y} \tag{2}$$

$$MAE = \frac{\sum_{i=1}^{n} |y_i - \hat{y}_i|}{n} \tag{3}$$

$$MSE = \frac{\sum_{i=1}^{n} (y_i - \hat{y}_i)^2}{n} \tag{4}$$

$$R^2(y, \hat{y}) = 1 - \frac{\sum_{i=1}^{n} (y_i - \hat{y}_i)^2}{\sum_{i=1}^{n} (y_i - \overline{y})^2} \tag{5}$$

Accuracy measures the proportion of correctly classified samples out of the total. In this case, the Pearson correlation coefficient measures the linear relationship between the real values of the target variable and the values calculated by the algorithms.

In the various sites analyzed, the models have been trained for a single diameter of those available since it is not the objective of this article to evaluate the effect of the different diameters.

In addition, only those perforations made less than 2 m from its associated penetrometer have been considered. In this way, an attempt is made to avoid errors due to changes in the terrain with the distance to the penetrometer. The effect of separation is evaluated later in this article. This limitation greatly affects the number of samples available (Table 5).

Table 5. Data at a distance ≤2 m from its associated penetrometer.

Site	Columns	Data Rows
1	10	1375
2	12	1030
3	6	536
4	6	549

The input variables are the drilling speed (m/h), rotation speed (rpm), rotation torque (t·m) and thrust (t). Therefore, the same parameters as in traditional formulas are considered. The variable to predict is the blow values per 20 cm of the penetrometers.

For every calculation, the data set is divided randomly into a train set with 70% of the data and a validation set with the remaining 30%.

7.1. Fully Developed Trees

As a first approximation, fully developed DTs are used, without restrictions on their growth. This is achieved with the default Scikit-learn parameters. Table 6 shows accuracy, correlation coefficient, MAE, MSE and R^2 values for a fully developed DT. These values vary slightly when you rerun the algorithm due to random effects even for the same train-test split. The algorithm has been tested on 500 different train-test splits. The mean values are shown in the tables with their standard deviation in brackets.

Table 6. Correlations obtained for a fully developed DT.

Site	Set	Accuracy	Correlation Coefficient	MAE	MSE	R^2
1	Test	0.15 (±0.02)	0.62 (±0.04)	4.10 (±0.20)	35.03 (±3.36)	0.28 (±0.08)
	Train	1.00 (±0.00)	1.00 (±0.00)	0.00 (±0.00)	0.00 (±0.00)	1.00 (±0.00)
2	Test	0.27 (±0.03)	0.42 (±0.07)	2.59 (±0.17)	17.06 (±2.38)	−0.20 (±0.16)
	Train	1.00 (±0.00)	1.00 (±0.00)	0.00 (±0.00)	0.00 (±0.00)	1.00 (±0.00)
3	Test	0.32 (±0.03)	0.57 (±0.09)	3.00 (±0.30)	22.48 (±3.70)	−0.09 (±0.20)
	Train	1.00 (±0.00)	1.00 (±0.00)	0.00 (±0.00)	0.01 (±0.00)	1.00 (±0.00)
4	Test	0.15 (±0.02)	0.24 (±0.08)	5.34 (±0.38)	59.72 (±7.55)	−0.38 (±0.20)
	Train	1.00 (±0.00)	1.00 (±0.00)	0.00 (±0.00)	0.00 (±0.00)	1.00 (±0.00)

The models overfit since a perfect correlation is achieved for the train set while the correlation is much lower in the test set. In addition, fully developing the trees has resulted in trees between 150 and 650 leaves, which makes their interpretation impossible. Furthermore, the R^2 values for the test set are very low. Although their deviation is also very high, such low values of R^2 seem to be caused by the outliers.

7.2. Hyperparameter Calibration Using GridSearchCV

In order to find the best hyperparameters, the GridSearchCV meta-estimator has been used. Although it would be possible to calibrate all hyperparameters, due to the nature of the data analyzed, not all of them are relevant. Table 7 shows the set of hyperparameters introduced in GridSearchCV.

Table 7. Hyperparameters used during calibration with GridSearchCV.

Hyperparameters	Criterion	Splitter	Max_depth	Min_samples_split	Min_samples_leaf	Max_leaf_nodes
Values	Gini, entropy	Best, random	1, 5, 10, 15, None	2, 4, 6, 8	1, 2, 3, 4, 5	None, 400, 200, 100, 50, 10

To perform this calculation, a random_state = 42 has been established. This way, GridSearchCV always works on the same data set, obtaining very similar results regardless of how many times the calculation is made (Table 8). However, it must be considered that different sets can give rise to different hyperparameters that may be better or worse than those obtained in this case.

Table 8. Best parameters obtained by GridSearchCV.

Site	1	2	3	4
criterion	gini	gini	gini	gini
splitter	random	random	random	random
max_depth	None	15	10	10
min_samples_split	2	8	2	4
min_samples_leaf	1	1	1	1
max_leaf_nodes	400	400	50	100

With the new calculated hyperparameters, there is less overfitting than for fully developed DTs. However, the predictability of the validation set is not significantly improved (Table 9). On the other hand, there is a reduction in the complexity of the model (fewer leaf nodes and less depth). This would make the use of hyperparameters calculated using GridSearchCV preferable (Table 10).

Table 9. Correlations obtained using GridSearchCV.

Site	Set	Accuracy	Correlation Coefficient	MAE	MSE	R^2
1	Test	0.12 (±0.02)	0.62 (±0.04)	4.08 (±0.26)	34.56 (±4.34)	0.29 (±0.09)
	Train	0.64 (±0.22)	0.88 (±0.08)	2.46 (±0.92)	21.24 (±7.96)	0.56 (±0.16)
2	Test	0.23 (±0.03)	0.49(±0.07)	2.52(±0.18)	15.99 (±2.34)	−0.12(±0.14)
	Train	0.53 (±0.21)	0.60 (±0.15)	1.40 (±0.55)	9.33 (±3.77)	0.35 (±0.26)
3	Test	0.32 (±0.02)	0.34 (±0.10)	2.99 (±0.29)	24.92 (±4.13)	−0.20 (±0.14)
	Train	0.60 (±0.08)	0.54 (±0.11)	2.63 (±0.35)	22.19 (±3.71)	−0.07 (±0.17)
4	Test	0.14 (±0.02)	0.24 (±0.11)	5.27 (±0.44)	61.56 (±9.19)	−0.41 (±0.17)
	Train	0.50 (±0.16)	0.54 (±0.18)	3.92 (±0.94)	47.35 (±12.41)	−0.09(±0.28)

Table 10. Complexity of fully developed trees compared to those obtained through GridSearchCV.

	Fully Developed Tree		GridSearchCV	
Site	**Depth**	**Leaf Nodes**	**Depth**	**Leaf Nodes**
1	20	681	20	400
2	22	437	15	177
3	16	180	10	50
4	20	269	10	100

7.3. Random Forests

Figure 10 shows the evolution of the accuracy as a function of the number of trees considered for site 1. It has been verified that with a number of trees in the order of 50, it is ensured that the maximum precision has been reached. The calculation has been performed with 75 trees.

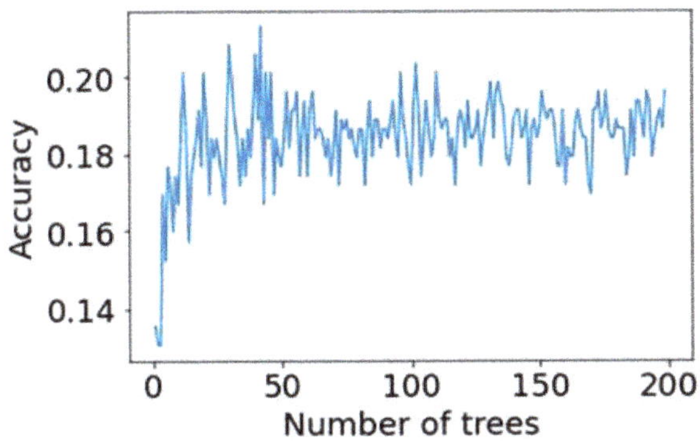

Figure 10. Accuracy as a function of the number of trees considered for site 1.

Table 11 shows the values of the metrics for RFs.

Table 11. Correlations obtained for RFs.

Site	Set	Accuracy	Correlation Coefficient	MAE	MSE	R^2
1	Test Train	0.18 (±0.02) 1.00 (±0.00)	0.76 (±0.02) 1.00 (±0.00)	3.45 (±0.02) 0.00 (±0.00)	26.81 (±0.02) 0.00 (±0.00)	0.45 (±0.02) 1.00 (±0.00)
2	Test Train	0.31 (±0.03) 1.00 (±0.00)	0.54 (±0.06) 1.00 (±0.00)	2.15(±0.13) 0.00 (±0.00)	12.82(±1.84) 0.00 (±0.00)	0.10(±0.11) 1.00 (±0.00)
3	Test Train	0.38 (±0.03) 1.00 (±0.00)	0.57 (±0.07) 1.00 (±0.00)	2.60 (±0.24) 0.00 (±0.00)	20.03 (±3.05) 0.01 (±0.00)	0.04(±0.13) 1.00 (±0.00)
4	Test Train	0.18 (±0.02) 1.00 (±0.00)	0.28(±0.07) 1.00 (±0.00)	4.83 (±0.36) 0.00 (±0.00)	54.30 (±7.00) 0.02 (±0.00)	−0.26(±0.16) 1.00 (±0.00)

The results obtained with RF are better than those obtained in all the analyzed metrics.

7.4. Comparison with Compound Parameters

Table 12 shows the values of the correlation coefficient between the blows of the penetrometers and the results of both the analytical formulas and the algorithms used. The values obtained by linear regression are also shown, in order to compare with the simplest possible ML algorithm.

Table 12. Correlation coefficient for the compound parameters.

Compound Parameter	Site			
	1	2	3	4
Penetration resistance	−0.45	−0.09	−0.40	−0.24
Somerton index	0.40	−0.07	0.42	0.12
Drilling specific energy	0.52	0.13	0.44	0.34
Specific energy	0.44	0.02	0.48	0.32
Alteration index	0.49	0.14	0.39	0.29
DT	0.62	0.42	0.57	0.24
DT (GridSearch CV)	0.62	0.49	0.34	0.24
RF	0.76	0.54	0.57	0.28
Linear regression	0.56	0.39	0.49	0.36

As can be seen, the ML algorithms show a higher correlation with the penetrometers than the compound parameters in all sites except for site 4, where similar values of the correlation coefficient are obtained. DT and RF outperform linear regression in three of the four sites.

7.5. Representation of Data as a Function of Depth

Unlike other machine learning problems, in our case it is possible to represent the data with ease, showing the blow values obtained versus depth.

Figures 11–14 show real and calculated values versus depth for the test subset. The first graph of each figure corresponds to the real blow values per 20 cm, the second are the values obtained from a fully developed DT and the third is the values for RF. These images quickly show the ability of the algorithms to reproduce the penetrometric profile of the soil.

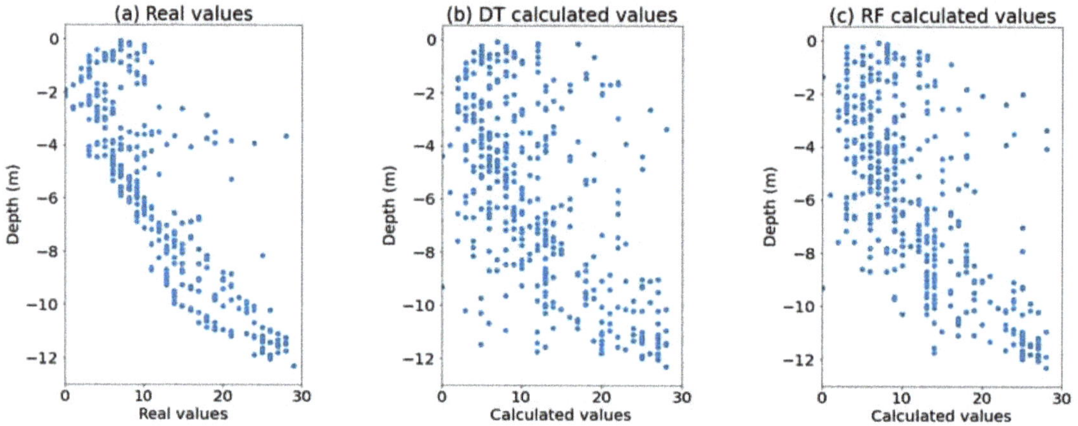

Figure 11. Real and calculated values for site 1.

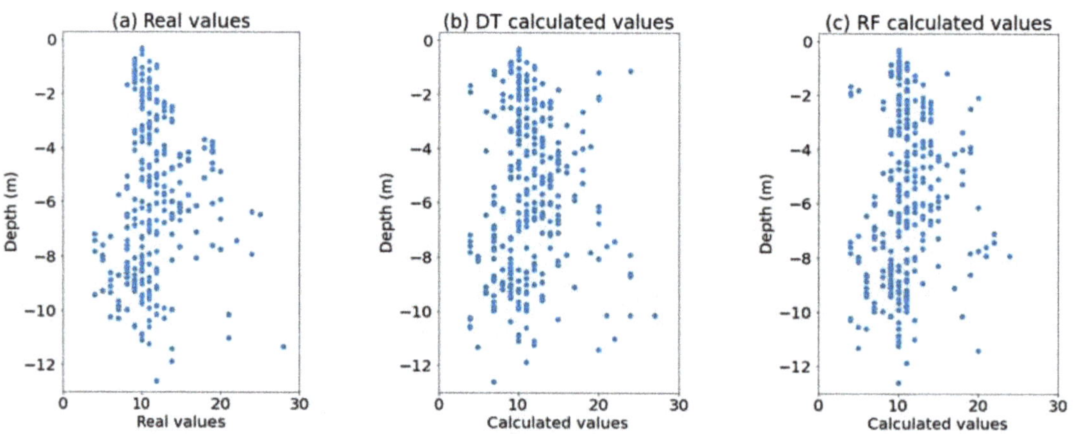

Figure 12. Real and calculated values for site 2.

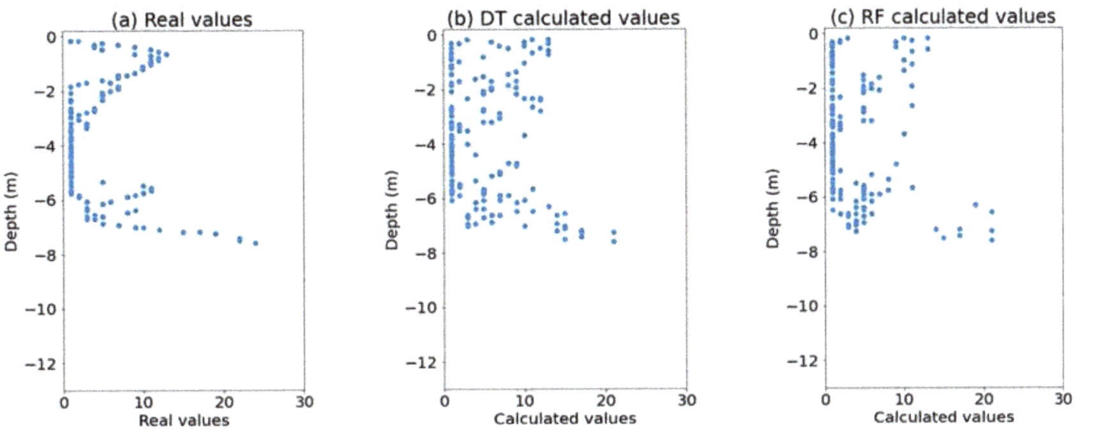

Figure 13. Real and calculated values for site 3.

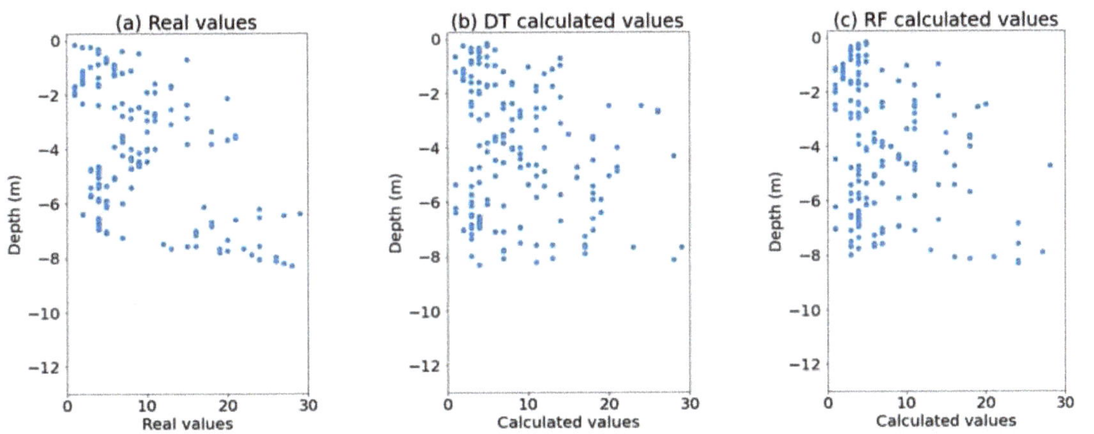

Figure 14. Real and calculated values for site 4.

7.6. Advantage of Taking Depth into Account

The variables used in the previous sections (drilling speed, rotational speed, torque and thrust force) have been the same as those considered in the classical formulas.

However, when proposing a machine learning algorithm, it may be interesting to also use depth. In this way, the algorithm should be able to detect different layers on the ground and give values more in line with reality. As can be seen in Tables 13 and 14, the increase in prediction precision is very important. Figures 15–18, show that the profile obtained with the models, considering the depth, is practically identical to the real one.

Table 13. Correlations obtained for fully developed DTs with depth.

Site	Set	Accuracy	Correlation Coefficient	MAE	MSE	R^2
1	Test	0.42 (±0.03)	0.87 (±0.18)	1.97 (±0.15)	11.98 (±1.74)	0.76 (±0.04)
2	Test	0.30 (±0.03)	0.61 (±0.07)	1.70 (±0.16)	9.81(±2.12)	0.31(±0.15)
3	Test	0.64 (±0.73)	0.94 (±0.24)	0.65 (±0.15)	2.32 (±1.01)	0.89 (±0.05)
4	Test	0.35 (±0.04)	0.62 (±0.07)	3.10 (±0.33)	29.06 (±5.80)	0.32 (±0.15)

Table 14. Correlations obtained for RFs with depth.

Site	Set	Accuracy	Correlation Coefficient	MAE	MSE	R^2
1	Test	0.38 (±0.03)	0.90 (±0.01)	1.66 (±0.11)	8.82 (±1.21)	0.82 (±0.03)
2	Test	0.38 (±0.04)	0.74 (±0.05)	1.35 (±0.12)	6.54 (±1.56)	0.54 (±0.09)
3	Test	0.50 (±0.06)	0.91 (±0.03)	0.90 (±0.13)	3.74 (±1.17)	0.82 (±0.05)
4	Test	0.33 (±0.03)	0.62 (±0.50)	2.70 (±0.27)	22.68 (±4.72)	0.48 (±0.10)

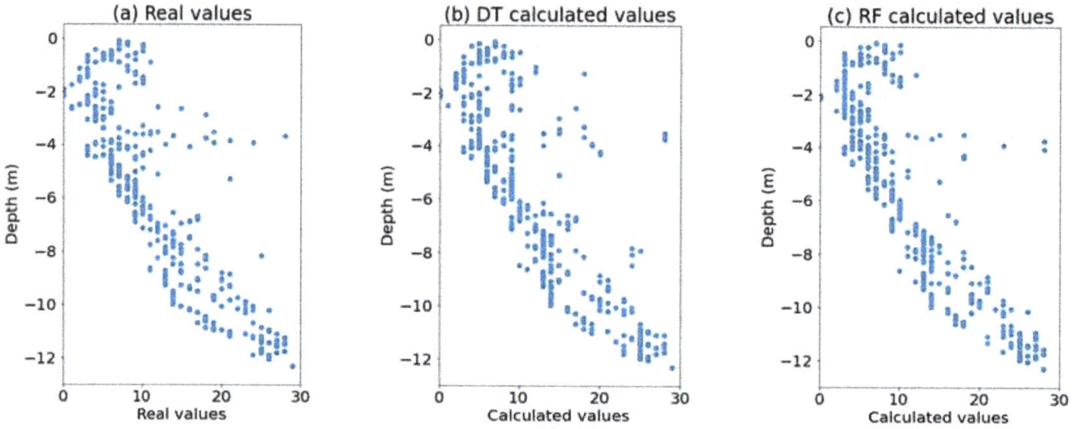

Figure 15. Real and calculated values for site 1 considering depth.

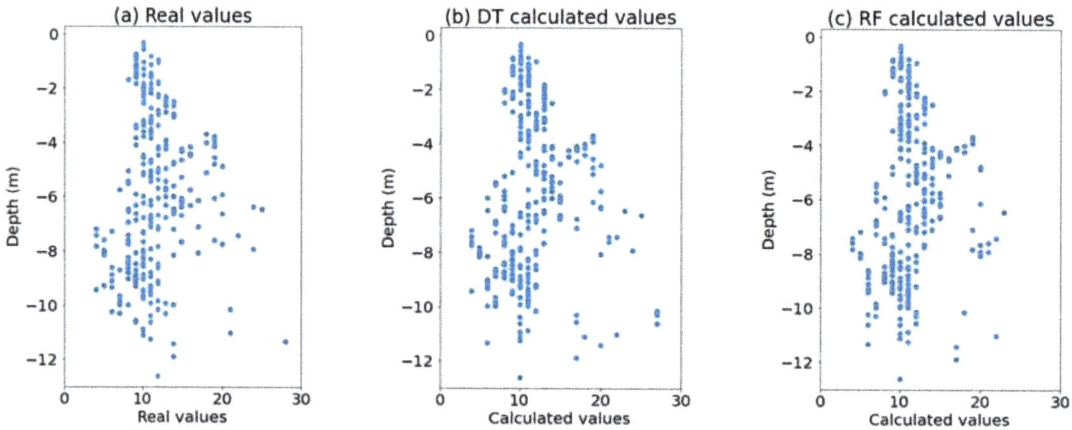

Figure 16. Real and calculated values for site 2 considering depth.

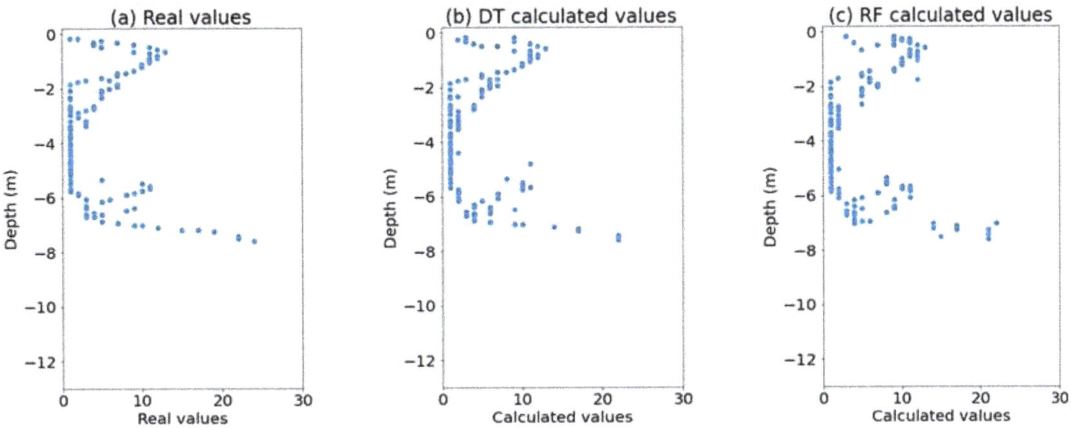

Figure 17. Real and calculated values for site 3 considering depth.

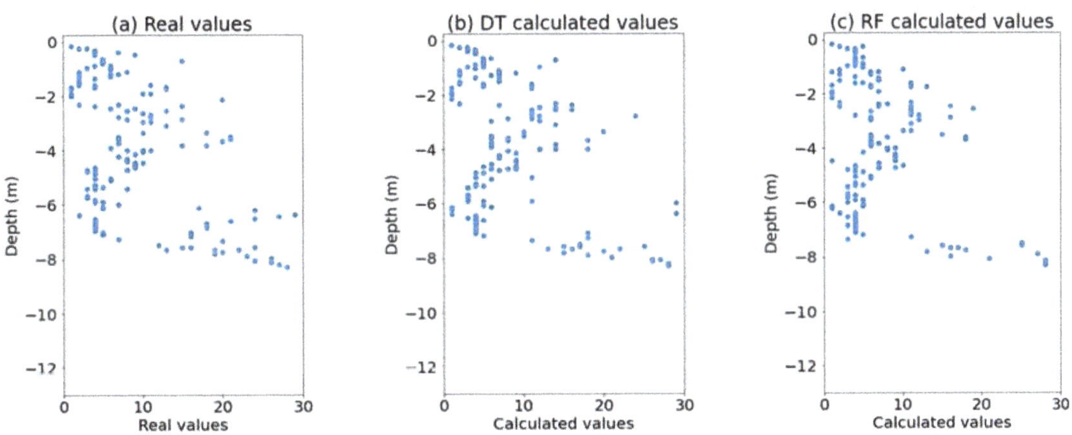

Figure 18. Real and calculated values for site 4 considering depth.

7.7. Increase in Error with Distance to Perforations

Normally, only a few penetrometers are carried out and, subsequently, there will not be too many columns near such tests. This greatly limits the relevant data, resulting in a handicap when using certain algorithms.

The calculations performed up to this point have only considered data that was within 2 m or less of its associated penetrometer.

Through an iterative calculation, the increase in error has been verified, based on the F-score and the correlation coefficient as the distance to the perforation increases. It must be borne in mind that this error depends on the type of soil. With a homogeneous layer distribution (with horizontal continuity) in the soil, the error should not be large, but it can be important in highly variable terrain.

The results obtained for the different sites (for fully developed DTs and taking depth into account) are shown in Figure 19.

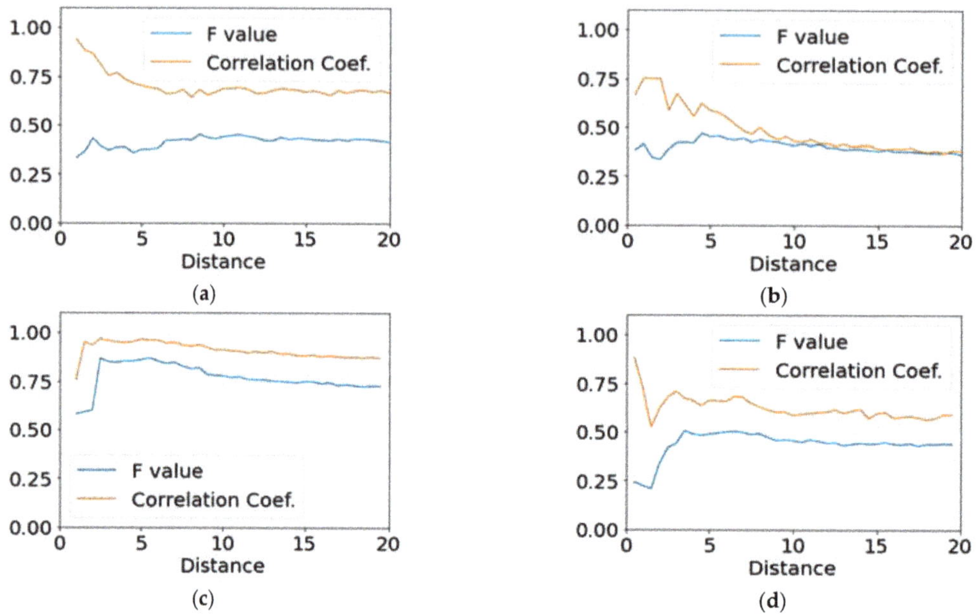

Figure 19. F-score and correlation coefficient as a function of distance for the different sites. (**a**) Site 1. (**b**) Site 2. (**c**) Site 3. (**d**) Site 4.

As can be seen, the precision is maximum when distances between columns and the penetrometer are small (in the order of 2 m). From that point, there is a rapid decline in the first few meters which later slows down.

In sites 3 and 4, the point of maximum precision does not occur with the minimum distance but instead, a little further. This may be due to the bias that would occur when considering very small samples.

According to the values obtained and considering the high degree of uncertainty commonly associated with geotechnical works, the increase in error with distances to penetrometers greater than 2 m can be considered acceptable when making predictions if there are not enough values in closer distances.

7.8. Application to a Set of Sites

As a culmination of the analysis carried out, the operation of the algorithm has been verified with a set of sites. Many are the factors that can vary from one job to another in terms of soil behavior, equipment performance and so on.

The same calculations shown in the previous sections have been carried out for sites 1, 2 and 3. Site 4 has been omitted since the penetrometers executed were Borro type instead of DPSH, as in the rest of the sites.

When applying the algorithms to the data set of the three sites, the results are equally promising, as reflected in Table 15 and Figure 20.

The precision obtained is similar to the average of the three sites separately. It therefore seems possible to analyze all the sites without producing a significant increase in error. It should be noted, however, that the materials of the works studied are of a similar nature. It will be necessary to evaluate in the future the effect of materials of a different composition.

Table 15. Correlations obtained for the set of sites 1, 2 and 3.

Site	Model	Set	Accuracy	Correlation Coefficient	MAE	MSE	R^2
1, 2, 3	Fully developed DT with depth	Test	0.35 (±0.03)	0.82 (±0.02)	1.92 (±0.10)	11.84 (±0.18)	0.69 (±0.03)
	RF with depth	Test	0.37 (±0.02)	0.88 (±0.01)	1.52 (±0.07)	7.77 (±0.83)	0.80 (±0.02)

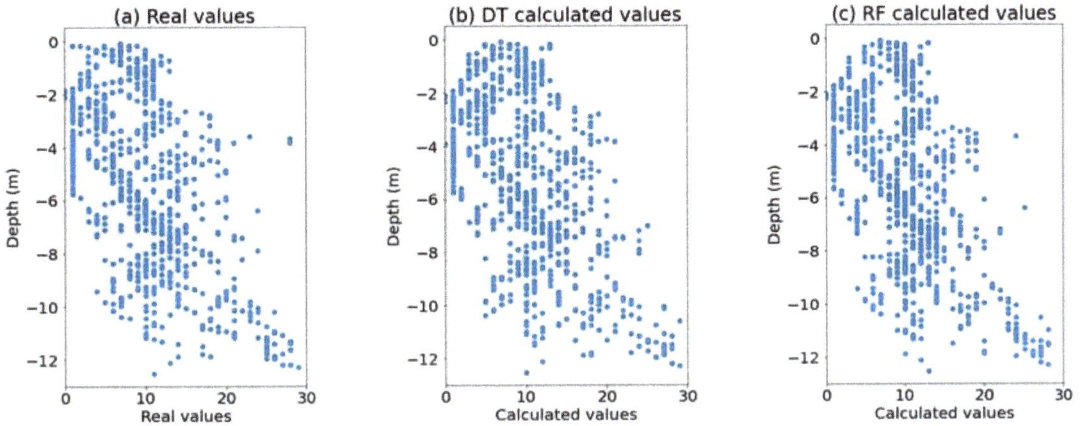

Figure 20. Real and calculated values for site 1, 2 and 3 considering depth.

8. Discussion and Conclusions

In this article, the application of machine learning algorithms to correlate data obtained from MWD with dynamic penetrometer values has been studied.

Decision tree algorithms and random forests have been used to predict penetrometer values, unlike in other studies where neural networks have been used. These algorithms have a simpler implementation as few parameters must be adjusted and these are easy to interpret. The graphical representation of the results is even possible, although in this case, it has not been practical due to the large number of resulting nodes. This simplicity, together with the existence of libraries such as scikit-learn [45], allow an easier application.

The study has been carried out from real execution data. It is not common to have data from sites, as most of the previous studies use laboratory data or historical cases from the bibliography. In addition, many of the existing cases with MWD data are from rock drilling and not soils, as in this article.

The results obtained have shown that, for the four sites studied, a fully developed decision tree is capable of achieving great precision. The importance of taking depth into account in the calculations has also been shown since it significantly increases the degree of correlation. It should be noted, however, that the number of penetrometer tests available is low and giving excessive importance to depth could mask terrain of a difference in nature if the soil profile is not homogeneous within the site.

By means of a reasonably simple mechanism to implement, a very high level of security is achieved in terms of the soil traversed. This would allow the execution to be adapted to the real conditions of the terrain, as well as to detect possible anomalies. The results obtained, being penetrometer values, are easy to interpret by any geotechnical engineer as this is one of the most common tests and there are various accurate correlations with other soil parameters.

Moreover, this methodology is a more flexible tool than the classical analytical formulas shown in Table 1. In addition, it is automatically adapted to the case under study, making it much more precise than said formulas.

The sites analyzed have been drilled with soil displacement. However, the methodology should be applicable to any type of drilling with parameter registration.

The methodology has shown equally good results when analyzing a set of sites. Although the application to different sites requires further analysis, these results seem to show that, maintaining the conditions of the equipment and with a sufficient amount of data, it would be possible to train a general algorithm applicable to any site.

For future developments, another possible approach to this problem is to use a hybrid method combining supervised and unsupervised learning, initially dividing the data into clusters and from there predicting the target variable [46–48].

Author Contributions: Conceptualization, E.M.G., A.A.A.Á. and M.G.A.; methodology, E.M.G., A.A.A.Á. and M.G.A.; software, E.M.G.; validation, A.A.A.Á. and M.G.A.; investigation, E.M.G.; data curation, E.M.G.; writing—original draft preparation, E.M.G.; writing—review and editing, E.M.G., A.A.A.Á. and M.G.A.; visualization, E.M.G., A.A.A.Á. and M.G.A.; supervision, A.A.A.Á. and M.G.A. All authors have read and agreed to the published version of the manuscript.

Funding: This research was funded by Universidad Politécnica de Madrid through the Innovation Project code IE22.0407.

Data Availability Statement: Restrictions apply to the availability of these data. Data was obtained from Menard España and are available from the authors with the permission of Menard España.

Acknowledgments: The authors gratefully acknowledge Menard Spain for supplying the data used in this study They also offer their gratitude to the Ministry of Economy and Competitiveness of Spain by means of the Research Fund Project PID2019-108978RB-C31 and Calle 30 for supporting the Enterprise University Chair "Cátedra Universidad Empresa Calle30-UPM".

Conflicts of Interest: The authors declare no conflict of interest.

Nomenclature

The following abbreviations are used in this manuscript:

Abbreviation	Meaning
AI	artificial intelligence
MWD	Measurement-While-Drilling
ML	Machine learning
SPT	Standard Penetration Test
NN	neural networks
CPT	Cone penetration test
CAPWAP	Case Pile Wave Analysis Program
GP	Gaussian regression process
SVM	Support vector machines
RMR	rock mass rating
SDE	Drilling specific energy
DT	Decision trees
RF	Random forests
API	application programming interface
CART	Classification and Regression Trees
DPSH	Dynamic Probing Super Heavy
MAE	Mean absolute error
MSE	Mean squared error
R^2	Coefficient of determination

Symbol	Meaning
$(t)_{dz=0.2m}$	Time for drilling 0.2 m
C_R	rotation torque
E_n	Normalized energy
E_s	Specific energy
F	Frequency
I()	Indicator function
I_a	Alteration index
L(z)	Entropy of S
P_E	effective weight on the bit
P_M	Hammer power
P_O	thrust force
R_p	Penetration resistance
S	Señal considerada
S_0	drilling area
S_d	Somerton index
V_A	drilling speed
V_R	rotation speed
$\bar{y}$	Mean value of the samples
y_i	True value of the i-th sample
$\hat{y}_i$	Predicted value of the i-th sample
$\alpha, \beta, \gamma, k_0, k_1$	Coefficients
σ_X	Standard deviation of X
σ_{XY}	Covariance
σ_Y	Standard deviation of Y

References

1. Morgan, J.N.; Sonquist, J.A. Problems in the Analysis of Survey Data, and a Proposal. *Am. Stat. Assoc. J.* **1963**, *58*, 415–434. [CrossRef]
2. Ricci, F.; Rokach, L.; Shapira, B. *Recommender Systems Handbook*; Springer Nature: Berlin, Germany, 2011; ISBN 9780387858203.
3. Sckit-Learn 7. Dataset Loading Utilities. Available online: https://scikit-learn.org/stable/datasets.html (accessed on 10 April 2022).
4. Bukkapatnam, S.T.S.; Afrin, K.; Dave, D.; Kumara, S.R.T. Machine learning and AI for long-term fault prognosis in complex manufacturing systems. *CIRP Ann.* **2019**, *68*, 459–462. [CrossRef]
5. Jang, H.; Topal, E. A review of soft computing technology applications in several mining problems. *Appl. Soft Comput. J.* **2014**, *22*, 638–651. [CrossRef]
6. Salazar, F.; Oñate, E.; Toledo, M.A. *A Machine Learning Based Methodology for Anomaly Detection in Dam Behaviour*; Universitat Politècnica de Catalunya: Barcelona, Spain, 2016.
7. Zhou, Y.; Li, S.; Zhou, C.; Luo, H. Intelligent Approach Based on Random Forest for Safety Risk Prediction of Deep Foundation Pit in Subway Stations. *J. Comput. Civ. Eng.* **2019**, *33*, 1–14. [CrossRef]
8. Vlahogianni, E.I.; Karlaftis, M.G.; Golias, J.C. Short-term traffic forecasting: Where we are and where we're going. *Transp. Res. Part C Emerg. Technol.* **2014**. [CrossRef]
9. Puri, N.; Prasad, H.D.; Jain, A. Prediction of Geotechnical Parameters Using Machine Learning Techniques. *Proc. Procedia Comput. Sci.* **2018**, *125*, 509–517. [CrossRef]
10. Juwaied, N. Applications of artificial intelligence in geotechnical engineering. *ARPN J. Eng. Appl. Sci.* **2018**, *13*, 2764–2785.
11. MolaAbasi, H.; Saberian, M.; Khajeh, A.; Li, J.; Jamshidi Chenari, R. Settlement predictions of shallow foundations for non-cohesive soils based on CPT records-polynomial model. *Comput. Geotech.* **2020**, *128*, 103811. [CrossRef]
12. Hu, J. A new approach for constructing two Bayesian network models for predicting the liquefaction of gravelly soil. *Comput. Geotech.* **2021**, *137*, 104304. [CrossRef]
13. Shuku, T.; Phoon, K.K.; Yoshida, I. Trend estimation and layer boundary detection in depth-dependent soil data using sparse Bayesian lasso. *Comput. Geotech.* **2020**, *128*, 103845. [CrossRef]
14. Zhao, T.; Wang, Y. Interpolation and stratification of multilayer soil property profile from sparse measurements using machine learning methods. *Eng. Geol.* **2020**, *265*, 105430. [CrossRef]
15. Rai, P.; Schunesson, H.; Lindqvist, P.-A.; Kumar, U. An Overview on Measurement-While-Drilling Technique and its Scope in Excavation Industry. *J. Inst. Eng. Ser. D* **2015**, *96*, 57–66. [CrossRef]
16. Kadkhodaie-Ilkhchi, A.; Monteiro, S.T.; Ramos, F.; Hatherly, P. Rock Recognition From MWD Data: A Comparative Study of Boosting, Neural Networks, and Fuzzy Logic. *IEEE Geosci. Remote Sens. Lett.* **2010**, *7*, 680–684. [CrossRef]

17. Beattie, N. Monitoring-While-Drilling for Open-Pit Mining in a Hard Rock Environment. Ph.D. Thesis, Queen's University, Kingston, ON, Canada, 2012.
18. Zhou, H.; Monteiro, S.T.; Hatherly, P.; Ramos, F.; Nettleton, E.; Oppolzer, F. Spectral feature selection for automated rock recognition using Gaussian Process classification. In Proceedings of the 2009 Australasian Conference on Robotics and Automatio 2009, Sydney, Australia, 2–4 December 2009.
19. Zhou, H.; Hatherly, P.; Monteiro, S.T.; Ramos, F.; Oppolzer, F.; Nettleton, E.; Scheding, S. Automatic rock recognition from drilling performance data. In Proceedings of the 2012 IEEE International Conference on Robotics and Automation, Saint Paul, MN, USA, 14–18 May 2012; pp. 3407–3412.
20. Leung, R.; Scheding, S. Automated coal seam detection using a modulated specific energy measure in a monitor-while-drilling context. *Int. J. Rock Mech. Min. Sci.* **2015**, *75*, 196–209. [CrossRef]
21. Goh, A.T.C. Back-propagation neural networks for modeling complex systems. *Artif. Intell. Eng.* **1995**, *9*, 143–151. [CrossRef]
22. Goh, A.T.C. Empirical design in geotechnics using neural networks. *Géotechnique* **1995**, *45*, 709–714. [CrossRef]
23. Flaate, K. An investigation of the validity of three pile driving formulae in cohesionless material. *Nor. Geotech. Inst. Publ.* **1964**, 11–22.
24. Lee, I.-M.; Lee, J.-H. Prediction of pile bearing capacity using artificial neural networks. *Comput. Geotech.* **1996**, *18*, 189–200. [CrossRef]
25. Meyerhof, G.G. Bearing capacity and settlement of pile foundations. *J. Geotech. Geoenvironmental Eng.* **1976**, *102*, 197–228. [CrossRef]
26. Teh, C.I.; Wong, K.S.; Goh, A.T.C.; Jaritngam, S. Prediction of Pile Capacity Using Neural Networks. *J. Comput. Civ. Eng.* **1997**, *11*, 129–138. [CrossRef]
27. Rausche, F.; Moses, F.; Goble, G.G. Soil Resistance Predictions From Pile Dynamics. In *Current Practices and Future Trends in Deep Foundations*; American Society of Civil Engineers: Reston, VA, USA, 2004; pp. 418–440.
28. Diaz, M.B.; Kim, K.Y.; Shin, H.S.; Zhuang, L. Predicting rate of penetration during drilling of deep geothermal well in Korea using artificial neural networks and real-time data collection. *J. Nat. Gas Sci. Eng.* **2019**, *67*, 225–232. [CrossRef]
29. Pal, M.; Deswal, S. Modelling pile capacity using Gaussian process regression. *Comput. Geotech.* **2010**, *37*, 942–947. [CrossRef]
30. Galende-Hernández, M.; Menéndez, M.; Fuente, M.J.; Sainz-Palmero, G.I. Monitor-While-Drilling-based estimation of rock mass rating with computational intelligence: The case of tunnel excavation front. *Autom. Constr.* **2018**, *93*, 325–338. [CrossRef]
31. Martínez García, E. Registro de parámetros y control de ejecución de las columnas de módulo controlado. In *18ª Ses. SEMSIG-AETESS Control e Instrumentación en Obras Geotécnicas*; AETESS: Madrid, Spain, 2018.
32. Laudanski, G.; Reiffsteck, P.; Tacita, J.L.; Desanneaux, G.; Benoit, J. Experimental study of drilling parameters using a test embankment. In Proceedings of the Fourth International Conference on Geotechnical and Geophysical Site Characterization, Pernambuco, Brazil, 6 September 2012; Volume 1, pp. 435–440.
33. Möller, B.; Bergdahl, U.K.E. Soil-rock sounding with MWD—A modern technique to investigate hard soils and rocks. In Proceedings of the 2nd International Conference on Site Characterization, Porto, Portugal, 19–22 September 2004.
34. Somerton, W.H. A Laboratory Study of Rock Breakage by Rotary Drilling. *Trans. AIME* **1959**. [CrossRef]
35. Teale, R. The concept of specific energy in rock drilling. *Int. J. Rock Mech. Min. Sci.* **1965**, *216*, 92–97. [CrossRef]
36. Pfister, P. Recording Drilling Parameters in Ground Engineering. *Gr. Eng.* **1985**, *18*, 16–21.
37. Nishi, K.; Suzuki, Y.; Sasao, H. Estimation of soil resistance using rotary percussion drill. In Proceedings of the First International Conference on Site Characterization, Atlanta, Georgia, 19–22 April 1998.
38. Duchamp, J. Apport Des Techniques Statistiques Pour L'exploitation Des Diagraphies Instantanées en Génie civil. Ph.D. Thesis, University of Bordeaux, Bordeaux, France, 1988.
39. Louppe, G. Understanding Random Forests: From Theory to Practice. *arXiv* **2014**, arXiv:1407.7502.
40. Pedregosa, F.; Varoquaux, G.; Gramfort, A.; Michel, V.; Thirion, B.; Grisel, O.; Blondel, M.; Prettenhofer, P.; Weiss, R.; Dubourg, V.; et al. Scikit-learn: Machine Learning in Python. *J. Mach. Learn. Res.* **2011**, *12*, 2025–2830.
41. Breiman, L.; Friedman, J.H.; Olshen, R.A.; Stone, C.J. *Classification and Regression Trees*; Routledge: London, UK, 2017; ISBN 9781315139470.
42. Breiman, L. Random forests. *Mach. Learn.* **2001**, *45*, 5–32. [CrossRef]
43. Menard Menard CMC. Available online: https://www.menard-group.com/en/techniques/controlled-modulus-columns/ (accessed on 10 April 2022).
44. Dinh, D.T.; Huynh, V.N.; Sriboonchitta, S. Clustering mixed numerical and categorical data with missing values. *Inf. Sci.* **2021**, *571*, 418–442. [CrossRef]
45. Buitinck, L.; Louppe, G.; Blondel, M.; Pedregosa, F.; Mueller, A.; Grisel, O.; Niculae, V.; Prettenhofer, P.; Gramfort, A.; Grobler, J.; et al. API design for machine learning software: Experiences from the scikit-learn project. *arXiv* **2013**, arXiv:1309.0238.
46. Astolfi, D.; Pandit, R. Multivariate wind turbine power curve model based on data clustering and polynomial lasso regression. *Appl. Sci.* **2022**, *12*, 72. [CrossRef]
47. Dinh, D.-T.; Fujinami, T.; Huynh, V.-N. *Estimating the Optimal Number of Clusters in Categorical Data Clustering by Silhouette Coefficient*; Springer: Singapore, 2019; ISBN 9789811512094.
48. Märzinger, T.; Kotík, J.; Pfeifer, C. Application of hierarchical agglomerative clustering (Hac) for systemic classification of pop-up housing (puh) environments. *Appl. Sci.* **2021**, *11*, 11122. [CrossRef]

applied sciences

MDPI

Article

An Experimental Study of Nailed Soil Slope Models: Effects of Building Foundation and Soil Characteristics

Mahmoud H. Mohamed, Mohd Ahmed * and Javed Mallick

Civil Engineering Department, College of Engineering, King Khalid University, Abha 61421, Saudi Arabia; m_hussein71@yahoo.com (M.H.M.); jmallick@kku.edu.sa (J.M.)
* Correspondence: mall@kku.edu.sa; Tel.: +966-172-418-439

Abstract: A soil nailing system is a proven effective and economic method used to stabilize earth slopes from the external (factors increasing the shear stress) and internal (factors decreasing material strength) failure causes. The laboratory models with scales of 1:10 are used to study the behavior of nailed soil slope with different soil and building foundation parameters. The models consist of Perspex strips as facing and steel bars as a nailing system to increase the stability of the soil slope. The models of sand beds are formed using an automatic sand raining system. Devices and instruments are installed to monitor the behavior of soil-nailed slope during and after construction. The effect of the soil type, soil slope angle, foundation width and position on the force mobilized in the nail, lateral displacement of the slope, settlement of the foundation and the earth pressure at the slope face, under and behind the soil mass at various foundation pressures, has been observed. It is found that the increase of soil density reduces both slopes facing displacement and building foundation settlements. The slope face displacement and footing settlement will increase with an increase in the width of the foundation and foundation position near the crest of the slope.

Keywords: soil nailing system; soil characteristics; soil slope; foundation pressure; nailed soil

Citation: Mohamed, M.H.; Ahmed, M.; Mallick, J. An Experimental Study of Nailed Soil Slope Models: Effects of Building Foundation and Soil Characteristics. *Appl. Sci.* **2021**, *11*, 7735. https://doi.org/10.3390/app11167735

Academic Editor: Giuseppe Lacidogna

Received: 9 June 2021
Accepted: 18 August 2021
Published: 22 August 2021

1. Introduction

The failure of slopes leads to extensive social and economic consequences and is accompanied by the degradation of the natural environment. A soil nailing system is a proven effective and economic method used to stabilize earth slopes from the external (factors increasing the shear stress) and internal (factors decreasing material strength) failure causes. A review of the literature related to the susceptibility of slope failure is presented by Pourghasemi et al. [1]. The comprehensive review of ground improvement techniques using earth nailing systems, including system installation and construction, failure modes of soil-nailed structures and effects of various construction parameters on the nailed soil slope design method, is presented by Sharma et al. [2]. The guidelines are suggested by Yelti [3] to improve the design procedures for soil-nailed structures considering the lateral pressures generated by soil expansion. The soil nailing system is used to stabilize earth slopes by installing closely spaced nails in the ground in situ. A soil-nailed slope essentially consists of three elements, namely, the soil mass, the slope facing and steel reinforcement (nails). The deformations behavior of the soil nailing system depends on the composite interactions between the soil nailing system elements. The slope facing provides the local stability of the ground between the nails and limits its decompression in the soil nailing system [4]. Bridges et al. [5] have proposed a measurement procedure for the crest displacement of the soil-nailed wall and concluded that effective soil cohesion has the greatest effect on crest displacement. Kotake and Sato [6] have studied the influence of bending stiffness of facing material, installed on a nailed slope surface of dense sand, on deformation and bearing capacity characteristics by conducting the tests on rigid and flexible models of footings. Various experimental and numerical models of nailed soil structures have been studied for the evaluation of structural performance and the development of reliable design methods for nailed soil structures. Ceccato et al. [7] have developed numerical models based on the material point method to investigate the application of plate anchors for landslide stabilization. The numerical model has been validated with the results of some small-scale laboratory tests. Sahoo et al. [8] have carried out the shaking table tests to study the seismic behavior of nailed soil slopes. A simplified limit equilibrium method is presented by Deng et al. [9] to analyze the stability of slopes reinforced with anchor cables considering nonlinear Mohr–Coulomb strength criterion for the shear failure of the slip surface. The laboratory models have been developed by Mahmoud et al. [10] to study the effects of surcharge loading and nails characteristics in Nailed Soil Slope behavior. They found that the increase in length and inclination of soil nails decreases the vertical, horizontal stress and footing settlement, while the increase of spacing of nails increases the vertical and horizontal stress behind the soil mass.

Babu and Singh [11] have studied the lateral displacements, nail forces and failure modes of nailed soil vertical slope under both static and seismic conditions. Numerical analysis of stabilizing mechanisms of loose-fill slopes using hybrid nail arrangement has been carried out by Cheuk et al. [12], who studied the influence of hybrid nail orientations on the behavior of the ground nailing system and concluded that a hybrid nail arrangement would enhance the robustness of the ground nailing system. Taule [13] has applied Bishop's Simplified method to evaluate and rank the sensitivity of the soil cohesion, unit weight and internal friction angle with respect to the global factor of safety using Monte Carlo simulation in a soil nailing design. He has inferred that the soil internal friction angle is the most sensitive parameter, while cohesion and unit weight are less sensitive with respect to the global safety factor. Villalobos et al. [14] present the re-assessment stability analysis of soil nailing design and construction in a heavily weathered granite (residual soil) using a limit equilibrium sliding block method (bi-linear failure surface). They found that even after a maximum acceleration of 0.63 g of a stronger earthquake, the nailed wall did not show any damage, probably due to the use of un-drained shear strength parameters. Alhabshi [15] has studied the hybrid mechanically stabilized earth and soil nail structure and has proposed the optimum length of nails required to reduce the structure displacement. A slope stability analysis of a combined system of soil nails and stabilization piles on loess soil is presented by Wu et al. [16]. Benayoun et al. [17] have proposed the genetic algorithm-based optimization of the soil-nailed structure design. Chen et al. [18] have developed the novel element nail pull-out test to describe the shearing mechanisms at the soil–nail interface in nailed cohesive soils. The stability and displacement characteristics of a composite silt–soil-nailed symmetrical foundation pit are numerically investigated by Han et al. [19]. They have found the critical values of inclination and spacing of the soil nails, the diameter and embedded depth of the mixing pile for the stability of the foundation pit. The soil slope deformation pattern is investigated by Sojoudi and Sharafi [20] using the numerical and experimental modeling of layered and homogenous piled stabilized earth slopes. They study the impact of pile position and surcharge distance from the slope crest on soil deformation patterns, the bearing capacity improvement ratio and slope stability improvement ratio.

Various parameters affect the behavior of the earth nailing systems, such as nail parameters, method of installation of nails, soil density, soil slope height and angle, foundation parameters and loadings, among others. It is clear from the literature survey that the effect of the building foundation parameter on nailed soil slope has not been given due attention. The present paper has investigated the performance of soil-nailed slopes laboratory models under varying soil and building foundation parameters. The effect of the type of soil, soil slope angle, footing width and position of footing on the force mobilized in the nail, lateral displacement of the slope, settlement of the footing and the earth pressure at the slope face, under and behind the soil mass at various foundation pressure have been observed. The results for lateral displacement and footing settlement are presented in terms of the percentage of slope height.

2. Laboratory Models

The three-dimensional small-scale laboratory models of a dry sand slope with a steel nailing system were developed to predict the performance of soil nailing system behavior. The physical models, their dimensions, state of stress and strain and enclosing walls are built and designed so that their performance perfectly simulates the prototype behavior. The scale factor of 10 is adopted in simulating the soil slope with consideration to material handling. The physical models under strip loading are assumed to develop plane strain conditions. The developed physical models have no influence of soil enclosure boundaries on the behavior of the soil nailing system elements, i.e., soil, nails, facing wall and the footing. The depth and width of the experimental model of nailed soil slope also fulfill the criteria that pressure isobars of strip loading laterally extend almost two times the footing width from the center line of the footing and vertically extend five times the footing width.

2.1. Simulation and Dimensional Analysis

The stiffness parameter for the facing wall and nails are maintained for both model and prototype to achieve the correct similarity as per the following relationship.

$$(H^4/EI)_{model} = (H^4/EI)_{prototype} \tag{1}$$

where: H = Slope height (m); E = Young's modulus of elasticity of wall or nail (kPa); I = Moment of inertia per unit length of the wall or the nail (m^4/m).

If we consider a scale factor (N) = ($H_{prototype}/H_{model}$) and in the case of using the same material for the model and prototype, the relation between the model and prototype becomes:

$$I_{model} = I_{prototype}\ (E_{prototype}/E_{model})(1/N)^4 \tag{2}$$

Considering the same unit weight and shearing resistance in the model and prototype soil, then the length and spacing between the nails, the distance of footing from the crest of the slope and applied footing pressure of the model as well as the resulting settlements are proportional to the prototype with the same ratio as the total excavation height ratio.

2.2. Materials, Testing Tank and Loading Frame

The main components of the nailed soil slope system are back-filling material, nails, facing unit and footings. The air-dried siliceous sand is used in the construction of the soil slope model. The model in layers of sand is formed using an automatic sand raining system to control the density. The properties of the sand were determined as per ASTM standards. The procedure for determining the properties of sand is discussed in the literature [21]. Three sand types are used in the model, namely, loose sand with a relative density of 35% and internal angle friction of 30°, medium sand with a relative density of 48% and internal angle friction of 34°, and dense sand with a relative density of 69% and angle internal friction of 40°. The physical properties of the soil material (sand) are given in Table 1. The facing unit in the nailed soil system only has a minor mechanical role, and the main function of the facing is to ensure local stability of the soil between the nails and to limit its decompression. The facing materials should be flexible enough to withstand ground displacement during excavation. Therefore, a Perspex plate with a thickness of 5.0 mm is selected to simulate the Shotcrete in the prototype soil nailing system of thickness 140.0 mm. The nails used in this study are circular cross-section steel bars with a length of 700 mm. The diameter of the model nails is 5 mm to simulate the actual nail diameter of 50 mm. A direct tension test is conducted to know the mechanical properties of the nail [21].

Table 1. Physical properties of soil.

Property	Value	Property	Value
% of clay	0.00	Coefficient of uniformity (C_u)	1.99
% of silt	1.33	Coefficient of gradation (C_u)	1.00
% of fine sand	39.17	Specific gravity (G_s)	2.62
% of medium sand	58.63	Minimum unit weight (γ_{min}) (kN/m^3)	15.30
% of coarse sand	0.87	Maximum unit weight (γ_{max}) (kN/m^3)	17.80
% of fine gravel	0.00	Minimum void ratio (e_{min})	0.472
Effective diameter (D_{10}) mm	0.126	Maximum void ratio (e_{max})	0.712

Considering the simulations criteria for the material and dimensions of the tank, an open-front Perspex box measuring 1760 mm × 850 mm × 1000 mm is chosen as the experimental model of nailed soil slope. The specially designed loading system is used in this work to prevent any disturbance to the nailed soil model and is shown in Figure 1. The loading system is made of a loading frame, loading frame base and hydraulic loading system. A rigid steel plate of dimensions 840 mm × 150 mm × 22 mm thick is used to act as a building foundation to exert a distributed load on the soil. The nailed soil model in the testing box has used a total of 2250 kg of dry sand and nine nails. The mechanical properties of nails, facing and building foundation are given in Tables 2 and 3.

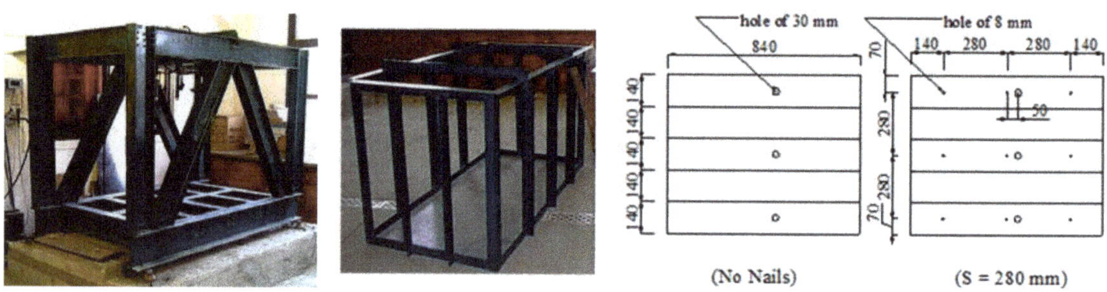

(a) (b) (c)

Figure 1. Testing loading frame, sand tank and slope facing panels; (**a**) loading frame; (**b**) sand tank; (**c**) slope facing panels.

Table 2. Nail properties.

Element	Length (mm)	Maximum Tensile Force (kN)	Strains at Ultimate Stress (με)	Maximum Tensile Strength (kPa)	Young's Modulus (kPa)	Flexural Rigidity, EI (kN.mm^2)	Normal Stiffness, EA (kN/m)
5 mm Steel Nails	700.0	17.4	7625	8.8×10^5	21.2×10^7	6.51×10^{-3}	4165.9

Table 3. Slope facing and foundation plate properties.

Plate Element	Material	Width (mm)	Thickness (mm)	Young's Modulus (kPa)	Bending Stiffness (kN.mm^2)	Axial Stiffness (kN)
Footing	Steel	150	22.0	21.2×10^7	188,186.6	4665.78
Facing	Perspex Plate	140	5.0	4200	0.04375	0.021

2.3. Experimental Program

2.3.1. Preparation of Sand Filled Tank with Reinforcement

1. The 150 mm thick first layer of the sand is formed using the sand spreader and leveled as desired to obtain the required predetermined density.
2. The bottom layer nails with tubes are fixed in their positions.
3. Then, the second layer is formed, and the layer is compacted to reach the required density.
4. In each layer, the horizontal and vertical pressure cells were installed in their predetermined positions.
5. Steps No. 2, 3, 4, were repeated until the tank is filled up and ready to begin the test.

2.3.2. Preparation of Sequence of Construction Phases

The experimental set-up is planned to match the sequence of slope construction in field, i.e., excavation of soil, construction of slope soil nailing system and subsequent placement of foundation pressure. To simulate the actual behavior of the prototype model, the test was divided into two stages. The first stage was the excavation stage to make sand slope, and the second was the loading stage.

1. The 140 mm thick layer of sand was excavated.
2. The first facing panel is placed that includes pressure cells in line with the middle of the panel, and this facing panel is held with two polyethylene strips (sticky tape) to the box side to prevent sand from dropping from the gap between the edges of the slope facing and the box sides.
3. Then, the installation of the first row of strain gauge fitted nails is carried out, which is tightened to the slope facing with nuts by pushing the nail into the sand. These nails are placed in their position using the tubes (black lines in Figure 2).
4. Dial gauges are fitted to record the displacements of the slope facing due to excavation of the second layer.
5. Another four layers of the sand, each 140 mm thick, are dug, and the facing panels are placed as before. At the third and fifth layers, a row of nails and pressure cells are fitted. The facing panel details are shown in Figure 1.
6. In each stage, displacement of the slope face, tensile force in the nails and pressure (horizontal and vertical) are recorded.
7. As the excavations are complete to a depth of 700 mm, the rigid footing was placed at the required location from the nailed slope crest at the top surface, leveled by a water bulb level and then loaded to the required pressure. The different stages of nailed soil slope constructed are depicted in Figure 2.
8. The settlement of the footing is measured using LVDTs.
9. The strain gauge reading is recorded simultaneously with the LVDTs and pressure cells for each foundation pressure case. The spreading of pressure in the nailed soil, horizontal stresses on the slope facing and the horizontal pressure behind the nailed soil are measured by means of five soil pressure transducers and six miniature strain gauge cells, namely, local cell pressure transducers.

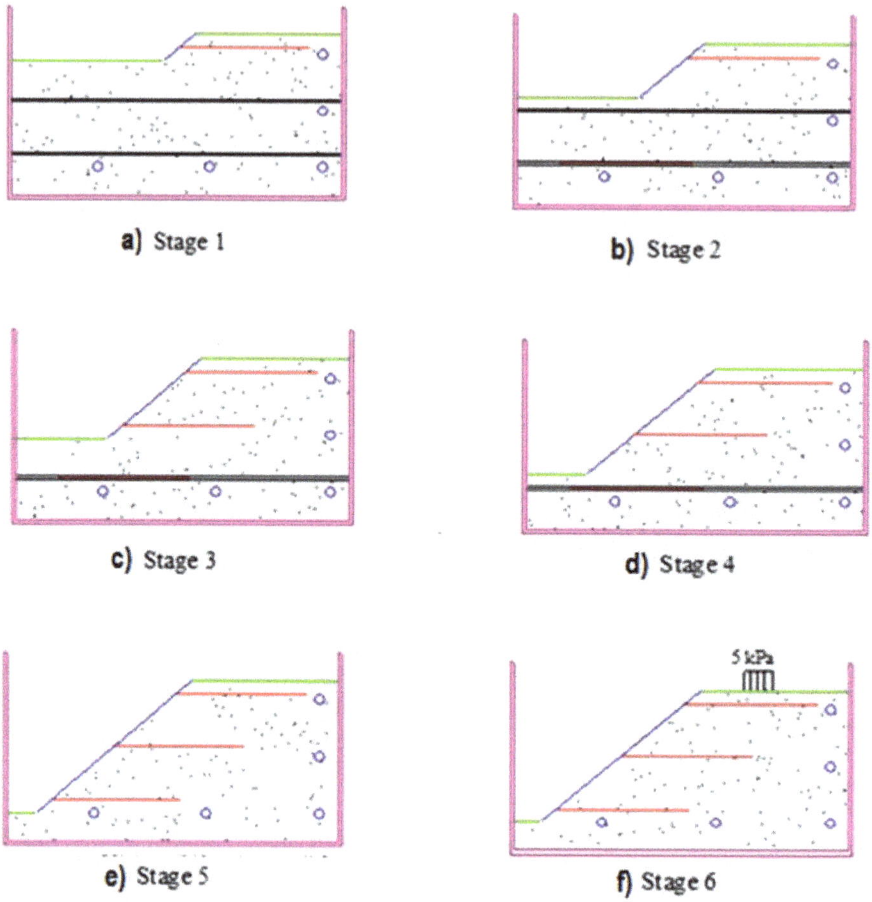

Figure 2. Slope-nailed system construction stages [21].

3. Results and Discussion

In the present study, the laboratory models are prepared with three soil types with different relative densities, i.e., 35%, 48%, 69%, with two angles of slope (θ), i.e., 40°, 45°, with two footing widths (B), i.e., 150 mm and 200 mm and with four positions of footing from upper slope corner (X), i.e., 75 mm (0.5 B), 150 mm (B), 225 mm (1.5 B), 300 mm (2 B). The displacement of the slope, force in the nail, settlement of the footing and the earth pressure in back-fill soil due to different soil and footing parameters (relative density, slope angle, footing width, footing position) are measured and obtained results are discussed below. The horizontal nails (δ = 0) with a length of slope height (Ln = 700 mm) are used in the models. The vertical (Sv) and horizontal (Sh) spacing of nails are provided as 0.4 times the slope height (0.4 *H*). The lateral displacement of the nailed slope facing are recorded with three dial gauges at three measurement points A, B and C at distances Z = 70 mm (0.1 *H*), 350 mm (0.5 *H*) and 630 mm (0.9 *H*) from the top surface. Pressure cell numbers are mounted flush with the back of the slope facing with a special arrangement to measure horizontal pressure, as shown in Figure 3 at points A, B and C. Readings are measured during construction of slope and after each increment of foundation pressure (q) from 5.0, 10.0, 20.0, to 30.0 kPa. The complete nailed soil models with various parameters are shown in Figure 3.

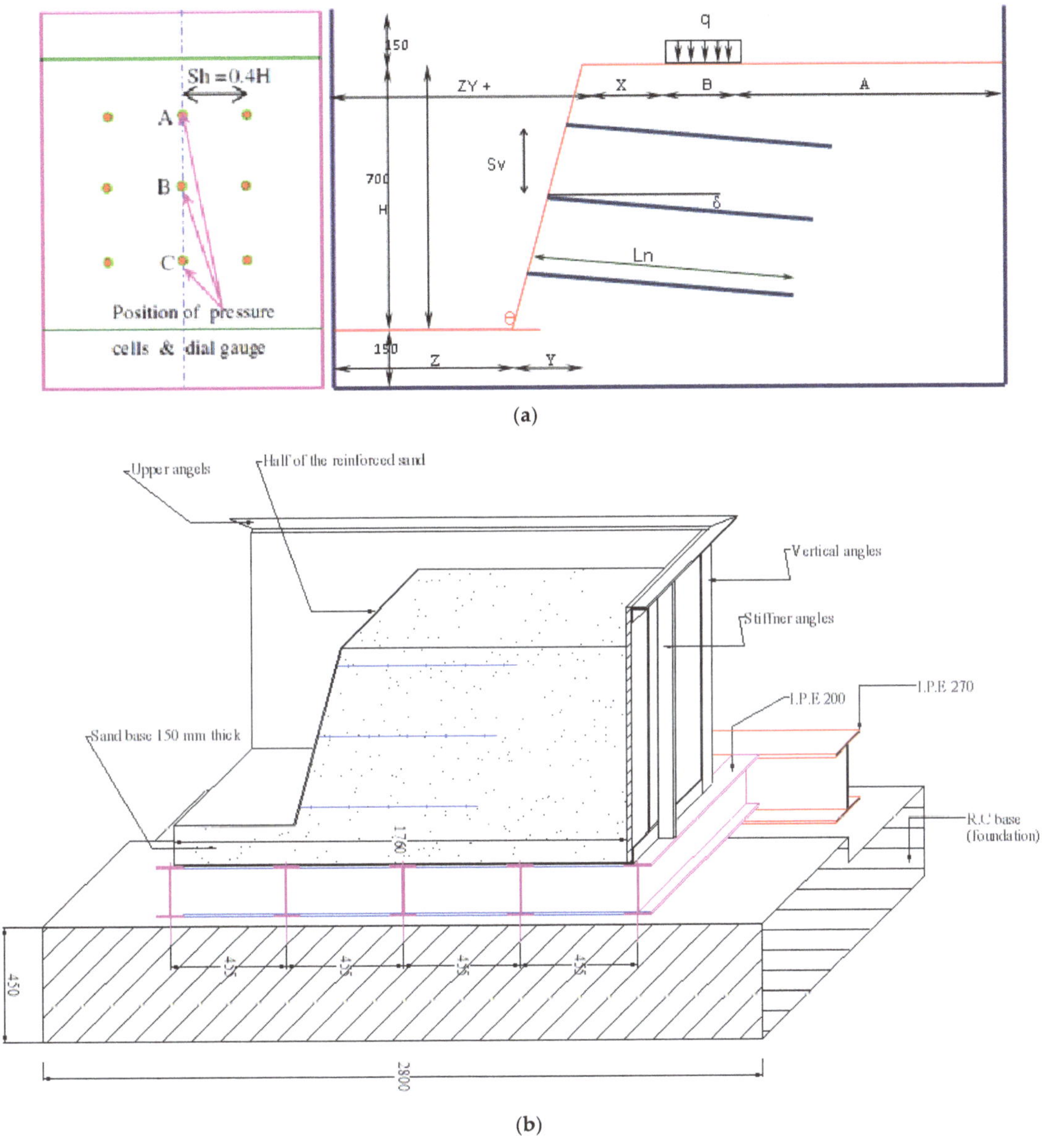

(a)

(b)

Figure 3. Laboratory model with different soil and building foundation parameters; (**a**) section and elevation of nailed soil slope model; (**b**) 3D model of nailed soil slope (all dimensions are in mm).

3.1. Effect of Soil Type

3.1.1. Horizontal Movement of the Soil Nailing Slope

The effect of the soil type on the horizontal movement of the slope face of the nailed soil system during the construction stage and loading stages is shown in Figure 4a. It is clear from the figure that the inclusion of reinforcement reduces the horizontal movement of the slope face. The figure also depicts that the horizontal movement of the slope in the middle third is higher than the slope top and bottom levels, and it has the lowest value at the top of the slope. This is attributed to the boundary condition of the soil slope. The top level of the slope with foundations and bottom portion will restrain the movement of the slope face.

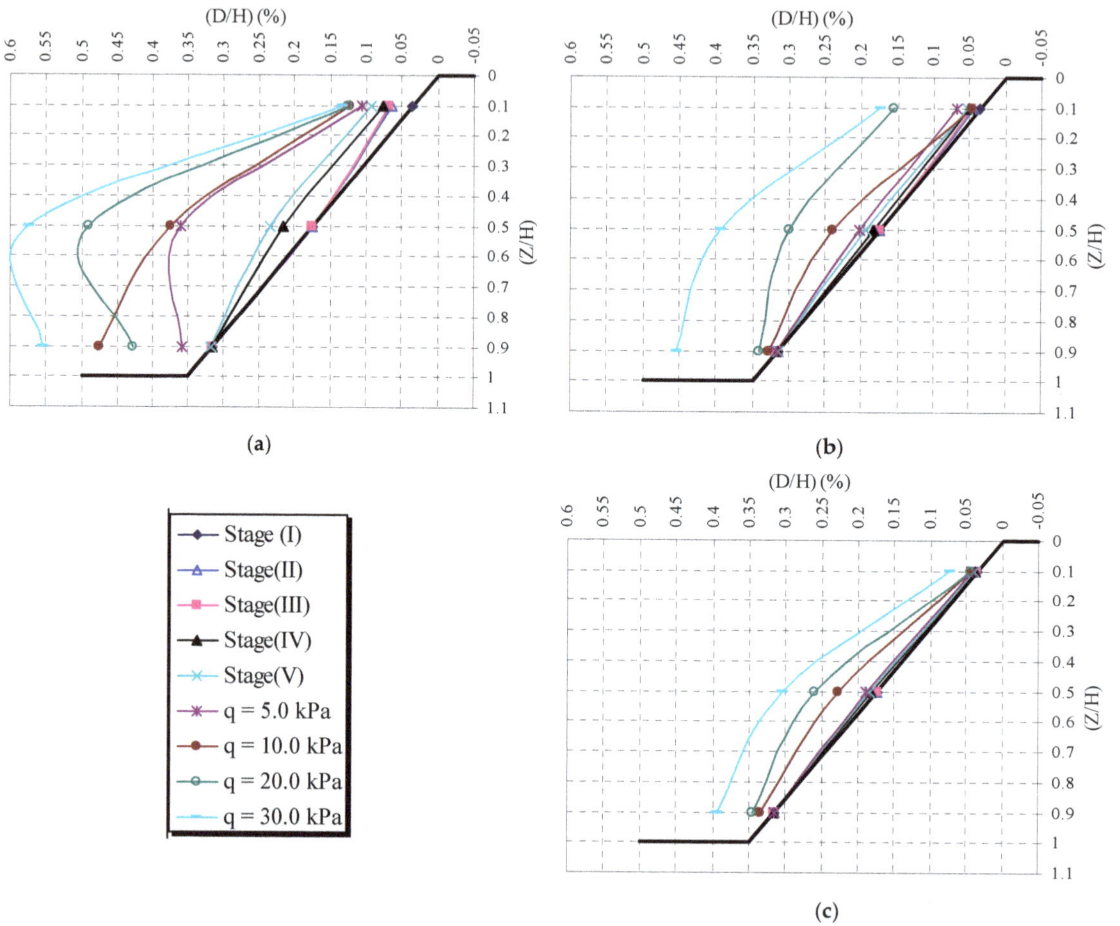

Figure 4. Influence of soil density on the horizontal movement of slope face in construction and loading stages; (**a**) loose sand; (**b**) medium sand; (**c**) medium sand.

Figure 5 shows the effect of the soil type on the horizontal movement of the slope face during the construction stage and loading stages of varying foundation pressure. It can be inferred from the results that the increase of the soil density reduces the lateral movement of the soil-nailed slope. Reduction in the slope movement may be because the friction between soil and reinforcement is strongly dependent on the density of the soil and restrained dilatants behavior in dense granular soil. The lateral movement of the slope face of the soil with 69% relative density is less than half of the lateral movement of the slope face of the soil with 35% relative density at higher foundation pressure.

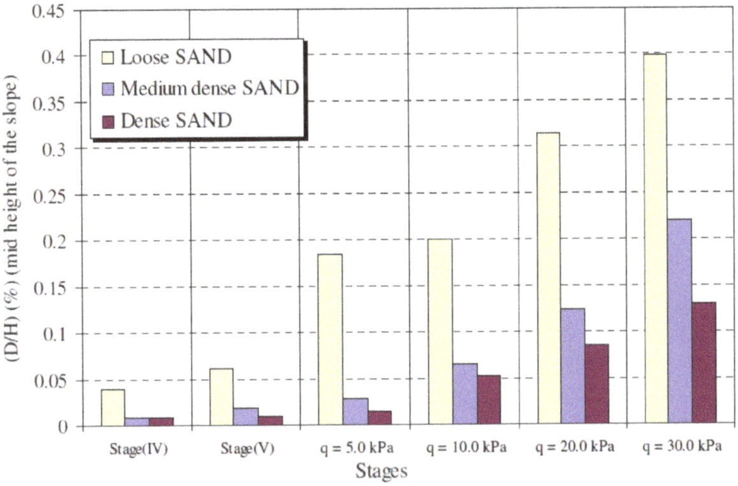

Figure 5. Influence of soil density on the horizontal movement at mid-height of slope.

3.1.2. Settlement of the Footing

The footing settlements for different soil densities at different footing pressures are presented in Figure 6. It is clear from the figure that the increase of the soil density reduces the footing settlement. This reduction of the settlement could be attributed to the reduction of the facing displacement. Furthermore, the increase of the soil density is accompanied by an increase in shearing resistance and thus a reduction of the footing settlement. The difference of footing settlement of loose sand and dense sand is more than twice at all stages of loading, but the difference of settlement of footing on medium dense sand nailed slope reduces at higher foundation pressure.

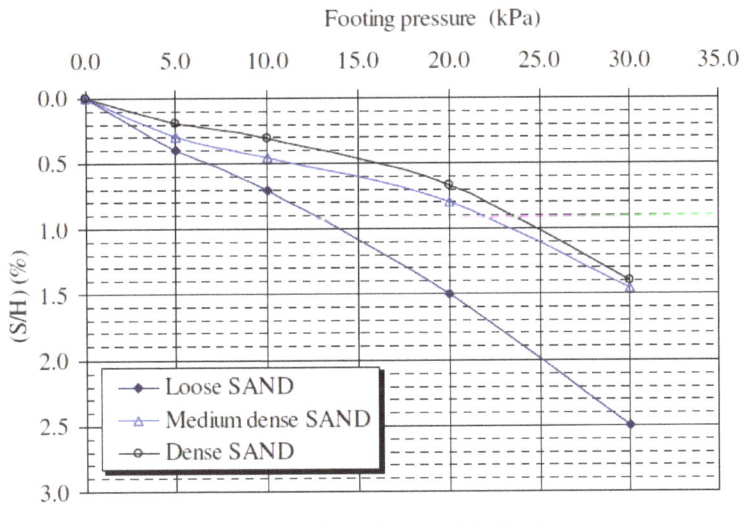

Figure 6. Influence of soil density on the settlement of the footing.

3.1.3. Force in the Nail

The effect of soil density on the maximum tensile force in the nails at different levels is shown in Figure 7. It is clear from the figure that as the density of the soil increases, the tensile force in nails both in construction stages and loading stages. This may be because the dense sand is more stable and mobilizes a smaller force than the force in a loose state. It is also clear that the lower nails have less tensile force than the upper nails, but the middle nails have the maximum tensile force. This may be due to the occurrence of maximum horizontal displacements at the mid-height of the slope.

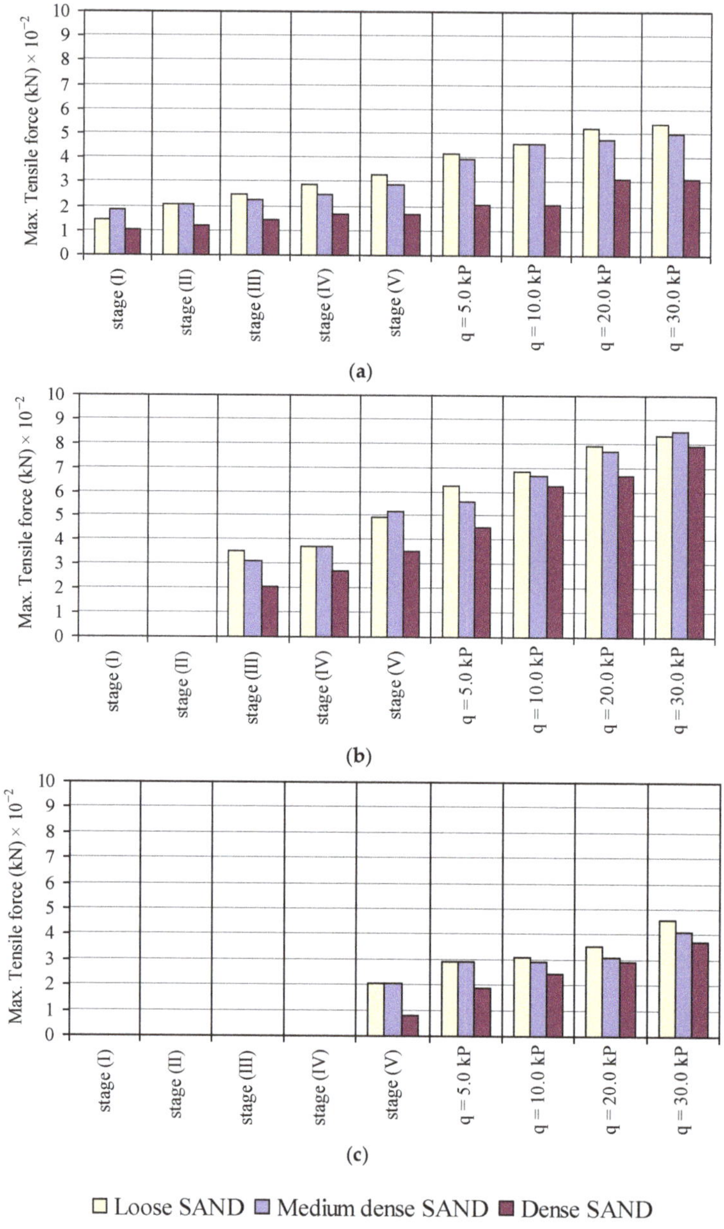

Figure 7. Influence of soil density on the maximum tensile force in the nails; (**a**) upper nail; (**b**) middle nail; (**c**) lower nail.

3.1.4. Vertical Pressure under Nailed Soil Mass

Figure 8 presents the effect of soil density on vertical stresses under the soil mass at various loading stages. It is evident from the figure that the effect of soil density on vertical stresses under the soil mass was directly proportional, i.e., the denser the soil, the higher the vertical stresses under the soil mass. This may be due to an increase in the unit weight.

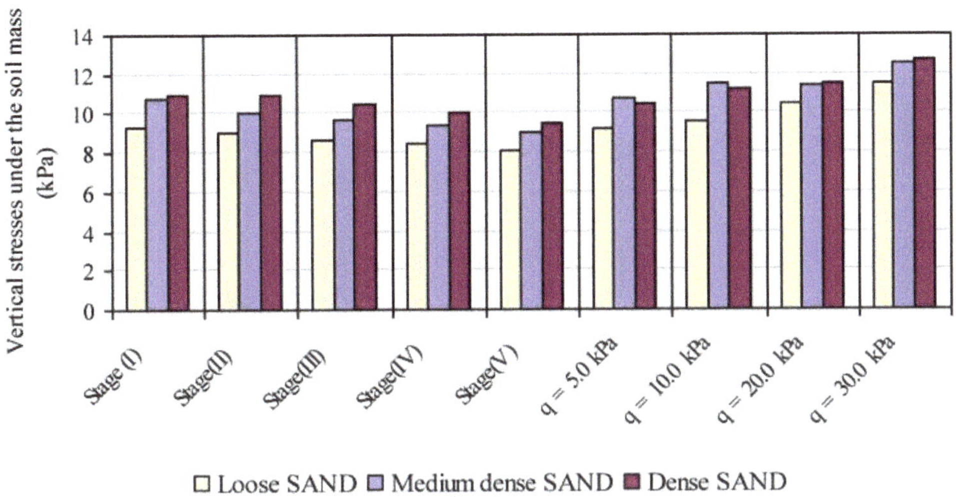

Figure 8. Influence of soil density on the vertical stresses under the nails.

3.1.5. Horizontal Stresses on the Slope Face

Figure 9 shows the influence of the soil density on the horizontal stresses at the slope face at the middle of the slope during construction stages and loading stages. The figure depicts that, in general, when the soil density increases, the horizontal stresses increase. The percentage increase in horizontal stresses at the slope face increases with the increase of surcharge loading. The percentage increase in horizontal stresses at the slope face subjected to higher surcharge loading is about 20% when the soil density increases from 35% to 69%.

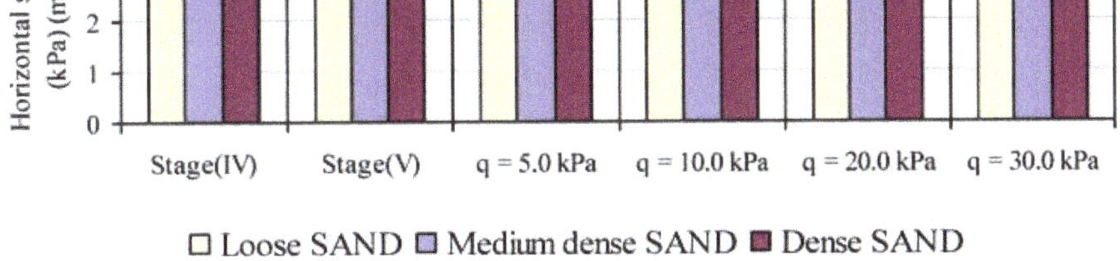

Figure 9. Influence of soil density on the slope face horizontal stresses at mid-height.

3.1.6. Horizontal and Vertical Stresses behind Reinforced Soil Mass

The horizontal and vertical stress behind the soil mass at different soil densities during construction and loading stages are given in Figures 10 and 11. It is clear from the figures that at construction stages, the vertical stresses behind the soil mass increase as the soil density increases. The behavior of horizontal stresses behind the soil mass with the soil density shows an inverse behavior, i.e., the horizontal stresses behind the soil mass decrease when the soil density increases. A similar trend of variation of horizontal and vertical stresses is also obtained at loading stages. The increase or decrease in horizontal stresses at various stages is more as compared to vertical stresses behind the soil mass at different soil densities.

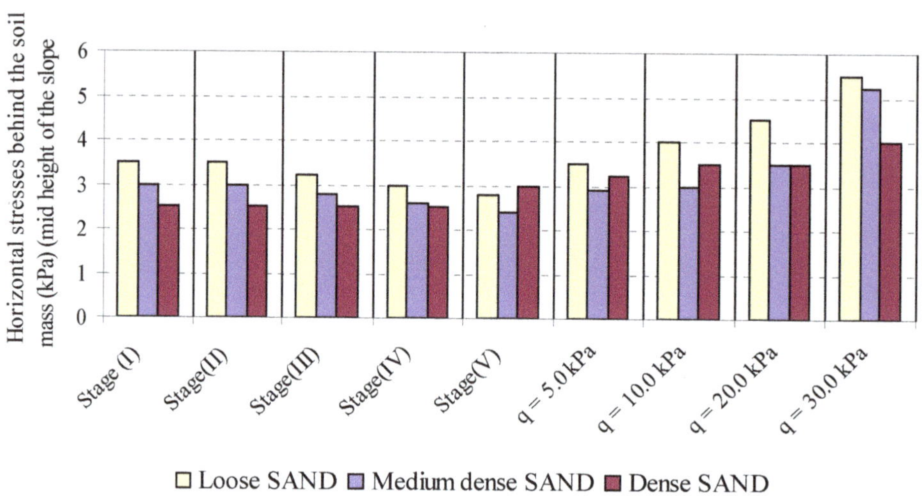

Figure 10. Influence of soil density on horizontal stresses behind the nailed soil mass at mid-height of slope.

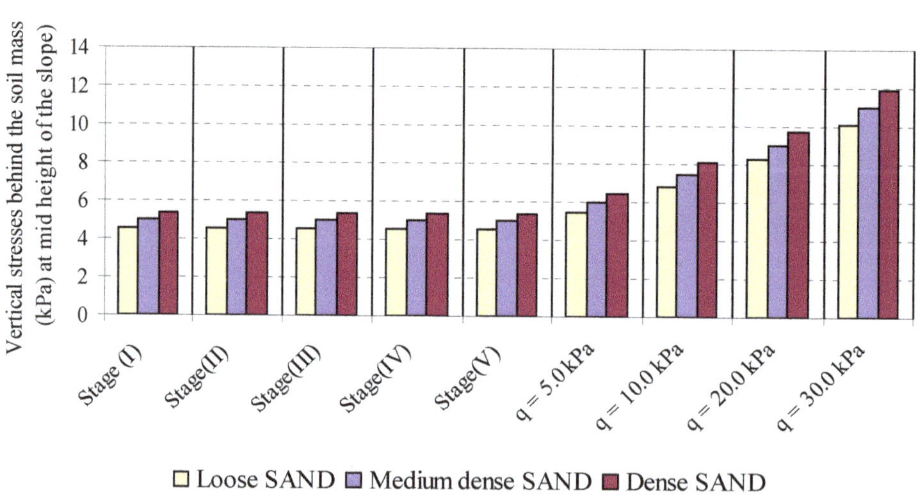

Figure 11. Influence of soil density on vertical stresses behind the nailed soil mass.

3.2. Inclination of the Soil Slope

3.2.1. Horizontal Movement of the Nailing Slope

Figure 12 represents the effect of soil inclination on the horizontal movement of the slope face during construction stages and loading stages. It is clear from the figures that there is no significant effect of slope inclination on the horizontal movement of the slope face except at heavy surcharge loading.

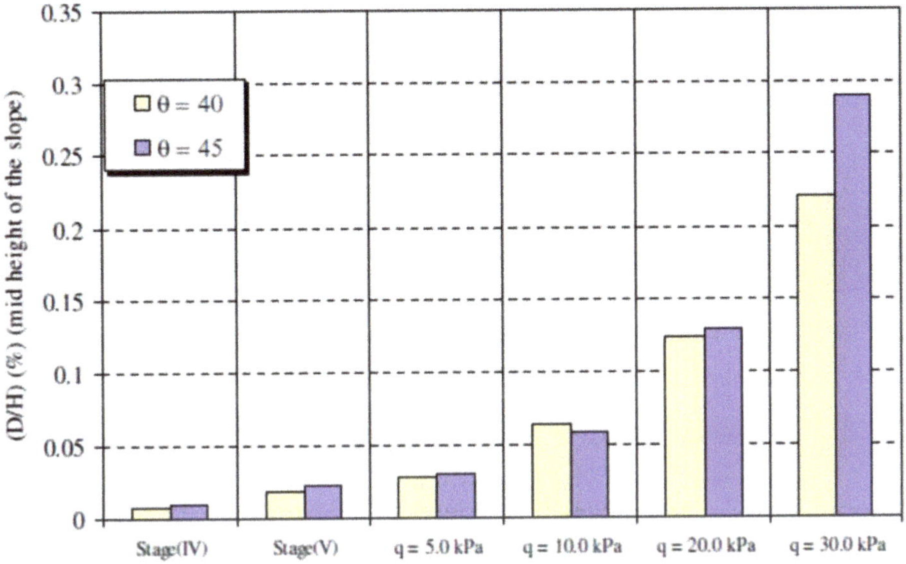

Figure 12. Influence of slope inclination on the horizontal movement at mid-height of slope.

3.2.2. Settlement of the Footing

Figure 13 shows the footing settlements for slope inclinations at various footing pressures. It is observed from the figure that the footing settlements have not reduced considerably with the increase of the slope inclination at various surcharge pressure. The reduction of footing settlements with an increase of the slope inclination by a ratio ranged from 20.0% to 37.9%.

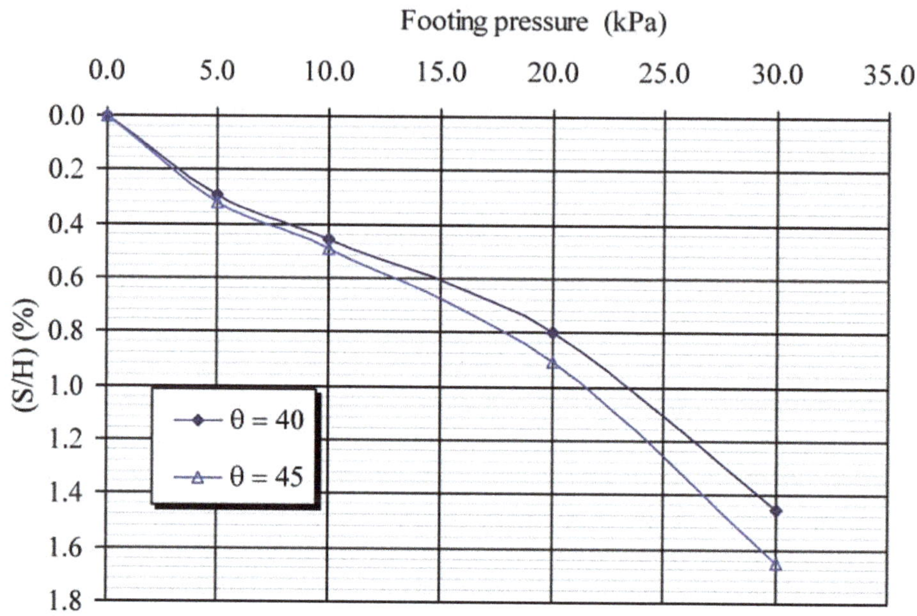

Figure 13. Influence of slope inclination on the settlement of the footing.

3.2.3. Force in the Nail

The influence of the slope inclination on the maximum tensile force in the nails at different levels is shown in Figure 14. It is evident from the variation of tensile force that as the slope inclination increases, the maximum tensile force of middle nails decreases in construction stages but increases in tensile force loading stages. The variation of tensile force in the upper and lower nails is different, and the maximum tensile force increases with the increase of slope inclination at all stages. It is also clear that the lower nails have lesser tensile force than the upper nails, and the middle nails have the maximum tensile force. This may be attributed to the maximum horizontal displacements at mid-height of the slope.

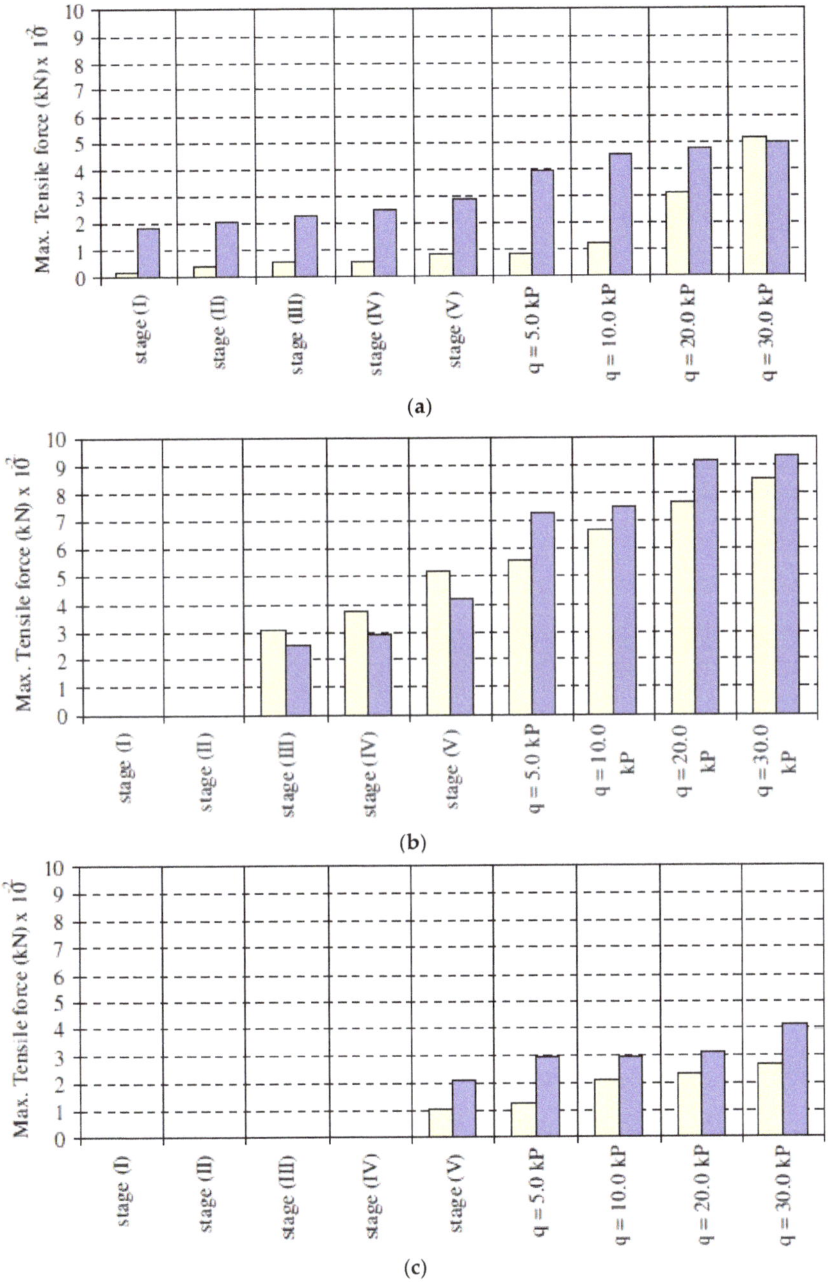

Figure 14. Influence of slope inclination on the maximum tensile force of the footing; (**a**) upper nail; (**b**) middle nail; (**c**) lower nail.

3.2.4. Horizontal Stress on the Slope Face

Figure 15 represents the influence of the slope inclination on the horizontal stresses at the slope face at the middle point of the slope during construction stages and after applying footing pressure. It is clear from the figure that as the slope inclination increases, the horizontal stress at the slope face increase in both construction and loading stages. The percentage increases in the horizontal stresses at the slope face range from 16% to 43%.

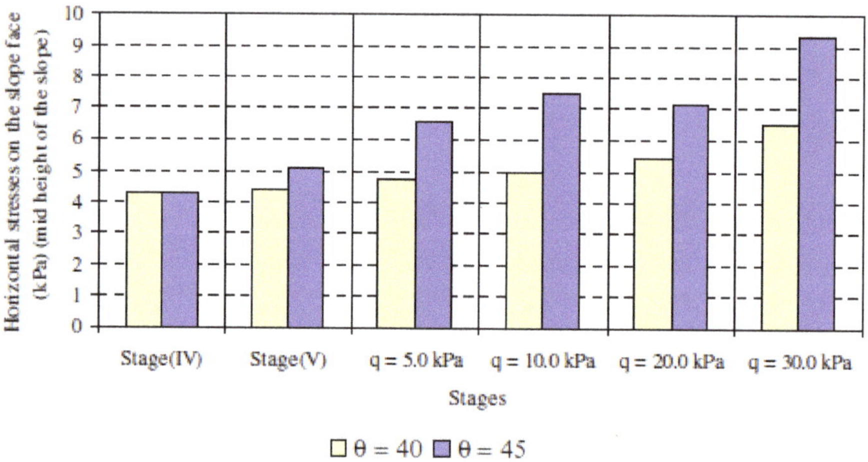

Figure 15. Influence of slope inclination on the slope face horizontal stresses at mid-height of slope.

3.2.5. Vertical Pressure under Nailed Soil Mass

The influence of the slope inclination on the vertical stresses under the nailed soil mass during construction stages and under surcharge loading is presented in Figure 16. It can be concluded from the figure that the effect of slope inclination on vertical stresses under the soil mass was directly proportional in loading stages, i.e., the higher the inclination angle, the higher the vertical stress under the soil mass. The vertical pressure during the slope construction will be the same at different angles of slope.

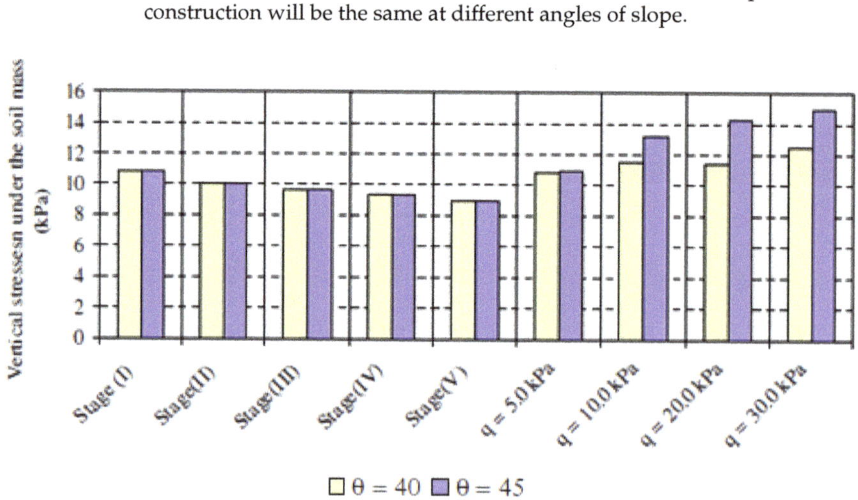

Figure 16. Influence of slope inclination on vertical stresses under the nailed soil mass.

3.2.6. Horizontal and Vertical Stresses behind Nailed Soil Mass

Figures 17 and 18 show the horizontal and vertical stresses behind the slopped soil mass in construction and loading stages measured at the middle of the slope. The figures depict that during excavation stages, the horizontal stresses behind the soil mass increase as the slope inclination decreases, whereas it decreases as the slope inclination decreases. The vertical stresses behind the soil mass decrease as the slope inclination decreases at all stages. The horizontal and vertical stresses behind the soil mass were increased when the surcharge load is increased.

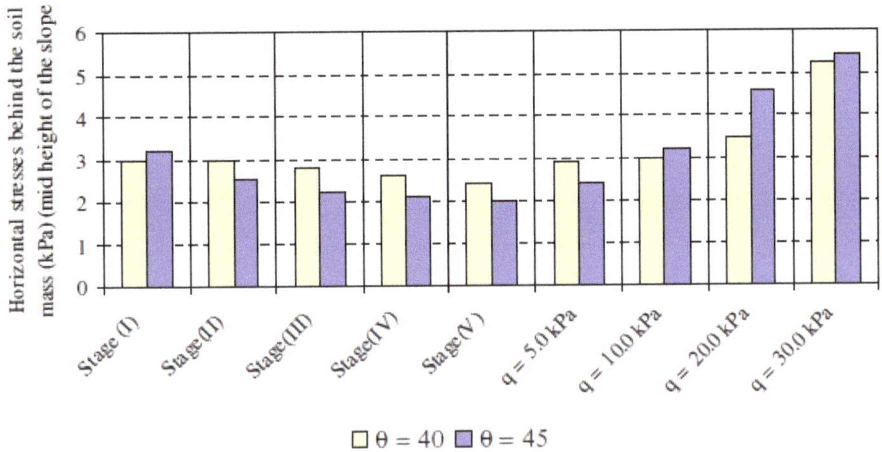

Figure 17. Influence of slope inclination on horizontal stresses behind the nailed soil mass at mid-height of slope.

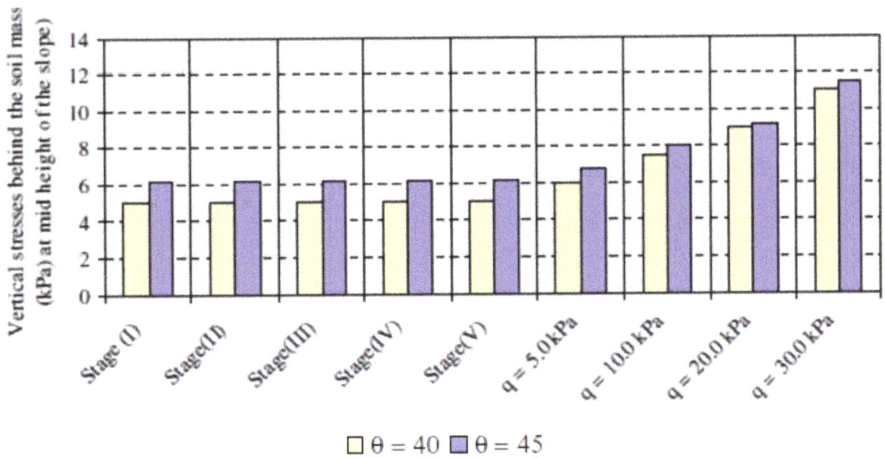

Figure 18. Influence of nail inclination on vertical stresses behind the nailed soil mass at mid-height of slope.

3.3. Width of the Footing

3.3.1. Lateral Displacement of the Soil-Nailed Slope

Figure 19 shows the effect of the footing width on the horizontal movement of the slope face during loading stages. The results indicate that at lower surcharge loading, there is no influence of footing width on the lateral movement of the soil-nailed slope, but with an increase of surcharge loading, lateral movement of the soil-nailed slope is increased. It can also be concluded that the lateral movement of the facing increases with an increase of footing width.

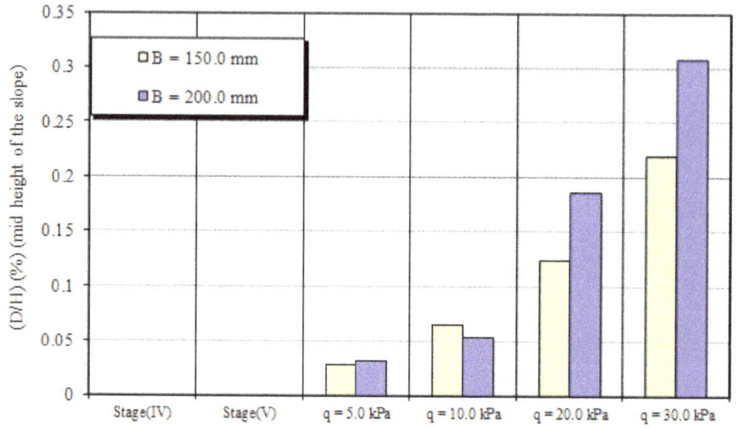

Figure 19. Influence of footing width on the horizontal movement at mid-height of slope.

3.3.2. Settlement of the Footing

The footing settlements for different footing widths at various footing pressures are given in Figure 20. It can be seen from the figure that the footing settlement increases when the footing width increases, and the percentage increase in footing settlement ranges from 20% to 38%. The increase of the footing settlement could be attributed to the increase of the lateral movement of the slope.

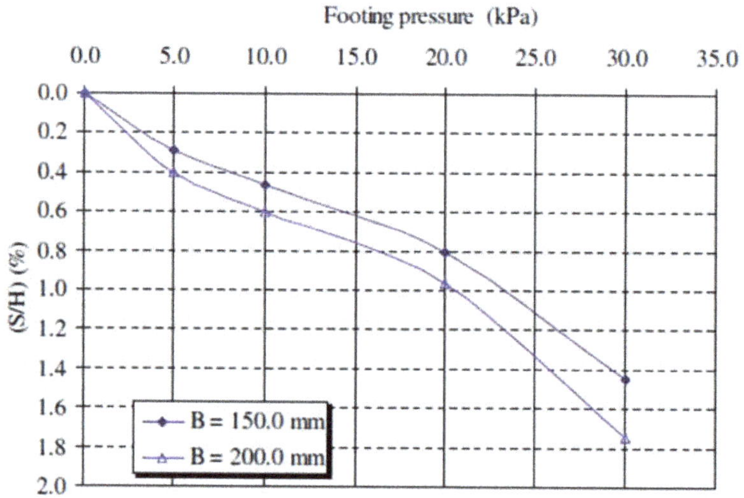

Figure 20. Influence of footing width on the settlement of the footing.

3.3.3. Force in the Nail

The influence of footing width on the maximum tensile force mobilized in the nails at different levels is presented in Figure 21. It can be concluded that the lower nails have lesser tensile force as compared to the upper nails at all stages of loading, and the tensile force increases in all nails as the footing pressure increases with different footing widths. The increase in footing width decreases the tensile force at lower surcharge (5 kPa), but with further increase of surcharge loadings, the tensile force in the nails is increased with the increase of footing width because the stability of the slope at high footing width needs more tensile force mobilization.

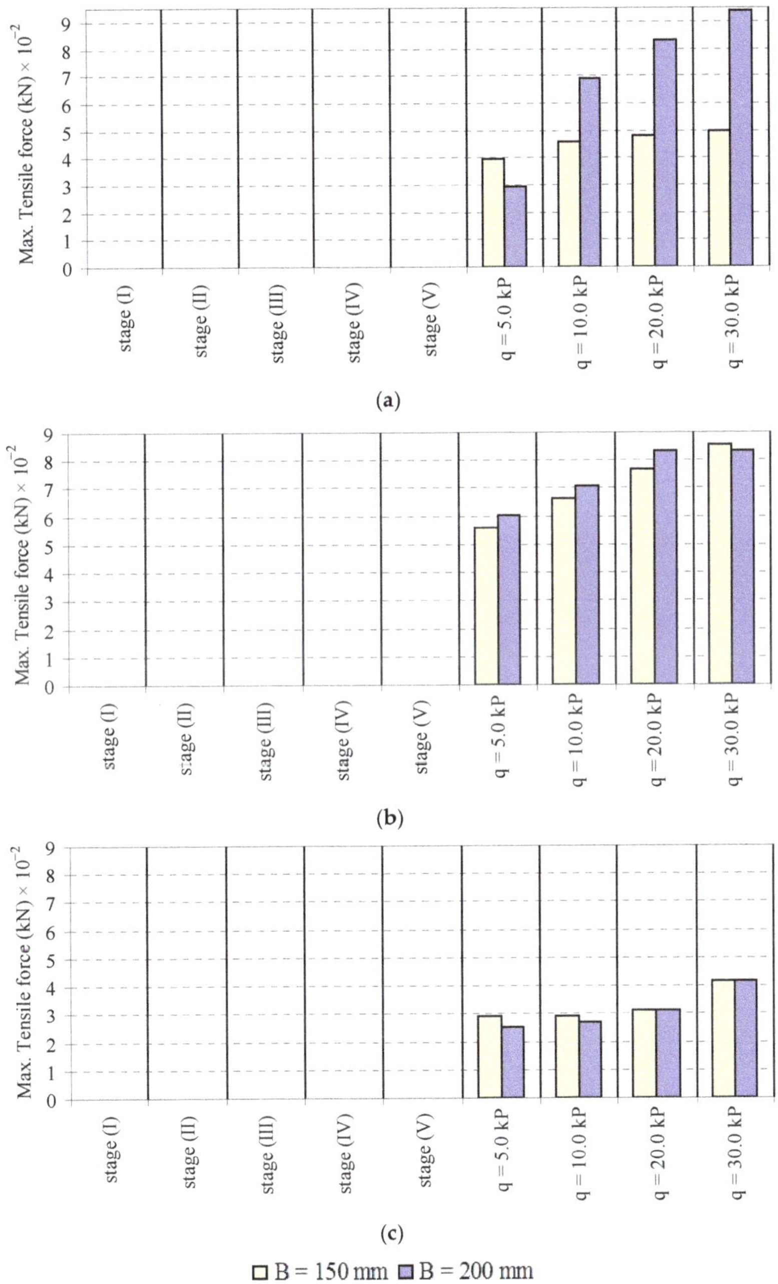

Figure 21. Influence of footing width on the maximum tensile force of the footing; (**a**) upper nail; (**b**) middle nail; (**c**) lower nail.

3.3.4. Horizontal Stresses on the Slope Face

Figure 22 gives the slope face horizontal stresses at mid-height of the slope with the different footing widths during the loading stages. The figure shows that the horizontal stresses at the slope face increase with the increase of the footing width and the footing pressure. The influence of the footing width is more at higher footing pressure. The percentage increase in the slope face horizontal stress with a 35% increase in footing width at the highest surcharge loading is considered to be about 20%.

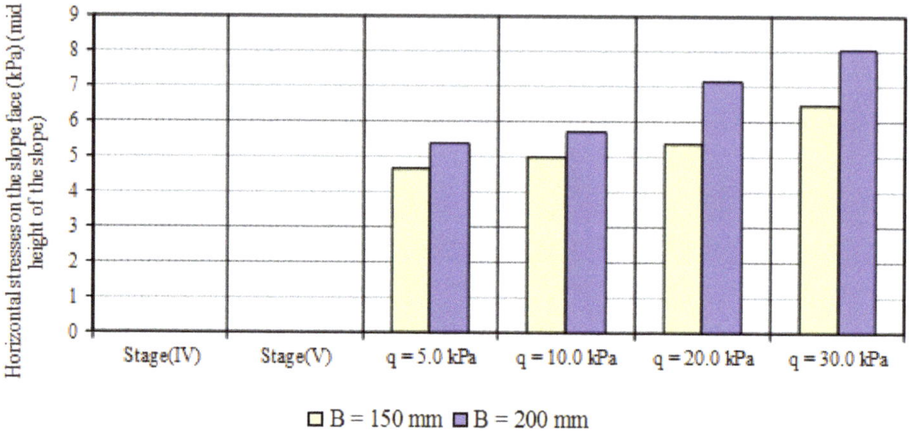

Figure 22. Influence of footing width on the slope face horizontal stresses at mid-height of slope.

3.3.5. Vertical Pressure under the Nailed Soil Mass

The vertical stresses under the nailed soil mass with the different footing widths under different stages are given in Figure 23. It can be seen from the figure that the effect of footing width on vertical stresses under the nailed soil mass was directly proportional, i.e., the higher the footing width, the higher the vertical stresses, and at different footing widths, increasing the surcharge load leads to increase in vertical stresses under the nailed soil mass.

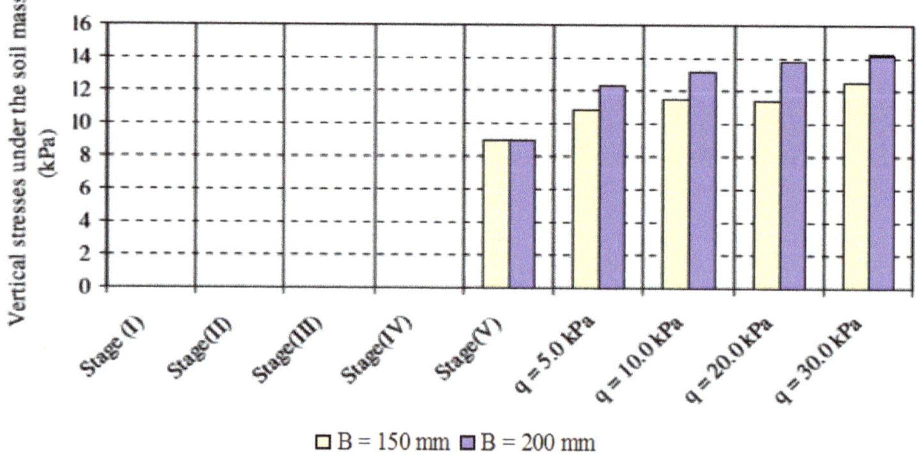

Figure 23. Influence of footing width on the slope face vertical stresses under the nails at mid-height of slope.

3.3.6. Horizontal and Vertical Stresses behind Nailed Soil Mass

Figures 24 and 25 present the effect of footing width on the horizontal and vertical stresses behind the soil mass in loading stages in the central portion of soil mass. The figures show that increasing footing width results in increasing the horizontal and vertical stresses behind the soil mass. The increase in the horizontal and vertical stresses behind the soil mass ranges from 0% to 21% and 0% to 57%, respectively, with a 35% increase in footing width and when surcharge load is increased from 5 kN/m^2 to 30 kN/m^2.

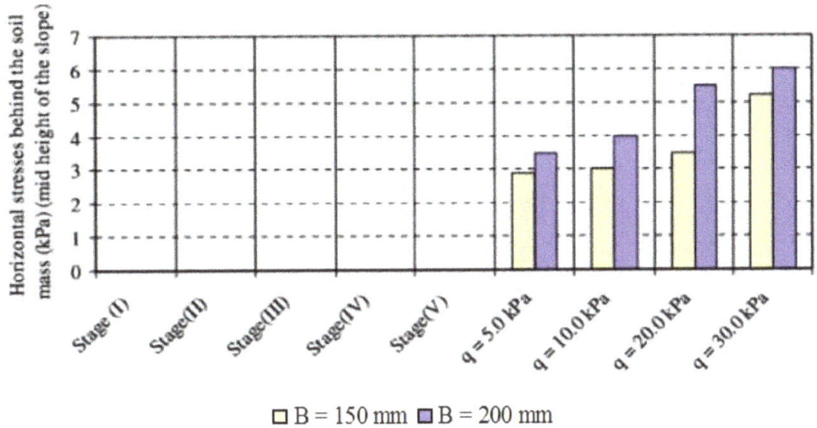

Figure 24. Influence of footing width on horizontal stresses behind the nailed soil mass at mid-height of slope.

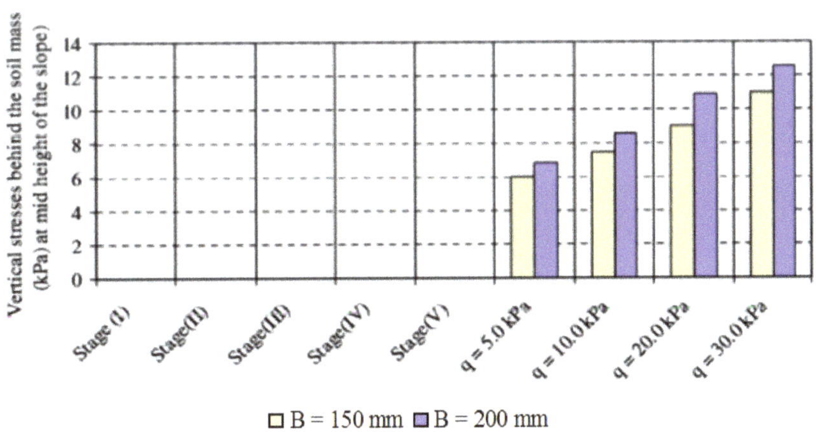

Figure 25. Influence of footing width on Vertical stresses behind the nailed soil mass at mid-height of slope.

3.4. Footing Position from Slope Face

3.4.1. Horizontal Displacement of the Soil Nailing Slope

Figure 26 shows the effect of the footing location on the horizontal displacement of the slope face under the loading stage. It can be observed from the results that at the loading stage of footing pressure 20.0 kPa, with an increase of the footing distance from the slope face, the horizontal movement of the slope face at the mid-row of nails decreases. There is a sharp decrease in lateral movement of slope up to a footing distance of 1.5 B, and then a percentage decrease in slope movement is reduced. The reduction in slope-facing displacement is about 82%, with an increase in the footing distance of twice the footing width.

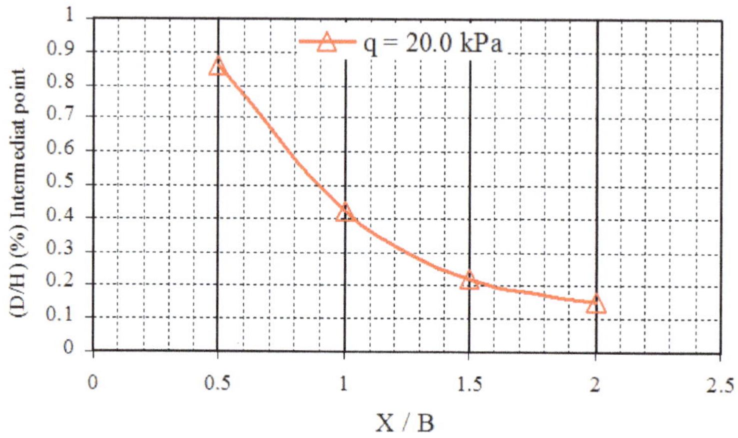

Figure 26. Influence of footing position on the mid-height slope horizontal movement.

3.4.2. Settlement of the Footing

Figure 27 presents the variation of the footing settlements for different footing locations under different footing pressures. From the figure, it can be seen that the footing settlement decreases as the footing distance from the slope face increases. A similar variation is obtained at all surcharge loads. This could be attributed to the fact that increasing the footing distance leads to a more stable slope.

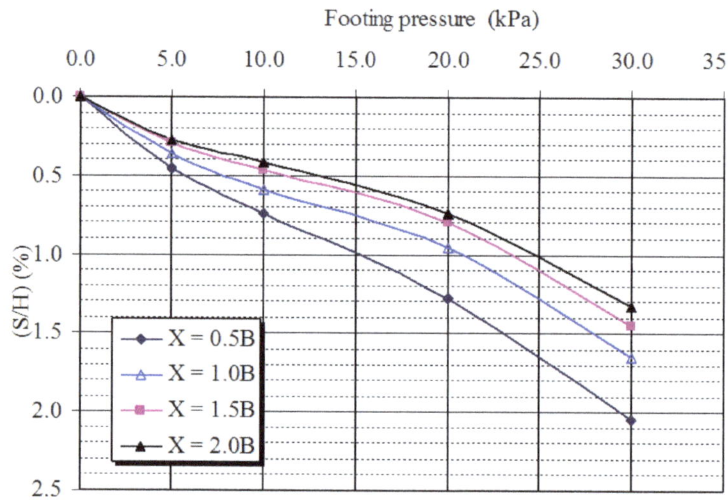

Figure 27. Influence of footing position on the settlement of the footing.

3.4.3. Force in the Nail

The variation of the footing position on the mobilized maximum tensile force in the nails at different levels is shown in Figure 28. The influence of footing position is small in the upper and lower nails. As the footing distance from the slope face increases, the maximum tensile force of the middle nails decreases. This may be attributed to the maximum horizontal displacements occurring at the mid-height of the slope.

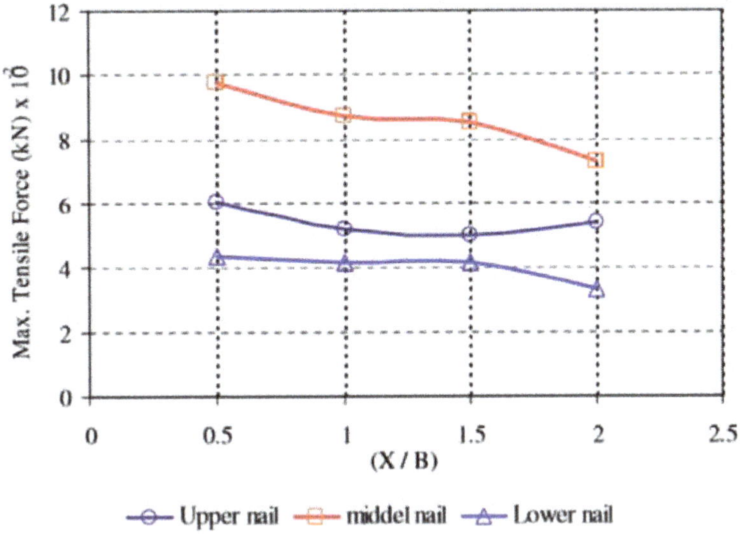

Figure 28. Influence of footing position on the maximum nail tensile force at levels.

3.4.4. Horizontal Stresses on the Slope Face

The variation of the footing position on the slope face horizontal stress at the middle of the slope during loading stages is shown in Figure 29. From the figure, it can be inferred that there are consider horizontal stresses present at the slope face, and the horizontal stresses measured at the same location of footing increase with the increases of footing pressure. As the footing distance from the slope face increases, the horizontal stresses at the slope face decrease as expected. By placing the footing to twice the footing width, the slope face horizontal stresses reduce to less than half of the original value.

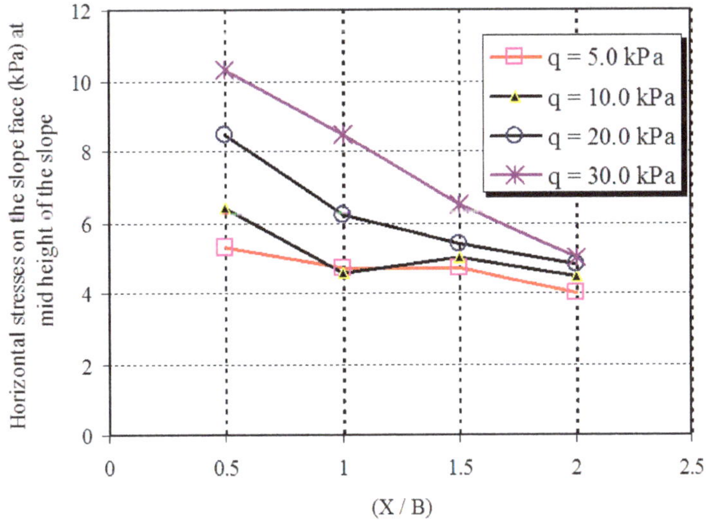

Figure 29. Influence of footing position on the slope face horizontal stresses under the nails at mid-height of slope.

3.4.5. Vertical Pressure under Nailed Soil Mass

Figure 30 presents the variation of the footing locations on the vertical stress under the soil mass with different footing pressure. From this figure, it can be seen that the effect of footing distance on vertical stresses under the soil mass is fluctuating, i.e., increasing the footing distance from slope face leads to an increase in vertical stresses under the soil mass up to a certain distance (x = 1.50 B) and a further increase of footing distance decreases the vertical stresses.

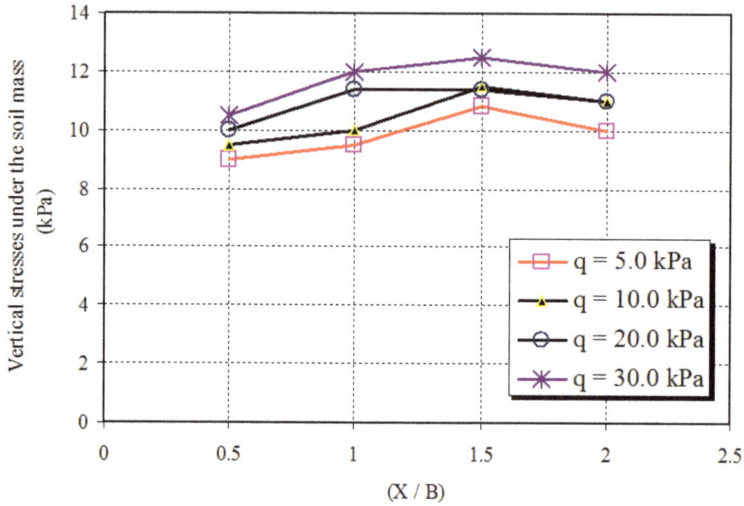

Figure 30. Influence of footing position on vertical stresses under the nailed soil mass.

3.4.6. Horizontal and Vertical Stresses behind Nailed Soil Mass

Figures 31 and 32 give the variation of the horizontal and vertical stresses behind the soil mass with different footing locations over the surface of the nailed slope with different foundation pressures. The influence on the horizontal and vertical stresses behind the soil mass with different footing locations is generally small at lower footing pressure. It is also clear from the figures that the horizontal stresses and vertical stresses behind the soil mass increase as the footing distance from the slope increases.

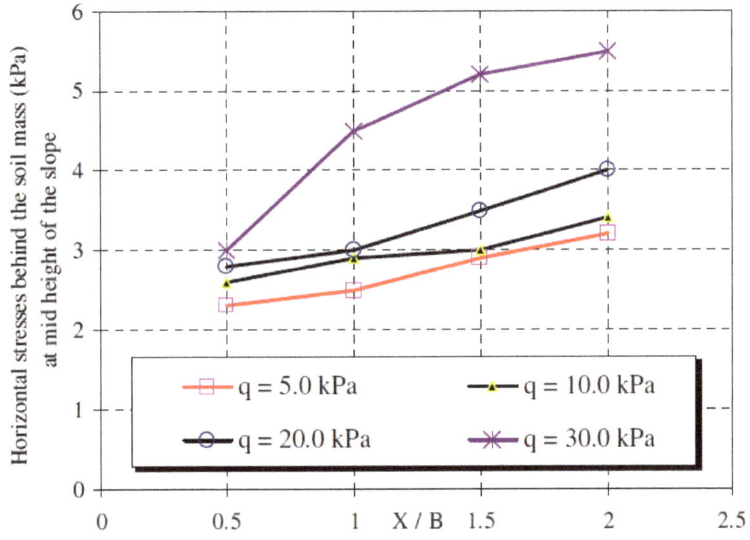

Figure 31. Influence of footing position on horizontal stresses behind the nailed soil mass at mid-height of slope.

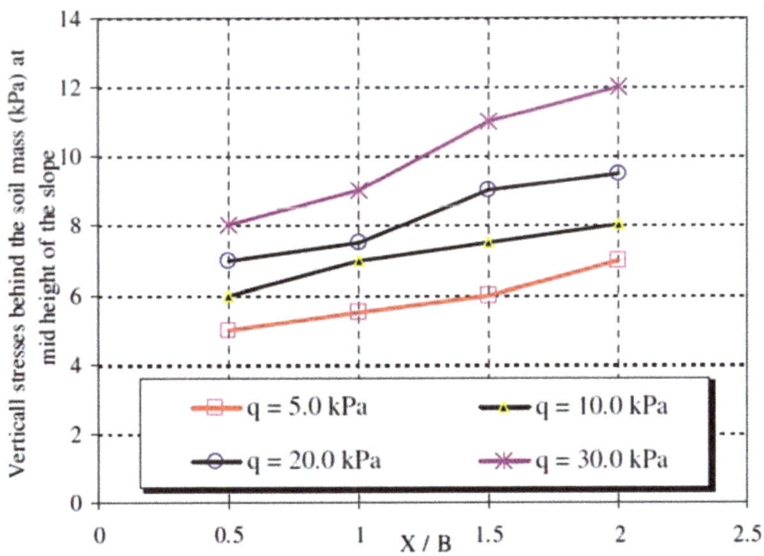

Figure 32. Influence of footing position on vertical stresses behind the nailed soil mass at mid-height of slope.

4. Numerical Modelling: Validation

The numerical model of earth slopes based on the finite element method is developed to ascertain the accuracy of laboratory model results. The PLAXIS-2D code is used to develop computational model of nailed soil slope considering plane strain conditions of soil slope. Four types of elements are used to model the sand, steel and Perspex glass material. A nonlinear elastic (Hyperbolic) model is used to simulate the sand material. In this model, the stress–strain curve is hyperbolic, and the soil modulus is a function of confining stress and the shear stress that a soil is experiencing. Depending on the stress state and stress path, three soil moduli are required, namely, the initial modulus, the tangential modulus and the unloading–reloading modulus.

The geometry of the model is taken as 850 mm in height and 1760 mm in length, the inner dimensions of the sand bed used in the experimental model. The other nailed soil slope geometry parameters used in the model are taken from Figure 3. The geometry of the model is divided to meet the requirement of the construction stages and loading phases. The construction sequence is modeled in stages, as explained in Section 2.3.2 above. The 5 mm diameter steel bars with a length of 700 mm are modeled as the reinforcement, and a rigid steel plate of dimensions 840 mm × 150 mm × 22 mm is used to simulate the footing. Perspex plate with a thickness of 5.0 mm is selected to simulate the facing material, which is flexible enough to withstand the ground displacement during excavation. The strength parameters of nails, facing and footing materials are taken from Tables 2 and 3. The soil parameter used in the material model is given in Table 4. The discretized nailed soil slope model with boundary conditions is shown in Figure 33.

Table 4. Soil parameters of sand used in material model.

Soil Type	Density (kN/m^3)	Mohr-Coulomb Model				Plastic Straining Due to Primary Comp./Deviator Loading $(E_{oed})^{ref}$ (kPa)	Elastic Unloading/Reloading		Stress-Dependent Stiffness (m)
		φ	C (kPa)	ψ	υ		$(E_{ur})^{ref}$ (kPa)	υ_{ur}	
Loose sand (RD = 34%, e = 0.631)	16.06	30	0.2	0	0.33	1327	3981	0.2	0.5
Medium sand (RD = 48%, e = 0.597)	16.41	34	0.2	4	0.31	1959	5877	0.2	0.5
Dense sand (RD = 68%, e = 0.549)	16.92	40	0.2	10	0.26	2632	7896	0.2	0.5

φ—friction angle; C—cohesion; ψ—dilatancy angle; υ—Poisson's ratio; e—void ratio.

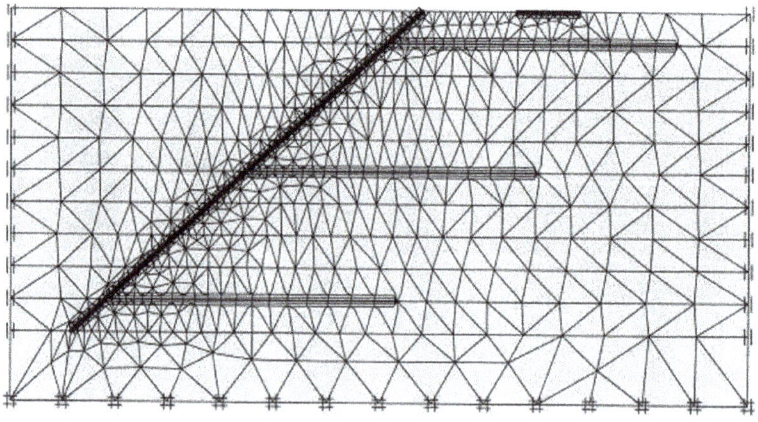

Figure 33. Triangular element discretization of nailed soil slope model.

To verify the accuracy of the laboratory results, a numerical model of nailed soil slope of height (*H*) 700 mm is developed with a length of horizontal nails as *H* and vertical spacing of nails as 0.4 *H*. The footing settlements (S/*H*, %) at different soil densities are computed using the developed numerical model. The plots of computational and experimental results of sand having three different relative densities (i.e., 35%—Fine sand, 48%—Medium sand and 68%—Dense sand) at different surcharge pressure, i.e., at the excavation stage and loading stage, are depicted in Figure 34. It is clear from the figure that the maximum footing settlement occurred at a surcharge load exerting a foundation pressure of 30.0 kN/m^2 with both models. Moreover, the minimum footing settlement occurred at a surcharge load exerting a foundation pressure of 5.00 kN/m^2. It can be observed that the footing settlements obtained from the numerical model were overestimated compared to those obtained from the laboratory model for all cases of sand used. The percentage of difference in the settlement between experimental and numerical models ranged from 4.37% to 18.72% in the case of loose sand, whereas it ranged from 13.37% to 40.59% in the case of medium sand and it ranged from 0.39% to 35.02% in case of dense sand.

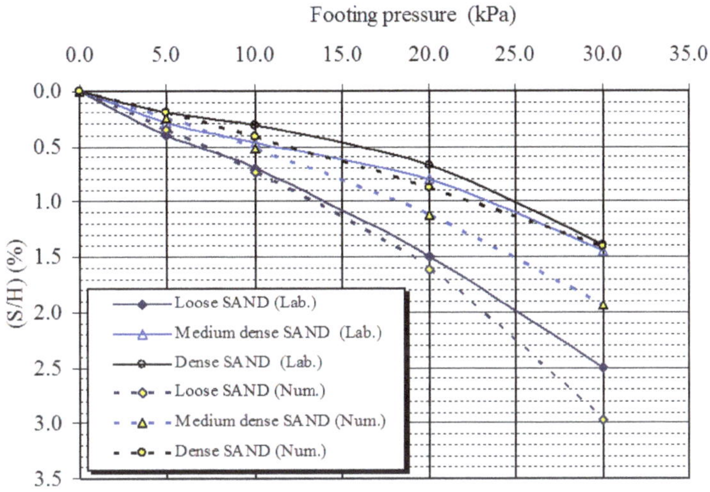

Figure 34. Footing settlement from laboratory and numerical models of nailed soil slope.

Figure 35 presents the vertical stresses under the nailed soil mass at excavation and loading stages obtained using the laboratory and numerical models at different soil densities. It is inferred from both model results that the denser the soil, the higher the vertical stresses under the soil mass. However, the vertical stresses are overestimated in numerical models as compared to laboratory models.

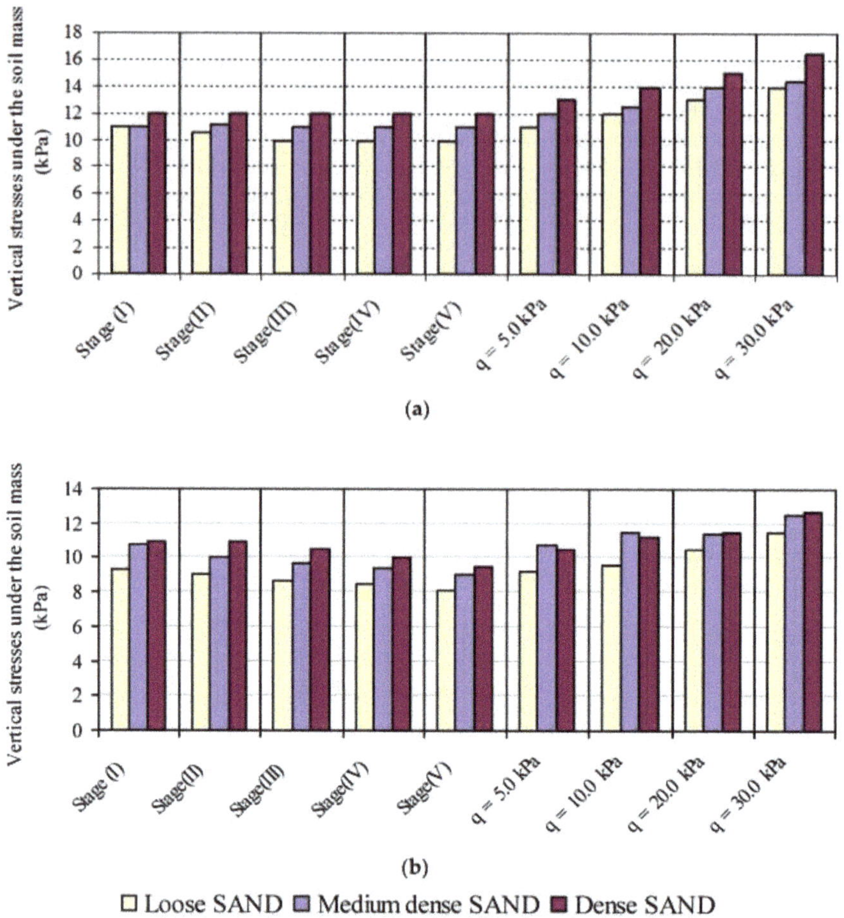

Figure 35. Effect of soil density on the vertical stresses under the nailed soil mass at mid-point; (**a**) numerical model; (**b**) laboratory model.

5. Conclusions

Soil nailing system is a ground improvement method used to stabilize the soil slopes. The study presents the laboratory investigation of earth nailing system models of non-cohesive soil having various soil and building foundation parameters. The laboratory models are prepared to have three soil types with different relative densities, i.e., 35%, 48%, 69%, two angles of slope, i.e., 40°, 45°, two footing widths, i.e., 150 mm and 200 mm and four positions of building foundation from the upper slope corner, i.e., 75 mm (0.5 B), 150 mm (B), 225 mm (1.5 B), 300 mm (2 B). The displacement of the slope, force in the nail, settlement of the footing and the earth pressure in back-fill soil due to different soil and footing parameters (relative density, slope angle, footing width, footing position) are measured.

The following conclusions may be drawn from the present study:

- The maximum lateral displacement and slope face horizontal pressures occur in the middle third and minimum lateral displacement, and slope face horizontal pressures occur in the lower third of slope with different soil and footing parameters under various footing pressures.

- The lateral movement of slope and footing settlements reduces with an increase of soil density and decreases with an inclination of soil slope. The increase in footing width increases the slope face displacement and footing settlement. The increase in strip footing distance from slope crest reduces the slope face displacement and footing settlement.
- The vertical and horizontal pressures behind the soil mass are unaffected during the construction process but begin to increase with the increase in the footing pressure. The distribution of vertical stress under the soil mass away from the slope part remains constant during various construction stages, and under different loading stages, the vertical stress under the soil mass is increased with an increase in foundation pressure.
- A decrease in soil density leads to an increase in maximum tensile force mobilized in the nails. Increasing the soil slope results in increasing the maximum tensile force of nails at the middle third of the soil slope, while soil slope increase results in a decrease in the maximum tensile force of nails at the lower and upper third of the soil slope. The mobilized tensile forces in nails are increased due to an increase in soil density and a decrease in footing distance from the slope face.
- The horizontal stresses at the slope face decrease with the decrease of relative density of soil while they are increased due to an increase in soil slope angle. The horizontal stresses at the slope face increase with the increase of footing width and decrease of footing distance from the slope crest.
- The vertical stress under the soil mass is more in high slope soil, denser soil, larger footing and greater distance from slop. The increase in soil density and slope angle results in an increase in vertical stress behind the soil mass and a decrease in the horizontal stress behind the soil mass with the increase of soil density and slope angle.
- Increasing the footing width leads to an increase in the vertical and horizontal stress behind the soil mass. The vertical and horizontal stresses behind the soil mass decrease with the decrease of footing distance from the slope crest.

Author Contributions: Conceptualization, M.H.M. and M.A.; methodology, M.H.M. and M.A.; resources, J.M.; writing—original draft preparation, M.H.M. and M.A.; writing—review and editing, J.M.; project administration, M.A. and J.M.; funding acquisition, M.A. All authors have read and agreed to the published version of the manuscript.

Funding: Funding for this research was given under award numbers R.G.P2/73/41 by the Deanship of Scientific Research; King Khalid University, Ministry of Education, Kingdom of Saudi Arabia.

Institutional Review Board Statement: Not Applicable.

Informed Consent Statement: Not Applicable.

Acknowledgments: The authors extend their appreciation to the Deanship of Scientific Research at King Khalid University for funding this work through a General Research Project under grant number (R.G.P2/73/41).

Conflicts of Interest: The authors declare no conflict of interest.

References

1. Pourghasemi, H.R.; Yansari, Z.T.; Panagos, P.; Pradhan, B. Analysis and evaluation of land slide susceptibility: A review on articles published during 2005–2016 (periods of 2005–2012 and 2013–2016). *Arab. J. Geosci.* **2018**, *11*, 193. [CrossRef]
2. Sharma, M.; Samanta, M.; Sarkar, S. Soil Nailing: An Effective Slope Stabilization Technique. Landslides: Theory, Practice and Modelling. In *Advances in Natural and Technological Hazards Research*; Pradhan, S., Vishal, V., Singh, T., Eds.; Springer: Cham, Switzerland, 2019; Volume 50.
3. Yelti, N. Analysis and Design of Soil Nail Walls in High Plasticity Clays. Master's Thesis, Texas Tech University, Lubbock, TX, USA, 2011.
4. Sanvitale, N.; Simonini, P.; Bisson, A.; Cola, S. Role of the facing on the behavior of soil-nailed slopes under surcharge loading. In Proceedings of the 18th International Conference on Soil Mechanics and Geotechnical Engineering, Technical Committee 207, Paris, France, 2–6 September 2013; pp. 2019–2094.

5. Bridges, C.A.; Kizil, M.; Winter, M.G.; Smith, D.M.; Eldred, P.J.L.; Toll, D.G. The determination of deformation of soil nailed structures by a simplified method. In *Geotechnical Engineering for Infrastructure and Development, Proceedings of the XVI European Conference on Soil Mechanics and Geotechnical Engineering, Edinburgh, UK, 13–17 October 2015*; ICE: London, UK, 2015; pp. 913–918.
6. Kotake, N.; Sato, E. Bearing capacity of a flexible plastic plate for soil nailing. *Int. J. Phys. Model. Geotech.* **2021**, *21*, 26–39. [CrossRef]
7. Ceccato, F.; Bisson, A.; Cola, S. Large displacement numerical study of 3D plate anchors. *Eur. J. Environ. Civ. Eng.* **2020**, *24*, 520–538. [CrossRef]
8. Sahoo, S.; Manna, B.; Sharma, K. Shaking Table Tests to Evaluate the Seismic Performance of Soil Nailing Stabilized Embankments. *Int. J. Geomech.* **2021**, *21*, 04021036. [CrossRef]
9. Deng, D.-G.; Li, L.; Zhao, L.-H. Stability analysis of slopes reinforced with anchor cables and optimal design of anchor cable parameters. *Eur. J. Environ. Civ. Eng.* **2019**, 1–16. [CrossRef]
10. Mohamed, M.H.; Ahmed, M.; Mallick, J.; Hoa, P.V. An Experimental Study of a Nailed Soil Slope: Effects of Surcharge Loading and Nails Characteristics. *Appl. Sci.* **2021**, *11*, 4842. [CrossRef]
11. Babu, G.L.S.; Singh, V.P. Numerical Analysis of Performance of Soil Nail Walls in Seismic Conditions. *ISET J. Earthq. Technol.* **2008**, *45*, 31–40.
12. Cheuk, C.Y.; Ho, K.K.S.; Lam, A.Y.T. Influence of soil nail orientations on stabilizing mechanisms of loose fill slopes. *Can. Geotech. J.* **2013**, *50*, 1236–1249. [CrossRef]
13. Taule, C.R. Soil Nailing: A Monte Carlo Simulation of Soil Parameters. Master's Thesis, Department of Civil, Environmental and Natural Resources Engineering, Luleå University of Technology, Luleå, Cape Verde, 2019.
14. Villalobos, F.A.; Villalobos, S.A.; Oróstegui, P.L. Observations from a parametric study of the seismic design of soil nailing. *Proc. Inst. Civ. Eng. Ground Improv.* **2018**, *171*, 112–122. [CrossRef]
15. Alhabshi, A. Finite Element Design Based Design Procedures for MSE/Soil—Nail Hybrid Retaining Wall System. Ph.D. Thesis, Department of Civil Engineering, Texas Tech University, Lubbock, TX, USA, 2006.
16. Wu, J.J.; Cheng, Q.G.; Liang, X.; Cao, J.L. Stability analysis of a high loess slope reinforced by the combination system of soil nails and stabilization piles. *Front. Struct. Civ. Eng.* **2014**, *8*, 252–259. [CrossRef]
17. Benayoun, F.; Boumezerane, D.; Bekkouche, S.R.; Bendada, L. Application of genetic algorithm method for soil nailing parameters optimization. In Proceedings of the 5th International Conference on New Advances in Civil Engineering (ICNACE), Kyrenia, Cyprus, 8–10 November 2019.
18. Chen, C.; Zhang, G.; Zornberg, J.; Xinxiu, Z. Element Nail Pullout Tests for Prediction of Soil Nail Pullout Resistance in Expansive Clays. *Geotech. Test. J.* **2019**, *42*, 20170431. [CrossRef]
19. Han, W.; Li, G.; Sun, Z.; Luan, H.; Liu, C.; Wu, X. Numerical Investigation of a Foundation Pit Supported by a Composite Soil Nailing Structure. *Symmetry* **2020**, *12*, 252. [CrossRef]
20. Sojoudi, Y.; Sharafi, H. Study of soil deformation pattern in earth slope stabilised with pile. *Eur. J. Environ. Civ. Eng.* **2021**, *25*, 205–225. [CrossRef]
21. Mohamed, M.H.; Ahmed, M.; Mallick, J. Pullout Behavior of Nail Reinforcement in Nailed Soil Slope. *Appl. Sci.* **2021**, *11*, 6419. [CrossRef]

applied sciences

MDPI

Article

Behaviour of the Steel Welded Grid during a Simplified Pullout Test in Fine Sand

Marcin Ćwirko [1] and Małgorzata Jastrzębska [2,*]

1 VALLA GROUP Marcin Ćwirko, Elizy Orzeszkowej 24/6, 41-103 Siemianowice Śląskie, Poland; m.cwirko@valla.pl

2 Department of Geotechnics and Roads, Faculty of Civil Engineering, Silesian University of Technology, Akademicka 5, 44-100 Gliwice, Poland

* Correspondence: malgorzata.jastrzebska@polsl.pl; Tel.: +48-32-237-1543

Featured Application: A potential application of the study is the use of the technology described in the construction of bridges, in particular for the construction of retaining structures within abutments.

Abstract: This study considered the possibility of using steel gabion baskets made of welded mesh for a soil-strengthening function. Examples of such applications are known for meshes made of wires with diameters from 6.3 to 12.7 mm and mesh openings from 125 to 225 mm. In the case of a welded grid, the pulling resistance of fine-grained soil consists of two factors: frictional resistance and bearing resistance. Therefore, for the purposes of this issue, a simplified laboratory pullout test was carried out with four types of welded steel grid (common in Europe) embedded in the fine sand. The geometry of the grid (opening size: 76.2 × 76.2 mm), the type of steel (low carbon steel, tensile strength from 500 to 700 MPa), the diameter of the wire (2.7–4.5 mm) and its cover (ZN + PVC or ZnAl) were taken into account during the analysis. It was unequivocally stated that as the stiffness of the steel grid itself increases, its strength increases during the pullout test, which is not so obvious in the case of popular steel woven meshes. In addition, it has been shown that steel welded meshes with wire diameters less than 6 mm are suitable for soil reinforcement in structures with gabion facing, and the determined apparent friction coefficient (μ_k = 0.39–1.47) takes values similar to the friction coefficient given in references for welded meshes of larger diameters. This is a positive premise for starting further research on the use of wires of smaller diameters for welded mesh production used as soil reinforcement.

Keywords: pullout test; retaining structures; gabions; bearing resistance; welded steel grid; woven steel grid; reinforcement of embankments

Citation: Ćwirko, M.; Jastrzębska, M. Behaviour of the Steel Welded Grid during a Simplified Pullout Test in Fine Sand. *Appl. Sci.* **2021**, *11*, 9147. https://doi.org/10.3390/app11199147

Academic Editor: Ana María Camacho López

Received: 7 August 2021
Accepted: 9 September 2021
Published: 1 October 2021

1. Introduction

Retaining structures are used in the case of construction of buildings where there is a difference in ground level. A special place for the use of such structures are infrastructure projects, such as the construction of roads, railroads, bridges, or airports. Examples of existing implementations in Poland are depicted in Figure 1. Another innovative solution is the use of gabion baskets in the design of a highly sustainable agricultural building (with a cellar), located in a farm in the south of the Tuscany region, Italy. Gabions and natural stones were used for the realization of the foundations, the ground retaining walls, and all other bearing walls, including the external wall, which remains entirely above ground [1].

Retaining structures made of gabion baskets, i.e., cuboidal steel mesh baskets filled with ballast material, constitute a method of direct reinforcement for the mechanical protection of slopes [2]. The history of their use in construction dates to the time of the pharaohs (c. 5000 B.C.), when reed and rock elements were used to build dams and to prevent erosion and strengthen the banks of the Nile [3]. Similar structures were erected in

China around 1000 B.C. along the Yellow River [4]. It is believed that the first modern wire mesh gabions were made in the Bologna area of Italy around 1879 [5].

(a) (b) (c) (d)

Figure 1. Examples of the use of gabion baskets in Poland (own pictures): (**a**) gravity retaining wall made of gabion baskets; (**b**) the slope of a motorway embankment reinforced with gabions; (**c**) gravity retaining wall supporting a motorway embankment; (**d**) an example of a gabion wall erected in harmony with the environment.

The use of gabions filled with stones is environmentally friendly (Figure 1d) [6]. The main reason for the limited impact on the environment is the reduction in the materials used compared to traditional reinforced concrete structures, especially steel and concrete, the manufacture of which requires the use of cement, which is a highly energy-consuming raw material to produce. It is true that the steel of the gabion baskets is also an energetically expensive material, both in terms of production and the recycling process. However, it has the feature of being able to be totally recycled an unlimited number of times. Even the zinc used for steel galvanization is fully recyclable [1]. Both natural and recycled materials or alternative materials—e.g., clothing and plastic—can be used as ballast material for gabion baskets [7,8]. The development of anticorrosion protection techniques has made steel wire structures durable. Their durability is estimated at over 100 years [9–11].

The fact that about 30–40% of the volume of gabions are air voids has a positive influence on the durability of the gabion structures, thanks to which they are characterized by excellent water permeability (hydraulic conductivity $k > 2.5 \times 10^{-3}$ m/s), which facilitates the drainage of water from behind the retaining structure and prevents its level from rising [10–16].

Phenomena related to the influence of water on retaining structures are often neglected or treated with insufficient care when analysing the surroundings of bridge structures, which leads to numerous minor failures that are difficult to eliminate due to the scale of earthworks. The importance of proper drainage system implementation is often underestimated, both at the design and execution stages [17]. A geotechnical engineer should always be consulted for design solutions in such situations, and it should be noted that gabion retaining structures, apart from their structural function, also drain water from the soil medium.

An additional advantage is the fact that the gabion structures are deformable, thanks to which they do not break, but adapt to changing external conditions [18], which allows the use of gabions in areas of mining damage (e.g., retaining structure by the motorway in Ruda Śląska, in Silesia, a vast area of mining activity in southern Poland [19]). Gabions also have the ability to absorb noise [20], which allows them to be used for sound-absorbing partitions (e.g., noise barriers on the railroad in The Hague, Netherlands). Furthermore, the gabion revetments have the capacity for vegetation restoration (it provides a good habitat for floral, faunal, and microbial growth) and resist periodic flood disturbance [21,22]. Gabion structures are easy to assemble and do not require extensive construction site facilities, and their landscape and architectural aesthetics allow them to be adapted to the surrounding natural terrain (e.g., retaining structures on Victoria Road in Cape Town, South Africa).

Gabion structures made for the purpose of soil reinforcement technology deserve special attention. The whole structure is made of woven mesh, more precisely double-woven mesh (Figure 2a) [23]. Gabion retaining structures made using welded meshes (Figure 2b) appear mainly as gravity structures (e.g., Figure 1a,c) or as structures made of soil reinforced with geosynthetics (e.g., Figure 1b) [24]. The use of welded steel meshes as backfill reinforcement seems to be the next step in the development of this technology. A diagram of such a solution is presented in Figure 3.

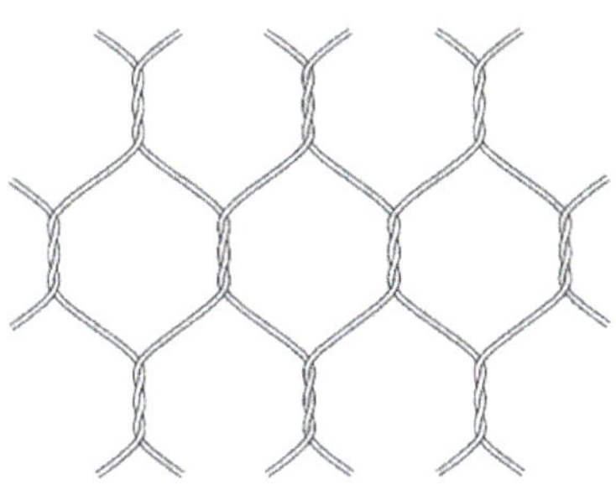

(a)

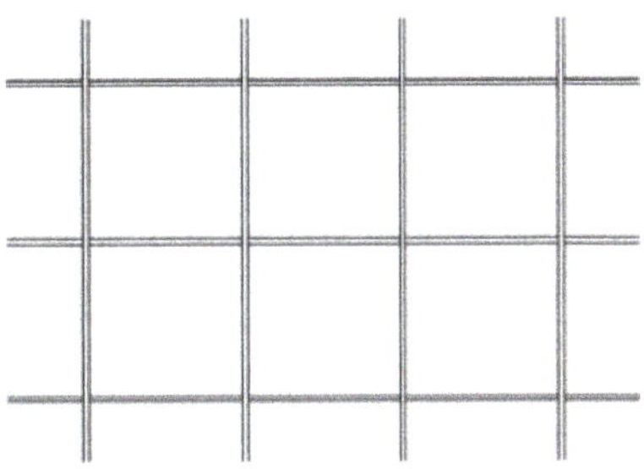

(b)

Figure 2. Comparison of the mesh opening: (**a**) woven (double-woven) mesh; (**b**) welded mesh.

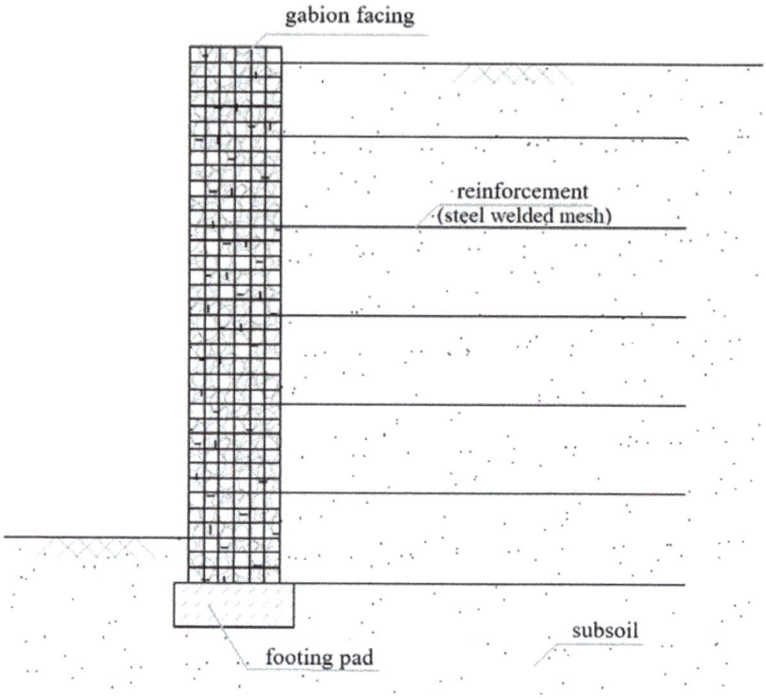

Figure 3. Diagram of a gabion retaining structure for soil reinforcement technology.

In addition, lower demand for steel mesh, the considerably lower consumption of ballast material for filling the gabions, the reduction of the necessary transportation, as well as the lower labour demand constitute other advantages of such gabion structures. Structures using soil reinforcement technology are ultimately 25% to 60% cheaper to build than massive structures [24–28]. Ciomcia [25] carried out a meticulous analysis of costs and construction time on the example of the construction of a typical bridge abutment, showing that the construction of the bridge abutment using soil reinforcement technology with steel elements is 35% to 37% cheaper and takes from 38% to as much as 63% less time, compared to the construction of the abutment in the traditional monolithic form. Jayasree [26] analysed the costs of erecting retaining structures with a height of 1 to 10 m, showing that the construction of gabions, both massive and with soil reinforcement, is about 30% cheaper than the construction of RCC (reinforced cement concrete) retaining walls. For higher structures from 5.5 to 10.0 m, the savings range from 48% to 52% for massive structures and from 70% to 78% for reinforced soil gabion structures).

Gabion baskets made of woven or welded mesh are always selected on the basis of their deformability, which is a function of wire thickness, mesh size, type, and the grain size of the ballast material, as well as the construction of the gabion basket itself. The same parameters determine the selection of an appropriate gabion mattress solution for the regulation of riverbeds depending on the speed of flowing water [11].

In turn, the additional use of steel meshes as soil reinforcement requires the determination of their deformability after building in the soil medium, the force pulling them out of the soil, and the friction coefficient between the reinforcement and the soil, which also directly affects the length of the anchorage. Woven mesh, when stretched in the air (without building into the ground), tends to narrow, which is a result of the geometry of the mesh openings. This phenomenon drastically increases the tensile elongation of the woven mesh. It does not occur when stretching welded mesh with rectangular openings, because the force is transmitted along the wires and their cross-sectional area directly affects the mesh elongation during its stretching. It turns out that after building in the soil medium, both types of mesh (woven and welded) are characterized by similar values of relative deformation. Pullout tests are dedicated to the determination of tensile forces, and consist of building a reinforcement mesh in a box with soil, applying a vertical load, simulating the weight of the entire geotechnical structure, and then slowly ejecting the reinforcement through the slot in the box, while simultaneously measuring both force and displacement. In this type of research, woven meshes exhibit a different mechanism of cooperation with the ground from welded meshes [29].

All these premises were the starting point for this research paper, which focuses on the observation of the behaviour of welded steel mesh during a simplified pullout test in fine sand. Pullout tests are performed to determine the coefficient of friction between the reinforcement and the soil, the value of which is essential for the correct and safe design of reinforced soil structures [30].

In construction site practice, woven and welded meshes are commonly used in the construction of gabion retaining structures. In the case of retaining structures with soil reinforcement, a double-woven mesh is typically used if the baskets are also made of it, or geosynthetics, regardless of the material of the gabion basket. Gabion welded mesh for soil reinforcement was not used. Although welded mesh has been successfully used to reinforce soil in light retaining structures since the 1980s, it is made of wires with a much larger diameter than those used for gabion basket production.

The essential issue in the design, performance, and durability of the use of a structure with soil reinforcement is the connection of the wall facing with the soil reinforcement. This connection is often the weakest point in terms of strength, effectively limiting the use of the total load-bearing capacity of the soil reinforcement, and is the site of potential failure during use. Integrating the facing with the reinforcement (e.g., extending the base of the gabion to such a length that it is also the reinforcement of the soil) makes it possible to eliminate the problematic connection, as is currently the case in structures where woven meshes are used to make gabions and reinforce soil.

The use of welded steel meshes as backfill reinforcement seems to be, of course, the next step in the development of this technology. This is important because welded meshes can achieve higher tensile strength than braided meshes due to the technological possibility of using thicker wire for their production.

2. Materials and Methods

Soil reinforcement with steel welded meshes identical or similar to those used for gabion structures is a field for discussion among researchers dealing with these issues. There are many indications that the use of this type of reinforcement is possible and justified [29]. In this study, it was decided to explore the subject of the cooperation of the soil reinforcement system in a simplified laboratory pullout test. For this purpose, four types of welded steel mesh and one type of coarse-grained soil were used, and the mesh was built in that soil. The question to be investigated was how the different types of reinforcement in a particular soil type affect the force required to pull the reinforcement out of the soil during the test.

2.1. Fine Sand

During laboratory tests, dry fine sand (FSa) with the following parameters was used: grey-yellow colour, calcium carbonate content $CaCO_3 < 1\%$, angle of internal friction $\varnothing = 24.0^\circ$, cohesion $c = 0.9$ kPa, coefficient of uniformity $C_u = 2.57$, coefficient of curvature $C_c = 0.71$, optimum water content (OWC) $w_{opt} = 8.6\%$, maximum dry density $\rho_{ds} = 1.76$ g/cm^3. A detailed description of studies identifying the fine sand under consideration was presented by Sułkowska [31].

2.2. Welded Steel Grid

Welded meshes are manufactured of steel wire with a tensile strength of 500–700 MPa in accordance with the PN-EN 10223-8 standard [32]. Wires with a diameter of 2.2 mm to 5.0 mm are used, and the edge wire may be the same or thicker than the filling wire. The opening is rectangular and its most common sizes are 50 × 50 mm, 50 × 100 mm, 75 × 75 mm or 100 × 100 mm, as well as 76.2 × 76.2 mm (3 × 3 inches).

A commonly used anticorrosion protection of welded meshes is used to cover the wire, with one of three types of coating according to the PN-EN 10244-2 [33] and 10245-2 [34] standards:

- type 1: thick zinc coating,
- type 2: zinc–aluminium alloy layer up to 350 g/m^2,
- type 3: a coating consisting of a pure zinc layer and an additional polymer layer, usually PVC.

Laboratory tests were conducted on four types of welded meshes, which differed with respect to wire diameter as well as the type of anticorrosion coating used during production. They had been chosen as they are widely used for the production of gabion baskets available on the European market. It is worth mentioning here that soil reinforcement with gratings made of thick steel rods (6.3–12.8 mm in diameter) is used in practice in the North American market [35]. However, soil reinforcement with both gratings made of steel rods and welded steel meshes made of thinner wire (such as those used in the production of gabions) is an innovative method on a European scale.

In the series of tests, sheets with dimensions of 762 × 1676 mm, consisting of 11 tensile longitudinal wires and 23 transverse wires, are used, and in the case of the largest diameter wire mesh, also sheets with dimensions of 305 × 1676 mm, consisting of 5 and 23 wires, respectively. The smaller number of wires was the result of the limitation of the maximum force that could be applied during the test. The choice of meshes with type 2 and type 3 coatings was based on the manufacturer's guaranteed long-term protection against corrosion under the conditions prevailing in the soil or in contact with the soil. This period is 80–100 years, and even more than 120 years respectively. The types of meshes are presented in Table 1.

Table 1. Basic geometrical parameters of the welded meshes used in the test.

Designation	Wire Diameter	Mesh Opening	Coating
Zn + PVC 4.30 mm	4.30 mm (ext.)/3.80 mm (int.)	76.2 × 76.2 mm	Type 3
Zn + PVC 3.20 mm	3.20 mm (ext.)/2.70 mm (int.)	76.2 × 76.2 mm	Type 3
ZnAl 4.50 mm	4.50 mm	76.2 × 76.2 mm	Type 2
ZnAl 3.00 mm	3.00 mm	76.2 × 76.2 mm	Type 2

3. Test Procedure

The basic pullout type test is carried out in accordance with the EN 13738 standard [36] in a large-scale pullout apparatus with a box with internal dimensions of 1600 mm × 600 mm × 360 mm (length × width × height). The test consists of pulling the reinforcement uniformly out of the soil at a speed not exceeding 10 mm/min, at a given vertical tension (up to 200 kPa). The displacement of the reinforcement points inside the box is monitored by linear variable differential transformers (LVDT) with the corresponding value of the pullout force. Both values are recorded continuously in the computer system [37]. Originally, the pullout test was dedicated to geosynthetics, but eventually it also came to be used to study the interaction between soil and reinforcement in the form of welded or woven steel mesh. As a result, many researchers conduct tests that are called simplified ones, in a suitably modified apparatus and according to a procedure adapted to the situation (e.g., [38–41]). In this study, the authors carried out a simplified pullout test. This test is consistent with the concept of the standard method; however, the simplification is related to the way force and displacement were measured. The pullout force was set and not measured, and the displacements were measured at the extreme nodes of the reinforcement and mounting brackets, which were not embedded in soil. Finally, the test was conducted in accordance with the following procedure:

1. fill the steel box with soil in layers of approximately 100 mm,
2. add an appropriate amount of water to the soil and mix evenly if it is necessary to obtain a water content close to the optimum water content (OWC),
3. compact the soil using an electric plate compactor,
4. the box should be filled according to points 1–3 until the height at which the slot in the box is reached Figure 4
5. insert the steel mesh and the mounting flat into the clamping jaw and screw the two parts of the jaw gently together,
6. remove the gap in the clamping jaws manually by pulling the mesh towards the steel box,
7. make sure that the welded mesh does not touch any part of the steel box,
8. screw the two parts of the mounting jaws tightly together,
9. continue filling the steel box according to points 1–3, until a height of approximately 45 mm below the top edge of the steel box is reached,
10. level the upper soil surface and install the cover, the force gauge, and hydraulic actuator No 2,
11. apply a vertical load,
12. place the hydraulic actuator No 1 in the load frame,
13. check the correct position of the welded mesh in the box slot,
14. install the electronic dial displacement sensors and reset them,
15. apply the horizontal load gradually every 3.1 kN and after each load is applied and the reading stabilizes on the jaw displacement sensors, note indications of these sensors,
16. conduct the test until the damage of the welded mesh or until the moment at which the indication on the hydraulic actuator No 1 stops increasing despite the mesh continuing to move out of the ground (reaching the maximum anchoring force) or the maximum extension of this hydraulic actuator is reached,
17. recoil the hydraulic actuator, disassemble all elements of the stand, pull the two parts of the loading frame towards each other using a lashing strap to the original position, and start another test from point 1 of this procedure.

Figure 4. Mesh installed on the level of the slot height in the box.

During the test, a total of four tests were carried out according to the above procedure, one for each type of welded steel mesh.

3.1. Test Stand

For this research, a stand was constructed (Figure 5). The test stand was originally intended for the tests of geosynthetics (geogrids) carried out by Kawalec [42]. Then it was adapted for pullout tests by the authors of this study. It consisted of typical pullout test equipment such as two hydraulic actuators, a force gauge, four electronic dial displacement sensors with a range of 0–100 mm, and a steel load frame (length 3260 mm × width 2160 mm), welding mesh mounting brackets, and a steel soil box with internal dimensions of 900 mm × 900 mm × 630 mm (length × width × height) and sturdy retaining trestles. The dimensions of the soil box were selected on the basis of the observations of Kim [43] and Prashanth et al. [39], who used boxes ranging from 400 to 2000 mm in length, 200 to 1100 mm in width and 100 to 1100 mm in height. The whole structure was fixed to a concrete floor, ensuring the sliding of the frame elements in the horizontal plane. The geometry of the frame, especially the position of the axis of the hydraulic actuator and the axis of the force that stretches the welded mesh in relation to the articulated joint of the frame, causes the force that stretches the mesh to be exactly twice as large as the force applied by the hydraulic actuator. Displacement sensors were applied to the clamping jaws in such a way that the average displacement for each jaw showed displacement both in the plane of the mesh itself and half the width of the mesh. This method of mounting the sensors facilitated their installation and did not require the installation of sensors inside the box with the soil.

The mounting brackets were constructed in such a way that there were two bolts on each side of each of the tensile wires of the welded mesh to twist the independent parts of the bracket. The tensile force from the bracket to the steel mesh was transmitted through a 30 × 5 mm steel flat bar, which ensured an even distribution of force across the whole width of the mesh.

The aforementioned retaining trestles made of steel channel bars allowed one of the arms of the loading frame to be anchored to the floor. Thanks to this solution, it was possible to expand only one side of the frame and thus to pull the steel mesh out of the soil in the central part of the test stand. Horizontal load was applied gradually every 3.1 kN and after each load was applied and the reading stabilizes on the jaw displacement sensors, the indication of these sensors was noted.

The vertical load was applied once to the value corresponding to the stress of approx. 60 kPa in the level of the reinforcement. The force was applied using a hydraulic actuator, which from the top rested against a very thick and durable reinforced concrete ceiling, and from below it transferred the force to the ground by means of a strongly ribbed steel cover, which ensured the even distribution of force over the entire surface of the box in the box. After the indication of the vertical force, measuring force was stabilized, the reinforcement was pulled out of the soil.

A view of the test stand is shown in Figure 5. To get as much information as possible from this test, it was also decided to install digital displacement dial sensors directly to the grid, but on the opposite side to the mounting jaws.

The test stand during the work is shown in Figure 6.

3.2. Test Conditions

The test was conducted in a dry and closed room at room temperature. The soil water content was controlled during its construction. If necessary, water was added in sufficient quantity to obtain the optimum water content. The soil was compacted manually using an electric plate compactor. An important factor to be taken into account when analysing the issue of meshes embedded in soil is the state of normal stress in the soil at the level of the tested reinforcement. The intention of the authors was to illustrate the stress induced in an embankment of approximately 3–4 m height. Using hydraulic actuator 2, a vertical stress at the level of the reinforcing steel mesh of about 60 kPa was obtained, which meets the assumptions adopted by the authors of this study.

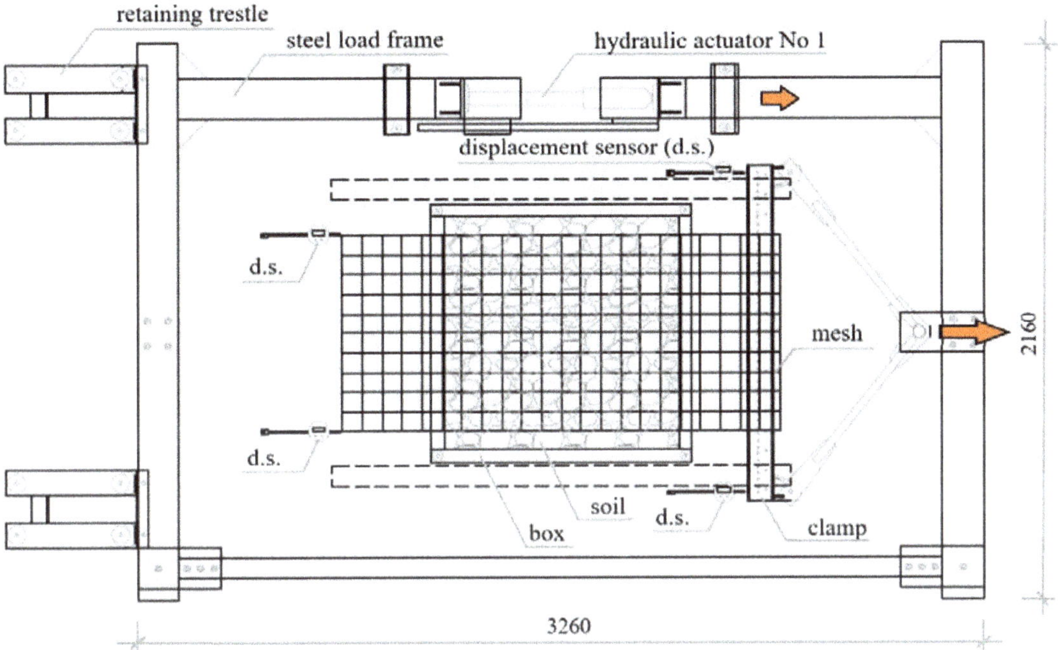

(a)

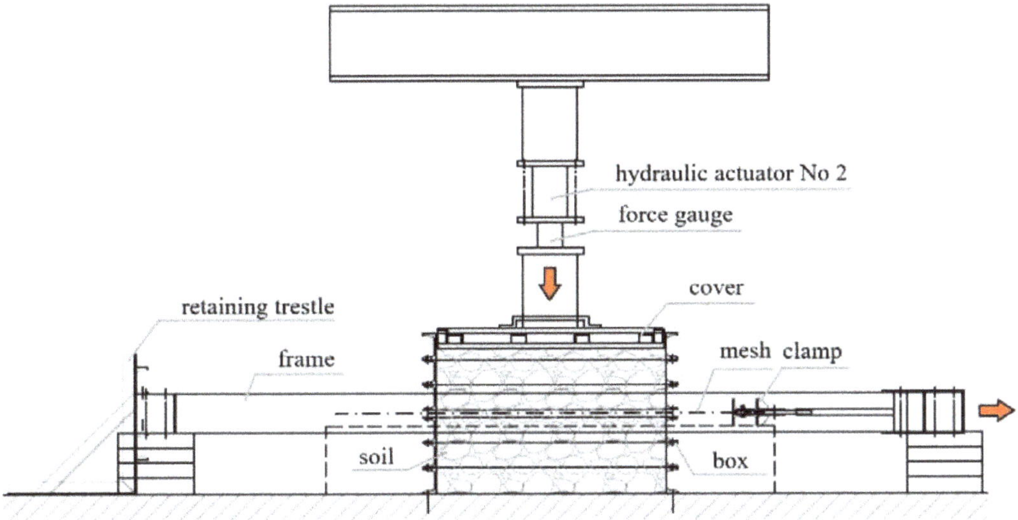

(b)

Figure 5. Schematic of the test stand for pulling meshes embedded in the soil: (**a**) top view; (**b**) longitudinal section.

Figure 6. The test stand for pulling meshes embedded in the soil.

4. Results and Discussion

During the test, the test with the Zn + PVC 3.20 mm mesh resulted in the destruction of the mesh in the form of its tearing outside the box with the soil. In other cases, ejection of the reinforcement from the fine sand was achieved. For easier interpretation, the results obtained from the test were converted to unit width reinforcement, and these are presented in Figure 7.

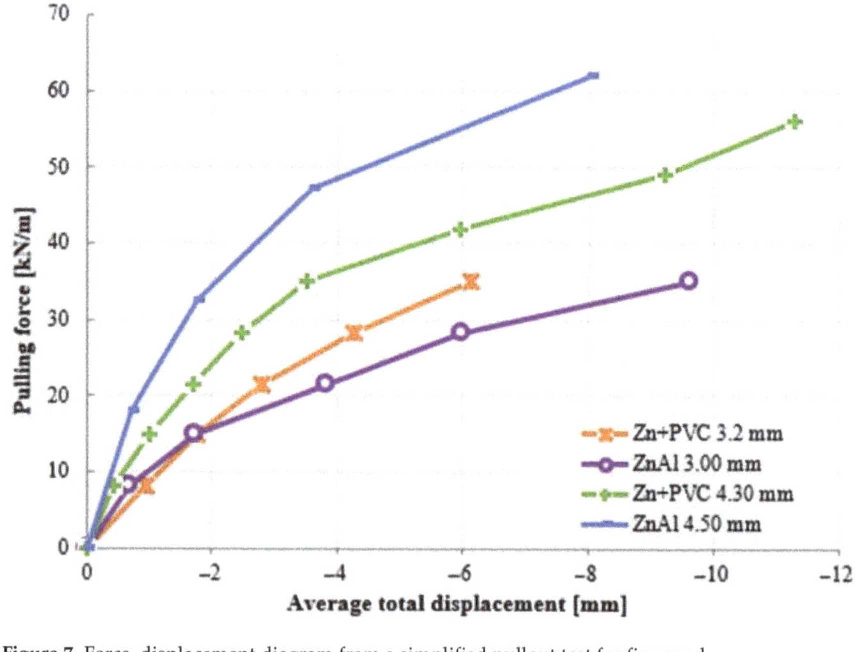

Figure 7. Force–displacement diagram from a simplified pullout test for fine sand.

On the basis of the test results, it can be unequivocally stated that in the case of the Zn + PVC 3.20 mm mesh, the key parameter is its low strength, as it was the destruction of the mesh that caused the termination of the test (Figure 8). In the case of the remaining three meshes, the condition for termination of the test was that the load bearing capacity of the soil reinforcement system was reached (i.e., the mesh ejection from the soil).

Figure 8. Damage of Zn + PVC 3.20 mm mesh after the pullout test—large displacements of the wires that were broken during the test are visible (the photo is greyed out for better visibility of the mesh against the floor).

The approximate values of the unit force necessary to damage the soil reinforcement system during a pullout test were also determined, and the results are presented in Table 2.

Table 2. List of forces pulling the reinforcement in the form of welded meshes.

Mesh Type	Pullout Force T_{max} [kN/m]
Zn + PVC 4.30 mm	55
Zn + PVC 3.20 mm	35
ZnAl 4.50 mm	60
ZnAl 3.00 mm	35

It has also been shown that as the stiffness of the reinforcement increases, the pullout resistance of this reinforcement increases as well, which is consistent with the observations of Yuan and Chua [44].

The values obtained for fine sand are similar to those presented in [44–47]. The authors of [44–47] also carried out similar tests in sands (or soil containing a lot of sand fraction in the case of [46]), with similar values of normal σ_n (the overall range is 2–140 kPa), but for different types of reinforcement, proving the validity of the conducted test. A summary of the test conditions is presented in Table 3.

Furthermore, the decrease in the displacements along the length of the reinforcement embedded in soil during pullout tests of whole meshes embedded in soil medium was analysed. In the tests performed in fine sand, it was found that the displacements recorded on the sensors behind the box (BACK) were smaller than those recorded on the sensors at the force application side (FRONT) by about 30–40%, as presented in Figure 9.

Table 3. Summary of selected test conditions.

Researchers	Type of Soil	Normal Pressure Box Dimensions	Reinforcement
Yuan & Chua [44]	fine to medium sand (FSa–MSa) $ø = 42°$ $c = 0$ kPa	σ_n = 5–35 kPa 760 mm × 710 mm × 610 mm	Geolon 200 (woven geotextile) Tensar SR2 (geogrid)
Lajevardi, Dias & Racinais [45]	medium sand (MSa) $ø = 36.5°$ $c = 2$ kPa	σ_n = 20–140 kPa 2000 mm × 1100 mm × 800 mm	Welded steel mesh (wire diameter from 8 to 12 mm, opening from 125 to 185 mm)
Bergado et al. [46]	lateric soil $ø$ = 15–66° c = 11–80 kPa	σ_n = 2–130 kPa 1300 mm × 800 mm × 500 mm	Welded steel mesh (wire diameter from 6.3 to 12.7 mm, opening from 152 to 225 mm)
Bergado & Teerawattanasuk [47]	silty sand (siSa) $ø = 30°$ $c = 10$ kPa	σ_n = 55–105 kPa 1270 mm × 760 mm × 508 mm	Woven steel mesh PVC coated (wire diameter 2.7 mm, opening size 80 × 100 mm)
Ćwirko & Jastrzębska (this study)	fine sand (FSa) $ø = 24°$ $c = 0.9$ kPa	σ_n = 60 kPa 900 mm × 900 mm × 630 mm	Welded steel mesh PVC or ZnAl coated (wire diameter from 2.7 to 4.5 mm, opening 76.2 × 76.2 mm)

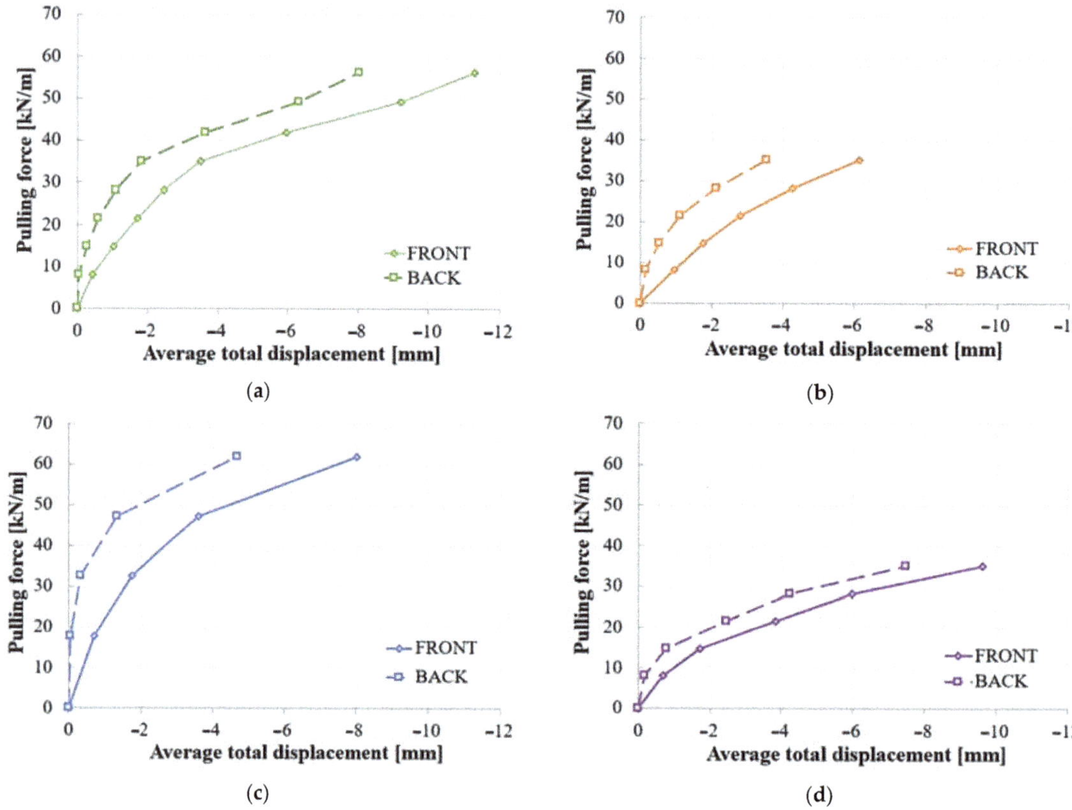

Figure 9. Recorded displacements at the front and back of the box for: (**a**) 4.30 mm Zn + PVC mesh; (**b**) 3.20 mm Zn + PVC mesh; (**c**) 4.50 mm ZnAl mesh; (**d**) 3.00 mm ZnAl mesh.

The relationship described above clearly indicates that the type of mesh affects the anchorage length in the passive zone. On the basis of the studies of Teerawattanasuk et al. [48], it is known that the displacements along the length of the reinforcement decrease in a curvilinear manner with increasing distance from the face of the tested structure. As the authors of this study only have the value of this displacement at two points, assuming its rectilinear slope, the anchorage length (for the maximum pullout forces applied in the tests) for the fine sand tested is approximately 2.3 to 3.0 m. These lengths are certainly underestimated, as the displacement value does not decrease in a linear way. Unfortunately, due to the lack of a third measurement point along the length of the reinforcement, the exact course of the curve describing this phenomenon cannot be determined.

It is worth noting that, on the basis of the research carried out, it is possible to easily determine the value of the apparent friction coefficient μ_k using Relationship (1):

$$\mu_k = T_{max}/(2 \cdot S \cdot \sigma_v), \tag{1}$$

where:

T_{max}—a maximum pulling force [kN] (according to Table 2),

S—a mesh area [m^2] (0.34 m^2 for ZnAl, 4.50 mm and 0.75 m^2 for others),

σ_v—a vertical stress at the level of the reinforcing steel mesh [kPa] (60 kPa in test).

The summary of the friction coefficients between the reinforcement and the soil μ_k calculated according to the above formula is presented in Table 4. The obtained values of the friction coefficient 0.39–1.47 are similar to those given by Lajevardi et al. [45], who reported values of the coefficient μ_k equal to 0.31–0.72. It is worth noting that in this study, for a steel mesh with a wire diameter of 4.5 mm, the friction coefficient (μ_k = 1.47) is much greater than that of Lajevardi et al. [45] for a steel grid with a wire diameter of 12 mm (μ_k = 0.72). This can be explained by the fact that the friction coefficient is determined not only by the diameter of the mesh wire (the larger it is, the greater the coefficient of friction is), but also the mesh opening (the smaller it is, the greater the friction coefficient is, because more wires are located on the same surface across the force action). In this study, we used 76.2 × 76.2 mm mesh opening, while Lajevardi et al. [45] examined grids with the mesh opening from 125 × 140 mm to 125 × 505 mm. On the other hand, Bergado et al. [46] showed coefficients of friction equal to μ_k = 1.04–1.78 for welded meshes with a wire diameter of 6.3–12.7 mm and a mesh opening from 152 × 152 mm to 152 × 225 mm. At the same time, it was confirmed that the coating type of the mesh wire affects the friction coefficient. The zinc-coated wires of the same diameter (or similar) to the polyvinyl chloride (PVC)-coated ones are characterized by a higher coefficient of friction, which is consistent with the results of the hexagonal wire mesh studies realised by Teerawattanasuk et al. [48]. By comparison, Bergado and Teerawattanasuk [47] demonstrated for woven meshes (with a similar geometry as in Lajevardi et al. [45] and Bergado et al. [40]) a coefficient of friction equal to μ_k = 0.2–0.30. This clearly shows that the welded meshes are more suitable for soil reinforcement than the commonly used woven meshes.

Table 4. Coefficient of friction between reinforcement and soil μ_k based on the presented tests.

Mesh Type	Friction Coefficients between the Reinforcement and the Soil μ_k [-]
Zn + PVC 4.30 mm	0.61
Zn + PVC 3.20 mm	0.39
ZnAl 4.50 mm	1.47
ZnAl 3.00 mm	0.39

5. Conclusions

This study deals with the interaction of welded steel mesh, used for the production of gabions, with the soil depending on (1) the soil type (in this case fine sand FSa) and

(2) the mesh parameters. The tested meshes (four types) were differentiated with respect to the diameter of the steel wire (2.7–4.5 mm) as well as the type of anticorrosion coating (Zn + PVC or ZnAl). Each of them had the same opening size: 76.2 × 76.2 mm. For the analysis, a proprietary stand was constructed (based on the geogrid test stand) to carry out a simplified pullout test.

Undoubtedly, it can be concluded that it is possible to make a retaining structure using soil reinforcement technology using steel welded meshes with a much smaller diameter (3.0, 3.8 and 4.5 mm) than 6–12 mm, used for the production of gabion baskets. This is the basic premise for the application of sustainable geotechnical principles in practice.

To confirm the above statement, a trial design replacement of the retaining structure with an extension of the abutment was performed using the parameters obtained during the tests described in this study. Originally, the retaining wall was designed using soil reinforced with flat steel bars with a facing of precast concrete elements. The wall has a height of 9.50 m from the top of the foundation to the highest point of the backfill. The replacement design of this retaining wall assumes that it will be made of gabions of a welded mesh made of Zn + PVC 4.30 mm wire. The facing made of gabion baskets is vertical. Facing basket reinforcement 1.0 m thick and 9.0 m long were adopted at a spacing of 0.75 m between the three lowest reinforcement layers and 1.0 m between the higher layers of soil reinforcement. A detailed description of the design process, the solutions adopted, and the geometry of the structure is presented in [29,49]. It is hereby shown that gabion retaining structures with welded meshes made using reinforced soil technology are possible for use in infrastructure construction applications.

Furthermore, on the basis of the conducted tests and the analysis of the obtained results, it can stated that the low deformability (hence greater stiffness) of the tensile welded meshes embedded in soil allows them to be used for soil reinforcement in all types of reinforced soil structures, including the abutments of engineering facilities (the maximum deformation in the elastic work phase was only equal to 0.35%). In the case of woven meshes, this type of use is not so obvious.

Another valuable observation is the dependence whereby the mesh with higher stiffness (e.g., made of a thicker wire, with smaller opening or stiff welded node), during the pullout test, obtained higher resistance to pulling from the soil compared to a less stiff mesh. This is a favourable phenomenon from the point of view of the use of retaining structures. The reinforcement effect can be explained by the favourable phenomenon of mesh wedging between soil grains. Taking into account that fine sand was considered in this study, it may be assumed that the effect will be greater in soils with a larger grain size. The authors plan to continue the initiated series of studies using other soils.

In addition, the obtained results are a good basis for further model tests both in the laboratory and in situ on welded meshes in function of reinforcing soil. The main objective is to propose all coefficients necessary for the design of the engineering structures mentioned above in the future. In this study, the first of such parameters was determined—the apparent friction coefficient μ_k. The next ones require more advanced research.

Author Contributions: Conceptualization, methodology, M.J. and M.Ć.; investigation, validation, research tests, M.Ć.; formal analysis, M.J. and M.Ć.; writing—original draft preparation, M.J. and M.Ć.; visualization, writing—review and editing, M.J.; funding acquisition, M.J. All authors have read and agreed to the published version of the manuscript.

Funding: This research received no external funding. The APC was partially funded by the professor's grant No 03/070/RGP19/0125 awarded by the Rector of the Silesian University of Technology (Gliwice, Poland).

Institutional Review Board Statement: Not applicable.

Informed Consent Statement: Not applicable.

Data Availability Statement: The data presented in this study is available in the source materials included in the References.

Acknowledgments: The authors would like to express their gratitude to the company BETAFENCE Sp. z o.o. which delivered the steel welded grids for this study free of charge, and the company CzekAmet PL Sp. z o.o. Sp. k. delivered clamping jaws for this study free of charge.

Conflicts of Interest: The authors declare no conflict of interest.

References

1. Conti, L.; Barbari, M.; Monti, M. Design of sustainable agricultural buildings. A case study of a wine cellar in Tuscany, Italy. *Buildings* **2016**, *6*, 17. [CrossRef]
2. Kania, M. Safe escarpment and slopes. *Geoinżynieria Drog. Mosty Tunele* **2013**, *42*, 28–33. Available online: https://inzynieria.com/geoinzynieria/artykuly/33008,bezpieczne-skarpy-i-zbocza (accessed on 20 February 2013). (In Polish)
3. Nevečerel, H.; Pavešić, K.; Murgić, I.; Papa, I.; Landekić, M.; Lepoglavec, K. Possibilities of gabion application. *Nova Meh. Sumar.* **2016**, *37*, 59–68. Available online: https://hrcak.srce.hr/index.php?show=clanak&id_clanak_jezik=256570 (accessed on 1 September 2021). (In Croatian).
4. Bouafia, A. *Conception et Calcul des Ouvrages Géotechniques*; Editions Pages Bleues Internationales: Bab Ezzouar, Algeria, 2011. (In French)
5. Surowiecki, A. About the designing of gabion structures in communication construction. *Drogownictwo* **2001**, *3*, 81–86. Available online: http://drogownictwo.sitkrp.org.pl/index.php?option=com_content&view=article&id=136:streszczenia-abstract-do-numeru-3-2001-r&catid=10&Itemid=105&lang=pl (accessed on 1 September 2021). (In Polish)
6. Toprak, B.; Sevim, Ö.; Kalkan, İ. Gabion walls and their use. *Int. J. Adv. Mech. Civ. Eng.* **2016**, *3*, 56–58. Available online: http://ijamce.iraj.in/paper_detail.php?paper_id=5392&name=Gabion_Walls_and_Their_Use (accessed on 1 September 2021).
7. Surowiecki, A.; Kozłowski, W. Gabion retaining walls stabilizing road embankments in the light of model studies. *Geoinżynieria Drog. Mosty Tunele* **2011**, *5*, 34–38. Available online: https://inzynieria.com/geoinzynieria/artykuly/27232,gabionowe-sciany-oporowe-jako-stabilizacja-nasypow-komunikacyjnych-w-swietle-badan-modelowych (accessed on 1 September 2021). (In Polish)
8. Craswell, T.; Akib, S. Reducing bridge pier scour using gabion mattresses filled with recycled and alternative materials. *Eng* **2020**, *1*, 188–210. [CrossRef]
9. Jarominiak, A. *Lightweight Retaining Structures*, 3rd ed.; WKiŁ: Warsaw, Poland, 2000; Available online: https://www.naukowa.pl/Ksiazki/lekkie-konstrukcje-oporowe-173495 (accessed on 1 September 2021). (In Polish)
10. Kosiński, T. Maccaferri solutions in the stabilization and reconstruction of landslides. *Geoinżynieria Drog. Mosty Tunele* **2013**, *42*, 34–38. Available online: https://inzynieria.com/geoinzynieria/artykuly/33009,rozwiazania-maccaferri-w-stabilizacji-i-rekonstrukcji-osuwisk (accessed on 1 September 2021). (In Polish)
11. Najder, T.; Najder, A. Gabions and gabion mattresses as elements of environmentally friendly construction. *Mater. Bud.* **2005**, *1*, 87–91. Available online: https://www.sigma-not.pl/publikacja-10410-gabiony-i-materace-gabionowe-jako-elementy-budownictwa-przyjaznego-%C5%9Brodowisku-materialy-budowlane-2005-1.html (accessed on 1 September 2021). (In Polish)
12. Balawejder, A.; Surowiecki, A.; Kozłowski, W. The embankment with gabion wall slope built-slope, damming up water. *Drog. Lądowe Powietrzne Wodne* **2011**, *6*, 36–43. (In Polish)
13. Kosiński, T.; Michalski, T.; Bosak, J. Terramesh as a source of savings in the reconstruction of road embankments. *Górnictwo i Geoinżynieria* **2010**, *34*, 405–413. Available online: https://journals.bg.agh.edu.pl/GORNICTWO/index.php?vol=2010-02 (accessed on 1 September 2021). (In Polish)
14. Kuc, M. Properties of gabions made of woven or welded mesh. *Mag. Autostrady* **2012**, *7*, 18–20. Available online: https://autostrady.elamed.pl/uploads/mau/articles/autostrady_artykul_2012_07_36125.pdf (accessed on 1 September 2021). (In Polish)
15. Kuc, M.; Mika, A. Gabion retaining wall in the area of mining damage. *Mater. Bud.* **2004**, *10*, 85–86. Available online: https://www.sigma-not.pl/publikacja-2767-gabionowy-mur-oporowy-na-terenie-szk%C3%B3d-g%C3%B3rniczych-materialy-budowlane-2004-10.html (accessed on 1 September 2021). (In Polish)
16. Pisarczyk, S. *Geoengineering-Methods of Soil Modification*, 3rd ed.; Oficyna Wydawnicza Politechniki Warszawskiej (OWPW): Warsaw, Poland, 2020; Available online: https://www.ibuk.pl/fiszka/221801/geoinzynieria-metody-modyfikacji-podloza-gruntowego.html (accessed on 1 September 2021). (In Polish)
17. Sobala, D.; Rybak, J. Role to be played by independent geotechnical supervision in the foundation for bridge construction. *IOP Conf. Ser. Mater. Sci. Eng.* **2017**, *245*, 022073. [CrossRef]
18. Campelo, N.; Filho, M.; Valadares, O.; Paes, M.; Aragao, A. Georeferenced monitoring of displacements of gabion walls. *Geotech. Eng.* **2018**, *171*, 64–77. [CrossRef]
19. *Construction Report 2014: Gabion Retaining Structure, A4 Motorway Ruda Śląska, Poland*; Maccaferri Company Poland: Warsaw, Poland, 2014. Available online: https://www.maccaferri.com/pl/produkty/gabiony-2/. (accessed on 1 September 2021). (In Polish)
20. Koussa, F.; Defrance, J.; Jean, P.; Blanc-Benon, P. Acoustic performance of gabions noise barriers: Numerical and experimental approaches. *Appl. Acoust.* **2013**, *74*, 189–197. [CrossRef]
21. Tang, V.T.; Fu, D.; Ngoc Binh, T.; Rene, E.R.; Sang, T.T.T.; Singh, R.P. An investigation on performance and structure of ecological revetment in a sub-tropical area: A case study on Cuatien River, Vinh City, Vietnam. *Water* **2018**, *10*, 636. [CrossRef]

22. Fan, Y.; Yang, Z.; Li, M.; Zhang, Z.; Li, D. Comparative analysis on environmental adaptability of two types of bank stabilization structures along the middle and lower reaches of the Yangtze River. *Sustainability* **2020**, *12*, 7991. [CrossRef]
23. Lin, Y.L.; Yang, G.L.; Li, Y.; Zhao, L.H. Engineering behaviors of reinforced gabion retaining wall based on laboratory test. *J. Cent. South Univ.* **2010**, *17*, 1351–1356. [CrossRef]
24. Kosiński, T. Retaining Structures and Embankments Made of Soil Reinforced with Steel Elements. In *Seminar Materials of IBDiM and PZWFS on Steel Structures in Geotechnics*; unpublished. (In Polish)
25. Ciomcia, B. Bridgeheads of the engineering structures made of a reinforced soil. *Mag. Autostrady* **2014**, *8-9*, 64–67. Available online: https://autostrady.elamed.pl/uploads/mau/articles/20956/64-67.pdf. (accessed on 1 September 2021).
26. Jayasree, P.K. Performance of Gabion Faced Reinforced Earth Retaining Walls. Ph.D. Thesis, Cochin University of Science and Technology, Kochi, India, 2008. Available online: http://dyuthi.cusat.ac.in/purl/2821 (accessed on 1 April 2008).
27. Koerner, R.M. *Designing with Geosynthetics*, 5th ed.; Pearson College Div: Prentice-Hall, NJ, USA, 18 April 2005. Available online: https://www.slideshare.net/AgungNoorsamsi/designing-with-geosynthetics-by-koerner-5th-edition-2005. (accessed on 1 September 2021).
28. Koerner, R.M.; Soong, T.Y. Geosynthetic reinforced segmental retaining walls. *Geotext. Geomembr.* **2001**, *19*, 359–386. [CrossRef]
29. Ćwirko, M. The Analysis of the Cooperation of Steel Welded Meshes with Soil in Retaining Structures of Gabion Walls Technology. Ph.D. Thesis, Silesian University of Technology, Gliwice, Poland, 1 September 2019. Available online: https://www.researchgate.net/publication/336141586_Analiza_wspolpracy_stalowych_siatek_zgrzewanych_z_gruntem_w_konstrukcjach_oporowych_wykonanych_w_technologii_murow_gabionowych (accessed on 1 September 2021). (In Polish)
30. Bilgin, Ö. Failure mechanisms governing reinforcement length of geogrid reinforced soil retaining walls. *Eng. Struct.* **2009**, *31*, 1967–1975. [CrossRef]
31. Sułkowska, A. The Model Tensil Tests Analysis of the Welded Meshes in the ZincAlu Coating or Coated with PVC, Embedded in a Fine Sand. M.Sc. Thesis, Silesian University of Technology, Gliwice, Poland, 2018. (In Polish)
32. PN-EN 10223-8:2014-03. Steel Wire and Wire Products for Fencing and Netting. In *Welded Mesh Gabion Products*; PKN: Warszawa, Poland, 2014. (In Polish)
33. PN-EN 10244-2:2010. Steel Wire and Wire Products. Non-ferrous Metallic Coatings on Steel Wire. In *Zinc or Zinc Alloy Coatings*; PKN: Warszawa, Poland, 2010. (In Polish)
34. PN-EN 10245-2:2011. Steel Wire and Wire Products. Organic Coatings on Steel Wire. In *PVC Finished Wire*; PKN: Warszawa, Poland, 2011. (In Polish)
35. Sampaco, C.L. Behavior of Welded Wire Mesh Reinforced Soil Walls from Field Evaluation and Finite Element Simulation. Ph.D. Thesis, Utah State University, Logan, UT, USA, 1996. Available online: http://rizal.library.ateneo.edu/index.php/node/9734. (accessed on 1 September 2021).
36. PN-EN 13738:2006. Geotextiles and Geotextile-Related Products. In *Determination of Pullout Resistance in Soil*; PKN: Warszawa, Poland, 2006. (In Polish)
37. Bolt, A.; Duszyńska, A. Pullout testing of geosynthetics embedded in soil in large scale pullout apparatus. *Zesz. Nauk. Politech. Śląskiej* **2003**, *1573*, 387–394. Available online: http://delibra.bg.polsl.pl/Content/40386/BCPS_44148_2003_Badania-oporu-geosyn.pdf. (accessed on 1 September 2021). (In Polish)
38. Ezzein, F.M.; Bathurst, R.J. A new approach to evaluate soil-geosynthetic interaction using a novel pullout test apparatus and transparent granular soil. *Geotext. Geomembr.* **2014**, *42*, 246–255. [CrossRef]
39. Prashanth, V.; Murali Krishna, A.; Dash, S.K. Pullout tests using modified direct shear test setup for measuring soil–geosynthetic interaction parameters. *Int. J. Geosynth. Ground Eng.* **2016**, *2*, 10. [CrossRef]
40. Park, J.B.; Kim, D.K.; Yang, S.B.; Kim, J.H. Pullout characteristics of geosynthetics reinforced earth using multilayer spreading pullout test. *Adv. Mater. Sci. Eng.* **2017**, *2017*, 9485826. [CrossRef]
41. Rahmaninezhad, S.; Han, J.; Kakrasul, J.; Weldu, M. Stress distributions and pullout responses of extensible and inextensible reinforcement in soil using different normal loading methods. *Geotech. Test. J.* **2019**, *42*, 1606–1623. [CrossRef]
42. Kawalec, J. Aggregate–Geogrid Interaction in Stabilisation Applications. How far Standard Laboratory Index Test Results are from Reality. In In Proceedings of the 11th International Conference on Geosynthetics 2018 (11ICG), Seoul, Korea, 16–21 September 2018; Korean Geosynthetics Society: Seoul, Korea, 2018; Volume 4, pp. 3142–3149. Available online: https://www.researchgate.net/publication/328289757_Aggregate-geogrid_interaction_in_stabilisation_applications_How_far_standard_laboratory_index_test_results_are_from_reality (accessed on 1 September 2021).
43. Kim, J. Development and Applicability of Multi-layer Pullout Apparatus System. Ph.D. Thesis, Sunchon National University, Suncheon, Korea, 2008.
44. Yuan, Z.; Chua, K.M. Numerical evaluation the pullout box method for studying soil-reinforcement interaction. *Transp. Res. Rec.* **1990**, *1278*, 116–124. Available online: http://onlinepubs.trb.org/Onlinepubs/trr/1990/1278/1278-015.pdf. (accessed on 1 September 2021).
45. Lajevardi, S.H.; Dias, D.; Racinais, J. Analysis of soil-welded steel mesh reinforcement interface interaction by pull-out tests. *Geotext. Geomembr.* **2013**, *40*, 48–57. [CrossRef]
46. Bergado, D.T.; Macatol, K.C.; Amin, N.U.; Chai, J.C.; Alfaro, M.C.; Anderson, L.R. Interaction of lateritic soil and steel grid reinforcement. *Can. Geotech. J.* **2011**, *30*, 376–384. [CrossRef]

47. Bergado, D.T.; Teerawattanasuk, C. Analytical models for predicting the pullout capacity and interaction between hexagonal wire mesh and silty sand backfill. *J. Appl. Sci. Eng.* **2001**, *4*, 227–238. [CrossRef]
48. Teerawattanasuk, C.; Bergado, D.T.; Kongkitkul, W. Analytical and numerical modeling of pullout capacity and interaction between hexagonal wire mesh and silty sand backfill under an in-soil pullout test. *Can. Geotech. J.* **2003**, *40*, 886–899. [CrossRef]
49. Woźniak, L. A Retaining Wall Made of Reinforced Soil in the Technology of a Gabion Wall Reinforced with a Steel Welded Mesh. M.Sc. Thesis, Silesian University of Technology, Gliwice, Poland, 2019. (In Polish)

applied sciences

MDPI

Article

Recycling of End-of-Life Tires (ELTs) for Sustainable Geotechnical Applications: A New Zealand Perspective

Ali Tasalloti [1], Gabriele Chiaro [1,*], Arjun Murali [1], Laura Banasiak [2], Alessandro Palermo [1] and Gabriele Granello [1,3]

1 Department of Civil and Natural Resources Engineering, University of Canterbury, Private Bag 4800, Christchurch 8041, New Zealand; ali.tasalloti@canterbury.ac.nz (A.T.); arjun.murali@pg.canterbury.ac.nz (A.M.); alessandro.palermo@canterbury.ac.nz (A.P.); gabriele.granello@gmail.com (G.G.)
2 Institute of Environmental Science and Research Ltd., Christchurch 8041, New Zealand; laura.banasiak@esr.cri.nz
3 School of Engineering, The University of Edinburgh, Edinburgh EH8 9YL, UK
* Correspondence: gabriele.chiaro@canterbury.ac.nz

Abstract: End-of-life tires (ELTs) are tires, unusable in their original form, which go into a waste management scheme (for recycling and energy recovery purposes), or otherwise are disposed. In New Zealand, the annual disposal of 3.5 million ELTs is posing critical environmental and socio-economic issues, and the reuse of ELTs through large-volume recycling engineering projects is a necessity. In this study, gravel and recycled granulated rubber were mixed to explore the possibility of obtaining synthetic granular geomaterials (with adequate geotechnical and environmental characteristics) that are suitable as structural fills for geotechnical applications including foundation systems for low-rise light-weight residential buildings. Moreover, an original framework with a set of geo-environmental criteria is proposed for the acceptance of gravel–rubber mixtures (GRMs) as structural fills. It is shown that when gravel-size like rubber particles are used, GRMs with volumetric rubber content of 40% or less have adequate strength ($\phi' > 30°$), low compressibility ($\varepsilon_v \leq 3\%$), excellent energy adsorption properties, and acceptable leachate metal concentration values (e.g., Zn < 1 mg/L), making them ideal synthetic structural fill materials for many sustainable geotechnical applications.

Keywords: end-of-life tires; recycling; gravel–rubber mixtures; sustainable geotechnical applications; foundation systems

Citation: Tasalloti, A.; Chiaro, G.; Murali, A.; Banasiak, L.; Palermo, A.; Granello, G. Recycling of End-of-Life Tires (ELTs) for Sustainable Geotechnical Applications: A New Zealand Perspective. *Appl. Sci.* **2021**, *11*, 7824. https://doi.org/10.3390/app11177824

Academic Editor: Daniel Dias

Received: 19 July 2021
Accepted: 24 August 2021
Published: 25 August 2021

1. Introduction

Tire recycling is the process of converting unwanted end-of-life tires (ELTs)—that can no longer be re-grooved or re-treaded—into materials that can be utilized in new products or applications [1]. While in many countries, ELTs are a controlled waste under stringent environmental protocols, currently no national regulations are in place in New Zealand to properly manage ELT recycling. As a result, rising environmental and socio-economic concerns are commanding the reuse of ELTs by means of large-volume recycling civil engineering schemes. Below, the issues, challenges, and possible solutions to the ELT disposal problem are described with reference to the New Zealand context.

1.1. Issues

Currently, over 5 million ELTs are produced yearly in New Zealand (i.e., one per capita), including four million passenger vehicle tires and one million truck tires. Such numbers are expected to grow over time with increasing volume of vehicles on roads. It is estimated that only 30% of such ELTs are exported or recycled, with the remaining 70% disposed of in stockpiles, landfills, illegal dumping, or otherwise unaccounted for [2,3].

A typical example of inadequate ELT disposal practice in New Zealand is shown in Figure 1. The dumping of scrap tires into landfills is certainly the least appropriate option for the disposal of ELTs. Such practice is unsustainable, causing significant environmental, socio-economical, and health problems such as inappropriate use of valuable land (up to 75% void space), harbor for pests and insects that may spread contagious and unknown diseases [4], potential water and soil contamination due to leaching of metals and other chemicals contained in the scrap tires [1,5–7], and the likelihood of uncontrolled fires of stockpiled tires [8–12]. Therefore, there are major benefits in moving away from ELT disposal and implementing sustainable recycling schemes.

Figure 1. A typical ELT dumping practice seen in New Zealand.

1.2. Challenges

Tire recycling may be challenging, but it is not impossible to achieve. In Europe, USA, Japan, Canada, and many other countries where strategic environmental procedures have been put in place to effectively make use of recycled ELTs, their disposal has been reduced to 20% or less [2,4,13].

As shown in Figure 2, reduce, reuse, recycle, and energy recovery are the four integrated basic options that should always be considered in dealing with waste management problems [14]. Similar to any other waste, also in the case of ELTs, the highest priority should be to avoid/reduce the generation of waste. This is obviously impractical due to the increasing volume of tired vehicles on roads. The next most preferred options should be reuse (as many times as possible and without further processing) and recycle (making new products) of ELTs. This will keep ELTs in the productive economy and benefits the environment by lessening the need for new materials and waste management. When extra recycling is not achievable, it could be possible to recover the energy from ELTs [15] (but only if environmentally adequate). Finally, only if ELTs could be safely recycled and direct treatment is not feasible, their disposal could become an eventual management alternative.

While avoidance/reduction is currently impracticable, ELT reuse and recycling are certainly feasible opportunities, and should be without a doubt preferred to energy recovery and disposal. In this regard, the most promising solution would be to use ELT-derived products as construction materials in sustainable large-scale civil engineering projects.

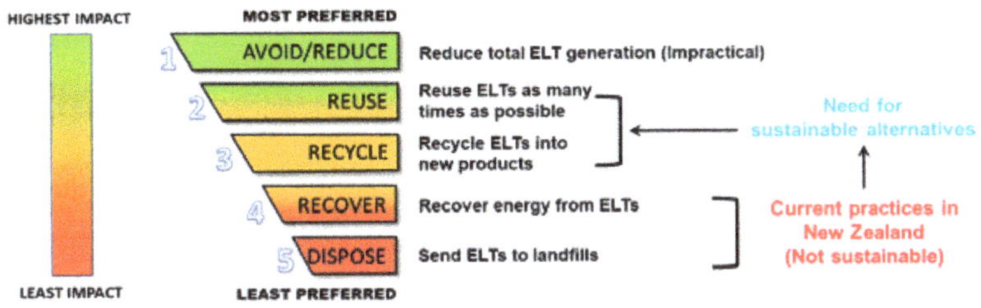

Figure 2. Waste management hierarchy applied to ELTs in the New Zealand context (adopted from [16]).

1.3. Opportunities

Reuse and recycling of industrial by-products, commercial wastes, and construction and demolition materials in geotechnical engineering applications are progressively required in Australasia as it provides important benefits in terms of increased sustainability and reduced environmental impacts [17]. In this setting, coal wash and steel slag mixtures have been reused as structural fills for a port reclamation development [18–20], recycled rubber mixed with coal wash and steel slag has been used in rail tracks [21], recycled glass and recycled concrete compounds have been characterized as pavement base [22], and recovered plastic and demolition wastes blends have been engineered as a capping layer for railway applications [23].

Given the above background, and aimed at facilitating the use of ELT-derived products as construction materials in sustainable civil engineering projects in New Zealand, a geo-environmental-structural engineering experimental research program—funded by the Ministry of Business, Innovation and Employment (MBIE)—has been jointly carried out by researchers of the University of Canterbury and the Institute of Environmental Science and Research Ltd. (ESR), Christchurch, New Zealand. Thus far, the main effort of such interdisciplinary research has been the development of "eco-rubber geotechnical seismic-isolation (ERGSI) foundation systems" for low-rise light-weight residential buildings [24–26]—readers can refer to the following website for full details of the project: https://sites.google.com/view/ecorubberfoundation/publications (accessed on 19 July 2021). Nevertheless, a series of experimental, numerical, and field investigations have been designed so that they would provide an in-depth understanding of key factors affecting the engineering behavior of soil–rubber mixtures and rubberized concrete, which must be taken into account in the design of such synthetic materials, hence, expediting their adoption not only into ERGSI foundation systems, but also in many other civil/geotechnical engineering applications.

In this paper, the results of the geotechnical and environmental investigations carried out to identify optimum energy-adsorption granular soil-recycled rubber mixtures, possessing excellent mechanical properties (e.g., compaction, compressibility, strength, dynamic properties etc.) and least leaching attributes are presented and discussed.

The results of the structural engineering laboratory tests (to design fiber-reinforced rubberized concrete structural elements, e.g., foundation raft, with satisfactory material and structural performance) and numerical investigations (i.e., DEM and FEM), physical models, and field trials (to verify the concept and assess the mechanical performance of geotechnical and structural elements individually and integrated into systems under both static and seismic load) will be presented elsewhere in due course.

2. Soil-Rubber Mixtures: Practical Implications and Material Suitability

Typically, ELT-derived aggregates (in the form of chips, crumbs, granules and shreds—ASTM [27]) assorted with cohesionless granular soil (mainly sand) have found use as lightweight backfill materials for embankments and retaining walls, drainage systems, slope remediation, and landfill construction [28,29]. However, more recently, due to their superior strength and dynamic properties, soil–rubber mixtures have been proposed as free-draining energy-adsorption backfill material for retaining walls, underground horizontal and vertical layers for liquefaction mitigation [30,31] and geotechnical seismic-isolation systems for residential buildings [32–35].

As reported in a comprehensive literature review undertaken by Tasalloti et al. [36], previous research has dealt primarily with the physical and mechanical characterization of sand–rubber mixtures. Generally, sand–rubber mixtures have good strength, low-shear modulus, and high damping properties. However, from a practical viewpoint, their high compressibility may result in low bearing capacity and undesirable settlement in the short- and long-term [34], limiting their adoption in many geotechnical applications. Moreover, in the selection of the soil type and recycled rubber size to form soil–rubber mixtures for use in geotechnical applications, the availability and the cost efficiency of both materials should be carefully considered [37,38]. Essentially, to avoid intrinsic segregation of binary assortments made of small and large particles [39–41], the recycled rubber should be cut into smaller (sand size-like) particles when mixed with sands. This, in turn, will unavoidably increase the implementation costs. Hence, the use of gravel–rubber mixtures (GRMs) instead of sand–rubber mixes has progressively been recommended.

Taking into consideration that in New Zealand it is a common practice to replace the topmost problematic soil layers (e.g., liquefiable sandy soils and compressible deposits) with well-compacted gravelly soil layers as part of the foundations for residential buildings and other geotechnical works, the adoption of GRMs in such applications seems the most appropriate. However, compared to sand–rubber mixtures, GRMs have been poorly characterized. Therefore, as part of the feasibility study reported in this paper, the physical properties, compaction characteristic, mechanical behavior, dynamic properties, and environmental aspects of different GRMs were evaluated by means of detailed laboratory investigations as described henceforth.

3. Experimental Study

In this study, to evaluate the combined effects of rubber content by volume and the aspect ratio (i.e., the ratio of median particle sizes of the rubber and the gravel, $AR = D_{50,R}/D_{50,G}$ [38–42]) on the physical properties and mechanical response of GRMs, a poorly-graded rounded gravel (G) and two coarse-sized recycled granulated rubber types—namely, large rubber (RL) and small rubber (RS)—were tested. Their particle size distribution (PSD) along with photos are shown in Figure 3, while their index properties are reported in Table 1. The specific gravity (G_s) was measured as 2.71 (G), 1.15 (RL), and 1.14 (RS)—the rubber was free of steel wires and fiber reinforcements. The aspect ratio of G-RL was AR = 0.67 and that of G-RS was AR = 0.33.

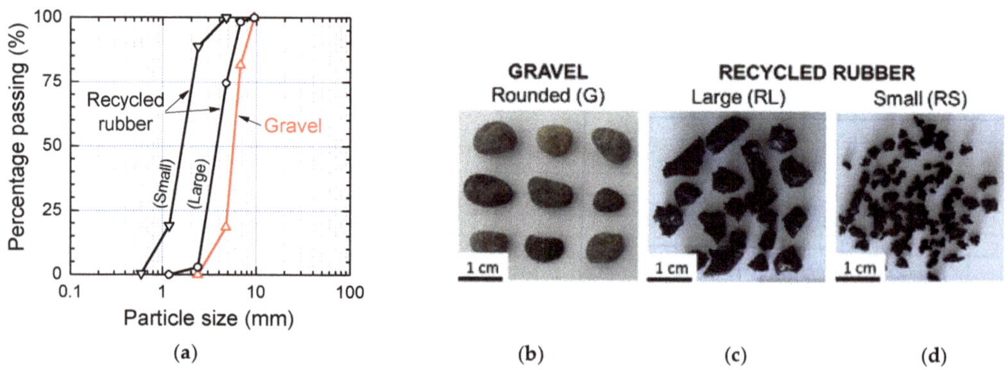

Figure 3. Material tested in this study: (**a**) Particle size distribution; and physical aspect of (**b**) gravel, (**c**) large rubber particles, and (**d**) small rubber particles (adopted from [16]).

Table 1. Index properties of the tested materials.

Material	Symbol	Grain Size (mm)		Specific Gravity G_s	Dry Unit Weight, γ_d (kN/m^3)	
		D_{max}	D_{50}		$\gamma_{d,min}$	$\gamma_{d,max}$
Gravel	G	10	6	2.71	15.4	17.2
Large rubber	RL	8	4	1.15	5.0	6.4
Small rubber	RS	4	2	1.14	4.2	5.9

Several mixtures were prepared at various volumetric rubber content (*VRC*) of 0%, 10%, 25%, 40%, and 100% by mixing RL and RS with G. Note that, *VRC* is defined as the ratio of the rubber particle volume (V_R) to the total volume of solid particles ($V_{TOT} = V_R + V_G$, where V_G is the volume of gravel grains), according to Equation (1):

$$VRC = \frac{V_R}{V_R + V_G}\,(\times 100) \tag{1}$$

Dry specimens were prepared by tamping method at a degree of compaction of 90% or above (based on standard Proctor tests—refer to details in Section 4.1). Segregation in the specimens was avoided by minimizing any vibration and preventing granular flow.

The one-dimensional (1-D) compressibility of GRMs was evaluated up to 500 kPa vertical stress by means of a medium-size (inner diameter 250 mm; specimen height from 150 mm) compression cell. Alternatively, the shear strength was estimated by using a medium-size (100 mm × 100 mm in cross-section and 53 mm in height) direct shear box under 30, 60, and 100 kPa normal stress levels (the horizontal displacement rate was 1 mm/min). Moreover, to evaluate the small-strain stiffness, shear modulus degradation, and damping ratio of GRMs, two series of bender element and small-strain drained cyclic triaxial tests (specimen size: diameter = 70 mm; height = 170 mm) were also carried out on dry specimens. Additionally, batch leaching tests besides conductivity, pH, total organic carbon (TOC), and inductively coupled plasma mass spectrometry (ICP-MS) analyses were performed to assess key environmental aspects of GRMs. A summary of the geo-environmental tests performed in this study is reported in Table 2.

Table 2. Summary of geo-environmental tests carried out in this study.

Materials	*VRC* (%)	Test Type				
		1-D Compression #	Direct Shear #	Bender Element #	Cyclic Triaxial #	Batch Leaching +
G	0	•	•	•	•	○
G-RL	10	•	•	•	•	○
G-RL	25	•	•	•	•	○
G-RL	40	•	•	•	•	•
RL	100	•	•	N/A	N/A	○
G-RS	10	•	•	N/A	N/A	○
G-RS	25	•	•	N/A	N/A	○
G-RS	40	•	•	N/A	N/A	•
RS	100	•	•	N/A	N/A	○

Tests on dry specimens prepared at 90% degree of compaction by dry tamping; + Materials tested in deionized water up to 28 days; • Completed; ○ In progress; N/A: Not available.

4. Experimental Results

4.1. Compaction Properties of GRMs

For satisfactory performance of structures made of soil–rubber assortments, it is necessary to properly control the compaction and correctly evaluate the physical properties of compacted materials. For the GRMs investigated in this study, vibratory table tests were found to be ineffective due mainly to the low moduli and energy absorption ability of the rubber aggregates. In contrast, Proctor compaction tests resulted in a suitable testing procedure, even if the moisture content was not a controlling factor. On the other hand, the minimum unit weight was simply attained by cautiously pouring the materials in the compaction mold with nearly zero depositional height. Typical results are reported in Figure 4a in terms of dry unit weight variation with *VRC*.

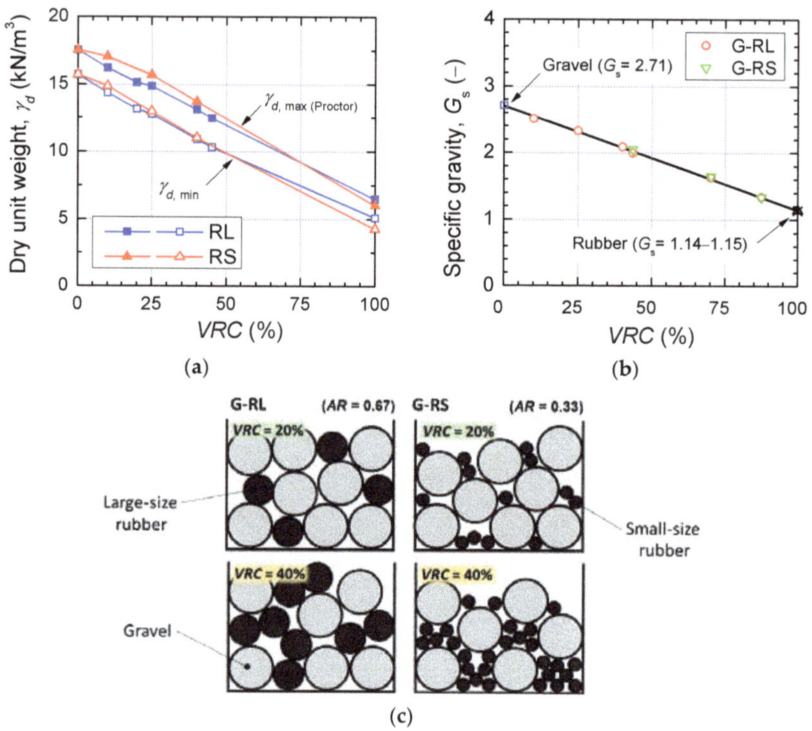

Figure 4. Density characteristics of tested GRMs: (**a**) minimum and maximum dry unit weight; (**b**) specific gravity; and (**c**) idealized illustration of volume of solids-biased density packing (adopted from [16]).

As the rubber particles are much lighter than the gravel ones (Figure 4b), both the minimum and maximum dry unit weights of GRMs decrease almost linearly by increasing *VRC*. It is worth mentioning that, at $VRC \leq 40\%$, the dry unit weight values of G-RS were slightly above the linear trends. This is because at lower *VRC* values, small rubber particles can easily occupy the large voids between gravel grains (Figure 4c), which results in an increase in the density state of the mixtures [39–41]. In contrast, because the size of RL particles is almost similar to that of the gravel, the rubber particles cannot fit in the voids between the gravel grains, even at lower *VRC*, but rather replace the gravel grains in the mixtures. For completeness, a schematic illustration showing idealized packing density arrangements between gravel and RL and RS particles is reported in Figure 4c for $VRC =$ 20% and 40%.

4.2. 1-D Compressibility of GRMs

Material compressibility (or the capability to increase/decrease in volume when subjected to an applied load) is one of the most important factors required in design considerations. For conventional granular soils with rigid particles (i.e., sand and gravel), any change in volume is due to the rotation, movement, and rearrangement of noncompressible grains [41]. The compressibility of GRMs, however, is completely different to that of granular soils, due to significant differences in the elastic modulus of the rigid particles of the host gravel and that of the soft rubber particles. Furthermore, under normal stress conditions, not much volume change is associated with individual particles of rubber (i.e., this is because the Poisson's ratio of pure rubber is $\nu \approx 0.5$), and distortion is the key phenomenon that takes place in pure rubber specimens.

1-D compression test results with creep loading are reported in Figure 5. In the tests, a 2-h creep (which, under the adopted testing conditions, was sufficient to attain negligible settlement under sustained vertical stress) was applied at different incremental loading stages and the measured vertical stress values were corrected for the loss of transferred load (due mainly to soil-wall friction) that was experimentally determined.

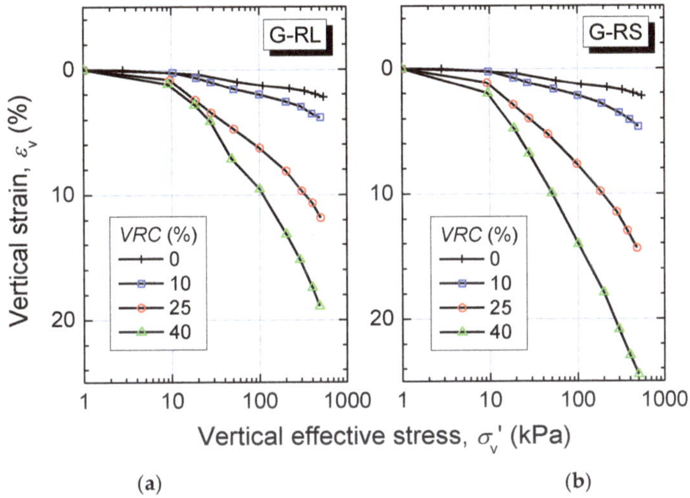

Figure 5. Compressibility of GRMs in 1-D compression tests with 2-h creep. (**a**) G-RL and (**b**) G-RS mixtures.

For any given vertical shear stress level, the trends reported in Figure 5 clearly show an increase in 1-D compression with increasing *VRC*. Furthermore, it can be seen that GRMs with *VRC* $\leq$ 10% performed better than those with *VRC* $\geq$ 25% (i.e., much smaller vertical strains (ε_v) developed under the same applied vertical effective stress (σ_v') level). Additionally, in the range of *VRC* tested, it can be observed that the GRMs made of small rubber particles (*AR* = 0.33) were much more compressible than those consisting of large rubber particles (*AR* = 0.67).

4.3. Shear Strength of GRMs

Alike conventional soils, shear strength is one of the most important characteristics contributing to the performance for GRMs. The shear strength is the result of friction and interlocking between particles and likely bonding at particle contacts. Due to interlocking, GRMs may contract or expand in volume as they are subject to shear strains.

The stress–strain–volumetric behavior of G-RL and G-RS attained by direct shear tests at 60 kPa normal stress (σ_n) is compared in Figure 6. It can be seen that, as *VRC* increased, the GRM response progressively changed from dilative with a clear peak shear state (gravel-like behavior) to contractive without a peak shear state (rubber-like behavior). Similar tendencies were observed at σ_n = 30 and 100 kPa.

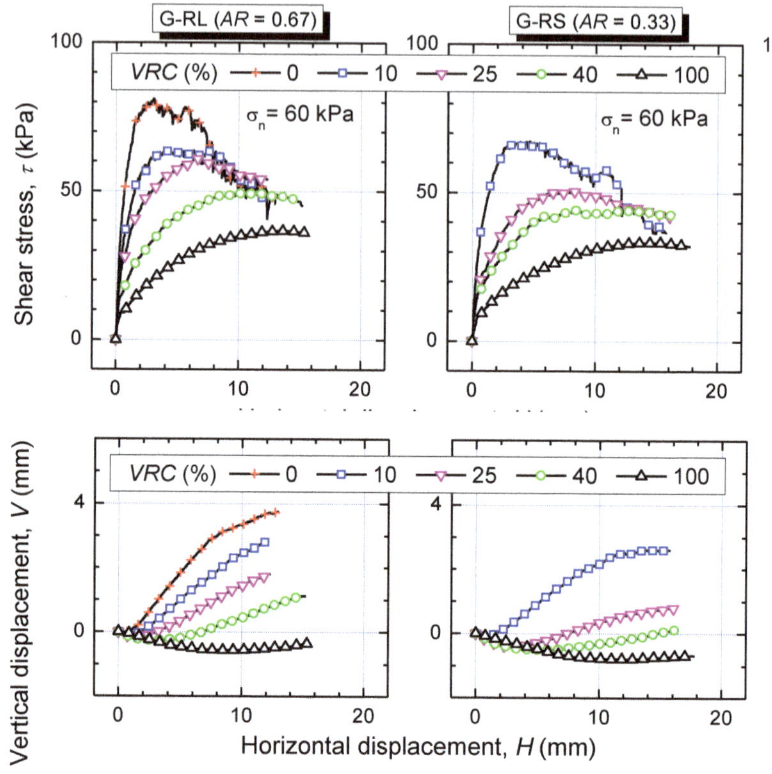

Figure 6. Direct shear behavior of GRMs at 60 kPa normal stress: (**a**) G-RL; and (**b**) G-RS mixtures.

The summary plot reported in Figure 7a shows the change in maximum shear stress (τ_{max} that was evaluated at peak state for dilative materials or at large deformation for contractive materials) with increasing *VRC* for all mixtures sheared at σ_n = 30, 60, and 100 kPa. It can be seen that the trends were very similar for both G-RL and G-RS, indicating that the effect of *AR* on the strength of tested GRMs was almost insignificant. Alternatively, for any given σ_n value, τ_{max} decreased considerably with increasing *VRC*. Moreover, such decay was more significant at higher σ_n. For example, at σ_n = 100 kPa, τ_{max} was approximately 137 kPa at *VRC* = 0% (gravel) and 73 kPa at *VRC* = 40%, corresponding to a reduction of 1.9 times. At σ_n = 30 kPa, τ_{max} is approximately 45 kPa at *VRC* = 0% and 28 kPa at *VRC* = 40%, corresponding to a reduction of 1.6 times.

Figure 7b reports the values of the Mohr–Coulomb effective friction angle (ϕ') for all mixtures. Essentially, ϕ' decreased significantly with increasing *VRC* (and only slightly with *AR*) from about 54° (gravel) to 29° (graduated rubber). Notably, excluding G-RS with *VRC* > 85% and G-RL with *VRC* > 95%, most of the GRMs had a high strength (i.e., ϕ' > 30°) regardless of the *VRC* and rubber particle size, making them potentially suitable structural fill materials for many geotechnical applications [17,19].

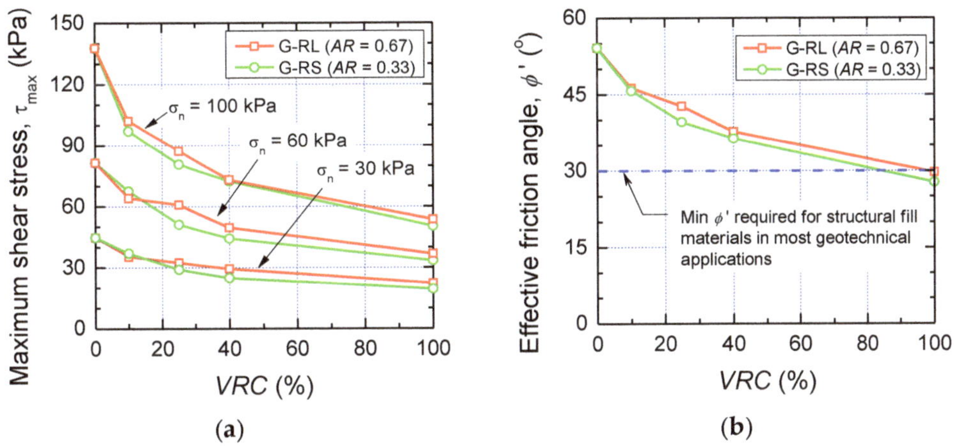

Figure 7. Strength characteristics of GRMs: (**a**) variation of maximum shear stress with normal stress and *VRC*; and (**b**) variation of effective friction angle with *VRC*.

The packing properties of GRMs described earlier (Section 4.1) may provide valuable information on their mechanical behavior (refer to Figures 5–7). Nonetheless, the load-transfer mechanism within the mixtures depends predominantly on the skeleton material [39–41], which is formed when particles of the same material are in contact with each other and are able to transfer loads. The material forming the skeleton becomes the matrix material that governs the general mechanical behavior of a mix. As shown by Figure 8, DEM-based micro-mechanical strong-force network analyses conducted by Chiaro et al. [43] have indicated that two matrix materials can be expected for GRMs: (i) a gravel matrix, leading to a stiffer gravel-like response of GRMs ($VRC \leq 30\%$); and (ii) a rubber matrix, producing a softer rubber-like behavior of GRMs ($VRC \geq 60\%$). In between, an intermediate gravel–rubber behavior will likely take place ($30\% < VRC < 60\%$). Such distinct behavioral zones provide a framework explaining the gradual change in behavior from stiff gravel-like to soft rubber-like, as observed in the direct shear tests.

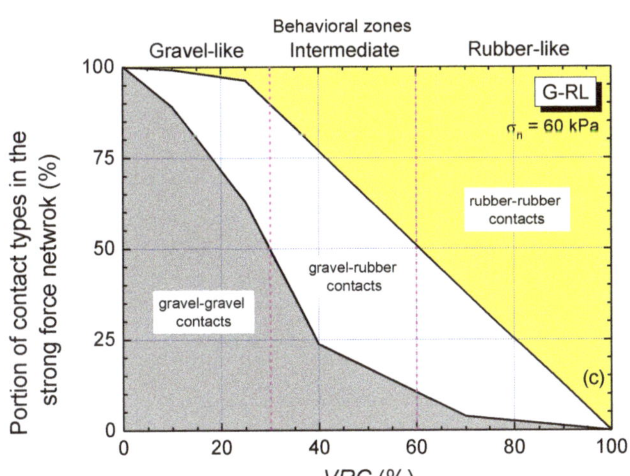

Figure 8. Behavioral zones for GRMs in direct shear tests evaluated by DEM strong-force network analysis.

4.4. Stiffness, Shear Strain Degradation and Damping Properties

The design of geotechnical structures made of GRMs subjected to cyclic shear loading conditions (e.g., earthquake or traffic loads) will necessitate the estimation of the dynamic and cyclic behavior of the mixtures including the small-strain shear stiffness (G_{max}), shear modulus (G) degradation, and damping ratio (D).

As shown in Figure 9, the results of bender element tests indicate that G_{max} of GRMs decreased with increasing *VRC* and increased with confining pressure. This is essentially consistent with previous studies on sand–rubber mixtures [28]. However, for any given confining pressure value, it has been noticed that the G_{max} of GRMs is generally greater than that of sand–rubber mixtures having the same *VRC* [39,41].

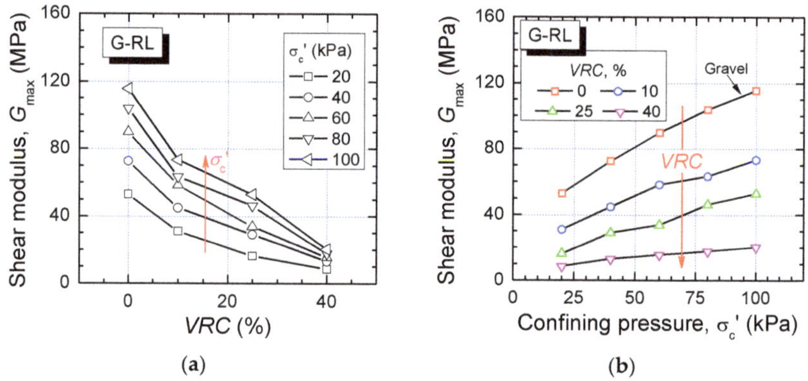

Figure 9. Variation of small-strain shear modulus (G_{max}) of G-RL with: (**a**) *VRC*, and (**b**) confining pressure (adopted from [16]).

On the other hand, the results of small-strain drained cyclic triaxial tests revealed that the addition of rubber aggregates into GRMs delays the G degradation on one side and increases D on the other (Figure 10). This is an important outcome demonstrating that GRMs will perform better than gravel only under cyclic loading, maintaining their initial stiffness and dynamic properties over a larger range of shear strains. Clearly, this is due to the ability of soft rubber particles to rebound and dissipate energy.

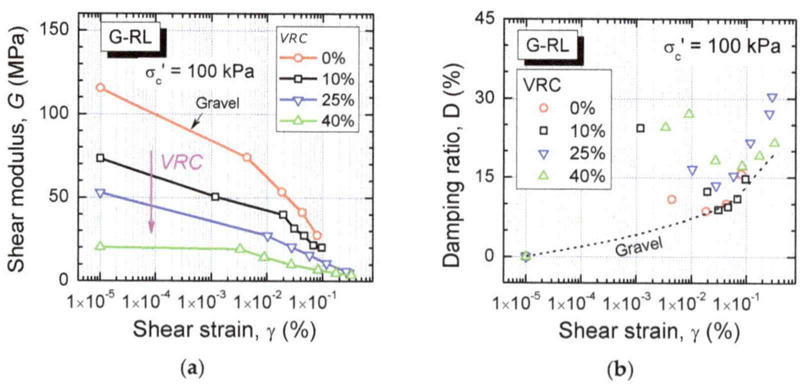

Figure 10. Dynamic properties of GRMs evaluated by small-strain drained cyclic triaxial tests at 100 kPa confining pressure: (**a**) shear modulus degradation and (**b**) damping ratio.

4.5. Leaching Characteristics of GRMs

Tires are largely composed of carbon black, vulcanized rubber, rubberized fabric containing reinforcing textile cords, antioxidants, silica, pigments, process and extender oils, accelerators, and steel wire [4] and may contain petroleum residues acquired through use [44]. In this study, due to the recycled rubber's proposed use in geotechnical applications, clarification was needed to understand whether and to what extent metals and inorganics could leach into the environment. This is to ensure that there are no negative, long-term impacts to the environment. Factors expected to affect the rate and concentration of tire leachate include the rubber particle size and content, contact time, and the aquatic or soil environments [6].

Within this scope, in this feasibility study, six pilot 28-day batch leaching tests were conducted on GRMs with *VRC* = 40% that were placed in deionized water (additional tests on GRMs with *VRC* = 0, 10, 25% are in progress and the results will be presented elsewhere). Samples were taken on days 0, 1, 2, 4, 7, 10, 14, 17, 21, 25, and 28 to track leaching characteristics, which were determined by ICP-MS analyses. The metals and organics tested for included aluminum (Al), arsenic (As), calcium (Ca), cadmium (Cd), cobalt (Co), chromium (Cr), copper (Cu), iron (Fe), potassium (K), magnesium (Mg), manganese (Mn), phosphorus (P), lead (Pb), sodium (Na), nickel (Ni), and zinc (Zn). Note that pH, conductivity, and TOC analyses were also performed and the details are reported in Banasiak et al. [7].

Table 3 summarizes the total average mass of predominant constituents leached for G-RL and G-RS mixtures, while Table 4 provides the maximum concentrations of all the tested elements. It is expected that the Ca, K, and Na contents are attributed to the gravel and the Zn and Mg to the tires. As an example, Zn concentrations measured during the entire 28-day leaching test period are reported in Figure 11.

Table 3. Total mass of elements leached at 28 days (average values from 3 tests).

GRMs	Total Mass (mg)					
	Ca	Na	K	Zn	Mg	Others
G-RL (*VRC* = 40%)	117	48	27	16	10	<2
G-RS (*VRC* = 40%)	424	55	55	57	28	<2

Table 4. Maximum concentration of leachate elements obtained by ICP-MS analyses (average values from three tests).

GRMs	Maximum Concentration (mg/L)															
	Al	As	Ca	Cd	Co	Cr	Cu	Fe	K	Mg	Mn	P	Pb	Na	Ni	Zn
G-RL	0.05	0.00	4.25	0.00	0.00	0.00	0.01	0.05	1.07	0.37	0.03	0.01	0.00	1.46	0.00	0.45
G-RS	0.03	0.00	15.0	0.00	0.01	0.00	0.00	0.02	1.58	1.08	0.08	0.01	0.00	1.61	0.00	1.76

The test results were compared to New Zealand Drinking Water Standards [45] and New Zealand Landfill Waste Acceptance Criteria [46] for class A and B leachability concentration limits. Class A landfills "provide a degree of redundancy for leachate containment" to "reduce the potential for adverse environmental effects" whereas class B landfills do not have such controls. Guideline values for Zn, Mn, Fe, and Al were exceeded. These values mean that there is an increased discoloration, taste, and staining risk.

From a practical viewpoint, these results highlighted that the leachate from smaller tire particles (G-RS) had a higher content of metals, implying that particle size and surface area influence the concentration of elements in GRM leachate. Additionally, the concentration of Zn in G-RS tests exceeded the leachability limit for class B landfills. Hence, G-RS mixtures would need to undergo pre-treatment if used in geotechnical works.

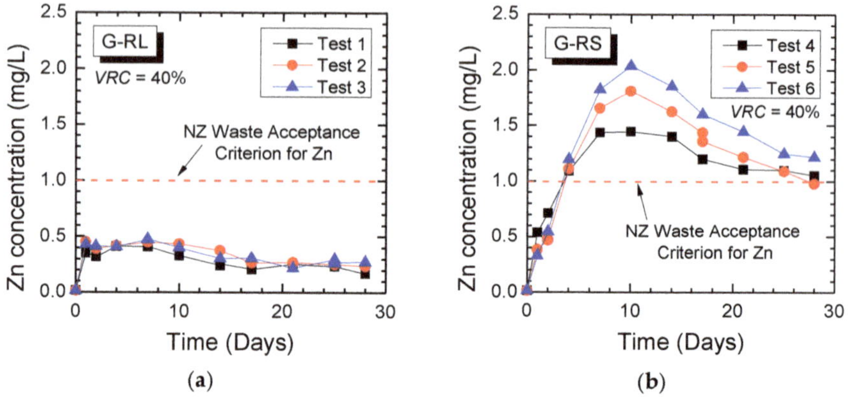

Figure 11. Concentration of zinc from 28-day batch leaching tests on GRMs (*VRC* = 40%) placed in deionized water: (**a**) G-RL; and (**b**) G-RS mixtures.

5. Design Considerations

GRMs are essentially lightweight materials (Figure 4). As a result, one of the benefits of using GRMs would be that the in situ overburden stresses associated with the self-weight of typical geomaterials could be greatly reduced. For instance, the reduction would be 6% for *VRC* = 10%, 13% for *VRC* = 25% and 23% for *VRC* = 40%. This is expected to lead to reduced design performance requirements, in terms of bearing capacity and settlement, for natural soil deposits that underlie a GRM layer installed as part of a residential structure foundation system or a geo-structure. Moreover, it would induce less settlement underneath embankments placed on compressible soil deposits or reduce the earth pressure behind retaining structures [16].

This study has demonstrated that GRMs have an adequate strength (i.e., $\phi' > 30°$) irrespective of the *VRC* and rubber particle size, making them suitable structural fill materials for many geotechnical applications. However, the ultimate adoption of GRMs as structural fills in geotechnical application would also depend on their compressibility under sustained loads (creep). As displayed in Figure 5, it is clear that the higher the vertical effective stress applied on GRMs, the higher the vertical strain developed, and the lower the *VRC* that may be accepted in the mixtures to satisfy compressibility requirements. Moreover, *AR* also plays a key role; in fact, for any combination of *VRC* and vertical effective stress, it can be observed that the mixtures with smaller rubber particles (G-RS) experience greater vertical strain compared to those with large rubber particles (G-RL). Therefore, to have GRMs with reduced compressibility, gravel and rubber particles with similar grain size should be used.

Permeability of GRMs can be related to the porosity of the mixes (which in turns depends on the particle size of rubber and gravel) as well as the vertical stress applied. Experimental evidence from relevant studies reported in the literature have shown that usually the hydraulic conductivity coefficient of soil–rubber mixtures with large rubber particles varies between that of typical gravelly soils at low confining stress levels and that of sandy soil at higher confining stress levels [36]. For mixtures of small rubber particles and *VRC* > 10%, the hydraulic conductivity coefficient was similar to that of sandy soils regardless of the confining stress applied [36]. In the case of foundations, retaining walls, and embankments, this means that GRMs will function as a free-draining fill material ensuring the rapid dissipation of excess pore water pressure. Remarkably, Hazarika et al. [30] reported that the use of GRMs (with *VRC* < 40%) placed on the top of liquefiable sandy soils can considerably reduce the pore water pressure generation in the sandy soil deposit, the reasons being: (1) GRMs themselves are not liquefiable and, therefore, can be conveniently used to replace the top most portion of liquefiable soil layers, and (2) having an hydraulic conductivity higher than the sand, GRMs provide a preferential pathway for water to easily flow upward and quickly dissipate pore water pressure.

The introduction of recycled ELT-derived materials in geotechnical applications may have benefits in terms of cost reductions and increased performance. However, it is essential to ensure that such innovations do not result in long-term negative impacts on the environment (e.g., through the leaching of toxic chemicals into the surrounding soil environment, groundwater, and surface water). The results of the leaching tests on GRMs have indicated that the leachate from smaller tire particles (RS) had a higher content of metals (e.g., Zn) compared to the large rubber one (RL), implying that particle size and surface area influence the concentration of elements in the GRM leachate. To minimize the leaching of metals, therefore, the use of larger rubber particle size is desirable. Otherwise, smaller rubber particles would need to undergo pre-treatment before use.

5.1. Proposed Acceptance Criteria for GRM Fill Materials (Static Loads)

In the case of most granular soils, design criteria based on frictional shear strength, bearing capacity, and permeability are used to evaluate their suitability as structural fill materials for geotechnical applications. It is often required that fills should possess a friction angle greater than or equal to 30° (and/or a California Bearing Ratio (CBR) > 10%), to guarantee a satisfactory shear resistance and to minimize post-construction settlement [17]. In addition, it is also recommended that fill material should have a permeability coefficient similar to that of sandy fills to ensure rapid dissipation of excess pore water pressure and to minimize internal erosion phenomena.

Because of the critical aspects identified in this study such as long-term creep compressibility and leaching of heavy metals, design criteria simply based on strength and permeability may not be sufficient to fully judge whether or not a GRM meets all the geotechnical requirements for structural fills. To overcome this issue, a modified framework with additional geotechnical design criteria including three levels of acceptance is proposed for GRMs (Figure 12):

- Level 1: *Frictional shear resistance/bearing capacity*. When used as structural fills, GRMs should possess adequate shear strength and bearing capacity to guarantee a satisfactory stability and reduced post-construction settlement. Specifically, GRMs should have a friction angle $\phi' \geq 30°$ and/or a CBR $\geq$ 10%. Otherwise, they should only be recommended as general fills.
- Level 2: *Compressibility (most critical level)*. Since compressibility could continue for a long period of time, it is suggested that GRMs be accepted as structural fill only if the volumetric strain (ε_v) is $\leq$ 3%. Otherwise, they should only be recommended as general fills.
- Level 3: *Permeability (not a critical level)*. In order to guarantee rapid dissipation of excess pore water pressure (i.e., maintaining relatively free-draining), it is recommended that GRM structural fills should have a permeability coefficient similar to that of sandy/gravelly fills (i.e., 1×10^{-6} cm/s $\leq k \leq 1 \times 10^{-4}$ cm/s). If $k > 1 \times 10^{-4}$ cm/s, GRMs could be used as general fills.

In addition to the above proposed geotechnical criteria, GRMs must also meet environmental standards (e.g., New Zealand Drinking Water Standards and New Zealand Landfill Waste Acceptance Criteria). Alternatively, pre-treatment or confinement of leached chemicals may be required.

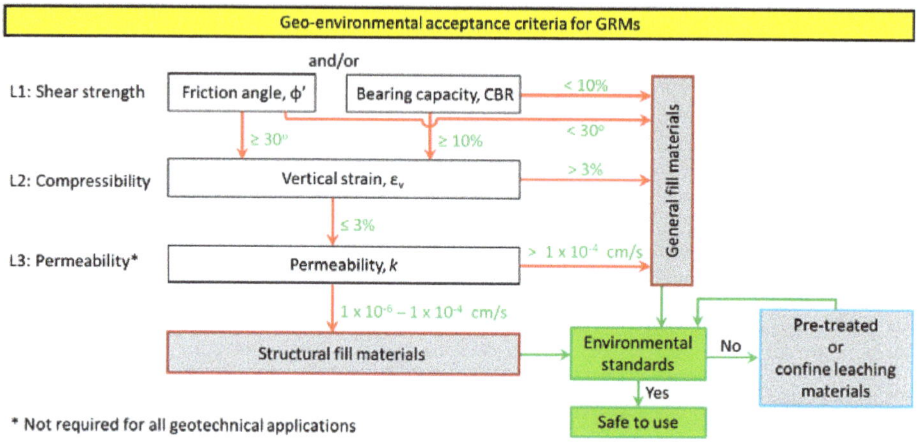

Figure 12. Proposed geo-environmental acceptance criteria for use of GRMs in geotechnical applications.

5.2. Practical Application: Seismic Design of ERGSI Foundation Systems

"Eco-rubber geotechnical seismic-isolation foundation systems" or in short "ERGSI foundation systems" are integrated systems comprising two key elements (Figure 13):

- a seismic energy-absorption shallow horizontal layer of GRMs; and
- a flexible fiber-reinforced rubberized concrete raft foundation.

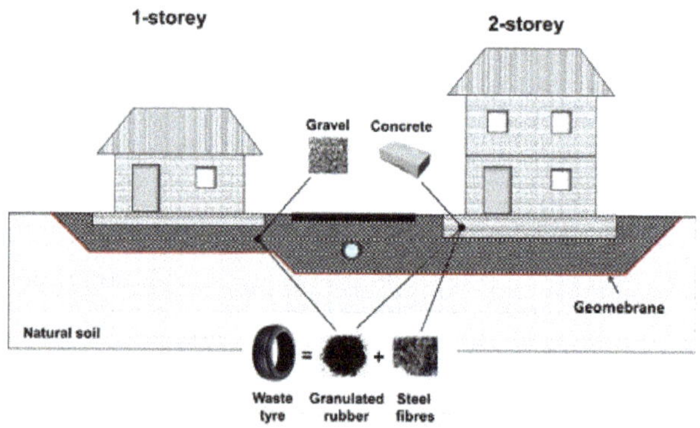

Figure 13. Schematic illustration of the "Eco-rubber geotechnical seismic-isolation" (ERGSI) foundation systems (adopted from [16]).

The geotechnical design of the GRM dissipative layer of a ERSGI foundation system requires considering not only the strength and compressibility criteria shown in Figure 12 (permeability is expected to be fulfilled given the fact that the GRMs are coarse granulated free-draining materials), but also the seismic performance of these synthetic materials.

Typically, in New Zealand, 1- to 2-storey timber-framed residential buildings transfer a vertical effective stress (σ_V') from the structure to the foundation of 7–10 kPa and 10–15 kPa [47,48], respectively. Considering only G-RL (due to its acceptable leaching characteristics, Figure 11) from Figure 7b, it is evident that mixtures with $VRC \leq 95\%$ would have a friction angle much greater than 30°, implying a good stability under static load conditions ($\sigma_V' \leq$ 10–15 kPa) for both 1-storey and 2-storey buildings.

On the other hand, to guarantee the serviceability during and after the superstructure construction (i.e., to avoid damage to pipelines and other essential services that may be embedded in gravel-rubber foundation layer), and meet the compressibility criteria ($\varepsilon_V \leq 3\%$), it appears that only the G-RL mixtures with *VRC* ≤ 40–45% would be suitable for 1-storey buildings and those with *VRC* ≤ 35% would be suitable for 2-storey buildings (Figure 14).

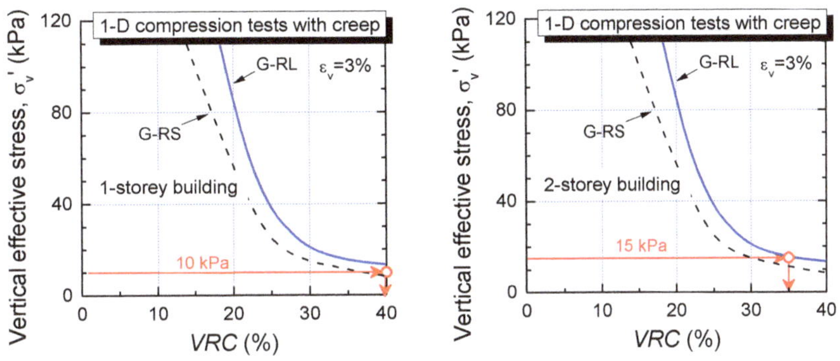

Figure 14. Selection of suitable GRMs for foundation applications based on maximum allowable vertical strain (ε_V = 3% contours are derived from the 1-D compression test results shown in Figure 5): (**a**) 1-storey building; and (**b**) 2-storey building.

The seismic performance of ERGSI foundation systems was assessed by small-scale impact load tests conducted on GRMs [49]. In such pilot tests, a foundation prototype was placed on a 60 cm thick layer (made of mixtures with *VRC* = 0, 10, 25, and 40%). As shown in Figure 15, it was found that as more rubber aggregates were added into the mixtures, the natural frequency of the foundation system shifted toward a smaller value (base-isolation effect) while the amplitude of the output peak acceleration decreased (i.e., damping effect).

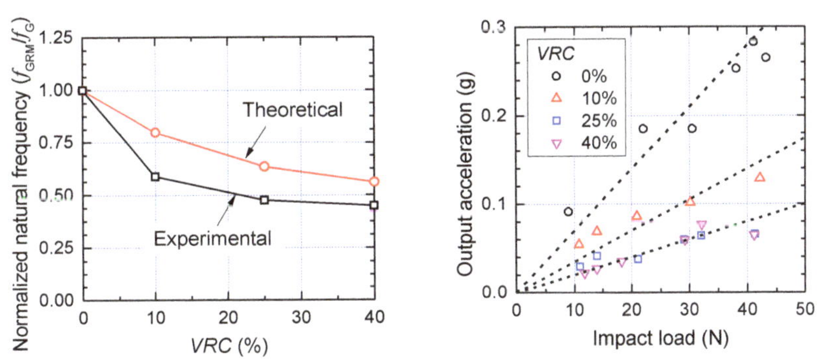

Figure 15. Dynamic properties of G-RL mixtures from impact load tests: (**a**) decay of natural frequency with increasing *VRC*; and (**b**) reduction in output acceleration with increasing *VRC*.

It is also evident that even a small amount of rubber (i.e., *VRC* = 10%) contributes significantly in the dissipation of seismic energy. On the other hand, *VRC* of 25% and 40% seem to provide the same dissipative effects.

Based on the above, it is recommended that the range of suitable GRMs for ERGSI foundation systems should be limited to *VRC* ≤ 40% and *VRC* ≤ 35% for 1-storey and 2-storey buildings, respectively. However, to achieve the best seismic performance in terms of seismic isolation and energy dissipation, it was eventually suggested to use 25% ≤ *VRC* ≤ 35–40%.

6. Conclusions

In New Zealand, every year, approximately 3.5 million ELTs are legally or illegally disposed of through landfills and stockpiles. However, ELTs cannot be considered as a simple waste to be disposed of. Large-scale sustainable recycling initiatives are crucial to tackle this nationwide problem. A promising solution is to reuse ELTs in the form of recycled granulated rubber as construction materials in civil engineering applications. In this context, this paper has presented: (i) the results of geotechnical and environmental investigations assessing the potential use gravel–rubber mixtures (GRMs) as structural fills for geotechnical applications (including seismic-isolation foundation systems), and (ii) a framework with a set of geo-environmental criteria for the acceptance of GRMs as geo-structural fills.

The following main conclusions can be drawn from this study:

- the compressibility of GRMs is highly influenced by the rubber content and the rubber size. The higher the *VRC* and the smaller the rubber size, the more compressible the mixtures;
- irrespective of the applied normal stress level and rubber particle size (or aspect ratio, *AR*), the direct shear response of GRMs changed rapidly from dilative-like to contractive-like with the addition of rubber in the mixtures; nevertheless, independently from *AR*, the effective friction angle (ϕ') was found to be between 54° (gravel) and 29° (rubber);
- from an environmental viewpoint, the use of GRMs with larger gravel-size like rubber particles (free from steel wires) is desirable since it would minimize the leaching of toxic metals (e.g., Zn) from the granulated rubber;
- GRMs with volumetric rubber content (*VRC*) $\leq$40% and gravel-size like rubber particles have satisfactory strength (i.e., friction angle >30°), low compressibility (long-term vertical strain $\leq$3%) and minimal leaching characteristics that make it possible to use GRMs as structural fills in many sustainable geotechnical applications; and
- specifically to the case of foundation systems, it was demonstrated that GRMs with $25\% \leq VRC \leq 40\%$ have excellent seismic-isolation and energy dissipation properties. As a result, their use can be particularly beneficial in applications aimed at enhancing the seismic resilience of low-rise light-weight residential buildings.

Future works. Note that the results reported in this paper are based on the use of uniformly-graded gravel and granulated rubber with maximum particle diameters of approximately 10 mm that could be tested in the laboratory using conventional devices. However, as at the Geotechnical Laboratory of the University of Canterbury, the capability of conducting tests on geomaterials with larger particles has become possible, the authors plan to conduct further geotechnical investigation on GRMs with larger gravel particles (i.e., D_{max} = 40 mm) that are commonly used in foundation applications and other geotechnical works in New Zealand. This will make it possible to verify some of the findings reported in this paper, and if required, refine the proposed acceptance criteria for GRMs as geo-structural fills.

To complement the batch leaching test results reported in this paper, additional leaching tests considering acid and alkaline solutions are intended. Moreover, a series of column tests will be conducted on 1-D consolidated GRMs with the aim of mimicking as much as possible the groundwater conditions and material properties underneath a foundation or geo-structure, and account for possible groundwater seasonal changes (i.e., repeated cycles of wet and drying) and the effect of 1-D compression on the leaching properties of GRMs.

Specifically for seismic-isolation foundation systems, the authors have planned to carry out detailed field testing on a large-scale prototype structure supported on ERGSI foundations. This is of paramount importance for several reasons: (i) to evaluate the response of the ERGSI foundation with realistic infinite boundary conditions (not feasible in the laboratory tests); (ii) to assess the effects that underlying site geology may have on the ERGSI foundation performance and design; (iii) to monitor and evaluate the long-term settlement and ensure that is within serviceable limits, and (iv) to use background seismicity to further assess the seismic performance of the foundation system.

Author Contributions: Conceptualization, G.C., A.P., L.B., and G.G.; Methodology, G.C. and L.B.; Investigation, A.T., G.C., A.M., and L.B.; Data curation, A.T., G.C., L.B., and G.G.; Writing—original draft preparation, A.T. and G.C.; Writing—review and editing, L.B., A.P., and G.G.; Project administration, G.C.; Funding acquisition, G.C., A.P., and L.B. All authors have read and agreed to the published version of the manuscript.

Funding: This research was funded by the Ministry of Business, Innovation, and Employment (MBIE) of New Zealand, MBIE Smart Ideas Endeavour Research Grant No. 56289.

Institutional Review Board Statement: Not applicable.

Informed Consent Statement: Not applicable.

Data Availability Statement: The data presented in this study are available on request from the corresponding author.

Acknowledgments: The laboratory assistance of Sean Rees is greatly appreciated. Figures adopted from [16] were republished with the permission of the New Zealand Geotechnical Society.

Conflicts of Interest: The authors declare no conflict of interest. The funders had no role in the design of the study; in the collection, analyses, or interpretation of data, in the writing of the manuscript, or in the decision to publish the results.

References

1. Basel Convention Working Group. *Technical Guidelines on the Identification and Management of Used Tyres*; Document No. 10; Basel Convention on the Control of Transboundary Movements of Hazardous Wastes and Their Disposal: Basel, Switzerland, 1999.
2. Ministry for the Environment. *Waste Tyres Economic Research*; Report 3; Ministry for the Environment: Wellington, New Zealand, 2015; p. 87.
3. Cann, G. NZ's Tyre Mountains Keep Growing in Absence of Recycling Scheme. *The Press*. Available online: https://www.stuff.co.nz/environment/92515136/millions-of-tyres-becoming-unrecyclable-in-absence-of-recycling-scheme (accessed on 8 July 2021).
4. Torretta, V.; Rada, E.C.; Ragazzi, M.; Trulli, E.; Istrate, I.A.; Cioca, L.I. Treatment and Disposal of Tyres: Two EU Approaches. A Review. *Waste Manag.* **2015**, *45*, 152–160. [CrossRef] [PubMed]
5. Lynch, P.; Gabolinscy, C.; Tyler, M.; Turner, N.; Caldwell, J.; Kim, N.; Latham, R.; Kirkham, A.; Middleton, N.; Weiss, S.; et al. *Guidance for Storage and Stockpiling End of Life Tyres for Local Government*; Report 5629; Waikato Regional Council: Hamilton, New Zealand, 2017; p. 14.
6. Banasiak, L.; Chiaro, G.; Palermo, A.; Granello, G. Recycling of End-of-Life Tyres in Civil Engineering Applications: Environmental Implications. In Proceedings of the WasteMINZ 2019 Conference, Hamilton, New Zealand, 23–26 September 2019; pp. 1–15.
7. Banasiak, L.; Chiaro, G.; Palermo, A.; Granello, G. Environmental Implications of the Recycling of End-of-Life Tires in Seismic Isolation Foundation Systems. *Lect. Notes Civ. Eng.* **2021**, *144*, 43–52.
8. Dangerfield, E. Cause of Massive North Canterbury Tyre Fire 'Unknown', Investigation to Begin. *The Press*. Available online: https://www.stuff.co.nz/the-press/news/101790124/hundreds-of-tyres-on-fire-in-north-canterbury (accessed on 8 July 2021).
9. Dangerfield, E. Community Asked to Call for Government to Combat Tyre Stockpiling Through Legislation. *The Press*. Available online: https://www.stuff.co.nz/the-press/news/north-canterbury/101908978/community-to-call-for-central-government-to-combat-tyre-stockpiling-through-legislation (accessed on 8 July 2021).
10. Dangerfield, E. Tyre Pile in Amberley, North Canterbury, Still a Risk a Year after Huge Fire. *The Press*. Available online: https://www.stuff.co.nz/national/110835486/tyre-pile-still-a-risk-a-year-after-huge-fire (accessed on 8 July 2021).
11. Dangerfield, E. Number of Stockpiled Tyres Underestimated by Environment Canterbury. *The Press*. Available online: https://www.stuff.co.nz/the-press/news/north-canterbury/111551658/number-of-stockpiled-tyres-underestimated-by-environment-canterbury (accessed on 8 July 2021).
12. Rowhani, A.; Rainey, T.J. Scrap Tyre Management Pathways and Their Use as a Fuel—A Review. *Energies* **2016**, *9*, 888. [CrossRef]
13. Pehlken, A.; Essadiqi, E. *Scrap Tire Recycling in Canada*; CANMET Materials Technology Laboratory: Hamilton, ON, Canada, 2005.
14. Waste Avoidance and Resource Recovery Act 2001 No 58. Available online: https://www.legislation.nsw.gov.au/#/view/act/2001/58 (accessed on 1 March 2021).
15. Constant, D.; Gosmann, D. Tire Derived Fuel Use in Cement Kilns. *GCI Tech. Notes* **1997**, 3. Available online: https://gcisolutions.com/library/gci_tech_notes/gcitn199709.html (accessed on 19 July 2021).
16. Chiaro, G.; Tasalloti, A.; Banasiak, L.; Palermo, A.; Granello, G.; Rees, S. Sustainable Recycling of End-of-Life Tyres in Civil (Geotechnical) Engineering Applications: Turning Issues into Opportunities in the New Zealand Context. *N. Z. Geomech. News* **2020**, *99*, 38–47.

17. Arulrajah, A.; Narsilio, G.; Kodikara, J.; Orense, R.P. Key Issues in Environmental Geotechnics: Australia-New Zealand. *Environ. Geotech.* **2015**, *6*, 326–330. [CrossRef]
18. Chiaro, G.; Indraratna, B.; Tasalloti, S.M.A.; Rujikiatkamjorn, C. Optimisation of Coal Wash–Slag Blend as a Structural Fill. *Ground Improv.* **2015**, *168*, 33–44. [CrossRef]
19. Rujikiatkamjorn, C.; Indraratna, B.; Chiaro, G. Compaction of Coal Wash to Optimize its Utilization as Water-Front Reclamation Fill. *Geomech. Geoeng.* **2013**, *8*, 36–45. [CrossRef]
20. Tasalloti, S.M.A.; Indraratna, B.; Rujikiatkamjorn, C.; Heitor, A.; Chiaro, G. A Laboratory Study on the Shear Behavior of Mixtures of Coal Wash and Steel Furnace Slag as Potential Structural Fill. *Geotech. Test. J.* **2015**, *38*, 361–372. [CrossRef]
21. Qi, Y.; Indraratna, B.; Twak, M. Use of Recycled Rubber Elements in Track Stabilisation. *Geotech. Spec. Publ.* **2020**, *GSP319*, 49–59.
22. Ghorbani, B.; Arulrajah, A.; Narsilio, G.; Horpibulsuk, S.; Bo, M.W. Dynamic Characterization of Recycled Glass-Recycled Concrete Blends Using Experimental Analysis and Artificial Neural Network Modeling. *Soil Dyn. Earthq. Eng.* **2021**, *142*, 106544. [CrossRef]
23. Arulrajah, A.; Naeini, M.; Mohammadinia, A.; Horpibulsuk, S.; Leong, M. Recovered Plastic and Demolition Waste Blends as Railway Capping Materials. *Trans Geotech.* **2020**, *22*, 100320. [CrossRef]
24. Chiaro, G.; Palermo, A.; Granello, G.; Banasiak, L. Direct Shear Behaviour of Gravel-Granulated Tyre Rubber Mixtures. In Proceedings of the 13th Australia New Zealand Conference on Geomechanics, Perth, Australia, 1–3 April 2019; p. 6.
25. Chiaro, G.; Palermo, A.; Granello, G.; Tasalloti, A.; Banasiak, L. Reuse of Waste Tires to Develop Eco-Rubber Seismic-Isolation Foundation Systems: Preliminary Results. *Lect. Notes Civ. Eng.* **2021**, *144*, 159–169.
26. Hernández, E.; Palermo, A.; Granello, G.; Chiaro, G.; Banasiak, L.J. Eco-rubber Seismic-Isolation Foundation Systems: A Sustainable Solution for the New Zealand Context. *Struct. Eng. Int.* **2020**, *30*, 192–200. [CrossRef]
27. American Standards for Testing Materials (ASTM). *Standard Practice for Use of Scrap Tires in Civil Engineering Applications*; American Society for Testing and Materials: West Conshohocken, PA, USA, 2017.
28. Senetakis, K.; Anastasiadis, A.; Pitilakis, K. Dynamic Properties of Dry Sand/Rubber (SRM) and Gravel/Rubber (GRM) Mixtures in a Wide Range of Shearing Strain Amplitudes. *Soil Dyn. Earthq. Eng.* **2012**, *33*, 38–53. [CrossRef]
29. Mashiri, M.; Vinod, J.; Sheikh, M.N.; Tsang, H.-H. Shear Strength and Dilatancy Behaviour of Ssand-Tyre Chip Mixtures. *Soils Found.* **2015**, *55*, 517–528. [CrossRef]
30. Hazarika, H.; Pasha, S.M.K.; Ishibashi, I.; Yoshimoto, N.; Kinoshita, T.; Endo, S.; Karmokar, A.K.; Hitosugi, T. Tire-Chip Reinforced Foundation as Liquefaction Countermeasure for Residential Buildings. *Soils Found.* **2020**, *60*, 315–326. [CrossRef]
31. Balaji, P.; Dashti, S.; Liel, A.B. In-Ground Gravel–Rubber Panel Walls to Mitigate and Base Isolate Shallow-Founded Structures on Liquefiable Ground. *J. Geotech. Geoenviron. Eng.* **2020**, *146*, 04020087. [CrossRef]
32. Tsang, H.-H.; Lo, S.H.; Xu, X.; Sheikh, M.N. Seismic Isolation for Low-to-Medium-Rise Buildings Using Granulated Rubber-Soil Mixtures: Numerical Study. *Earthq. Eng. Struct. Dyn.* **2012**, *41*, 2009–2024. [CrossRef]
33. Tsiavos, A.; Alexander, N.A.; Diambra, A.; Ibraim, E.; Vardanega, P.J.; Gonzalez-Buelga, A.; Sextos, A. A Sand-Rubber Deformable Granular Layer as a Low-Cost Seismic Isolation Strategy in Developing Countries: Experimental Investigation. *Soil Dyn. Earthq. Eng.* **2019**, *125*, 105731. [CrossRef]
34. Dhanya, J.S.; Boominathan, A.; Banarjeee, S. Performance of Geo-Base Isolation System with Geogrid Reinforcement. *Int. J. Geomech.* **2018**, *19*, 04019073. [CrossRef]
35. Pitilakis, D.; Anastasiadis, A.; Vratsikidis, A.; Kapouniaris, A.; Massimino, M.R.; Abate, G.; Corsico, S. Large-Scale Field Testing of Geotechnical Seismic Isolation of Structures Using Gravel-Rubber Mixtures. *Earthq. Eng. Struc. Dyn.* **2021**, *50*, 2712–2731. [CrossRef]
36. Tasalloti, A.; Chiaro, G.; Murali, A.; Banasiak, L. Physical and Mechanical Properties of Granulated Rubber Mixed with Granular Soils—A Literature Review. *Sustainability* **2021**, *13*, 4309. [CrossRef]
37. Hazarika, H.; Abdullah, A. Improvement Effects of Two and Three Dimensional Geosynthetics Used in Liquefaction Countermeasures. *Jpn. Geotech. Soc. Spéc. Publ.* **2016**, *2*, 2336–2341. [CrossRef]
38. Tasalloti, A.; Chiaro, G.; Banasiak, L.; Palermo, A. Experimental Investigation of the Mechanical Behaviour of Gravel-Granulated Tyre Rubber Mixtures. *Constr. Build. Mater.* **2021**, *273*, 121749. [CrossRef]
39. Lee, J.-S.; Dodds, J.; Santamarina, J.C. Behavior of Rigid-Soft Particle Mixtures. *J. Mater. Civ. Eng.* **2007**, *19*, 179–184. [CrossRef]
40. Pasha, S.M.K.; Hazarika, H.; Yoshimoto, N. Physical and Mechanical Properties of Gravel-Tire Chips Mixture (GTCM). *Geosynth. Int.* **2019**, *26*, 92–110. [CrossRef]
41. Kim, H.-K.; Santamarina, J.C. Sand-rubber mixtures (large rubber chips). *Can. Geotech. J.* **2008**, *45*, 1457–1466. [CrossRef]
42. Tasalloti, A.; Chiaro, G.; Palermo, A.; Banasiak, L. Effect of Rubber Crumbs Volumetric Content on the Shear Strength of Gravelly Soil in Direct Shear Apparatus. *Geotech. Spec. Publ.* **2020**, *319*, 259–266. [CrossRef]
43. Chiaro, G.; Tasalloti, A.; Chew, K.; Vinod, J.S.; Allulakshmi, K. Macro- and Micro-Scale Engineering Response of Rigid-Soft Gravel-Rubber Inclusions: Insights from Detailed Laboratory and DEM Numerical Investigations. *Lect. Notes Civ. Eng.* **2021**, 16, in press.
44. Edil, T.B. A Review of Environmental Impacts and Environmental Applications of Shredded Tire Scraps. In Proceedings of the International Workshop IW-TDGM 2007, Yokosuka, Japan, 23–24 March 2007; Taylor & Francis Group: London, UK, 2007.
45. Ministry of Health. *Drinking Water Standards for New Zealand*; Ministry of Health: Wellington, New Zealand, 2018.

46. Ministry for the Environment. *New Zealand Landfill Waste Acceptance Criteria—Module 2: Hazardous Waste Guidelines*; Ministry for the Environment: Wellington, New Zealand, 2004.
47. Thurston, S.J. *Base Isolation of Low Rise Light and Medium-Weight Buildings*; BRANZ Study Report No. 156; 2006; BRANZ Ltd.: Judgeford, New Zealand.
48. Jampole, E.; Deierlein, G.; Miranda, E.; Fell, B.; Swensen, S.; Acevedo, C. Full-Scale Dynamic Testing of a Sliding Seismically Isolated Unibody House. *Earthq. Spectra* **2016**, *32*, 2245–2270. [CrossRef]
49. Tasalloti, A.; Chiaro, G.; Young, J.; Ross, O.; Palermo, A.; Granello, G. Experimental Seismic Characterisation of Gravel-Granulated Tyre Mixtures and Design Implications. In Proceedings of the 2021 New Zealand Society of Earthquake Engineering Annual Conference, Christchurch, New Zealand, 14–16 April 2021; p. 10.

applied sciences

MDPI

Article

The Influences of Local Glacitectonic Disturbance on Overconsolidated Clays for Upland Slope Stability Conditions: A Case Study

Kamil Kiełbasiński [1,*], Paweł Dobak [1], Łukasz Kaczmarek [2] and Sebastian Kowalczyk [1]

[1] Faculty of Geology, University of Warsaw, Żwirki i Wigury 93 St., 02-089 Warsaw, Poland; p.dobak@uw.edu.pl (P.D.); s.kowalczyk@uw.edu.pl (S.K.)

[2] Faculty of Building Services, Hydro and Environmental Engineering, Warsaw University of Technology, Nowowiejska 20 St., 00-653 Warsaw, Poland; lukasz.kaczmarek@pw.edu.pl

* Correspondence: k.kielbasinski@uw.edu.pl

Featured Application: Spatial development planning, construction of high-rise buildings, construction with the use of deep excavations in areas near slopes (especially in urban areas), where the geological structure is complicated.

Abstract: Reliability of equilibrium state evaluation about settlement slopes in the context of natural and human-made hazards is a complex issue. The geological structure of the vicinity of the upland slope in the urban environment of Warsaw is characterised by a significant spatial diversification of the layers. This is especially due to the glacitectonics in the Mio-Pliocene clays, which are located shallowly under the sandy tills' formations. With substantial variability in the clay roof surface, point recognition by drilling is often insufficient. The use of electrical resistivity imaging (ERI) in the quasi-3D variant provides accurate images of the real ground conditions, which is crucial in optimal geotechnical design. In forecasting the behaviour of the slope, it is necessary to quantify the impact of spatially differentiated systems of disturbed layers on changes in the safety factor (SF), which corresponds to the observed landslide activity of the Warsaw Slope. This study concerns numerous calculation model analyses of the optional clay position in the context of slope stability conditions. A wide range of soil properties variability was taken into account, resulting from both lithogenesis and subsequent processes disintegrating the original soil structure. Regarding the geological conditions of the slip surface, the use of classical computational methods and numerical modelling (FEM) was considered for comparative purposes. The results indicated that local changes in equilibrium conditions were affected by the different morphology of the clay roof surface of the slope and the alternation in strength characteristics on the slip surfaces. The findings of the study contribute to sustainable spatial planning of near-slope regions.

Keywords: safety factor; Neogene clay elevation; Warsaw Slope; finite element method; electrical resistivity tomography

Citation: Kiełbasiński, K.; Dobak, P.; Kaczmarek, Ł.; Kowalczyk, S. The Influences of Local Glacitectonic Disturbance on Overconsolidated Clays for Upland Slope Stability Conditions: A Case Study. *Appl. Sci.* **2021**, *11*, 10718. https://doi.org/10.3390/app112210718

Academic Editors: Jaroslaw Rybak, Małgorzata Jastrzębska, Krystyna Kazimierowicz-Frankowska and Gabriele Chiaro

Received: 11 October 2021
Accepted: 9 November 2021
Published: 13 November 2021

1. Introduction

Slope stability, especially among urban areas, is a subject of extensive engineering-geology research [1,2]. The scope of these studies includes improvement of geological structures' identification, especially glacitectonic ones, as well as optimising methods of forecasting slope stability conditions. The first aspect uses shallow geophysics, which allows more precise identification and mapping of the course of structural disturbances. The second aspect of the analysis concerns the optimisation of computational methods that are applied in forecasting the equilibrium state of the slopes. Such an approach allows the correlation of the results of the identification of shallowly located glacitectonic structures obtained with the use of different methods. It provides a possibility to apply the data acquired with the electrical resistivity imaging (ERI) method to classical slope stability model analyses and to the finite element method (FEM). Combined, the analysed issues can be proposed as a pattern of conducting slope stability research, especially in the case of the wide availability of various research and computational methods, that do not necessarily fit all conditions.

The electrical resistivity method can be helpful in determining a model of spatial variability of strata and slope stability. This method is marked by high prospection efficiency and is commonly used to characterise geological structures, e.g., course and depth of geological boundaries, including the position of clays. Vertical electrical sounding [3–5] and electrical resistivity imaging (ERI) are commonly applied techniques [6–8] to determine the position of the roof of Neogene clays in Poland. The ERI method, also known as the electrical resistivity tomography (ERT) or continuous vertical electrical sounding (CVES) method, is often used in the mapping of landslides and slope stability [9–13]. The theoretical basis of the electrical resistivity imaging method, as well as the history of its development and application, was described by [14,15]. In these studies, the results of an ERI survey were applied to determine the position of the roof of the Neogene clays.

The results of the quantitative assessment of the equilibrium state conditions are the subject of numerous analyses [16–18] and were developed in this work according to the conclusions of [19,20], among others. An important influence on the methodology development of stability analyses has the effect of studies of landslides at different scales and conditions (e.g., [16–24]). Models introduced in the analysed case study take into consideration different factors the spatial variability of the geological structure and periodic changes in soil strength properties caused by changes in moisture content and consistency on potential structural slip surfaces. Those changes can be caused by the seepage of rainwater, especially during increasingly occurring events of heavy rains and sewage system failures. The effects of such situations were analysed using multivariant, computer-aided modelling. A number of classical methods of slope stability prediction, based on various assumptions, are in use, although those methods typically do not take geological conditions into account while determining the location of potential slip zones. Another group of analyses uses the modelling of the slope stability conditions based on the finite element method. The comparative approach to the slope stability issue is significant in terms of recognition of optimal slope safety conditions, but also in terms of the evaluation of various prognostic methods. These issues are addressed by the study of the assumptions of both the LEM and the determination of optimal slope slip surfaces using FEM. The obtained results indicate the necessity to further develop an outcome comparison from these two groups of calculation methods [19,25].

Another aspect that is frequently discussed in research is the adoption of simplified plane strain models for spatial issues [21,22]. The presented methodological approach allows the assessment of the influence of various spatial forms on slope stability, and it also answers the question of whether the simplification of the plane-deformation analysis model will have a significant impact on the obtained results. This task was carried out with the use of a 3D model by finite element method (FEM) calculations as well as performing calculations in 2D using two traditional limit equilibrium methods (LEMs; namely, Bishop and Janbu methods) and also once again by FEM.

The FEM calculations were carried out by the ZSoil 2016 software with the implementation of the proportional reduction in strength parameters (c/ϕ reduction—SRM). The SRM is commonly used for slope stability analysis [16,17,26–28]. The beginnings of SRM application can be associated with the publication of Zienkiewicz et al. (1975) [29]. Based on previous experiments [23,30], the safety factor was determined with consideration of the successively reduced values of strength parameters for iterative solving of the boundary problem of statics using the finite element method. During the calculation process, the state of equilibrium was analysed with the assumption of the geostatic distribution of the stress state in the soil medium. At the initial stage, the value of the reduction factor SF = 1 was determined for which the equilibrium state was defined for the initially adopted soil parameters. This stage served as a reference level for subsequent iterations in which there was a discrete reduction in the strength parameters of the soil medium by increasing the SF coefficient. This process was conducted until the equilibrium state was lost, which is associated with the development of the slip zone. The last value of the reduction factor SF at which it was possible to obtain the equilibrium state corresponds to the value of the estimated safety factor. The previously mentioned method was applied to solve the problems of the stability of the soil medium in 2D and 3D deformation models.

2. Analysed Area and Materials—Structural Factors of Stability Conditions

The subject of the study presented in this paper is potential changes in the slope stability conditions in a selected survey area located on the upland in the vicinity of the Warsaw Slope. The survey area was previously investigated with the use of geophysical methods.

Warsaw Slope, which forms the border of the Vistula River Valley, was used as the study site. Warsaw Slope is located on Warsaw Upland, which is built by glacitectonically deformed clays of the Poznań clay formation (Figure 1), covered by layers of glacial till and fluvioglacial sediments.

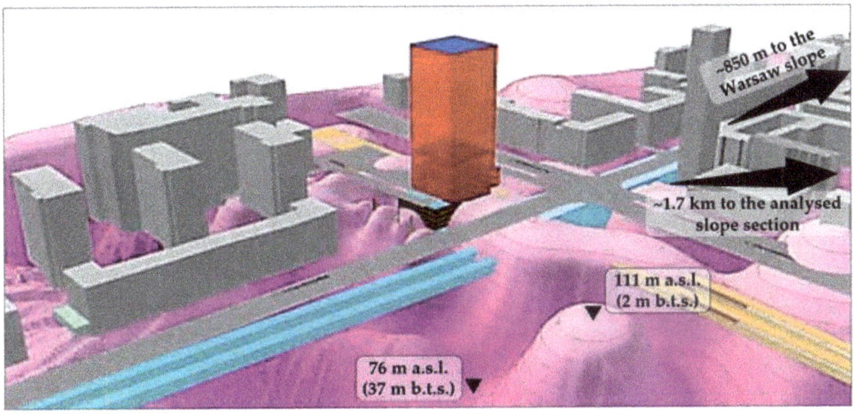

Figure 1. The Neogene clay roof within the Warsaw city centre (excluding soils lying over clay), modified from [31].

The spatially and genetically complex structural setting of Warsaw Slope has been carefully documented by many studies (e.g., [32,33]). Successive development of processes leading to the displacement of soils can be observed, among others, by comparison of historical iconography illustrating the panoramas of the Old Town, Krakowskie Przedmieście, and Żoliborz, seen from the side of Vistula River.

There are also historical records mentioning difficulties in placing building foundations, e.g., in the case of St Anne's Church, located directly on the Warsaw Slope. Archival materials, especially panorama paintings and shaded-relief maps, can be used to assess the nature and directions of changes in the Warsaw Slope and correlate them with a rich collection of results of contemporary research and studies. Lithological profiles of numerous boreholes, the results of site observation, i.e., condition of existing buildings, observations from research outcrops discovering old foundations, and data from recently conducted geophysical surveys are collected [9].

Comparisons on modern morphometry indicate that in the most densely developed downtown quarter, the mean height of the Warsaw Slope ranges from 15 to over 20 m, with a typical range of inclination of the slope ranging from 10 to 20 degrees. The highest inclination exceeds 30 degrees in the area of Oboźna Street and the Old Town. In the synthetic monograph prepared under the supervision of Wysokiński [32], the following geological areas were distinguished:

- Areas with active slope processes (with the simultaneous lack of structural predisposition to the formation of large landslides);
- Areas with low or periodic slope activity (with a predisposition to the formation of landslides);
- Areas with no traces of slope activity (determined by the structural predisposition for the formation of landslides).

In general, it can be stated that the status quo of development is maintained, which can be changed by periodically activated slope activity. However, the aforementioned classification does not apply in the areas where soil creeping processes are observed [30,32].

Due to the complexity of the spatial structure of the Warsaw Slope (Figure 2), classic methods based on locating the most favourable slip zones may not be accurate. Mechanistic optimisation related to the assumption of quasi homogeneity of the ground is difficult to accept. After all, the assumption stating that structurally conditioned slip surfaces are major factors in initiating and developing landslide processes is widely accepted. In the case of the Warsaw Slope, these circumstances can occur as a result of glacitectonic deformations of the Poznań clay formation, located in close proximity to the slope. Therefore, the cause of groundmass movements observed on various sections of the slope is often difficult to determine.

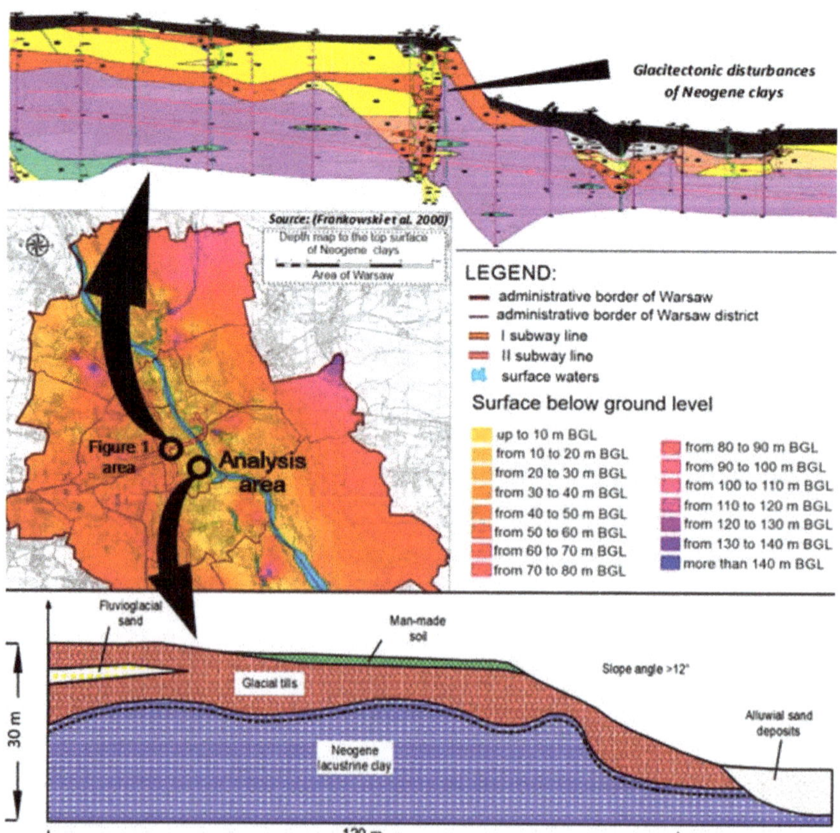

Figure 2. The roof of the Neogene clay within the borders of Warsaw with two exemplary sections crossing the Warsaw Slope (modified from [9,34]).

On this account, it is necessary to quantify the influence that clay roof deformations have on the stability of the slope. The location of the clay stratum noted in the drilling profiles varies significantly, which is documented in both cross section and map interpretations. Created images are highly dependent on the density of documented boreholes and on the methods of generalising the obtained information [30]. In earlier studies, the border of the clay formation was often captured as a mildly collapsing line parallel to the upland's slope. At the same time, the clay formation collateral to the edge of the upland is built of folds of various heights that create local elevations. The complexity of geological structure is determined by plastic, form-like deformations of clay, caused by the base of a dynamically moving ice sheet.

The denivelations of the clay roof are locally extensively documented (e.g., in the vicinity of the Świętokrzyska metro station [31,35]). Such situations may occur along the entire slope. For example, the illustrative cross section in the area of Dworkowa Street—about 2 km southward of the analysed area—and Sarmacka Street [36] shows a noticeable uplift of the clay stratum, the upper part of which lies in close vicinity of the upland's slope. The insight on mechanisms of multiplicative folds, along with the growing number of boreholes drilled in the intensively developed areas of Śródmieście and Mokotów districts, indicate that much more local, small elevations occur in the area. Data from newly drilled boreholes are successively added to the databases, making the image of the geological structure in the area more diverse and real. In this case, it is necessary to use geophysical methods, the possibilities of which, however, are hindered in urban areas by the presence of underground infrastructure interfering with the recorded fields.

An example of implementing geophysical prospection is the electrical resistivity tomography that was used to track the disturbances of the clay roof in the area of Piękna and Lenona streets in Warsaw.

3. Local Estimation of the Clay Roof Relief Based on Electrical Resistivity Prospection

ERI measurements were made with the Terrameter LS apparatus of the Swedish company ABEM, alongside three 70 m long measurement lines with 7 m intervals. The spacing of the electrodes on each of the lines was 1 m. Measurements were made by a gradient array with multiple current–electrode combinations, which yields solid outcomes when recognising/mapping horizontal variability [37]. Apparent resistivity data obtained from field measurements were processed with the Res2DInv software [38]. Acquired two-dimensional resistivity models of the surface level of the geological medium were used to perform the 3D interpretation in the Voxler program (Figure 3).

The determination of the top surface of the Neogene clays was the primary task for the ERI measurements. The location of the relief of the Neogene clay formation roof was assumed on the basis of the iso-resistivity line of 30 Ωm. Resistivity values below 30 Ωm are the most common resistivity values for clays in Poland [39]. In the Warsaw city region, the juxtaposition of borehole profiles with the distribution of low resistivity values shows that Neogene clays are assigned to soils with electrical resistivity value below 30 Ωm [9,13,40].

The results of the resistivity distribution in the study area were related to geological data from archive boreholes. Such an approach reduced uncertainty of interpretation and increased confidence in placing borders interpreted on the basis of resistivity distribution. Over the Neogene clays (resistivity below 30 Ωm), there is a layer of glacial till, with resistivity from 30 to 80 Ωm. Just above tills, there is a near-surface layer of high resistivity, with resistivity ranging from 80 to over 5000 Ωm. It can be identified as fluvioglacial sands or human-made soils. In this case, electrical resistivity tests do not allow for the differentiation of those deposits and the placement of clear borders between them.

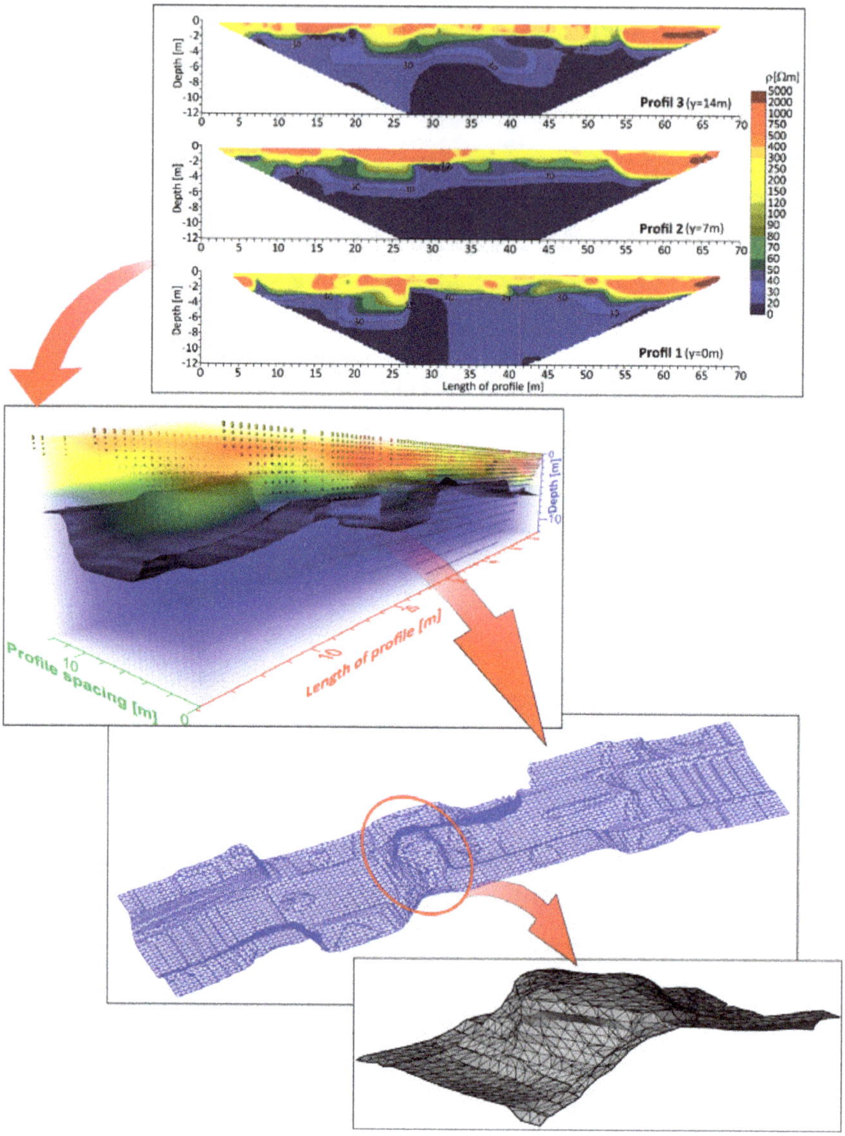

Figure 3. Workflow to select the elevation of the Neogene clay roof surface from the results of quasi 3D electrical resistivity tomography prospection.

4. Stability Analysis Results, Their Interpretation, and Discussion

The tests carried out by the ERI in the area of Piękna Street in Warsaw showed a local disturbance of the clay complex roof surface. The relative height of the displacement was determined on the basis of the iso-resistivity line of 30 Ωm and was about 3.5 m. The width of the base of the structure in the direction perpendicular to the slope was estimated at 7.5 m. As a result, the inclination of the Warsaw Slope reached 50 degrees in some places. The interpretation of the 3D ERI in the geophysical image indicates that the considered form of the roof surface had a local character. From a broader analysis of the geological area complexity, it can be assumed that underneath the base of the slope, a dike-like form parallel to the edge of the upland structure occurs. The obtained results were processed in the AutoCAD software and presented in the form of electrical resistivity distribution as a large-scale surface model (Figure 3).

The next stage of assessing the impact of clay roof surface deformation on the stability conditions in the slope area was the preparation of the upper-layer stratum sequence model. The architecture of the identified form was projected in the spatial model with a geometric approximation. Basic attributes such as the relative height, width, and slope gradient were maintained.

4.1. Stability Analysis by 3D Modelling

The investigated form is a disturbance with restricted spread, and therefore, the created solid model does not meet the criteria of plane strain analysis. Therefore, it proved necessary to perform a stability analysis, taking into account the three-dimensional stress analysis. In the first stage of the analysis, two generalised solid models were compared:

- Local elevation limited in the plan (3D analysis);
- Dike-like form (reduced to the 2D analysis).

In the calculation model used, the following was applied:

- Parameters of the strength of the soil medium based on the elastic–plastic constitutive Mohr–Coulomb model;
- The optional shape of the roof surface of Neogene clays.

In the stability calculations, variant values of the geotechnical parameters of clays were assumed, based on the results obtained on own research (from the analysed area adjacent to the slope) and literature data taking into account the glacitectonic disturbance of the structure [30,41,42].

As a starting point in computational iterations, the following assumptions were made:

- For non-glacitectonically disturbed layers, the maximum values of ϕ' and c′ were used based on archival data [41,42];
- For glacitectonically disturbed layers,
    - Archival cohesion values were reduced by 60% [42];
    - Values of ϕ' and c′ from TRX investigations; for soils showing the effects of structural disintegration, a reduction of 30% to 45% of the cohesion value and 10% to 30% of the angle of internal friction were applied [41,42];
    - Values of ϕ' and c′ from creep TRX's tests [30].

Section 4.2 presents the parameter value options used in the stability calculations in the following ranges: $\phi' = 9.7°–22°$; $c' = 0–55$ kPa.

The calculations were made with the following spatial assumptions: generally flat clay roof surface slightly inclined towards the slope (option S), which may include

- Point elevation with heights of 3.5 and 8 m (options PE3.5 and PE8);
- Parallel to the edge of the slope, a dike-like form disturbance with a height of 8 m (D8).

Option PE3.5 refers to the results of the local geophysical investigation, while the alternative variants represent other common types of geological settings found along the Warsaw Slope (Figure 2).

The results of the SF stability factor calculations in options S and PE3.5 were similar (SF(S) = 1.80; SF(PE3.5) = 1.83), which indicates the marginal impact of the 3.5 m disturbance on the slope stability.

In order to assess the impact of the disturbance size on the slope stability, the dike relative height was increased (up to 8 m) while maintaining the previous inclination. In this configuration, 7 options (PE8_1-PE8_30) differing in the extent of the disturbance (Figure 4) from 1 to 30 m, with an interval of 5 m, were analysed.

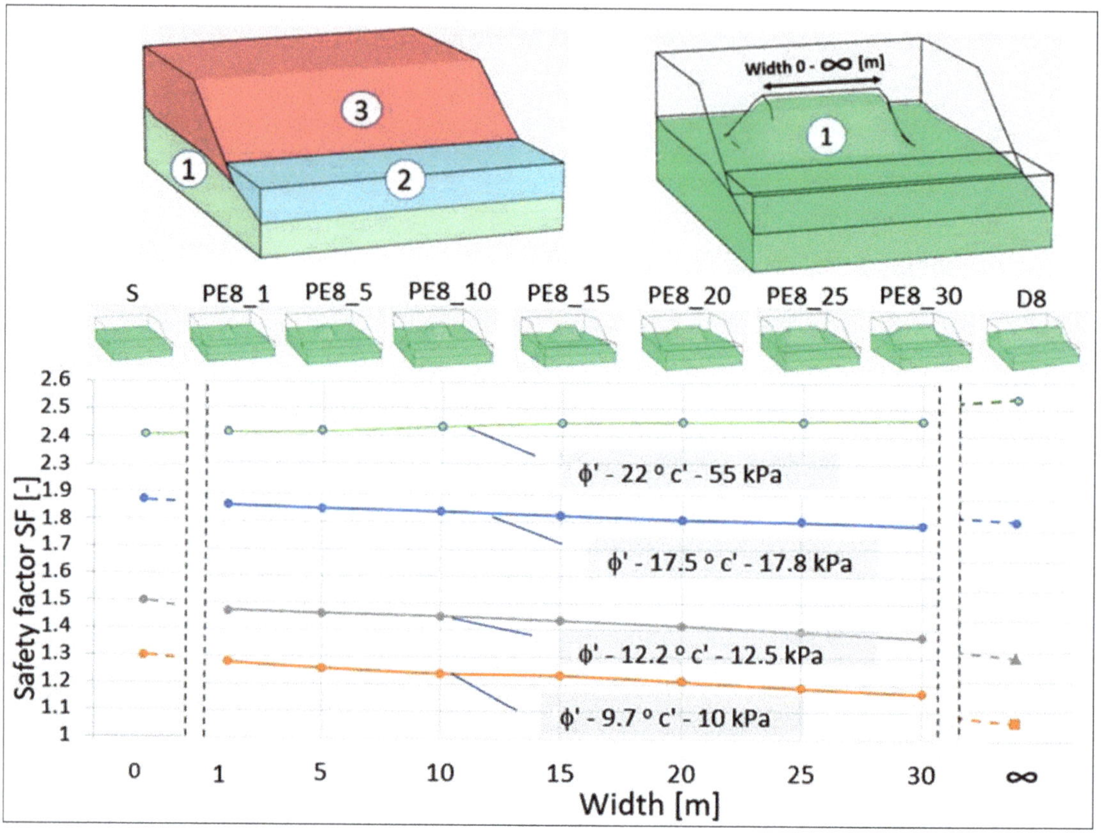

Figure 4. Results of slope stability 3D finite element modelling in different structural condition: (1) Cl/siCl—(clay, silty clay) strength parameters on above graph; (2) saCl/clSa—silty clay/clayey sand ϕ′—22°; c′—44 kPa; (3) MSa—(medium sand) ϕ′—30°; c′—0 kPa. Codification explanation: S—slightly inclined, PE—point elevation, D—dike-like form (type of disturbance with the height–disturbance width).

As a result, depending on the assumed strength parameters of the soil, the presence of disturbance in the clay roof surface over 20 m wide resulted in a reduction in the stability index by 3 to 7%m in comparison with option S. The greater presence of clays in the slip zone is a possible reason. The greatest decrease in the SF was observed with ϕ′= 9.7° and c′ = 10.0 kPa—the maximally reduced strength parameters of the clay. With the use of the maximum strength parameters (ϕ′—22° and c′—55 kPa), despite the presence of a disturbance in the clay roof surface, an enhancement in the stability conditions was achieved in relation to the S variant. The reason for the modelled slope strengthening was the fact that the assumed clay strength parameters were higher than those of the overlying glacial tills. The numerical simulations of the D8 option were carried out for 4 variants of the strength parameters of Neogene clays while maintaining the parameters of the remaining layers unchanged.

In further analysis, the results of stability modelling at point elevations, the width of which did not exceed 10 m (variant: PE8_10), were compared with the dike-like elevations. The sensitivity of models with an alternative method of shaping the clay roof surface depends on the considered ranges of strength parameters. The smallest difference in SF, amounting to approx. 6%, was observed for the clay parameters specified in TRX research. On the other hand, the largest difference in the stability indexes, amounting to 14%, was related to the variant taking into account the maximum reduced parameters of the clay ($\phi' = 9.7°$ and $c' = 10$ kPa). It should be emphasised that in each of the analysed variants, the slope stability with the dike elevation was lower than that of the slope with a point elevation of the clay roof surface in the basement. This allows the conclusion that the dike elevations provide an additional safety margin in relation to the point elevation models. Thus, it justifies simplifying the calculations for the plane strain analysis model, which provides an additional safety margin.

4.2. Stability Analysis by 2D Model (Plane Strain State)

The implementation of calculations in the plane deformation state allowed for additional comparative analyses using the limit equilibrium method (LEM), well established in engineering practice, and the 2D finite element method (FEM). Calculations made in a 2D environment require less computing power, compared with 3D systems.

The simplified Bishop method and the Janbu method were used in the LEM analysis. The simplified Bishop method is a strip method that does not take into account the vertical interaction forces between the stripes and the equilibrium condition of the horizontal forces. In the assumptions, it requires an approximation of the slip surface with a segment of a circle. Despite all these simplifications, it is very often used in practice and, in most cases, sufficient for the initial assessment of slope stability. The obtained estimation of the stability coefficient using this method does not differ from the accuracy obtained by more advanced methods such as the Morgenstern–Price method or the Spencer method [43].

The application of the circular slip surface method is possible mainly in the assessment of the stability of homogeneous structures or a horizontal arrangement of layers. In the case of soil with an inclined layer system or with a complicated course of layers, as in the analysed case, the use of this method may overestimate stability assessments. This fact is largely due to the mismatch between the idealised circular slip surface and the surface produced by the contact of inclined or folded layers. The aforementioned situation occurred in the analysed model. The wrinkled clay roof surface had the potential to initiate a slip. The complexity of the real surface relief, diverging from the cylindrical surface required in Bishop's method, resulted in a reduction in the accuracy of the safety factor estimation.

In the course of computational analyses, four models that differed in the size of the aforementioned discrepancies (caused by structural disturbances) were created. The initial relative height in the analyses of 3.5 m was increased optionally to 5, 6.5, and 8 m. This resulted in, inter alia, changes in the share of clay in the slip zone.

In the considered spatial models, the relief of the disturbed clay roof surface allowed only an approximate fit of the circular slip surface, which also intersected the underlying layers with higher parameters (Figure 5A). This situation had a direct impact on the result of the stability assessment and resulted in its overestimation. In order to assess the size of this overestimation, the results from Bishop's method were compared with the Janbu method, in which the shape of the slip line was adjusted to geological conditions. As a result, stability was analysed even with a more complex, broken slip surface, the course of which was adapted to the contact zone of soils of different genesis and properties (Figure 5B).

The FEM 2D analysis was carried out by ZSoil 2016 software. Geometric assumptions were identical as in the case of the LEM method (Figure 5C). As in the case of 3D analysis, the elastic–plastic constitutive Mohr–Coulomb model was also used to describe the behaviour of the soil medium.

In the plane strain analyses (LEM and FEM 2D), the thickness of the top layer of the clay complex was changed by 1 m relative to the set of parameters previously used in the 3D analysis. Calculation models were extended by several sets of strength parameters to assess their sensitivity and possible parameter changes. In total, 8 variants with different parameters of the clay layer were analysed. Additionally, the models differentiated the strength parameters of the deeper clay layer (non-weathered/non-disturbed) using 3 sets of parameters. The first corresponded to the parameters of the clay from the area of ul. Piękna [30], the second matched a glacitectonically disturbed clay [42], and the third matched the maximum documented strength parameters of these soils [42]. Parameter combination of the top clay and the deeper layers created 24 calculation variants. Results, in the form of a safety factor (SF), are summarised in Figure 6.

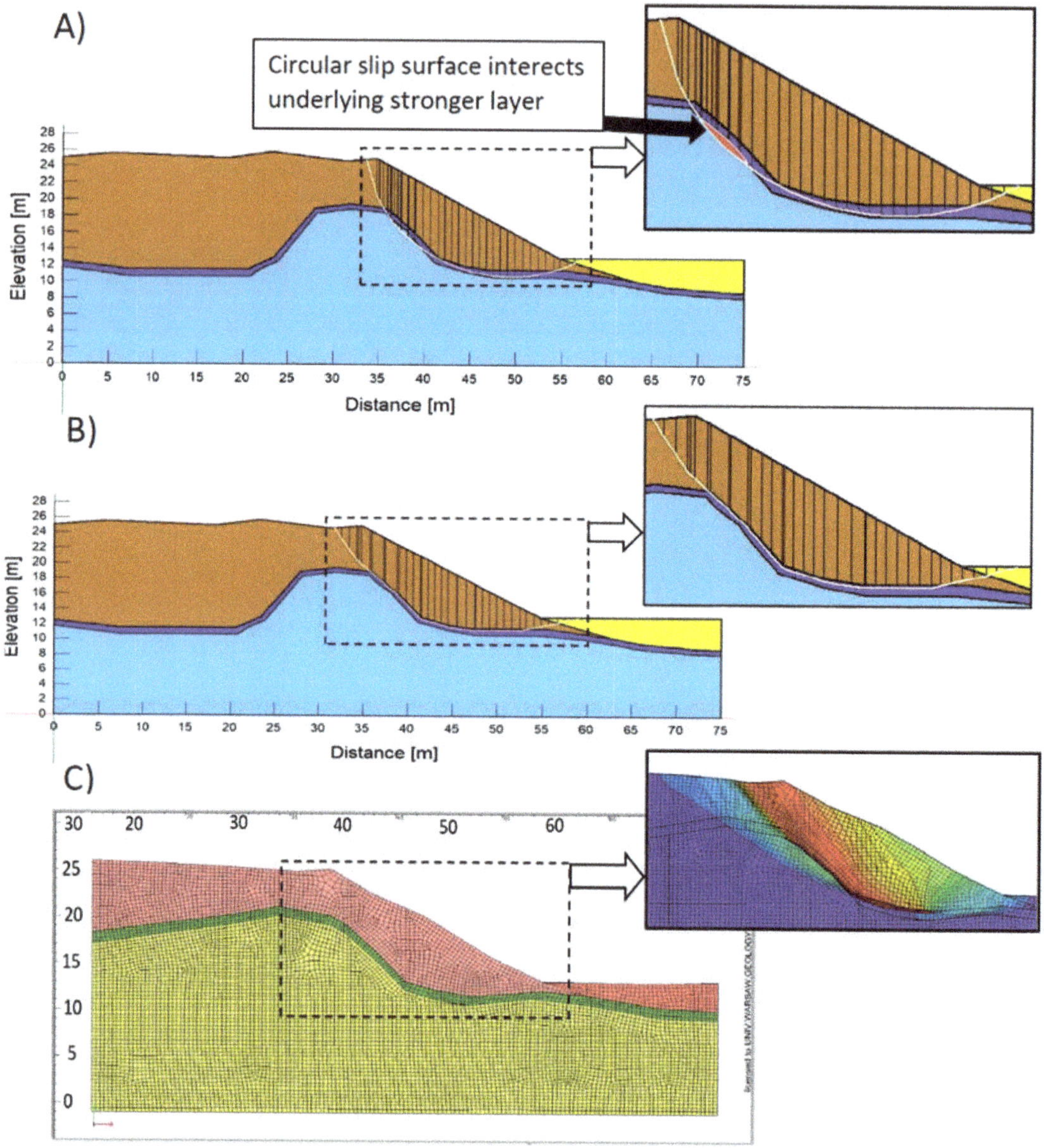

Figure 5. (**A**) Circular slip surface for Bishop's method; (**B**) fully specified slip surface for Janbu's method; (**C**) FEM 2D.

(Hight [m]/Cl*[%]/SaCl*[%])						8/71/29			6.5/65/35			5/58/42			3.5/50/50		
Method →						Bishop	Janbu	FEM 2D	Bishop	Janbu	FEM 2D	Bishop	Janbu	FEM 2D	Bishop	Janbu	FEM 2D
Week layer parameters			Clay strenght parameters														
Unit weight	Friction angle	Cohesion	Friction angle	Cohesion	Unit weight												
kN/m³	°	kPa	°	kPa	kN/m³	Safety Factor [-]											
17.6	9.7	10.0	17.5	17.8	20.1	1.12	0.91	<1	1.15	1.08	<1	1.21	1.18	<1	1.26	1.29	<1
18.6	12.2	9.8				1.19	0.99	<1	1.24	1.17	1.14	1.28	1.26	1.16	1.32	1.36	1.21
18.6	12.2	12.5				1.24	1.04	1.16	1.25	1.22	1.20	1.30	1.31	1.24	1.36	1.41	1.26
18.6	13.0	15.0				1.30	1.22	1.25	1.29	1.30	1.28	1.34	1.38	1.31	1.41	1.47	1.35
18.6	13.0	17.5				1.34	1.27	1.30	1.32	1.36	1.33	1.37	1.43	1.36	1.44	1.51	1.40
20.1	15.8	9.8				1.31	1.22	1.26	1.34	1.31	1.30	1.39	1.39	1.33	1.44	1.48	1.37
20.1	15.8	12.5				1.35	1.28	1.31	1.38	1.36	1.35	1.42	1.44	1.39	1.49	1.52	1.42
21	22.0	0.0				1.35	1.28	1.34	1.39	1.37	1.37	1.44	1.45	1.41	1.50	1.54	1.45
	-	-				1.38	1.46	1.47	1.42	1.51	1.49	1.48	1.59	1.51	1.55	1.65	1.55
17.6	9.7	10.0	22.0	22.0	21.0	1.18	0.91	<1	1.24	1.08	<1	1.28	1.18	<1	1.31	1.29	<1
18.6	12.2	9.8				1.27	0.99	<1	1.31	1.17	1.15	1.36	1.26	1.17	1.37	1.36	1.22
18.6	12.2	12.5				1.32	1.04	1.18	1.34	1.22	1.21	1.38	1.31	1.25	1.42	1.41	1.27
18.6	13.0	15.0				1.40	1.22	1.27	1.40	1.30	1.29	1.43	1.38	1.32	1.47	1.47	1.36
18.6	13.0	17.5				1.45	1.27	1.32	1.43	1.36	1.34	1.47	1.43	1.37	1.51	1.51	1.41
20.1	15.8	9.8				1.41	1.22	1.28	1.45	1.31	1.31	1.48	1.39	1.34	1.51	1.48	1.38
20.1	15.8	12.5				1.46	1.28	1.33	1.49	1.36	1.36	1.52	1.44	1.40	1.55	1.52	1.43
21	22.0	0.0				1.46	1.28	1.36	1.49	1.37	1.38	1.52	1.45	1.42	1.56	1.54	1.46
	-	-				1.67	1.73	1.75	1.71	1.77	1.79	1.75	1.82	1.84	1.79	1.86	1.86
17.6	9.7	10.0	22.0	55.0	21.0	1.34	0.91	<1	1.40	1.08	<1	1.43	1.18	<1	1.38	1.29	<1
18.6	12.2	9.8				1.43	0.99	<1	1.47	1.17	1.15	1.50	1.26	1.17	1.43	1.36	1.22
18.6	12.2	12.5				1.48	1.04	1.18	1.52	1.22	1.21	1.53	1.31	1.25	1.50	1.41	1.27
18.6	13.0	15.0				1.56	1.22	1.27	1.58	1.30	1.29	1.59	1.38	1.32	1.56	1.47	1.36
18.6	13.0	17.5				1.60	1.27	1.32	1.62	1.36	1.34	1.62	1.43	1.37	1.59	1.51	1.41
20.1	15.8	9.8				1.57	1.22	1.28	1.61	1.31	1.31	1.63	1.39	1.34	1.58	1.48	1.38
20.1	15.8	12.5				1.62	1.28	1.33	1.65	1.36	1.36	1.68	1.44	1.40	1.62	1.52	1.43
21	22.0	0.0				1.62	1.28	1.36	1.66	1.37	1.38	1.68	1.45	1.42	1.65	1.54	1.46
	-	-				2.22	2.17	2.24	2.21	2.17	2.23	2.20	2.17	2.23	2.18	2.17	2.22

Cl * - participation of the clay layer in the slip plane SaCl * - participation of the sandy clay layer in the slip plane

Risk of landslide accordnig ITB 424/2011 and ITB 304/1991			
< 1	high risk of landslide	(1.3 , 1.5]	low risk of landslide
(1 , 1.3]	risk of landslide	> 1.5	very low risk of landslide

Figure 6. Results of stability modelling (plane strain analysis).

Analysis of the obtained results indicated that in the Janbu method, differentiating the parameters of the clay layer below the weakened layer did not affect the SF value. In the case of FEM 2D modelling, the influence was also marginal, but with the application of the simplified Bishop method, change was significant. The reason for this is that the slip surface in the Bishop method passed through the deeper layer, while in the Janbu and FEM 2D method, it was limited to the weakened layer only. The FEM 2D modelling indicated a loss in slope stability with the maximally reduced value of the clay strength parameters in all analysed calculation variants. Convergent results were also obtained using the Janbu method but only for the variant taking into account the maximum assumed height of the disturbance (8 m). In other computational cases, despite the disturbance, the slope was stable.

The multivariate analysis of geometric factors and optional strength characteristics allowed presenting the three-dimensional dependencies of SF on the value of the internal friction angle and cohesion (Figure 7). They were mapped in the form of several surfaces corresponding to the different calculation methods used. In addition to the 2D modelling outcome, the results of the 3D FEM are presented. Different sensitivity of individual slope stability calculation methods to changes in strength properties of soils is represented by different inclinations of these surfaces. This influence was best seen in the Janbu method and 2D FEM modelling. Thus, concerning the Janbu method, the great importance of the real relief of the slip surface structural predispositions for the results of stability prediction could be observed. At the same time, the Janbu method obtained the relatively lowest SF values, making it possible to recommend this method as the safest method for geodynamic forecasting.

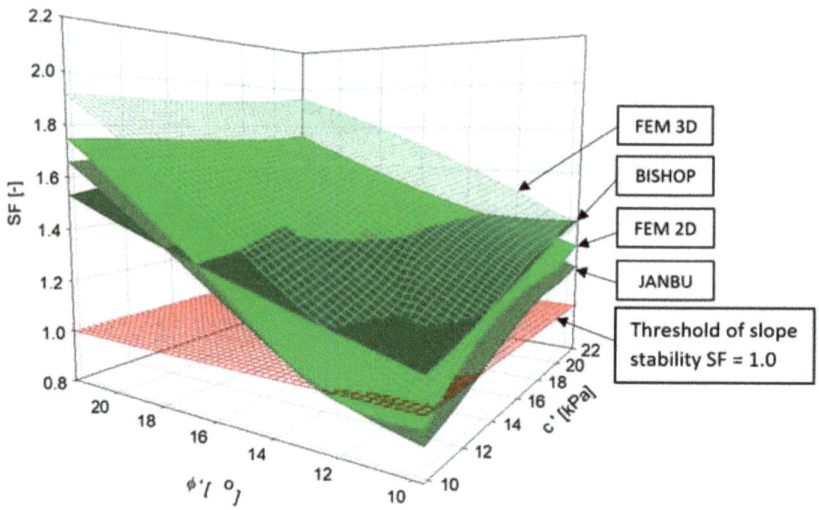

Figure 7. The three-dimensional dependencies of SF on the value of the internal friction angle and cohesion for different calculation methods.

The SF values obtained by 3D FEM modelling deviate from the results of other methods. In the range of low clay internal friction angle (9–11°), values of SF were comparable to the results obtained by the Bishop method. However, for ϕ′ values above 12°, the SF results were the highest of all the methods used. By assumption, the results of 3D modelling illustrate the natural geological situation better, especially taking into account the weaknesses in the slope [20]. Further, the 3D modelling outcomes differed from the results of 2D calculations. The SF values obtained in the 2D model were lower than the results of 3D modelling. The reason for the higher value of SF estimates may be the application of 3D models with a reduced density of the finite element mesh compared with the 2D variant. The increase in the finite element density causes a decrease in the predicted slope stability index value [26,27]. Without deciding which approach is closer to the real equilibrium state, it is worth noting that the consideration of lower estimated SF value by designers is treated as a safer procedure.

5. Conclusions

1. Structurally complex soil conditions in the area of the upland slopes (especially considering glacitectonic disturbances) require optional stability analyses, which take into account both the spatial variability of the soil structure and the variability of soil strength parameters.
2. For a more precise recognition of the spatial variability of the position of the soil layers, it is recommended to use quasi-3D electrical resistivity imaging (ERI; the so-called 2.5D electrical resistivity imaging). This method is necessary to obtain a 'continuous' image of the course of the disturbed roof of key structural surface (clays), which can determine the conditions of the equilibrium state of slopes, deep excavations, etc.
3. Identification of the Neogene clay elevation with the ERI method made performing a multivariant stability analysis possible. The results of the analysis allowed the determination of nomograms defining changes in SF depending on c′ and ϕ′.
4. The variability of soil properties caused by exogenous factors, in particular in the weak zones (e.g., on the clay roof), resulted in an alternation of strength parameters of the soil, which may change over time. For this reason, the use of parametric sensitivity analyses of computational models is recommended.

5. The analysed models indicated that the Neogene clay elevations were the main factor influencing the stability of the massif in the slope area. Changes in soils' strength parameters were the secondary factor in modifying the structural factors of stability. For example, according to applied models, changes in the strength parameters in the range of $\phi' \pm 5°$; $c' \pm 5$ kPa resulted in changes of SF $\pm$ 0.2. The value of internal friction angle was a more important factor influencing slope stability, as this parameter considerably depends on the stress state and indirectly on the geometry of the layers building the slope.
6. Stability predictions quite significantly depend on the applied calculation techniques, both analytical methods based on the conditions of moment equilibrium (e.g., Bishop) or forces (e.g., Janbu) and modelling using the finite element method (FEM). In calculations of the stability conditions performed in this paper, the influence of the weak layer was taken into account. Both Janbu and FEM 2D methods produced similar results. The values of the SF index calculated by the Bishop and FEM 3D methods were higher than those obtained with the other methods. The overestimation of the stability value in the case of the Bishop method was related to the use of a circular-cylindrical slip surface, which did not coincide with the probable slip surface related to the course of the weak layer (e.g., the weathered roof of Neogene clays, which is a screen for seeping groundwater). In addition, the highest values of SF results in the FEM 3D method require further research.
7. The conducted analyses showed the necessity of mapping the structural conditioning of stability and the necessity of deliberate selection of modelling method that takes into account the variability of both geometric factors and strength characteristics. It was proven by the usefulness of the use of geophysical prospection, which enriches the point characteristics obtained from traditional ground exploration by drilling. Thus, local disturbances and their influence on changes in slope stability can be taken into account.

Author Contributions: Conceptualisation and design: K.K. and Ł.K.; literature review: Ł.K. and S.K.; methodology and validation: K.K., Ł.K. and S.K.; formal analysis: K.K., Ł.K. and S.K.; investigation and data collection: K.K., Ł.K. and S.K.; data analysis and interpretation: K.K., Ł.K., P.D. and S.K.; writing—original draft preparation: K.K., P.D., Ł.K. and S.K.; writing—review and editing: P.D., K.K. and Ł.K.; supervision: P.D. All authors have read and agreed to the published version of the manuscript.

Funding: This research was funded by the University of Warsaw.

Conflicts of Interest: The authors declare no conflict of interest.

References

1. Carrión-Mero, P.; Briones-Bitar, J.; Morante-Carballo, F.; Stay-Coello, D.; Blanco-Torrens, R.; Berrezueta, E. Evaluation of Slope Stability in an Urban Area as a Basis for Territorial Planning: A Case Study. *Appl. Sci.* **2021**, *11*, 5013. [CrossRef]
2. El Kechebour, B. Relation between stability of slope and the urban density: Case study. *Procedia Eng.* **2015**, *114*, 824–831. [CrossRef]
3. Krawczyk, J.; Tkaczyk, A.M. Czwartorzędowe struktury rynnowe rejonu niecki wrocławskiej w świetle badań geofizycznych—elektrooporowych. *Przegląd Geol.* **1997**, *45*, 423–427. (In Polish)
4. Flieger-Szymańska, M.; Filipiak, P. Geophysical prospection problems in determining the presence and boundaries of hydrogeological structures of buried valleys geoelectric method. *Biul. Państwowego Inst. Geol.* **2011**, *445*, 97–106. (In Polish with English Summary)
5. Słodkowska, B.; Gałązka, D. Paleogene and Neogene sediments in mega-scale glaciotectonic structures of the Dylewskie Hills. *Biul. Państwowego Inst. Geol.* **2015**, *461*, 251–294. (In Polish with English Summary) [CrossRef]
6. Pacanowski, G.; Sokołowska, M.; Mieszkowski, R. Geoelectrical imaging of complex geological structures of the Morasko Hill in Poznań. *Przegląd Geol.* **2016**, *64*, 238–244. (In Polish with English Summary)
7. Kowalczyk, S.; Cabalski, K.; Radzikowski, M.; Jędrzejewski, J. Application of electrical resistivity imaging to ground recognition of the Warsaw Southern Ring-Road. *Przegląd Geol.* **2017**, *65*, 772–778. (In Polish with English Summary)
8. Lech, M.; Skutnik, Z.; Bajda, M.; Markowska-Lech, K. Applications of Electrical Resistivity Surveys in Solving Selected Geotechnical and Environmental Problems. *Appl. Sci.* **2020**, *10*, 2263. [CrossRef]

9. Kowalczyk, S.; Mieszkowski, R.; Pacanowski, G. The stability evaluation of Warsaw slope selected pieces based on Electrical Resistivity Tomography survey (ERT). *Przegląd Geol.* **2014**, *62*, 634–640.
10. Bellanova, J.; Calamita, G.; Giocoli, A.; Luongo, R.; Macchiato, M.; Perrone, A.; Uhlemann, S.; Piscitelli, S. Electrical resistivity imaging for the characterization of the Montaguto landslide (southern Italy). *Eng. Geol.* **2018**, *243*, 272–281. [CrossRef]
11. Hack, R. Geophysics For Slope Stability. *Surv. Geophys.* **2000**, *21*, 423–448. [CrossRef]
12. Moradi, S.; Heinze, T.; Budler, J.; Gunatilake, T.; Kemna, A.; Huisman, J.A. Combining Site Characterization, Monitoring and Hydromechanical Modeling for Assessing Slope Stability. *Land* **2021**, *10*, 423. [CrossRef]
13. Kaczmarek, Ł. Geomorphological and geophysical analysis of the Warsaw slope stability conditions in the Ursynów district. *Sci. Rev. Eng. Env. Sci.* **2014**, *65*, 215–226. (In Polish with English Summary)
14. Dahlin, T. The development of DC resistivity imaging techniques. *Comput. Geosci.* **2001**, *27*, 1019–1029. [CrossRef]
15. Loke, M.H.; Chambers, J.E.; Rucker, D.F.; Kuras, O.; Wilkinson, P.B. Recent developments in the direct-current geoelectrical imaging method. *J. Appl. Geophys.* **2013**, *95*, 135–156. [CrossRef]
16. Zhang, S.; Li, Y.; Li, J.; Liu, L. Reliability Analysis of Layered Soil Slopes Considering Different Spatial Autocorrelation Structures. *Appl. Sci.* **2020**, *10*, 4029. [CrossRef]
17. Pirone, M.; Urciuoli, G. Analysis of slope-stabilising piles with the shear strength reduction technique. *Comput. Geotech.* **2018**, *102*, 238–251. [CrossRef]
18. Azarafza, M.; Akgün, H.; Ghazifard, A.; Asghari-Kaljahi, E.; Rahnamarad, J.; Derakhshani, R. Discontinuous rock slope stability analysis by limit equilibrium approaches—A review. *Int. J. Digit. Earth* **2021**, 1–24, ahead-of-print. [CrossRef]
19. Tang, H.M.; Liu, X.; Xiong, C.R.; Wang, Z.Y.; Eldin, M. Proof of Nondeterministic Polynomial-Time Complete Problem for Soil Slope-Stability Evaluation. *Int. J. Geomech.* **2016**, *16*, 12. [CrossRef]
20. Xia, Y.-X.; Cheng, P.; Liu, M.-M.; Hu, J. Numerical Modeling of 3D Slopes with Weak Zones by Random Field and Finite Elements. *Appl. Sci.* **2021**, *11*, 9852. [CrossRef]
21. Peng, W.X.; Mo, J.J.; Xie, Y.J. Comparison for the Results from 2D and 3D Analysis for Slope Stability. *Appl. Mech. Mater.* **2011**, *90*, 255–259. [CrossRef]
22. Fredlund, M.D.; Fredlund, D.G.; Ivkovic, Z.; Lu, H. Extending slope stability analysis to multi-plane 2D and 3D limit equilibrium approaches. In Proceedings of the 70th Canadian Geotechnical Conference GeoOttawa 2017, Ottawa, ON, Canada, 1–4 October 2017.
23. Kaczmarek, Ł.; Popielski, P. Selected components of geological structures and numerical modelling of slope stability. *Open Geosci.* **2019**, *11*, 208–218. [CrossRef]
24. MacKillop, K.; Fenton, G.; Mosher, D.; Latour, V.; Mitchelmore, P. Assessing Submarine Slope Stability through Deterministic and Probabilistic Approaches: A Case Study on the West-Central Scotia Slope. *Geosciences* **2019**, *9*, 18. [CrossRef]
25. Lu, K.L.; Zhu, D.Y. A three-dimensional rigorous method for stability analysis and its application. *Bull. Eng. Geol. Environ.* **2016**, *75*, 1445–1457. [CrossRef]
26. Oberhollenzer, S.; Tschuchnigg, F.; Schweiger, H.F. Finite element analyses of slope stability problems using non-associated plasticity. *J. Rock Mech. Geotech. Eng.* **2018**, *10*, 1091–1101. [CrossRef]
27. Sazzad, M.M.; Rahman, F.I.; Mamun, M.A.A. Mesh Effect on the FEM Based Stability Analysis of Slope. In Proceedings of the International Conference on Recent Innovation in Civil Engineering for Sustainable Development, Gazipur, Bangladesh, 11–13 December 2015.
28. Tschuchnigg, F.; Schweiger, H.; Sloan, S.W. Slope stability analysis by means of finite element limit analysis and finite element strength reduction techniques. Part II: Back analyses of a case history. *Comput. Geotech.* **2015**, *70*, 178–189. [CrossRef]
29. Zienkiewicz, O.C.; Humpheson, C.; Lewis, R.W. Associated and nonassociated visco-plasticity and plasticity in soil mechanics. *Geotechnique* **1975**, *25*, 671–689. [CrossRef]
30. Kaczmarek, Ł.; Dobak, P.; Szczepański, T.; Kiełbasiński, K. Triaxial creep tests of glacitectonically disturbed stiff clay—structural, strength, and slope stability aspects. *Open Geosci.* **2021**, *13*, 1118–1138. [CrossRef]
31. Frankowski, Z.; Majer, E.; Sokołowska, M.; Ryżyński, G.; Ostrowski, S.; Majer, K. Engineering geological research conducted in the Polish Geological Institute during the last fifty years of activity. *Przegląd Geol.* **2018**, *66*, 752–768.
32. Wysokiński, L. *The Warsaw Slope in the Śródmieście District (from al. Jerozolimskie to the Sanguszki St.)*; Drukarnia Piotra Włodarskiego: Warszawa, Poland, 1999.
33. Kaczmarek, Ł.; Dobak, P. Stability conditions of the Vistula Valley attained by a multivariate approach—A case study from the Warsaw Southern Ring Road. *Geologos* **2015**, *21*, 249–260. [CrossRef]
34. Frankowski, Z.; Bażyński, J.; Zawadzki, R.; Lewkowicz, M.; Smagała, S.; Wysokiński, L.; Majer, E.; Łukasik, S.; Filipowicz, A.; Sobiech, J. *Engineering Geological Atlas of Warsaw on a Scale of 1: 10,000*; PIG-PIB: Warszawa, Poland, 2000.
35. Bogusz, W. Prediction of Tunneling-Induced Ground Movements. Ph.D. Thesis, ITB, Warsaw, Poland, 2021.
36. Sarnacka, Z. *Detailed Geological Map of Poland, Scale 1: 50,000, Warsaw East Sheet with Explanations*; PIG-PIB: Warszawa, Poland, 1980.
37. Dahlin, T.; Zhou, B. Multiple-gradient array measurements for multichannel 2D resistivity imaging. *Near Surf. Geophys.* **2006**, *4*, 113–123. [CrossRef]
38. Loke, M.H. *Tutorial: 2-D and 3-D Electrical Imaging Surveys*; Geotomo Software: Gelugor, Malaysia, 2002.
39. Stenzel, P.; Szymanko, J. *Metody Geofizyczne w Badaniach Hydrogeologicznych i Geologiczno-Inżynierskich (in Polish)*; Wydawnictwa Geologiczne: Warszawa, Poland, 1973.

gating the effectiveness of cement paste addition [24]. Preetham and Nayak [22] added granulated blast furnace slag (GBFS) and cement to the soil to investigate the geotechnical properties of mixtures, including physical properties, compaction characteristics, unconfined compression strength, and Unconsolidated Undrained (UU) triaxial compression test results [22]. Studies have focused on the strength related to hydraulic conductivity; Wu et al. [25] discussed that hydraulic conductivity could increase the uniformness of settlement and long-term safety showing the importance of hydraulic conductivity [25]. Wu et al. [26] also used cement to improve weak marine clay and found that strength is influenced by cement- and density-related factors. This indicates that the effect of density on strength is comparable to that of cement [26]. On the other hand, the moisture content used for preparing specimens applied in unconfined compression and triaxial compression tests affects results substantially. Several studies have applied high moisture content during specimens' preparation [20,27,28]. Ge et al. [28] applied moisture content levels of 1.8-, 2.0-, and 2.2-fold the liquid limit (LL) of kaolinite to prepare specimens [28]. Another study applied low moisture content to prepare specimens that primarily comprised ordinary Portland cement (OPC) and local lithomargic clay collected in the Dakshina Kannada district of India. In the study, clay was dried at 105 °C then screened and mixed by weight with varying proportions of cement (2.5%, 5%, 7.5%, and 10%). Water was added according to optimum moisture content (OMC). Thereafter, the obtained soil was put into a bag for curing for a given numbers of days before the specimen was removed for subsequent tests [29]. Similar methods were also proposed in other studies [15,16,22,30]. When we take practicality and on-site conditions into consideration, clay clods substantially affect engineering construction [24,31]. Soil gradation and the coefficient of uniformity (C_u) both affect soil static strength [32]. Furthermore, OMC percentage and the wet side of compaction tests can be adopted when hydraulic conductivity and the strength of compacted clay are taken into account [33]. Soil saturation also affects strength [34,35]. Thus, we applied multiple preparation methods, soil gradations, and dry density levels to prepare untreated samples in the study. Varying amounts of cement paste were added into samples subjected to compaction test treatment to increase the generalizability of the study. The relationships between elastic modulus and unconfined compression strength were discussed in many studies, with E_{50} recommended as the elastic modulus [19,26,33,36]. Jamsawang et al. [19] and Wu et al. [26] investigated and discussed the relationship between E_{50} and q_u, with favorable outcomes reported [19,26]. The present study used compression tests to clarify the relationship between elastic modulus and compression strength. The triaxial test included confined undrained (CU) and UU tests. Specimens used for the CU test were saturated and regarded as soaked samples, whereas those used for the UU test were unsaturated and regarded as unsoaked samples. The compression test only determined the soil strength q_u value. Soil shear strength parameters consisted of the c value and ϕ angle. According to the Mohr Coulomb theory, soil shear strength is calculated using the Equation (1).

$$\tau_f = c + \sigma \tan\phi, \quad (1)$$

where τ_f is shear strength, σ is normal stress, and c and ϕ represent cohesion value angle and internal friction angle, respectively. The swelling of soil can be assessed by identifying the effect of infiltration saturation and vertical stress on GMZ bentonite swelling indicated by Chen et al. [37]. The present study adopted the California bearing ratio (CBR) test. As clay, which is rich in smectite, possesses a high swelling potential and a low hydraulic conductivity, it is commonly used as a liner for landfills. The hydraulic conductivity of clay causes desiccation cracking and adverse effects. Therefore, one-dimensional swelling deformation apparatus introduced by Emmanuel and Anggraini [38] can be used as a research instrument. Swelling can also be determined through the soaking process of the CBR test [24,30]. The results of related studies indicate volume swelling. Clay clods and desiccation cracks also affect the mechanisms of hydraulic conductivity [18,29,31,39–42]. We used the indoor permeability test to clarify the hydraulic conductivity mechanism, while consolidation and compression were also intensively studied. Since the void/cement ratio is considered as a main factor that contributes to the strength and compression, clay-based lightweight aggregate concrete improved with cement addition [43,44].

In summary, (1) for the application of compaction test results, the moisture content of the specimens prepared in the present study was similar to the OMC level on the wet side; (2) soil strength was compared based on the CU and UU test results of the unconfined compression and triaxial compression tests, and the Mohr–Coulomb shear strength formula was used to determine the difference between mudstone and mudstone–cement samples; (3) after the compression test was completed, stress–strain curves were used to determine E_{50}, and the relationship between elastic modulus E_{50} and unconfined compression strength q_u was clarified; (4) the indoor permeability test was used to explore the factors related to on-site clods that could affect mudstone hydraulic conductivity; (5) the one-dimension consolidation test was used to explore the compression variation of mudstone before and after the addition of cement paste; and (6) the CBR test was used to clarify the swelling of mudstone after compaction and soaking.

2. Materials and Methods

2.1. Materials

Mudstone was collected from the Renwu district of Taiwan (Figure 1a). The district is located at the southernmost region of Taiwan where exposed mudstone can be found. Mudstone was obtained using excavators, with irrelevant on-site tree branches and stones removed. Mudstone was transported to the laboratory by trucks. The mudstone was soaked in water for days to allow for impurities to float to the water surface. Most of the water was poured out to facilitate air drying. The mudstone formed clods of various sizes, which were packed for future use. The chemical composition of mudstone from southwestern Taiwan is as follows: SiO_2 (63.49%), $Al_2O_3 \cdot Fe_2O_3$ (21.53%), and CaO (2.71%). According to Lee et al. (1996) and Lee et al. (2007), an X-ray scan of mudstone in Taiwan revealed the main mineral composition as follows: illite (35.4%), chlorite (28.70%), and quartz (28.45%). The secondary mineral composition of mudstone is as follows: feldspar, calcite, and kaolinite. The grain particle size distribution of weathered mudstone indicated that the silt-sized, clay-sized, and sand-sized particles accounted for 49%, 29%, and 22%, respectively, of the mudstone. OPC (type I) was purchased from Taiwan Cement Corporation and the basic chemical composition comprised SiO_2 (21.24%), Al_2O_3 (4.44%), Fe_2O_3 (3.44%), CaO (64.51%), MgO (2.35%), and SO_3 (2.10%). The fine modulus was 349 m^2/kg and the specific gravity G_s was 3.16. The initial and final setting times were 139 and 255 min, respectively, which is finished by the Vicat Needle method according to ASTM-C191-08 (2008).

2.2. Testing Methods

To explore the mechanical behavior of mudstone clods in which dry cement is added after compaction improvement, the present study conducted basic physical property, compaction, unconfined compression, triaxial compression, triaxial permeability, one-dimension consolidation, and CBR tests with the addition of 0%, 4%, and 8% of cement. The relevant regulations that are applied for the tests are as follows: ASTM D422 for grain size distribution curve [45]; ASTM D854 for G_s (specific gravity test) [45]; ASTM D423, D424, and D4318 for Atterberg limits [45]; ASTM D2487 for USCS [45]; ASTM D2166 for the unconfined compression test [45]; ASTM D4767 for the triaxial CU compression test [45]; ASTM D2850 for the triaxial UU compression test [45]; ASTM D2216 for soil moisture content [45]; ASTM D698 and ASTM D1557 for the Proctor compaction test [45]; ASTM D2434 for the indoor permeability test [45]; ASTM D2435 for one-dimension consolidation test [45]; and ASTM D1883 and ASTM D4429 for the CBR test [45].

2.3. Specimen Preparation

In 2007, Consoli et al. [16] discussed specimen preparation methods. On the basis of that study, the present study prepared specimens for unconfined and triaxial compression tests. Mudstone–cement samples were prepared by weighing soil and cement as per prior protocols before uniformization. The present study used cement content C (expressed in weight percentage) by referencing previous studies [16,24], with percentage levels of 0%, 4%, and 8% selected. The specimens were divided into three layers and placed in cylindrical copper split molds, layer by layer. Each layer was designed based on calculations. The final layer was slightly higher than the mold height. After the preparations were complete, the specimens were demolded. The specimens were 5 cm in diameter and 10 cm in height. Subsequently, the specimens were placed in wet rooms for curing and soaked in water tanks for 24 h to achieve saturation and minimize suction. The total curing time ranged from 7 to 28 days, with water temperature controlled at 25 °C. Before the tests were conducted, the specimens were removed from the water tank and the surface was dried with a water-absorbing cloth. Next, unconfined compression tests were conducted and the peak load of the specimens was recorded. The present study explored the effects of different compression methods on on-site civil engineering. Three methods, namely static, compaction, and kneading, were applied (Figure 2) to prepare the specimens. The mudstone soil specimens (without cement paste) were prepared using static, compaction, and kneading methods. This factor was used to facilitate extensive exploration. The compaction method was adopted for compaction tests and CBR tests. For the application of the compaction method, tests are required to determine how preset density can be achieved. Several studies have used the static method for specimen preparation, particularly the three-layer layered static method [16,30]. Studies have also used the kneading method for specimen preparation [19,46]. The kneading method was based on the JGS T821-1990 standard, and it can be difficult to implement for specimen preparation because prior tests are required to obtain the required density. Studies have used clods or soil gradation to investigate the hydraulic conductivity of clay [31,41] and identify the effect of clods on soil engineering characteristics [17,32,47]. On-site experience indicates that compression often leads to varying grain sizes. In contrast to laboratory-based research, massive on-site earthworks cannot be sifted by using just one sieve with a specific opening size. Thus, soil samples that passed through sieves that were size <#4 (4.75 mm), #4-3/4″ (19.05 mm), or larger than 3/4″ were used in further analyses. Figure 3 indicates that before unconfined compression tests were conducted using specimens from the three soil gradations, the specimens were removed, photographed, and subsequently destroyed. Specimens with added cement were prepared using the static method.

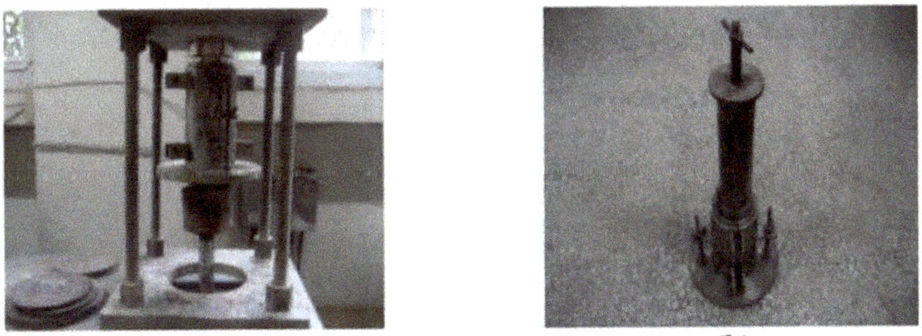

(a) (b)

Figure 2. *Cont.*

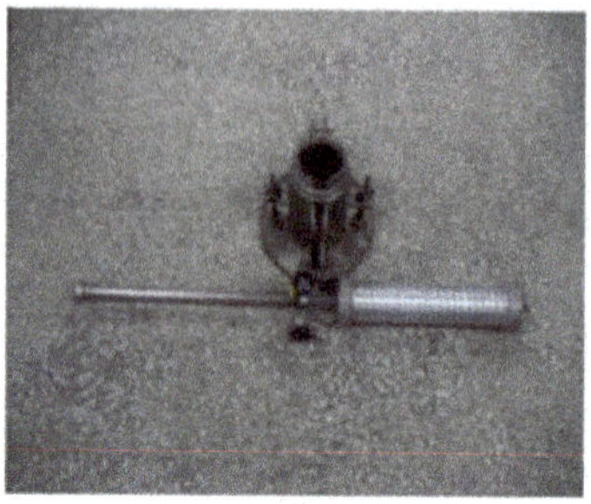

Figure 2. Three types of testing equipment for preparing soil samples: (**a**) static; (**b**) compaction; (**c**) kneading.

Figure 3. Sample appearance before failure: (**a**) compaction, kneading, and static (from left to right); (**b**) soil particles <#4, #4–3/4″, and >3/4″ (from left to right).

3. Testing Results and Analysis

3.1. Basic Physical Properties

The two particle size distribution curves of mudstone used in the present study are shown in Figure 4. One curve presents the relationship between the sieve analysis test curves and specific gravity test curves of the original mudstone soil that was not screened for particle size. Another curve relates to the soil samples that were passed through #4 sieves and underwent specific gravity testing before undergoing the sieve analysis test. Figure 4 indicates that approximately 77.8% of the original soil passed through the #4 sieve. The results of the basic soil physical property test revealed that the specific gravity of the grains was approximately 2.71. In addition to partial large grains, the grains were composed of sand-sized particles (approximately 27.3%), silt-sized particles (approximately 31.0%), and clay-size particles (approximately 19.5%). The Atterberg test results revealed that the LL, PL, and plasticity index of the soil was 26.8%, 18.1%, and 8.7%, respectively. As per the USCS, the original soil was classified as CL in Figure 5.

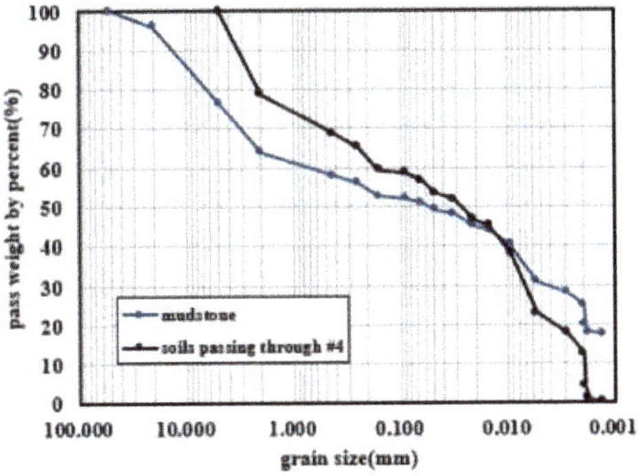

Figure 4. Particle size distribution curves of mudstone original soils and soils passing through #4.

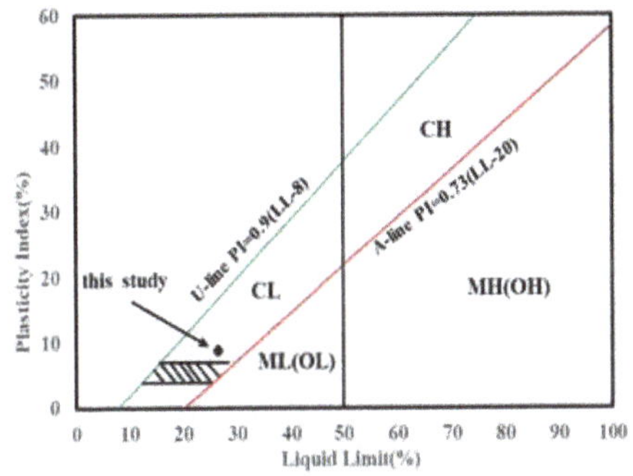

Figure 5. Testing soils in the paper located within the plastic chart according to the United Soils Classification System.

3.2. Compaction Test

As per ASTM D698 and ASTM D1557, the improved and standard soil compaction tests were conducted to determine the compaction curves, maximum dry density γ_{dmax}, and OMC. The present study distinguished clods with varying grain sizes and cement content to conduct improved and standard soil compaction tests. Curves of soil graduations for soil that passed through the sieves (<#4, #4-3/4", or larger than 3/4") were adopted to simulate clod size for tests involving the varying cement content levels of 0%, 4%, and 8%. Figure 6 presents the compaction curve results, with x- and y-axis denoting soil water content and dry density, respectively. Figure 6a indicates that larger clod grain sizes led to the rightward and downward movement of the compaction curves. The improved soil compaction curve was positioned to the upper-left side of the standard soil compaction curves. Figure 6b indicates that an increase in added cement caused the maximum dry density of both improved soil and standard soil compaction curves to move downward, indicating an increase in OMC and decrease in maximum dry density. The improved soil compaction test results revealed that the mudstone specimen OMC and γ_{dmax} were 10.1% and 19.9 kN/m^3, respectively. When the cement addition C was 4% and 8%, the OMC levels were 10.8% and 11.6%, respectively, and the maximum dry density levels were 19.6 kN/m^3 and 19.5 kN/m^3, respectively. For subsequent specimen preparation, compaction test results were employed as the basis, and a line was drawn approximately parallel to the zero air void curve (ZAVC) along the compaction curve on the wet side; this line represents the S_r = 91% curve (Figure 6). The curve was used as the basis for controlling moisture content for each dry density level during specimen preparation. The relevant moisture content levels for dry density levels of 17.5, 18.0, 18.5, 19.0, 19.5, and 20.0 kN/m^3 were

16.2%, 15.8%, 15.2%, 14.0%, 12.3%, and 10.1%, respectively. The moisture content controls for mudstone–cement specimens were identical.

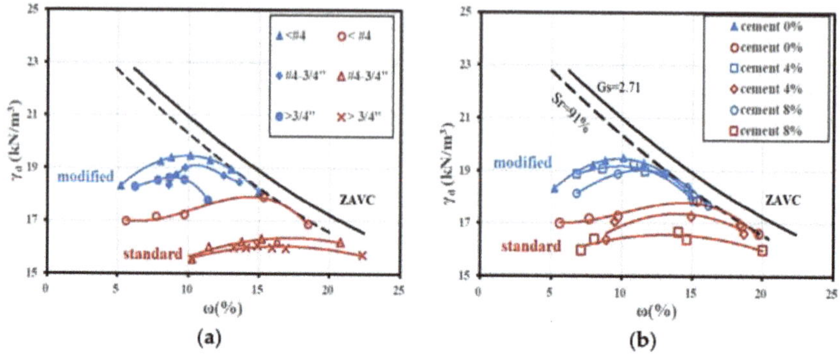

Figure 6. Compaction curves and zero air void curves (ZAVC) from the compaction tests for various mix conditions: (**a**) mudstone; (**b**) mudstone–cement.

3.3. Unconfined Compression Test

After dry cement is mixed with mudstone clods, the hydration of cement produces gels that fill a void. The gel increases the compactness of the clay framework and improves the clods. To understand the effectiveness of the implemented improvements, unconfined compression strength was used as the basic indicator of mudstone that was improved with the addition of cement. Unconfined compression tests were conducted per ASTM D2166. The tested items included specimen preparation methods, clods, number of curing days, cement addition, moisture content, and dry density variables. An explanation and analysis of the treated and untreated soil samples are provided. Figure 7a presents the curves for an untreated specimen clod with grains that passed through #4 sieves, with static methods used to prepare the specimens. Dry density ranged from 17.5 to 20.0 kN/m^3, the unconfined compression strength ranged from to 15.1 kPa, and the corresponding stain ranged from 0.83% to 1.74%. Unconfined compression strength appeared to increase with an increase in dry density. Similar trends were observed for the other two clod grain sizes. Figure 7b presents the treated specimens of various clods with a dry density of 19.5 kN/m^3. Grains that passed through a #4 sieve yielded an unconfined compression strength of 98.0 kPa and a corresponding strain of 1.53%. The unconfined compression strength of grains that were retained through the #4-3/4″ sieve was 95.1 kPa and the corresponding strain was 1.59%. The unconfined compression strength of grains that passed through or were retained on a 3/4″ sieve was 57.0 kPa and the corresponding strain was 1.28%. In summary, the unconfined compression strength was highest for clod grain that passed through the #4 sieve, followed by the #4-3/4″ sieve and >3/4″ sieve. Figure 8a presents the relationship between unconfined compression strength q_u and dry density γ_d of mudstone specimen clods. It indicates that the compaction method generated the highest unconfined compression strength, followed by the static and kneading methods. The treated specimens with C = 8% and grains that passed through the #4 and #4-3/4″ sieves after 28 days of curing exhibited higher unconfined compression strength q_u relative to the other specimens. In 2021, Wu et al. [26] proposed that the two factors cement and density influence the strength of marine clay that is improved through the use of cement. Figure 8a indicates that the mudstone specimens with six dry density levels yielded unconfined compression strength levels of between 10 and 400 kPa. However, after 8% cement paste was added to the specimens, which were then cured for 28 days, the specimens' unconfined compression strength increased to between 500 and 1200 kPa. Under the research conditions proposed by the present study, the addition of cement paste was more effective than an increase in dry density.

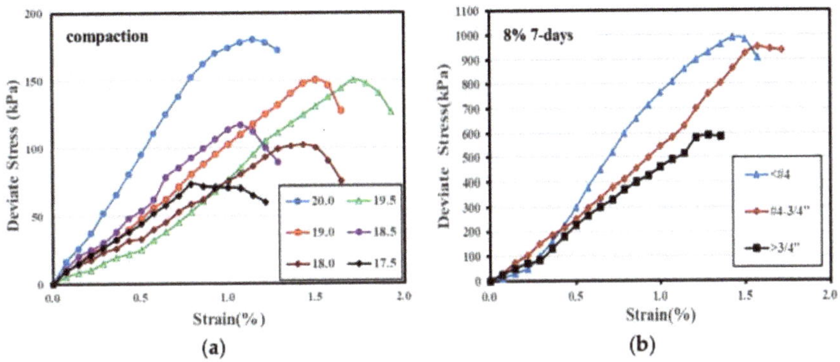

Figure 7. Deviate stress–strain curves of untreated and treated samples of unconfined compression test: (**a**) mudstone; (**b**) mudstone–cement.

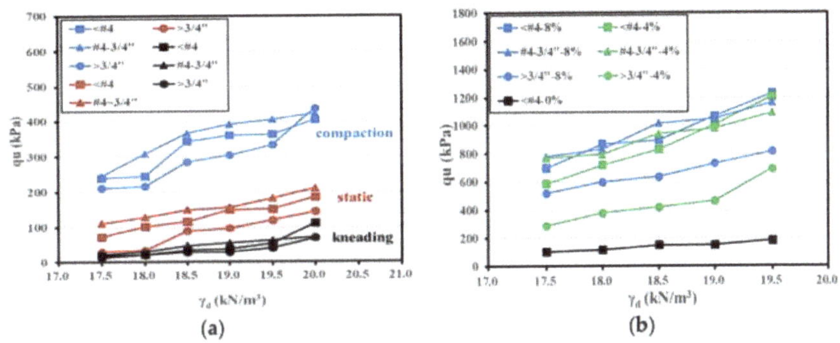

Figure 8. Unconfined compression strength with respect to dry density in soil samples for untreated and treated samples: (**a**) mudstone; (**b**) mudstone–cement.

Figure 9a,b presents the relationships between elastic modulus E_{50} and dry density γ_d in mudstone and mudstone–cement specimens, respectively. The trends presented by the aforementioned figures were identical to those in Figure 8. Among the mudstone specimens with six dry density levels, their elastic modulus E_{50} was between 1 and 24 MPa. After the specimens with added cement paste underwent 28 days of curing, their elastic modulus E_{50} was between 20 and 120 MPa. The relationships between elastic modulus E_{50} and unconfined compression strength q_u (Figures 8 and 9) are plotted in Figure 10. For the mudstone specimen formula, E_{50} = 47.699 q_u and R^2 = 0.8708, whereas for the mudstone–cement specimen formula, E_{50} = 69.397 q_u and R^2 = 0.9693. In 2017, Jamsawang et al. [19] used cement and fly ash to improve clay and reported that for the unsoaked sample, E_{50} = 90 q_u and R^2 = 0.9890, whereas for the soaked sample, E_{50} = 60 q_u and R^2 = 0.9894. In 2021, Wu et al. [26] indicated that the density control group had a formula of E_{50} = 131.1 q_u and R^2 = 0.89, whereas the cementation control group had a formula of E_{50} = 60.4 q_u and R^2 = 0.84. The difference in the formula coefficients was dependent on the added paste material and specimen preparation methods.

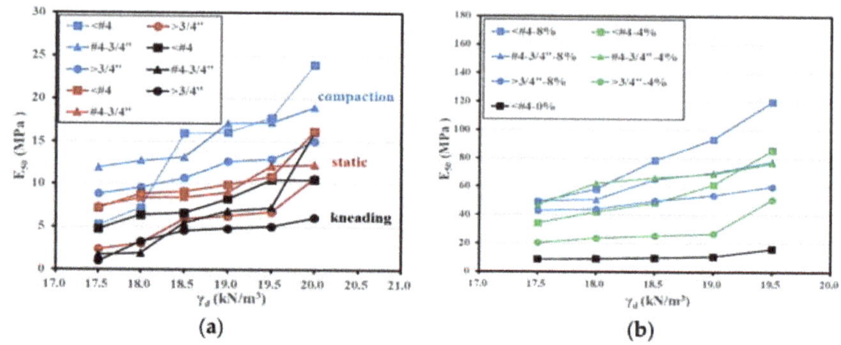

Figure 9. Elastic modulus E_{50} with respect to dry density for untreated and treated samples: (**a**) mudstone; (**b**) mudstone–cement.

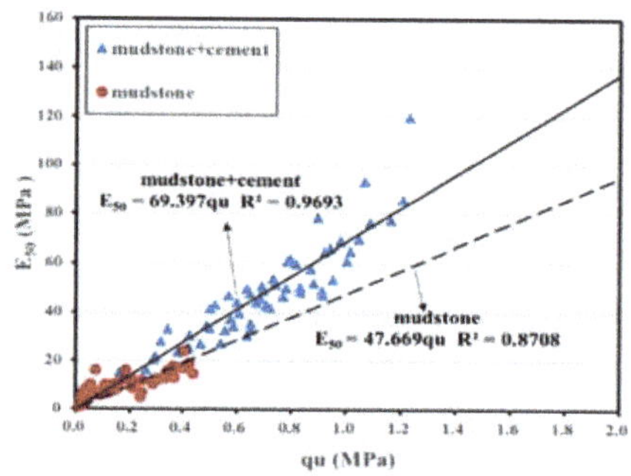

Figure 10. The relationships between q_u and E_{50} values based on mudstone and mudstone–cement samples.

3.4. Triaxial CU Test

In 2017, Das and Sobhan [34] proposed the use of triaxial CU test parameters to simulate load bearing-related engineering problems that are encountered shortly after completing the initial compaction of soil. The present study used ELE International's Digitest 50 system to conduct tests. The strain was controlled at 1.06 mm/min. The pneumatic loading system was connected to a measuring system, which considered load cells, LVDT, volume changes, and pore water pressure sensors. A set of data loggers was used to establish a connection with a computer for data storage and analysis. As per ASTM D2850, triaxial CU tests were conducted. The testing stages included specimen preparation, saturation, confining pressure, consolidation, and application of deviator stress. Specimen saturation requires the Skempton pore water pressure parameter B value to be at least 0.95. After the saturation stage, we can measure the B value by applying incremental confining pressure. The consolidation stage involved the complete exclusion of excess pore water pressure inside specimens. The present study used three grain sizes (grains that passed through #4, #4-3/4″, or >3/4″ sieves) to prepare mudstone specimens. The remolded specimens with added cement paste were prepared using static methods. The confining pressures of the specimens were 50, 100, and 200 kPa. Figure 11 presents the triaxial CU test results of original mudstone soil and treated soil samples. Figure 11a presents the original soil stress–strain curves. When the strain was between 10% and 15%, the deviator stress peaked and then gradually decreased. The pore water pressure peaked when the strain was between 0% and 5%, which dropped sharply afterward. When the deviator stress was at its peak, the pore water pressure was between −20 and −60 kPa. Figure 11b presents the triaxial CU test results of specimens with added cement. The results revealed that when the strain was between 1.5% and 2.0%, deviator stress peaked and then gradually decreased.

The pore water pressure peaked at a strain of 0.5% and decreased sharply afterward. When stress peaked, the pore water pressure was between −20 and −80 kPa.

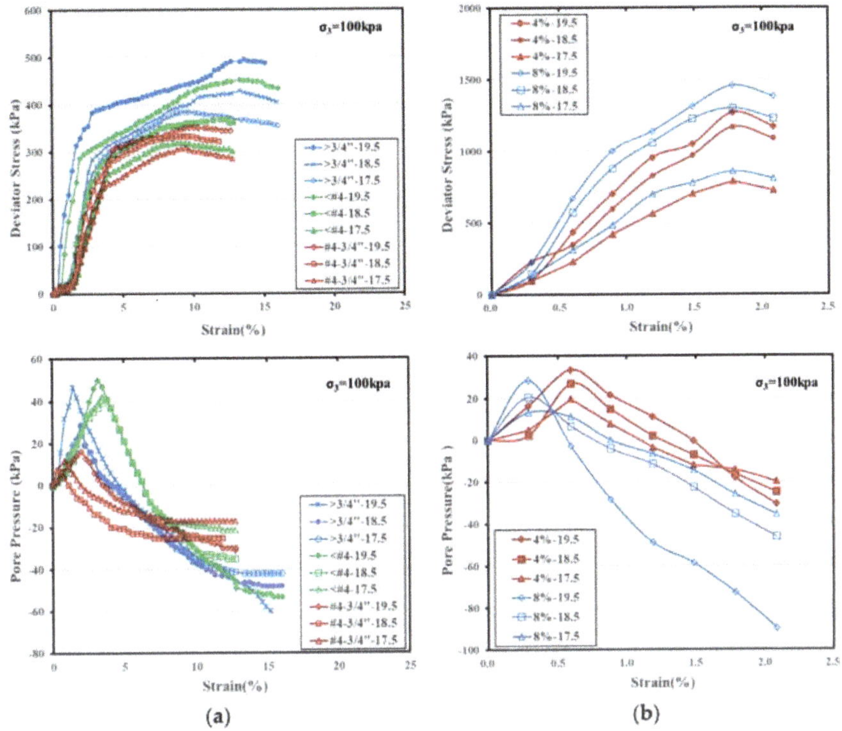

Figure 11. Typical triaxial Consolidated Undrained test results for untreated and treated samples: (**a**) mudstone; (**b**) mudstone–cement.

In all tests, shear strength increased with an increase in cement content and dry density. Figure 12 uses various confined compression Mohr circles and failure envelope curves to present the triaxial CU test results of the original mudstone soil and mudstone–cement soil samples. Because the specimens were compacted and contained cement paste, the pore water pressure was negative when the deviator stress was at its peak. At that point, effective stress and soil shear strength should increase. Consoli et al. [16] conducted saturated soil triaxial UU tests and reported negative pore water pressure. Figure 12 indicates that when the specimen γ_d was 19.5 kN/m^3 and cured for 28 days with C = 0%, 4%, or 8%, the c values were 85 kPa, 147 kPa, and 167 kPa, respectively, and the ϕ angles were 25.79°, 44.37°, and 44.35°, respectively. The cohesion force and internal friction angle increased when the amount of paste added was increased. Figure 13 presents the CU test results for mudstone and mudstone–cement specimens, including the total stress c values and ϕ angles. Figure 13a indicates that among the various clods, when γ_d increased, the ϕ angle increased notably to between 23.5° and 27.6°. However, the increase in the c value was not substantial, which could be related to the pore water pressure at specimen failure. The results of the mudstone–cement samples are presented in Figure 13b. C (i.e., amount of cement added) was 4% or 8% and the angle was between 37.1° and 44.8°.

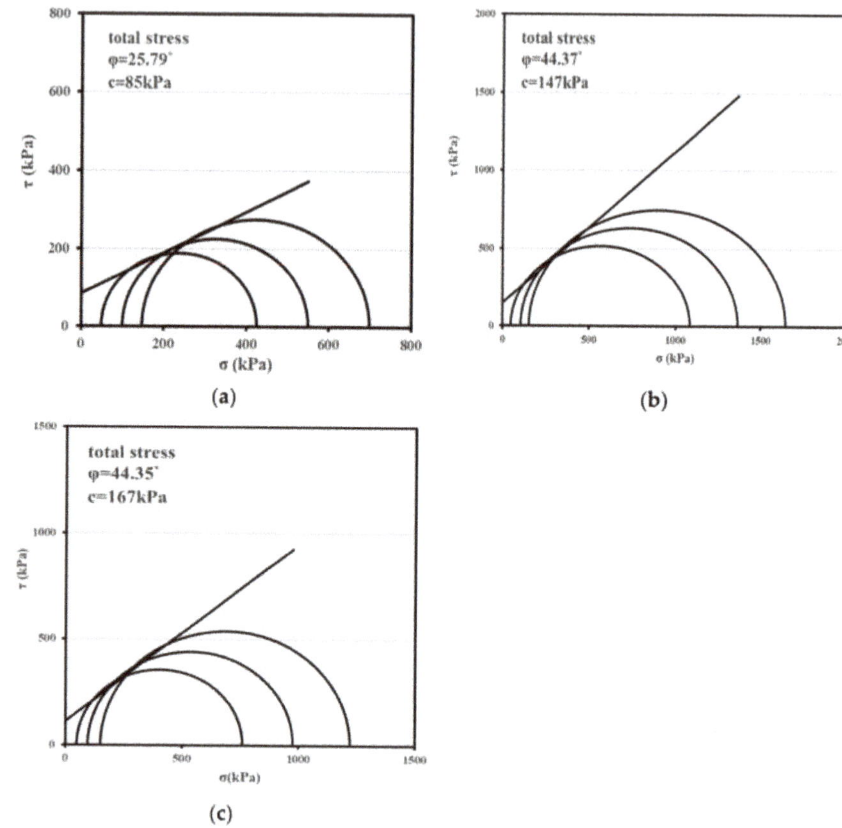

Figure 12. Mohr circle and shear strength parameters of the testing samples for triaxial Consolidated Undrained tests: (**a**) CU-0%; (**b**) CU-4%; (**c**) CU-8%.

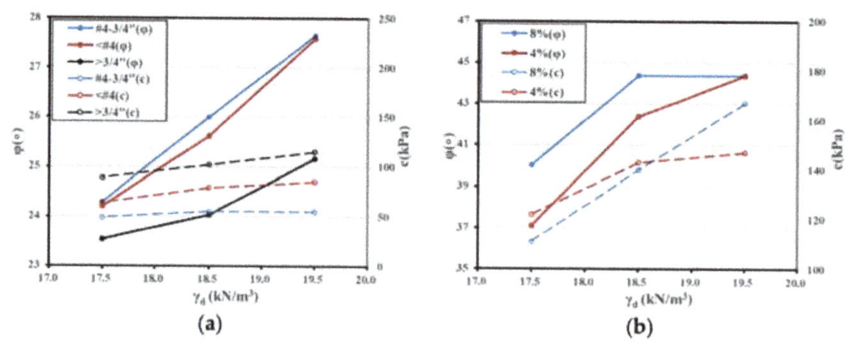

Figure 13. Cohesion values and internal friction angles of the samples for triaxial Consolidated Undrained tests: (**a**) mudstone; (**b**) mudstone–cement.

3.5. Triaxial UU Test

Das and Sobhan [34] proposed the use of the parameters of triaxial UU tests for simulating clay stability analysis and calculating bearing capacity when a construction load is high. To prepare remolded specimens, the present study used static triaxial UU tests. A discussion of mudstone and mudstone–cement specimens follows. Similar test methods were used and the specimens did not achieve saturation. Das [48] and Fredlund and Rahardjo [49] indicated that the curved envelope is sometimes approximated as a straight line as Equation (1) in unsaturated soils for design purposes. Care should be taken to not exceed the limits to which the envelope is applicable. Figure 14 indicates that when the specimens had a γ_d of 19.5 kN/m^3 and were cured for 28 days with C = 0%, 4%, and 8%, the c values were 70 kPa, 154 kPa, and 256 kPa, respectively, and the ϕ angles were 27.26°, 42.72°, and 43.13°, respectively. Cohesion and the internal friction angle increased with the amount of cement paste added. Clods with grains that passed through the #4 and #4-3/4″ sieves did not exhibit considerable differences. Specimens retained by the 3/4″ sieve exhibited greater clod grain sizes. After the completion of compaction, a large number of grains with varying sizes still existed in the specimens, preventing the complete compactness of the soil specimens. Therefore, specimens that were subjected to shearing forces tended to present dislocated grains and have reduced friction angles. Figure 15 presents the comparison diagrams for cement content, friction angle, and cohesion force. When the amount of cement paste added was increased, friction angle and cohesion force increased.

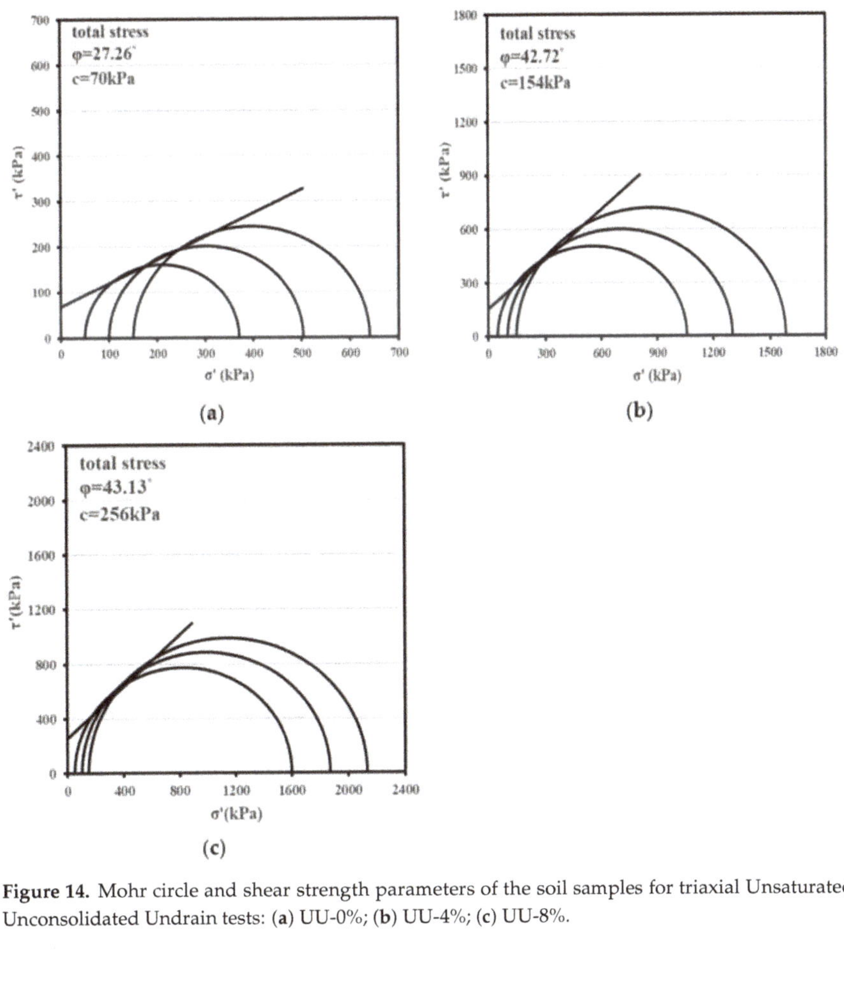

Figure 14. Mohr circle and shear strength parameters of the soil samples for triaxial Unsaturated Unconsolidated Undrain tests: (**a**) UU-0%; (**b**) UU-4%; (**c**) UU-8%.

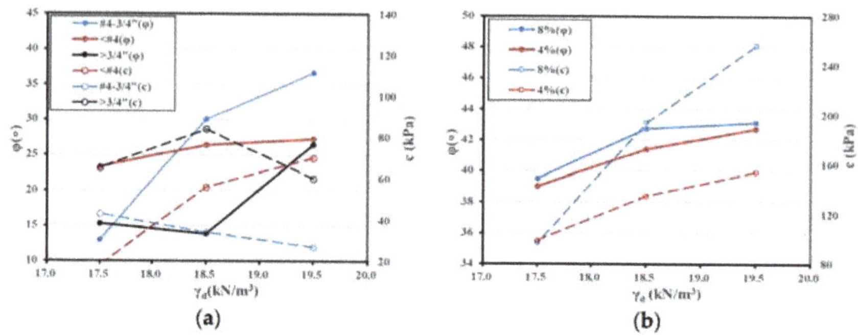

Figure 15. Cohesion values and internal friction circle of the samples for triaxial Unsaturated Unconsolidated Undrain tests: (**a**) mudstone; (**b**) mudstone–cement.

Unconfined compression strength measures the strength of specimens under unconfined conditions, with soil generally subjected to overburden pressure under the ground surface. Therefore, shear strength can be represented using Equation (1); that is, shear strength is calculated using the c value and ϕ angle parameters. The data presented in Figures 12 and 14 are organized in Table 1. For the parameter σ, a shallow overburden pressure of 40 kPa was applied. The triaxial CU test specimens were set as saturated and the triaxial UU test specimens were set as unsaturated. The mudstone specimens had a C of 0%, whereas the mudstone–cement specimens had a C of 4% and 8%. After 28 days of curing, the shear strength τ_f of saturated samples was 104.33 kPa, 186.13 kPa, and 206.10 kPa, while that of unsaturated samples was 90.61 kPa, 190.94 kPa, and 293.47 kPa. The relevant data are plotted in Figure 16, which indicates that unsaturated samples exhibited higher strength. The soaked samples generate negative pore water pressure because of compaction and the addition of cement under low strain conditions; thus, the specimen strength was not low. In Figure 8, the unconfined compression strength q_u of specimens with C values of 0%, 4%, and 8% after 28 days of curing was 151, 1207, and 1488 kPa, respectively. However, real geoengineering foundations are confined. Thus, the adoption of τ_f is more suitable for engineering applications.

Table 1. Shear strength calculated by CU and UU test results.

		CU test	
	mudstone	mudstone–cement	
	C = 0%	C = 4%	C = 8%
'	85	147	167
ϕ (°)	25.79	44.37	44.35
τf (kPa)	104.33	186.13	206.10
		UU test	
	mudstone	mudstone–cement	
	C = 0%	C = 4%	C = 8%
c (kPa)	70	154	256
ϕ(°)	27.26	42.72	43.13
τf (kPa)	90.61	190.94	293.47

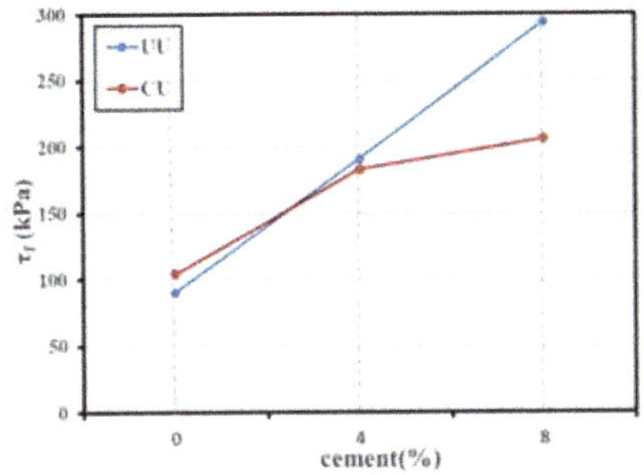

Figure 16. Comparison of shear strength of the samples for triaxial Consolidated Undrained and Unsaturated Unconsolidated Undrain test.

3.6. Triaxial Permeability Test

The indoor permeability test was based on the ASTM D2434 test and performed using a triaxial method. Infiltration was calculated using Darcy's law, that is,

$$q = kiA, \quad (2)$$

where i is the hydraulic gradient and A is the circulation area. The specimen was approximately 10 cm in length, and its upper and lower hydraulic head gap was 50 kPa, indicating that the hydraulic gradient i = 500/10 = 50. Each test required approximately 1 week for flow rate to stabilize before measurements were taken. Studies have explored the hydraulic conductivity of mudstone in Taiwan. They include a 1998 study by Sheu et al. [40], who reported that excessively high or low moisture content on both the dry and wet sides of compacted OMC resulted in high hydraulic gradient k values, and a 2004 study by Hsiao and Lin [34], who reported that reduced compaction energy and overly large grain sizes led to a substantial increase in the hydraulic gradient. In the present study, the triaxial permeability test was conducted to examine three clod grain sizes and two dry density levels, with the test revealing that the k values of the soil specimens ranged between 10^{-4} and 10^{-7} cm/s. Excluding the clods with grain sizes larger than 3/4″, other specimens exhibited low permeability or impermeability. These results corresponded to the impermeability characteristics of mudstone. A specimen with added cement paste (C = 4%) was used for testing. After 1 week, no notable water effluence was observed from the specimen, which was then regarded to be almost impermeable. Figure 17 presents the relationships between dry density and hydraulic conductivity for each specimen. When the size of clod grains increased, the hydraulic gradient k increased. Grain size was not directly associated with dry density. Although the soil was compacted, a considerable number of large-sized grains still existed in the clods, thereby leading to the formation of a substantial void among grains in the clods and greater hydraulic conductivity. Benson and Daniel [31] reported similar test results in their 1989 study.

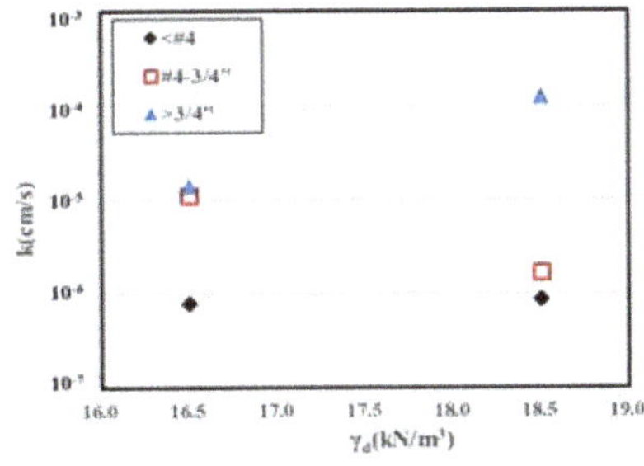

Figure 17. Permeability coefficient vs. dry density for different clods samples.

3.7. One-Dimension Consolidation Test

The one-dimension consolidation test was based on ASTM D2435. The loaded weights were 25, 50, 100, 200, 400, 800, and 1600 kPa, and the loading time was 24 h. Seven sets of tests were conducted to examine untreated soil specimens, treated specimens with 4% and 8% cement content, and three dry density levels. The test results are presented in Figure 18. Figure 18a indicates that the compaction level of mudstone soil was the highest. Specimens with 4% cement content exhibited flat straight lines of compaction when dry density was increased, suggesting low compaction levels. Figure 18b displays a similar trend. The results were used to calculate the compression index C_c and swelling index C_s. Figure 19a indicates that for specimens with C = 4% and dry density levels of 17.5, 18.5, and 19.5 kN/m^3, their corresponding C_c values were 0.034, 0.031, and 0.022, respectively. For specimens with C = 8% and dry density levels of 17.5, 18.5, and 19.5 kN/m^3, their corresponding C_c values were 0.028, 0.025, and 0.016, respectively. The C_c value of the soil specimen without cement content was 0.142. The results suggested that an increase in cement content and dry density reduced compression indexes. Thus, the addition of cement was effective in inhibiting soil consolidation settlement. Figure 19b indicates similar results for the swelling index C_s. In 2015, Zeng et al. [50] proposed that clay's initial moisture content and LL considerably affect remolded clay specimens' compression indexes. The present study used specimens with low moisture content and a LL of 26.8%; thus, the C_c value obtained in this study was not high. After cement paste was added, the compression index further decreased.

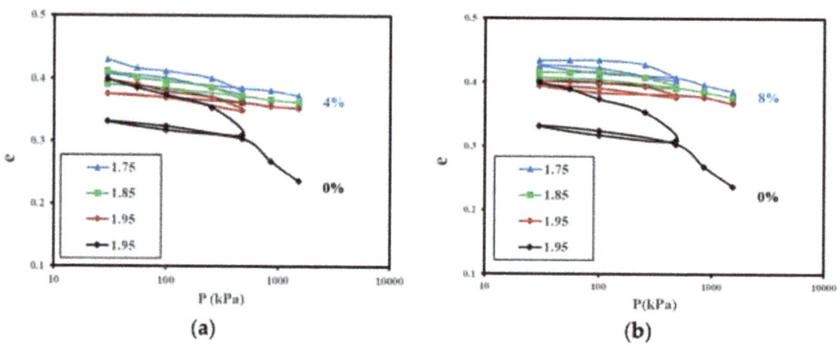

Figure 18. Compression curves of untreated and treated samples based on a soil consolidated test: (**a**) mudstone and mudstone–cement with C = 4%; (**b**) mudstone and untreated mudstone–cement with C = 8%.

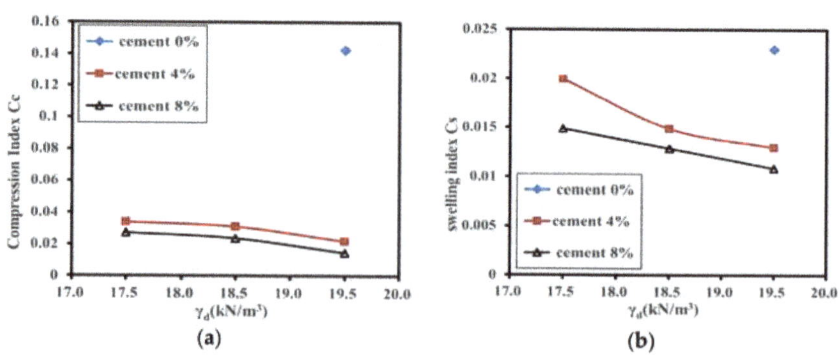

Figure 19. A comparison of compression index C_c and swelling index C_s versus testing soils dry density levels: (**a**) C_c versus γ_d; (**b**) C_s versus γ_d.

3.8. CBR Test

The CBR test was based on ASTM D1883 and D4429 and the indoor test was based on ASTM D1883. Each group consisted of specimens that were compacted 10 times, 25 times, and 56 times. The relationships between CBR and dry density were obtained. The CBR corresponding to the 95% γ_{dmax} of the compaction test was the true CBR. The test specimens were divided into 0%, 4%, and 8% added cement. Figure 20a presents the CBRs corresponding to a 95% maximum dry density. The CBRs of specimens with 0%, 4%, and 8% added cement were 15.6%, 93%, and 123%, respectively. The graph presents the relationship between dry density and CBR for three compaction patterns. When the number of compactions and cement content increased, the CBRs also increased. Thus, a concurrent increase in compaction energy and added cement led to a considerable increase in soil bearing capacity. Figure 20b indicates that when the number of compactions and amount of added cement increased, the swelling ratios decreased. In 1994, Lee et al. [3] suggested that that swelling of mudstone in Taiwan is related to clay minerals; specifically, the montmorillonite content of mudstone in Renwu is low.

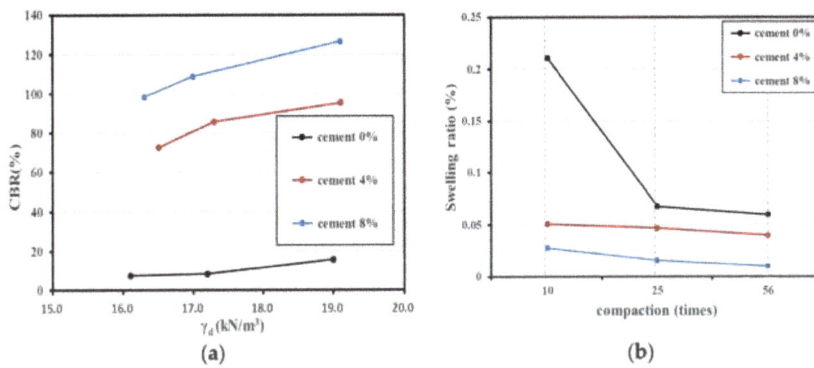

Figure 20. CBR values and swelling ratio for treated soils: (**a**) CBR versus γ_d; (**b**) Swelling ratio versus compaction (times).

4. Conclusions

The main objectives of the present study were to explore the multiple mechanical and engineering properties of mudstone improved with the increasing density or addition of 4% and 8% dry cement. Thus, mudstone clods with three grain sizes (grains that passed through #4, #4-3/4″, or >3/4″ sieves) were examined. Three methods (compaction, kneading, and static) and six dry density levels were used to prepare specimens. Two cement content levels were also applied. The following conclusions are drawn.

1. For the compaction tests, specimen moisture content was controlled and unconfined compression strength increased with γ_d. The compaction method produced specimens with the highest strength. In contrast, the clods with grain sizes larger than 3/4″ exhibited the lowest strength. The clod effect was similar with the results of the triaxial permeability test. The hydraulic gradient k values of mudstone clods ranged between 10^{-4} and 10^{-7} cm/sec, indicating low permeability or no permeability and that the hydraulic gradient k values of mudstone increased in relation to clods.
2. The mudstone specimen formula yielded $E_{50} = 47.699\ q_u$ and $R^2 = 0.8708$, while the mudstone–cement formula yielded $E_{50} = 69.397\ q_u$ and $R^2 = 0.9693$. By comparing our study with those that used formulas proposed by other researchers, we found that the differences between coefficients could be attributed to the material added and specimen preparation methods.
3. The triaxial CU and UU test results were used to calculate shear strengths through the application of the Mohr-Coulomb formula. For mudstone supplemented with

cement paste, its q_u values increased 7- to 10-fold and shear strength values increased 2- to 3-fold. Thus, the choice of the strength testing method is vital for geotechnical consideration.
4. The results from the one-dimension compression test indicate that the addition of cement for soil improvement is effective in inhibiting consolidation settlement. In addition, an increase in the number of compactions and amount of cement added led to a decrease in the swelling ratio according to the CBR test results. Therefore, an increase in compactions and cement addition are good ways to improve the property of mudstone.
5. Taken together, our study demonstrates the application of engineering techniques and the addition of small amounts of cement. This method not only adheres to low-carbon and environmental protection requirements, but also verifies the feasibility of using compacted mudstone as an engineering material.

Author Contributions: Project administration, D.H.; Experimental and analysis, D.H.; Supervision, D.H.; Writing—original draft preparation and editing, D.H; Validation; C.H.; Writing—review, C.H. All authors have read and agreed to the published version of the manuscript.

Funding: Not applicable.

Institutional Review Board Statement: Not applicable.

Informed Consent Statement: Not applicable.

Data Availability Statement: All data is available upon request.

Acknowledgments: CTCI Co. in Taiwan supplied materials to support the experiment's requirements throughout. All experimental works were performed in the geotechnical laboratory in NKUST, whose assistance is appreciated. The copy editing assistance provided by Hsiang-Jung Hsiao is also appreciated.

Conflicts of Interest: The authors declare no conflict of interest.

References

1. Goodman, R.E. *Engineering Geology-Rock in Engineering Construction*; Wiley: New York, NY, USA, 2002.
2. Wang, T.T.; Huang, T.H. An experience of tunnelling in mudstone area in southwestern Taiwan. *Tunn. Undergr. Space Technol.* **1993**, *17*, 425–436. [CrossRef]
3. Lee, D.H.; Jhin, Y.Y.; Tien, K.G. Characteristics of mudstone and the methods for slope protection. *Sino-geotechnics* **1994**, *48*, 35–47.
4. Lee, D.H.; Tien, K.G.; Juang, C.H. Full-scale field experimentation of a new technique for protecting mudstone slopes, Taiwan. *Eng. Geol.* **1996**, *42*, 51–63. [CrossRef]
5. Lee, D.H.; Yang, Y.E.; Lin, H.M. Assessing slope methods for weak rock slopes in Southwestern Taiwan. *Eng. Geol.* **2007**, *91*, 100–116. [CrossRef]
6. Higuchi, K.; Chigira, M.; Lee, D.H.; Wu, J.H. Pore-water chemistry and its influence on rock mechanical properties and hydrogeophysical processes in a mudstone slope in the southwestern Taiwan badlands. *Catena* **2020**, *190*, 104533. [CrossRef]
7. Zhao, C.; Zhao, D. Application of construction waste in the reinforcement of soft soil foundation in coastal cities. *Environ. Technol. Innov.* **2021**, *21*, 101195. [CrossRef]
8. Arulyajah, A.; Nikraz, H.; Bo, M.W. Assessment of marine clay improvement under reclamation fills by in-situ testing methods. *Geotech. Geol. Eng.* **2006**, *24*, 219–226. [CrossRef]
9. Bayesteh, H.; Sabermahani, M. "Field study on performance of jet grouting in low water content clay. *Eng. Geol.* **2020**, *264*, 105314. [CrossRef]
10. Wang, M.; Xu, X.; Li, J.; Shen, F.; Li, Y. An experiment study on stress relaxation of unsaturated lime-treated expansive clay. *Environ. Earth Sci.* **2017**, *76*, 241. [CrossRef]
11. Zhang, Y.S.; Guo, C.B.; Yao, X.; Qu, Y.G.; Zhou, N.J. Engineering geological characterization of clayey diatomaceous earth deposits encountered in highway projects in the Tengchong region, Yunnan, China. *Eng. Geol.* **2013**, *167*, 95–104. [CrossRef]
12. Chan, C.M. Influence of mix uniformity on the induced solidification of dredged marine clay. *Environ. Earth Sci.* **2014**, *71*, 1061–1071. [CrossRef]
13. Li, J.S.; Xue, Q.; Wang, P.; Liu, L. Influence of leachate pollution on mechanical properties of compacted clay: A case study on behaviors and mechanisms. *Eng. Geol.* **2013**, *167*, 128–133. [CrossRef]
14. Hou, J.; Li, H.; Liu, L. An experimental study on microstructure of leachate-polluted stabilized clay. *Environ. Earth Sci.* **2018**, *77*, 636. [CrossRef]

15. Miqueleiz, L.; Ramirez, F.; Seco, A.; Nidzam, R.M.; Kinuthia, J.M.; Abu Tair, A.; Garcia, R. The use of stabilised Spanish clay soil for sustainable construction materials. *Eng. Geol.* **2012**, *133–134*, 9–15. [CrossRef]
16. Consoli, N.C.; Foppa, D.; Festugato, L.; Heineck, K.S. Key parameters for strength control of artificially cemented soils. *J. Geotech. Geoenviron. Eng. ASCE* **2007**, *1332*, 197–205. [CrossRef]
17. Edil, T.B.; Acosta, M.; Benson, C.H. Stabilizing soft fine-grained soils with fly ash. *J. Mater. Civ. Eng.* **2006**, *182*, 283–294. [CrossRef]
18. Hsiao, D.H.; Phan, T.A.V. Experimental investigation into compressive strength of soft soil stabilization under curing stress conditions. *Electron. J. Geotech. Eng.* **2014**, *19*, 10429–10446.
19. Jamsawang, P.; Poorahong, H.; Yoobanpot, N.; Songpiriyakij, S.; Jongpradist, P. Improvement of soft clay with cement and bagasse ash waste. *Constr. Build. Mater.* **2017**, *154*, 61–71. [CrossRef]
20. Liu, L.; Zhou, A.; Deng, Y.; Cui, Y.; Yu, Z.; Yu, C. Strength performance of cement/slag-based stabilized soft clays. *Constr. Build. Mater.* **2019**, *211*, 909–918. [CrossRef]
21. Lorenzo, G.A.; Bergado, D.T. Fundamental parameters of cement-admixed clay—New approach. *J. Geotech. Geoenviron. Eng. ASCE* **2004**, *13010*, 1042–1050. [CrossRef]
22. Preetham, H.K.; Nayak, S. Geotechnical investigations on marine clay stabilized using granulated blast furnace slag and cement. *Int. J. Geosynth. Ground Eng.* **2019**, *5*, 28. [CrossRef]
23. Anagnostopoulos, C.A.; Papaliangas, T.; Manolopoulou, S.; Dimopoulos, T. Physical and mechanical properties of chemically grouted sand. *Tunn. Undergr. Space Technol.* **2011**, *26*, 718–724. [CrossRef]
24. Chen, L.; Lin, D.F. Stabilization treatment of soft subgrade soil by sewage sludge ash and cement. *J. Hazard. Mater.* **2009**, *162*, 321–327. [CrossRef]
25. Wu, J.; Deng, Y.; Zheng, X.; Cui, Y.; Zhao, Z.; Chen, Y.; Zha, F. Hydraulic conductivity and strength of foamed cement-stabilized marine clay. *Constr. Build. Mater.* **2019**, *222*, 688–698. [CrossRef]
26. Wu, J.; Liu, L.; Deng, Y.F.; Zhang, G.P.; Zhou, A.; Wang, Q. Distinguishing the effects of cementation versus density on the mechanical behavior of cement-based stabilized clays. *Constr. Build. Mater.* **2021**, *271*, 121571. [CrossRef]
27. Jongpradist, P.; Youwai, S.; Jaturapitakkul, C. Effective void ratio for assessing the mechanical properties of cement-clay admixtures at high water content. *J. Geotech. Geoenviron. Eng. ASCE* **2011**, *1376*, 621–627. [CrossRef]
28. Ge, L.; Wang, C.C.; Hung, C.W.; Liao, W.C.; Zhao, H. Assessment of strength development of slag cement stabilized kaolinite. *Constr. Build. Mater.* **2018**, *184*, 492–501. [CrossRef]
29. Nayak, S.; Sarvade, P.G. Effect of cement and quarry dust on shear strength and hydraulic characteristics of Lithomargic clay. *Geotech. Geol. Eng.* **2012**, *30*, 419–430. [CrossRef]
30. Lin, D.F.; Luo, H.L.; Hsiao, D.H.; Chen, C.T. Enhancing soft subgrade soil with a sewage sludge ash/cement mixture and nano-silicon dioxide. *Environ. Earth Sci.* **2016**, *75*, 619. [CrossRef]
31. Benson, C.H.; Daniel, D.E. The influence of clods on the hydraulic conductivity of a compacted clay Geotech. *J. Geotech. Eng.* **1989**, *1168*, 1231–1248. [CrossRef]
32. Monkul, M.M.; Etminan, E.; Şenol, A. Influence of coefficient of uniformity and base sand gradation on static liquefaction of loose sands with silt. *Soil Dyn. Earthq. Eng.* **2016**, *89*, 185–197. [CrossRef]
33. Daniel, D.E.; Benson, C.H. Water content-density criteria for compacted soil liners. *J. Geotech. Eng. ASCE* **1990**, *11612*, 1811–1830. [CrossRef]
34. Das, B.M.; Sobhan, K. *Principles of Soil Mechanics and Foundation Engineering*; Cengage Learning: Singapore, 2017.
35. Holtz, R.D.; Kovacs, W.D.; Sheaahan, T.C. *An Introduction to Geotechnical Engineering*, 2nd ed.; Pearson: Upper Saddle River, NJ, USA, 2011.
36. Yoobanpot, N.; Jamsawang, P.; Horpibulsuk, S. Strength behavior and microstructural characteristics of soft clay stabilized with cement kiln dust and fly ash residue. *Appl. Clay Sci.* **2017**, *141*, 146–156. [CrossRef]
37. Chen, Y.G.; Zhua, C.M.; Ye, W.M.; Cui, Y.J.; Chen, B. 2016 Effects of solution concentration and vertical stress on the swelling behavior of compacted GMZ01 bentonite. *Appl. Clay Sci.* **2017**, *124–125*, 11–20. [CrossRef]
38. Emmanuel, E.; Anggraini, V. Effects of desiccation-induced cracking and leachate infiltration on the hydraulic conductivity of natural and olivine-treated marine clay. *Int. J. Environ. Sci. Technol.* **2020**, *17*, 2259–2278. [CrossRef]
39. Kalkan, E. Influence of silica fume on the desiccation cracks of compacted clayey soils. *Appl. Clay Sci.* **2009**, *43*, 296–302. [CrossRef]
40. Sheu, C.; Lin, T.T.; Chang, J.E.; Cheng, C.H. The feasibility of mudstone material as a natural landfill liner. *J. Hazard. Mater.* **1998**, *58*, 237–247. [CrossRef]
41. Hsiao, D.H.; Lin, Y.C. A Study of Mudstone in Taiwan as a Landfill Material. In Proceedings of the 2004 Taiwan Rock Engineering Symposium, Tamsui, Taiwan, 21–22 October 2004; pp. 678–685.
42. Ray, H.M.T.; Yanful, E.K.; Fakher, A. Desiccation-induced cracking and its effect on the hydraulic conductivity of clayey soils from Iran. *Can. Geotech. J.* **2007**, *44*, 276–283. [CrossRef]
43. Horpibulsuk, S.; Miura, N.; Nagaraj, T.S. Clay–water/cement ratio identity for cement admixed soft clays. *J. Geotech. Geoenviron. Eng. ASCE* **2005**, *1312*, 187–192. [CrossRef]
44. Horpibulsuk, S.; Suddeepong, A.; Chinkulkijniwat, A.; Liu, M.D. Strength and compressibility of lightweight cemented clays. *Appl. Clay Sci.* **2012**, *69*, 11–21. [CrossRef]

45. ASTM. *Standard Practice for Classification of Soils for Engineering Purposes;* American Society for Testing and Materials: Conshohocken, PA, USA. Available online: https://www.globalspec.com/supplier/profile/ASTMInternational (accessed on 19 March 2019).
46. Takahashi, H.; Morikawa, Y.; Fujii, N.; Kitazume, M. Thirty-seven-year investigation of quicklime-treated soil produced by deep mixing method. *Ground Improv.* **2018**, *1713*, 135–147. [CrossRef]
47. Chen, M.; Wena, P.; Wang, C.; Chai, Z.; Gao, Z. Evaluation of particle size distribution and mechanical properties of mineral waste slag as filling material. *Constr. Build. Mater.* **2020**, *253*, 119183. [CrossRef]
48. Das, B.M. *Principles of Geotechnical Engineering, SI Edition*, 7th ed.; Cengage Learning: Stamford, USA, 2010.
49. Fredlund, D.G.; Rahardjo, H. *Soil Mechanics for Unsaturated Soils;* John Wiley & Sons INC.: Hoboken, NJ, USA, 1993.
50. Zeng, L.L.; Hong, Z.S.; Cui, Y.J. Determining the virgin compression lines of remolded clays at different initial water contents. *Can. Geotech. J.* **2015**, *52*, 1408–1415. [CrossRef]

applied sciences

MDPI

Article

Probabilistic Analysis as a Method for Ground Freezing Depth Estimation

Tomasz Godlewski [1,*], Łukasz Wodzyński [2] and Małgorzata Wszędyrówny-Nast [1]

1 Building Research Institute, 02-656 Warsaw, Poland; m.wszedyrowny@itb.pl
2 Institute of Civil Engineering, Warsaw University of Life Science, 02-776 Warsaw, Poland; lukasz_wodzynski@sggw.pl
* Correspondence: t.godlewski@itb.pl; Tel.: +48-22-56-64-163

Abstract: Accurate frost depth prediction is an important aspect in different engineering designs such as for pavements, buildings, bridge foundations, and utility lines. This paper presents a probabilistic method of assessment of the depth of soil freezing. Annual (winter) maxima of the position of the zero centigrade temperature measured in the soil were approximated by Gumbel probability distribution. Its parameters were estimated using maximum likelihood method. The results received on the basis of data from 36 meteorological stations in Poland and 50 years of observations, as characteristic values with 50-year return period, reflect the influence of the climatic conditions on the freezing depth. On the other hand, the soil structure and its conditions also play an important role in freezing. Nowadays they may be taken into account using correction coefficients. It is concluded that this method is more precise than a method using the air freezing index because through the use of direct measurements it takes into account additional factors affecting the actual depth of freezing. The obtained results are not the same as those given in the older Polish Standard which was based on the simplified and limited data. The results confirm the impact of climate change on ground freezing depth.

Keywords: soil freezing; probability; return period; reliability; correction coefficient; depth of ground freezing

Citation: Godlewski, T.; Wodzyński, Ł.; Wszędyrówny-Nast, M. Probabilistic Analysis as a Method for Ground Freezing Depth Estimation. *Appl. Sci.* **2021**, *11*, 8194. https://doi.org/10.3390/app11178194

Academic Editor: Daniel Dias

Received: 26 July 2021
Accepted: 30 August 2021
Published: 3 September 2021

1. Introduction

Freezing of the soil in wintertime is one of the climatic actions which must be considered when designing structures founded on soils. The actual frost depth is affected by the material type, soil thermal properties, soil water content, and climatic conditions such as temperature, wind speed, precipitation, and solar radiation. Frost depth can be estimated using numerical or analytical modeling technique (Gontaszewska [1], Konrad & Shen [2], Rajaei & Baladi [3], Venäläinen et al. [4],). In the past few years different numerical techniques have been used for modeling transient heat flow in pavement layers. These methods require a large number of input data [4,5]. In most cases the input data are either not available or just too expensive to collect. In the case of missing data, higher or similar accuracy can be obtained from simpler analytical and semi-empirical models to estimate frost depth prediction—for e.g., Stefan model, modified Berggren model or Chisholm and Phang empirical model [3,5]. In those methods, the depth of frost penetration has usually been defined using the so-called air freezing index. However, it is obvious that the frost penetration into the soil is a complex random process in which many factors affect the final results, which are often divergent from the observations [6]. Those are not only air temperature but also the depth and the thermal properties of snow cover, which is changing during winter, heat contained in the soil [7,8] and the soil properties, its structure and water content. All the climatic conditions have a random character and should be analyzed using probabilistic methods. Nowadays, the simplest approach is the probabilistic analysis of the annual (winter) maximum depth of the zero centigrade temperature. However, it should be emphasized that it is not the same as the depth of the soil freezing. The purpose of the paper is to present the probabilistic analysis of the maximum annual depth of the zero centigrade temperature of the soils, the implemented method and some examples of the obtained results.

Since 1955, the maps of the soil freezing depths have been given in Polish Standards. The values were calculated in centimeters according to the formula given in Soviet recommendations [9]:

$$h_z = 23 \cdot \sqrt{\sum m + 2} \tag{1}$$

where: m—an absolute value of the winter monthly mean negative air temperature; only negative values are summarized.

The numerical factor 23 means the type of soil. The value 23 is for loams and clays. Numbers for other types of soil were not given in the Polish Standards. According to Zavarina [10], this formula was derived on the basis of the experiments made at six Soviet meteorological stations. A similar formula but with more detail was elaborated at the US Army Cold Regions Research and Engineering Laboratory [6].

The results of the calculations according to the Formula (1) were implemented into the Polish Standard in 1955 (PN-55/B-03020) and its next editions with small modification in 1974. For the western half of Poland, the depth of freezing was 0.8 m, for the eastern part it was 1.0 m and for small area around Suwałki region it was 1.4 m.

The approach, based on the air freezing index, is still recommended [1]. However, now it is possible to define the depth of the zero centigrade temperature on the basis of direct measurements of the soil temperature.

2. Material and Methodology

2.1. Measurements of Soil Temperature

The Earth's surface is heated by the absorption of solar radiation. The variety of the substrate results in different heating mechanism. Snow and ice covering the ground reflects most of the radiation [11]. Water, due to its partial transparency, absorbs radiation throughout the layer to which it reaches, and the ground absorbs radiation in a thin surface layer [12]. Changes in ground temperature depend on its vegetation coverage, chemical composition, and to a large extent on water and air content. The source of heat in the ground can also be heat exchange with the atmosphere; thawing and freezing processes of water contained in the ground; precipitation heat and biological, chemical, and physical processes in the soil [13]. Soil, which is not covered with vegetation, is most strongly heated during the day in clear weather; at night in the same conditions, it intensely radiates heat. The size of the heat flux absorbed or given away by the soil depends on the temperature difference and the thermal capacity of the soil [14,15]. The thermal capacity of the soil is influenced by its mineral composition, water and air content. There is a very large impact of the snow cover on the depth of the zero isotherm position. The maximum positions of zero isotherms in the ground without snow cover are even more than twice as deep as in snow-covered soils [16,17].

Soil freezing (frost) depth can be determined using a number of direct approaches, for example, by direct measurement by thermometers installed in the ground. Another approach is to use the concept of 'degree days'—freezing degree days (FDDs) and thawing degree days (TDDs)—as a surrogate for frost depth. The advantage of the 'degree day' approach is that meteorological data is easier to obtain and less intrusive than the other described approaches. A limitation of the degree-day approach is the representativeness of temperature data taken off-site [18].

In Poland, the soil temperature is measured every day at the meteorological stations of the Institute for Meteorology and Water Management—State Research Institute (IMGW-PIB). The measurements were taken in the last 40 years using mercury-in-glass thermometers, at 6:00 and 18:00 UTC at the depths of 5 cm, 10 cm, 20 cm, and 50 cm and at the depth of 1.0 m but only at 12:00 UTC [19]. Measurements at this last depth have been made only since 1982. Currently, the measurements are automated, performed every minute with an accuracy of 0.1 °C. Measurements of the minimum temperature above the surface of the exposed soil and the ground temperature are made on the testing site. This measuring field is a part of the ground with area of 2 × 4 m; the longer sides of the plot are directed in the direction E-W. Experimental plot are free of the grass. The measurement technique for the soil temperature and the method of calculation of the zero centigrade depth were described in the instruction for meteorological measurements [20,21].

Freezing the ground is a physical process consisting of the freezing of water present in the ground and its molecules into a hard permafrost. The water contained in the soil is a solution of different salts of varying concentrations, the higher their concentration, the lower the ground freezing temperature. This temperature may drop to −2 °C and below. Freezing of the ground depends on the air temperature, the duration of the negative temperatures, the thickness and type of snow cover, the terrain profile, and the structure and humidity of the ground. The low air temperature that persists for a long time lowers the temperature of deeper and deeper layers of the ground, while strong but short-term temperature drops have less impact on the freezing depth of the ground. An important role is played by the snow cover: a thicker, less dense and even layer of snow inhibits the process of freezing the ground. Different types of soils are characterized by different ease of freezing. The process of ground freezing consists of solar radiation, the advection of warm air masses, and heat penetrating from deeper layers of soil. The timing of freezing of the first layers of soil, the length of permafrost persistence and the maximum freezing depth are of great importance in construction, agriculture, and water management. At weather stations measuring ground temperature at depths of 5, 10, 20, 50, and 100 cm, isotherm "0" in the ground is calculated. Isotherm (E_T) "0" is the plane connecting the points at 0 °C in the ground [19]. Depending on the course of the weather, there may be one or more "0" isotherms in the ground. On the basis of measured everyday values of soil temperature personnel of weather stations calculate the position of the zero centigrade temperature. Following formula is used:

$$Z = A + \frac{a \cdot \Delta T_A}{\Delta T} \tag{2}$$

where: A—the depth of the lowest place of thermometer which measured negative temperature; a—the distance between depths of thermometers measuring negative and positive temperature; ΔT_A—temperature at the level A; ΔT—the difference between temperature values at the levels with zero centigrade temperature between them.

An example is presented in Figure 1 below: A = 0.2 m, a = 0.5 − 0.2 = 0.3 m, ΔT_A = −0.4 °C, ΔT = −0.8 °C. Zero-degree temperature is at the depth of 0.35 m.

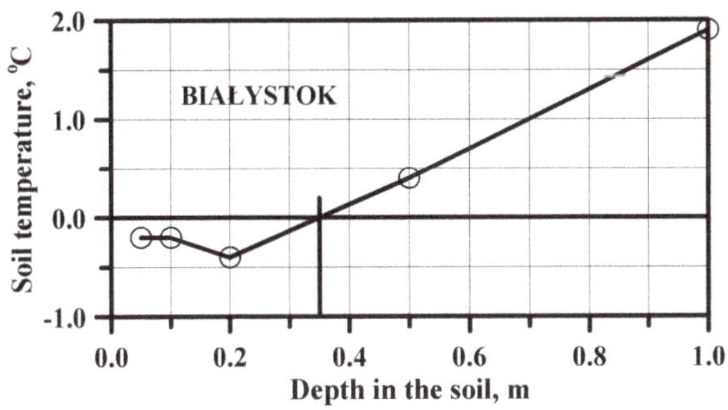

Figure 1. Example of temperature of the soil at the meteorological station in Białystok for one day (15 March 2005) adapted from [19].

The problem may occur when the negative soil temperature exceeds 1.0 m depth. In such a case it is recorded as below 1.0 m. The true value can be defined using extrapolation of values measured at smaller depths. An example is presented in Figure 2. However, it must be pointed out that such cases were not frequent.

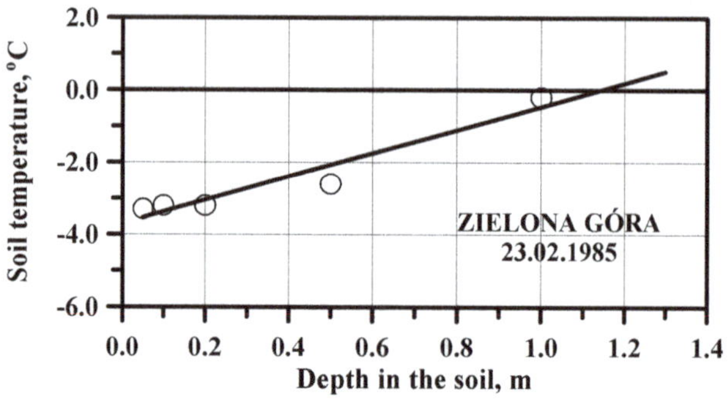

Figure 2. Example of extrapolation of the data on the soil temperature made in order to define maximal winter value adapted from [19].

When determining the position of a zero isotherm by extrapolation, it should be taken into account that as the temperature rises in the upper layers of the ground at an unchanged value at the depth of the largest measurement, the slope of the straight regression decreases. The use of such a set may lead to an error in the determination of the extrapolated value.

2.2. Methods of Probabilistic Analysis

The aim of the analysis is to find such values of the position of zero centigrade temperature which would be exceeded with the accepted probability in appropriate reference period. Nowadays, the commonly accepted annual probability of exceedance of climatic actions is 0.02 which means 50-year return period. Such an assumption may also be used for the probabilistic calculation of freezing depths. The probability that a value Z_k will be exceeded once in t years is:

$$P(Z > Z_k) = \frac{1}{t} \tag{3}$$

and the probability of not exceeding it is:

$$P(Z \leq Z_k) = 1 - \frac{1}{t} = F(Z_k) \tag{4}$$

In order to find Z_k one has to use the annual (winter) maxima of the position of the zero centigrade temperature and the appropriate probability distribution. The Gumbel distribution [22] has been preferred here. The Gumbel extreme—value distribution was used to analyze the data on locations of the zero isotherm. The purpose of this study was to fit an extreme-value distribution to a given set of winter maxima of soil freezing depth denoted by Z_i. The cumulative distribution function (CDF) is given as:

$$F(Z) = \exp\{-\exp[-\alpha(Z - U)]\} \tag{5}$$

where: α and U are parameters of the distribution estimated using the maximum likelihood method.

This distribution and procedure have been widely used to forecast wind velocity and snow load, e.g., [23]. It was also used to forecast soil freezing [24]. After being doubly logarithmized, it takes the form of a linear function with respect to α, U. The value Z_k of the depth of zero centigrade temperature and 50-year return period is calculated using the following equation:

$$Z = U - \frac{1}{\alpha}\ln[-\ln F(Z)] \tag{6}$$

The unknown Gumbel distribution parameters (α, U) were estimated using the least squares method, the maximum likelihood method, the method of moments, and the method described by Julius Lieblein in his work [25].

2.3. Estimation of Gumbel Distribution Parameters

In order to determine the parameters, the measurement results are depicted on the Gumbel probability grid depending on the position each observation occupies in the statistics order. The measurement results are treated as the values of the function which arguments have the form, where the ordinal number is in the ascending order of the n-element measurement data set [26].

2.3.1. Least Square Estimation

This method is relatively easy to use as double logarithm of Gumbel distribution results in the equation of a straight line. The parameters of the regression function are approximations of the Gumbel distribution parameters. Good approximation is obtained when noticeable linear relationship is present. The estimation method of Gumbel distribution parameters based on the least square method gives the best results as the coefficient of correlation $R \geq 0.95$. However, it is very sensitive to the cases when some points are located far outside the straight line. Such cases are often occurring, for example, in snow load on the ground. Such points are increasing the steepness of the regression line, which leads to the overestimation of the predicted values. In extreme cases, the linear regression do not fit the approximated empirical distribution at all, which then disqualifies this method.

The straight line was fitted to the points placed on the probability grid so that it runs as close as possible to all points, i.e., the sum of the squares of vertical distance of points from this straight line is as small as possible. The inverse of the slope coefficient of this line and its Z-intercept are the parameters of the Gumbel distribution—being the results of the least squares estimation (LSM).

2.3.2. Maximum Likelihood Method

Maximum likelihood estimation is recommended due to its many advantages. In this method, the assessment of the estimated parameters is based on values maximizing the credibility of the sample. Due to the lack of analytical solution for assessment of the distribution parameters, numerical methods are used. Estimators obtained with this method are consistent, asymptotically most efficient and they have normal asymptotical distribution. In practice, this method is rarely used due to calculation complexity, as the values of parameter distribution are assessed iteratively. The implementation of the maximum likelihood method (Max LM) consisted of seeking the maximum of the likelihood function given by the formula:

$$l = \prod_{i=1}^{n} \alpha \cdot \exp\{-\alpha(Z_i - U) - \exp[-\alpha(Z_i - U)]\} \quad (7)$$

The values α, U for which the Function (7) reaches its global maximum are the parameters of the Gumbel distribution estimated by the maximum likelihood method.

2.3.3. Method of Moments

The moment method is simpler. In order to determine the parameters (α, U) using the method of moments, certain sample moments referring to measured values Z_i were compared to the corresponding moments of the Gumbel distribution [27,28]. The arithmetic mean Z_m and variance σ^2 as properties of the Gumbel distribution are expressed by the following formulas:

$$Z_m = U + \frac{1}{\alpha} \cdot \gamma, \ \sigma^2 = \frac{\pi^2}{6 \cdot \alpha^2} \quad (8)$$

where: $\gamma = 0.5772$ is the Euler constant.

Comparing the Formulas (8) to corresponding quantities related to the n-element sample of winter maxima of freezing depth, we obtain the values α, U of the Gumbel distribution parameters estimated by the method of moments (MM).

2.3.4. Lieblein Method—BLUE

Julius Lieblein [25] proposed a method of estimating Gumbel distribution parameters based on order statistics, requiring the sample elements Z_i to be placed in ascending order. The estimators of the investigated distribution parameters take the form of a linear function with respect to positional statistics:

$$\hat{b}_{(n)} = \frac{1}{\hat{\alpha}_{(n)}} = \sum_{i=1}^{n} b_i Z_i,\ \hat{U} = \sum_{i=1}^{n} a_i Z_i \tag{9}$$

where: Best Linear Unbiased Estimators (BLUE) are with minimal variance. The coefficients b_i, a_i were collated in the paper [25] for sample size of $n = 1, 2, \ldots, 16$. The aforementioned work also describes a method of estimating parameters b, U for arbitrarily large sample with a size of n having given coefficients for an m-element sample if $m < n$. For an n-element sample, the coefficients b'_i, a'_i can be calculated using the following formulas:

$$\begin{gathered} b'_i = \sum_{j=1}^{m} b_j \cdot (j/i) p(n,m,i,j), \\ i = 1,2,\ldots,n \\ a'_i = \sum_{j=1}^{m} a_j \cdot (j/i) p(n,m,i,j), \end{gathered} \tag{10}$$

where:

$$p(n,m,i,j) = \binom{i}{j}\binom{n-i}{m-j} \Big/ \binom{n}{m} \tag{11}$$

is the hypergeometric probability function. The formulas below allow the calculation of "good" estimators $\hat{b}'_{(n)}$, $\hat{U}'_{(n)}$ for an n-element sample:

$$\hat{b}'_{(n)} = \frac{1}{\hat{\alpha}'_{(n)}} = \sum_{i=1}^{n} b'_i \cdot Z_i,\ \hat{U}'_{(n)} = \sum_{i=1}^{n} a'_i \cdot Z_i \tag{12}$$

The samples analyzed in this work contain significantly more than 16 measurement results; therefore, a set of BLUE coefficients b_j, a_j for $m = 16$ was used in the calculations.

The evaluation of the results obtained with four methods proved to be more difficult than the estimation of the Gumbel distribution parameters. As we have four Gumbel distributions, we face the question which of them best approximates the empirical distribution and can be used to determine the characteristic value Z_k of the location of the zero isotherm in the ground [26]. The test statistics of chi-squared, Kolmogorov–Smirnov and Cramer von Mises were adopted as a measure of adjustment of the theoretical to the empirical distribution. Low values of these statistics indicate that the given theoretical distribution truly reflects the empirical distribution.

Amid the known compliance tests, the chi squared test is mentioned first. It requires the decomposition of the Z axis into $r+1$ intervals $I_1, I_2, \cdots, I_{r+1}$ by the numbers $-\infty = g_0 < g_1 < \cdots < g_r < g_{r+1} = \infty$. The difference in the value of the cumulative distribution function F at the ends of the interval is equal to the probability p that the random variable Z will take a value belonging to that interval:

$$p_j = F(g_j) - F(g_{j-1}),\ j = 1,2,\ldots,r+1 \tag{13}$$

The chi squared test statistic for a measurement sequence consisting of n maximum freezing depths is given by the formula:

$$\chi^2 = \sum_{j=1}^{r+1} \frac{(n_j - np_j)^2}{np_j} \tag{14}$$

where: n_j—the number of measured values of Z within I_j, while np_j—is the expected number of measurement results which, according to the considered distribution, should be in I_j [29]. An examination of distribution compliance by means of statistics (14) leaves the choice of the number of intervals I_j into which the axis Z is divided and the points g_j constituting the limits of these intervals. In this work because of the ambiguity mentioned chi-squared test was implemented in two variants. The first one assumes the division of the Z variable axis into one-element intervals, the limits of which are arithmetic means of two values adjacent in the order statistics. The only exception here are points g_0, g_{r+1} equal to minus and plus infinity, respectively. In the second variant, the division into five intervals with equal probability p_j was used. Point g_1 was set as the arithmetic mean of the fifth and sixth elements in a series of ascending order data: $g_1 = (Z_5 + Z_6)/2$. The value of the cumulative distribution function at point g_1 is equal to the probability of the first interval: $F(g_1) = p$. The same probability of all intervals is ensured by the next boundary points g_j being defined as:

$$g_j = U - \frac{1}{\alpha}\ln\left[-\ln\left(p + F\left(g_{j-1}\right)\right)\right], \ldots j = 1, 2, \cdots, r \tag{15}$$

except the last: $g_{r+1} = \infty$.

Other tests refer to hypothetical cumulative distribution function and verify its compliance with empirical cumulative distribution function. Kolmogorov–Smirnov statistics D_{n1} verifies the quality of the adjustment on the basis of the largest absolute value of the hypothetical $F(Z)$ and empirical $F_n(Z)$ cumulative distribution function difference:

$$D_{n1} = \sup_{z}|F_n(Z) - F(Z)| \tag{16}$$

In Formula (16), unlike in [30], the $\sqrt{n}$ factor is omitted, because only equally numerous sets of values $F_n(Z)$ and $F(Z)$ are compared with each other and without any reference to samples of a different size n. Another variation of Kolmogorov–Smirnov statistics found in the literature was also used, namely:

$$D_n^+ = \max_{1 \le i \le n}\left|\frac{i}{n} - F(Z_{i:n})\right|,\ D_n^- = \max_{1 \le i \le n}\left|F(Z_{i:n}) - \frac{i-1}{n}\right| \tag{17}$$

from where:

$$D_{n2} = \max\{D_n^+,\ D_n^-\} \tag{18}$$

The above Formulas (16)–(18) define the maximum discrepancy between the postulated Gumbel distribution and the measured depths of the zero isotherm location. So, the idea arose to do a similar check for medium differences of hypothetical and geometrical cumulative distribution function. The subsequent tests are equivalent to statistics (16)–(18) and check not the maximum, but the average distance of the hypothetical and empirical cumulative distribution function. According to this reasoning, statistics (16) representing the average spacing of the cumulative distribution function is given by the formula:

$$D_{n1s} = \sum_{i=1}^{n}\frac{|F_n(Z_i) - F(Z_i)|}{n} \tag{19}$$

Whereas the equivalent of statistics (18) in relation to the average value is expressed by the formulas:

$$D_n^+ = \frac{1}{n}\sum_{i=1}^{n}\left|\frac{i}{n} - F(Z_{i:n})\right|,\ D_n^- = \frac{1}{n}\sum_{i=1}^{n}\left|F(Z_{i:n}) - \frac{i-1}{n}\right| \tag{20}$$

and next:

$$D_{n2s} = \frac{1}{2}\left(D_n^+ + D_n^-\right) \tag{21}$$

Another test used in this work, comparing the cumulative distribution function with order statistics of measurement results, is Cramer von Mises's statistics [31] in the form:

$$W_{CM} = \frac{1}{12n} + \sum_{i=1}^{n} \left(F(Z_i) - \frac{2i-1}{2n} \right)^2 \tag{22}$$

The best estimation method was considered to be the one indicated by most of the mentioned tests. For example, if the chi squared statistic is the smallest for the Gumbel distribution parameters α, U determined by the maximum likelihood method, it means that the chi squared test indicates the maximum likelihood method as the best of four applied methods.

3. Results

3.1. Estimation Models

From all recorded values of the position of zero centigrade temperature from measurements during every winter, one can extract the maximum depth and use in probabilistic calculations. Using the above mentioned Gumbel distribution, its parameters were estimated and the values of the predicted position of the zero isotherm in the ground were calculated (Figures 3–6). Calculations were made for data from years 1980–2011.

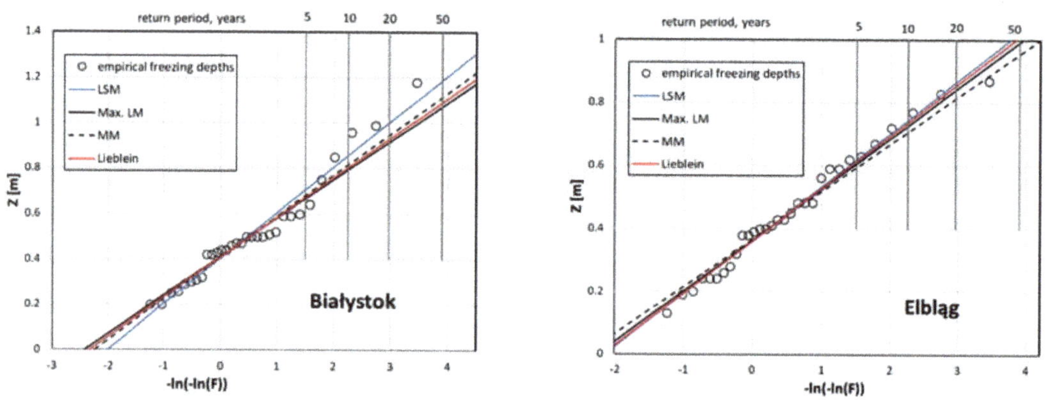

Figure 3. Estimation models for results from Białystok (**left**) and Elbląg (**right**) station—predicted position of zero isotherm for return period.

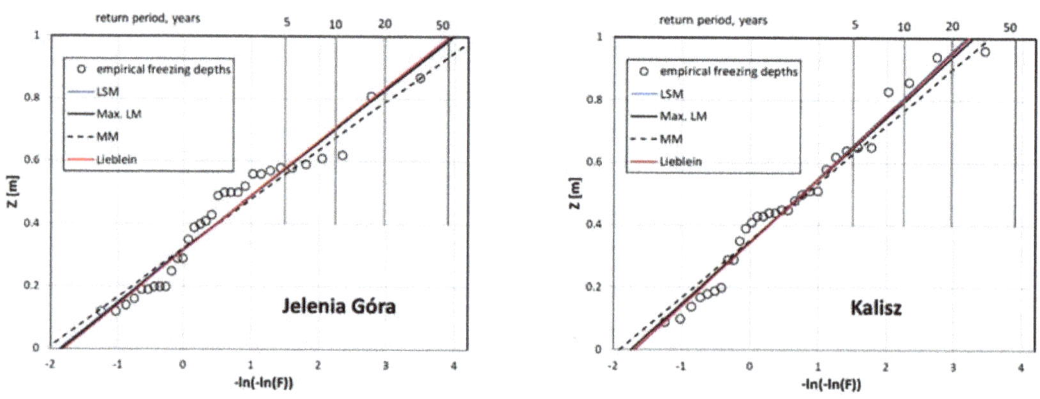

Figure 4. Estimation models for results from Jelenia Góra (**left**) and Kalisz (**right**) station—predicted position of zero isotherm for return period.

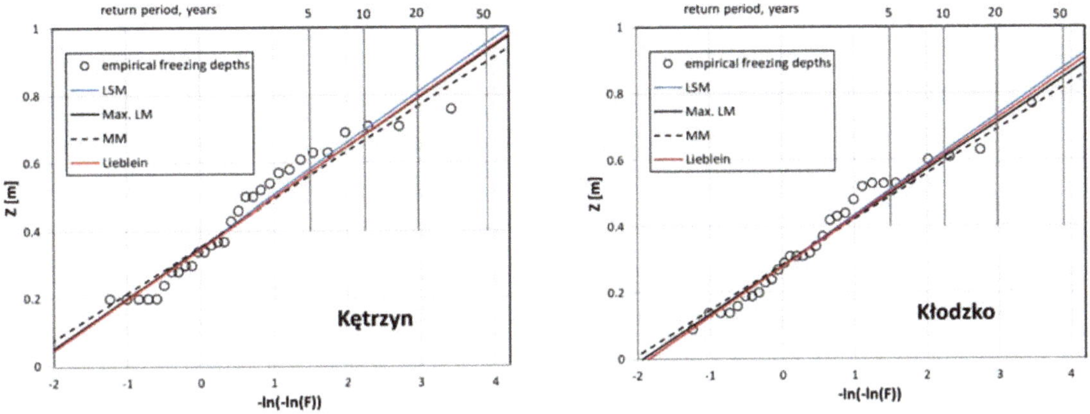

Figure 5. Estimation models for results from Kętrzyn (**left**) and Kłodzko (**right**) station—predicted position of zero isotherm for return period.

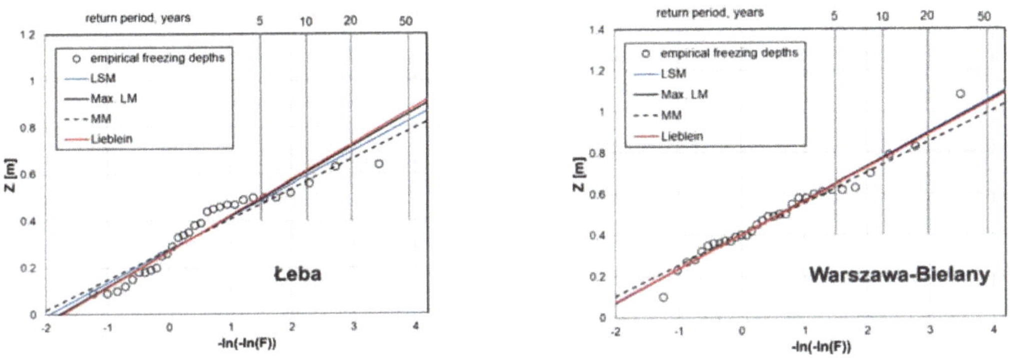

Figure 6. Estimation models for results from Łeba (**left**) and Warszawa-Bielany (**right**) station—predicted position of zero isotherm for return period.

This results in the location of the zero isotherm for the ground conditions of the 36 weather stations (first approach) where the measurements were made—Table 1.

3.2. Validation of Chosen Model of Estimation (Lieblein)

It was assumed that, as in the case of climatic interactions, the characteristic values of the zero isotherm position should have a return period of 50 years—as required reference level to ensure the reliability in design.

For the next step of calculation, only one model of Gumbel's method of analysis was used, with Lieblein's best linear unbiased estimators (BLUE). These estimators are recommended in European Standard for climate actions, e.g., derivation of wind speeds from measurements at meteorological stations. To assess the correctness of the predictions indicated in the first approach, a recalculation was performed for the enlarged dataset. The new data related to freezing depth measurements, obtained from 2011 to 2020 (black points on graphs), were added. On this basis, the previous forecasts have been compared with the values based on current observations (Figures 7–10). A new trendline has been set for the new data (1980–2020). With regard to the previous prediction, the change in the empirically determined freezing depth for the assumed return period (50 years) can thus be assessed.

Table 1. Results of calculation for all localizations.

No.	Localization	Gumbel Distribution Parameters for Chosen Estimation Models: LSM		Max. LM		MM		Lieblein		Correlation Coefficient R^2 for Lieblein
		α	u	α	u	α	u	α	U	
1.	Białystok	4.980	0.404	5.895	0.410	5.515	0.407	5.717	0.408	0.980
2.	Bielsko-Biała	8.958	0.212	10.479	0.214	10.019	0.214	10.133	0.213	0.984
3.	Chojnice	4.460	0.400	5.334	0.407	5.005	0.405	5.130	0.404	0.990
4.	Elbląg	5.902	0.363	6.182	0.362	6.617	0.366	5.999	0.360	0.992
5.	Gorzów	4.664	0.302	5.647	0.307	5.150	0.305	5.555	0.306	0.977
6.	Jelenia Góra	5.831	0.320	5.826	0.316	6.385	0.322	5.733	0.315	0.971
7.	Kalisz	4.864	0.347	5.014	0.346	5.419	0.351	4.894	0.343	0.985
8.	Katowice	7.291	0.201	8.759	0.204	8.199	0.204	8.584	0.203	0.986
9.	Kętrzyn	6.506	0.352	6.726	0.349	7.178	0.354	6.643	0.347	0.973
10.	Kielce	4.645	0.399	5.351	0.404	5.157	0.402	5.191	0.402	0.979
11.	Kłodzko	6.541	0.281	6.865	0.280	7.284	0.284	6.663	0.278	0.985
12.	Koło	5.973	0.336	5.809	0.334	6.658	0.339	5.687	0.332	0.983
13.	Koszalin	6.065	0.313	6.669	0.316	6.817	0.317	6.449	0.313	0.991
14.	Kraków-Balice	6.670	0.261	6.721	0.258	7.328	0.263	6.565	0.256	0.972
15.	Łeba	7.097	0.275	6.643	0.268	7.709	0.276	6.498	0.268	0.958
16.	Legnica	4.592	0.327	4.952	0.328	5.136	0.332	4.790	0.325	0.987
17.	Lesko	7.026	0.234	8.079	0.237	7.875	0.237	7.897	0.235	0.986
18.	Leszno	5.208	0.350	5.459	0.349	5.826	0.354	5.294	0.347	0.992
19.	Lublin	5.682	0.333	6.244	0.335	6.364	0.337	6.099	0.332	0.991
20.	Łódź	4.699	0.386	4.905	0.386	5.284	0.391	4.736	0.382	0.992
21.	Mikołajki	6.024	0.311	6.656	0.312	6.718	0.315	6.510	0.311	0.986
22.	Nowy Sącz	8.851	0.202	8.788	0.199	9.776	0.203	8.526	0.198	0.977
23.	Opole	5.006	0.324	5.689	0.327	5.607	0.328	5.475	0.325	0.988
24.	Resko	5.888	0.245	6.623	0.247	6.539	0.248	6.537	0.246	0.982
25.	Rzeszów	8.658	0.351	7.514	0.344	9.415	0.351	7.498	0.344	0.959
26.	Sandomierz	5.560	0.434	6.095	0.436	6.141	0.436	6.014	0.434	0.974
27.	Słubice	5.857	0.298	6.476	0.300	6.544	0.301	6.290	0.297	0.991
28.	Suwałki	6.276	0.442	6.343	0.441	6.969	0.445	6.236	0.439	0.982
29.	Świnoujście	5.232	0.433	5.322	0.432	5.860	0.437	5.205	0.429	0.988
30.	Szczecin	4.253	0.282	4.616	0.279	4.622	0.284	4.633	0.280	0.961
31.	Tarnów	6.189	0.322	6.494	0.321	6.884	0.325	6.323	0.319	0.984
32.	Toruń	4.528	0.446	5.095	0.450	5.079	0.450	4.978	0.447	0.989
33.	Warszawa	5.986	0.401	6.066	0.402	6.647	0.404	6.109	0.401	0.985
34.	Włodawa	3.726	0.507	4.476	0.515	4.188	0.513	4.334	0.512	0.992
35.	Zakopane	5.450	0.378	5.050	0.372	6.008	0.380	4.998	0.370	0.975
36.	Zielona Góra	3.553	0.421	3.589	0.416	3.915	0.425	3.516	0.413	0.972

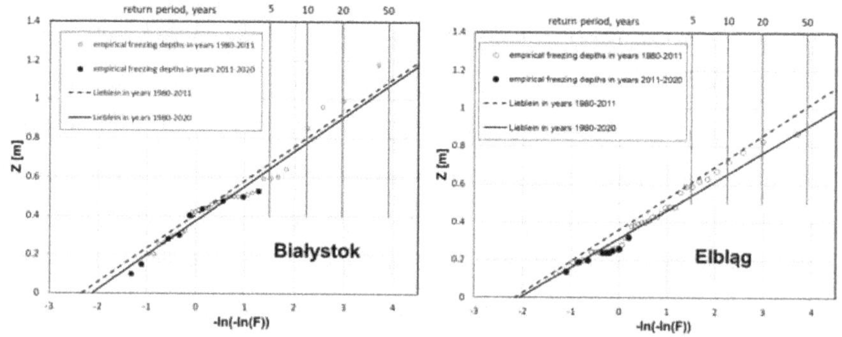

Figure 7. Comparison of predicted and measurement value for chosen location: Białystok (**left**) and Elbląg (**right**) station.

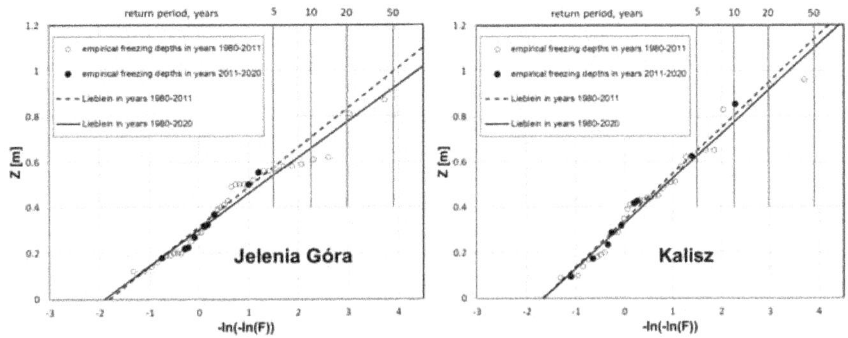

Figure 8. Comparison of predicted and measurement value for chosen location: Jelenia Góra (**left**) and Kalisz (**right**) station.

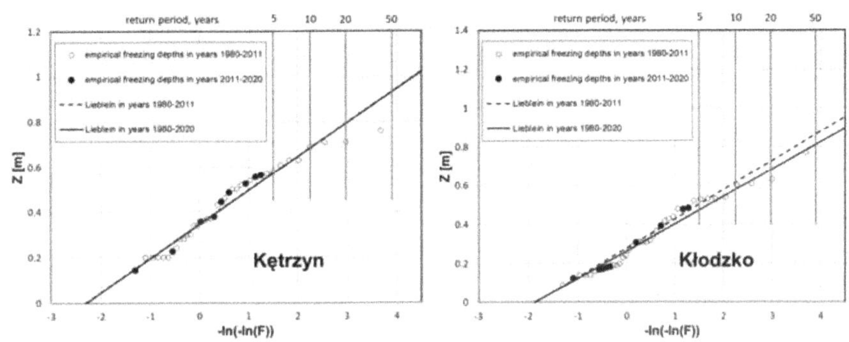

Figure 9. Comparison of predicted and measurement value for chosen location: Kętrzyn (**left**) and Kłodzko (**right**) station.

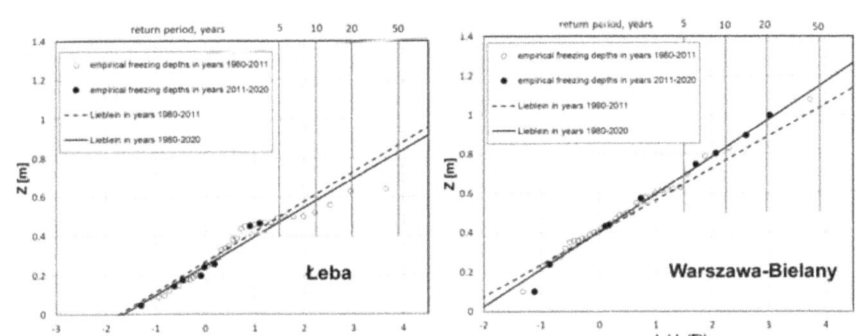

Figure 10. Comparison of predicted and measurement value for chosen location: Łeba (**left**) and Warszawa-Bielany (**right**) station.

4. Discussion and Conclusions

The use of the described method allows for a safe and reliable prediction of the depth of freezing of the ground. Real measurements over many years and an analysis of trends show that these values are not constant and subject to change. This is mainly due to the increasingly intense climate change associated with global warming. For the chosen locations, new predictions show that the empirical depth of freezing is general descending. This is one symptom of global warming.

The presented method of analyzing the depth of soil freezing takes into account actual soil temperature as a function of all external climatic actions with their coincidence and duration: air temperature, snow cover and precipitations. For this reason, it is more precise than the method using the so called air freezing index [16,26].

A comparison of calculation results related to the location of zero isotherm, conducted based on ground temperature analysis and freezing index (according to Gontaszewska [1]) leads to a conclusion that the second method has a lower accuracy. It does not account for snow cover, leading to much deeper freezing depths. Only in two cases (Leszno, Koszalin), almost the same results (difference of <2%) were obtained. The largest differences occurred in the mountains, where snowfall is the biggest (Jelenia Góra and Zakopane; values based on AFI are larger by 86%) and in North-East Poland (Białystok, 48%). Significant differences were also obtained in Warsaw (difference of 49%) and Kraków (77%).

Moreover, the probabilistic approach permits us to analyze the soil freezing in the terms of the reliability of structures. As the influence of climate can be easily analyzed using probabilistic approach, the influence of soil structure on the depth of freezing is much more complicated. It is due to the heterogeneity of soil and its structure. The depth of freezing is influenced by the type of the soil, its state of saturation, porosity, and mineral composition, and the arrangement of layers (Xu et al. [32], Żurawski & Godlewski [33], Yu, et al. [30]). This problem can be solved now only on the basis of the available literature using correction coefficients. One of the sources is available to authors now: former Soviet correction coefficients to the Formula (1) (Ickiewicz, Pogorzelski [34]): 23 for clays and silts, 28 for clay-sands, dusts and fine-grained sands, 30 for coarse-grained sands, and 34 for gravels. On the base of these values, it is possible to assume that gravel is a reference soil and the ratio of above mentioned values to 34 will be correction coefficients. In order to reduce the values based on different types of soil the inverse values of these coefficients will be used. An example is given below.

At Suwałki meteorological station the soil can be classified as silts and sands. In this case the actual depth of freezing reduced to the gravels as a standard reference soil can be received as a product of the calculated value 1.05 m for the 50-year return period and a factor of $34/28 = 1.21$. If the soil in Suwałki region is assumed as a reference gravel, then the freezing depths will be $1.05 \times 1.21 = 1.27$ m. On the contrary, if we want to calculate the depth for clays and foams, we will receive $1.27 \times (23/34) = 0.86$ m.

Also, the lack of snow cover under conditions of reduced temperature can significantly increase the freezing range. Snow cover greatly inhibits freezing depths in soils, such that freezing depths can vary from year-to-year or spatially within a given time period, solely dependent on the presence or absence of a snow cover. The implications of a future climate, without or an increasingly intermittent snow cover, are greater freezing depths or varying depths, respectively. The currently observed snowless winters (in Poland and central part of Europe), with low rainfall (hydrological drought) and possible sudden waves of frost can lead to an increase in the range of freezing of the ground, which is already observed in the results of measurements and presented predictions.

At present, new analysis is currently being done in order to update the old provisions presented in Polish Standards. The influence of the soil structure will be taken into account using correction coefficients, as listed above. It should be noted here that the new prediction ranges and the depths of the zero isotherm are to be considered as design values representative for reliability analysis for the return period of 50 years.

Further research on the influence of the soil structure on its freezing is necessary. Such investigations should be carried out for different types of soil. They should also examine the difference between the depth of the position of zero centigrade temperature and actual frost penetration into the soil. Meanwhile, the correction coefficients will be used as presented above.

Author Contributions: Conceptualization, T.G. and Ł.W.; methodology, T.G., Ł.W. and M.W.-N.; software, T.G. and Ł.W.; validation, Ł.W., and M.W.-N.; formal analysis, T.G. and M.W.-N.; investigation, Ł.W. and M.W.-N.; resources, T.G. and M.W.-N.; data curation, Ł.W.; writing—original draft preparation, Ł.W. and M.W.-N.; writing—review and editing, T.G.; visualization, Ł.W. and M.W.-N.; supervision, T.G.; project administration, T.G.; funding acquisition, T.G. All authors have read and agreed to the published version of the manuscript.

Funding: This research received no external funding.

Institutional Review Board Statement: Not applicable.

Informed Consent Statement: Not applicable.

Data Availability Statement: Not applicable.

Acknowledgments: We appreciate the soil temperature data provided by the Institute for Meteorology and Water Management—State Research Institute.

Conflicts of Interest: The authors declare no conflict of interest. The authors do not have any competing financial, professional, or personal interests from other parties.

References

1. Gontaszewska, A. *Thermophisical Properties in Soils in the Freezing Aspects*; Uniwersytet Zielonogórski: Zielona Góra, Poland, 2010. (In Polish)
2. Konrad, J.-M.; Shen, M. 2-D frost action modeling using the segregation potential of soils. *Cold Reg. Sci. Technol.* **1996**, *24*, 263–278. [CrossRef]
3. Rajaei, P.; Baladi, G.Y. Frost Depth: General Prediction Model, Transportation Research Record. *J. Transp. Res. Board* **2015**, *2510*, 74–80. [CrossRef]
4. Venäläinen, A.; Tuomenvirta, H.; Heikinheimo, M.; Kellomäki, S.; Peltola, H.; Strandman, H.; Väisänen, H. Impact of climate change on soil frost under snow cover in a forested landscape. *Clim. Res.* **2001**, *17*, 63–72. [CrossRef]
5. *Arctic and Subarctic Construction Calculation Methods for Determination of Depths of freeze and Thaw in Soils*; Technical Manual: TM5-852-6/AFM; Departments of the Army and the Air Force: Washington, DC, USA, 1966.
6. Aldrich, H.P., Jr.; Paynter, H.M. *Analytical Studies of Freezing and Thawing of Soils*; First Interim Report; U.S. Army Corps of Engineeris, Airfields Branch, Engineering Division, Military Construction, ACFEL TR42: Boston, MA, USA, 1953.
7. Kersten, M.S. Thermal properties of soils. In Proceedings of the 30th Annual Meeting of the Highway Research Board, Highway Research Board Special Report: Frost Action in Soils: A Symposium, Washington, DC, USA, 9–12 January 1951; Volume 2, pp. 161–166.
8. Edwards, A.C.; Cresser, M.S. Freezing and its Effects on Chemical and Biological Properties of Soil. *Adv. Soil Sci.* **1992**, *18*, 59–79.
9. Wiłun, Z.; Piasecki, A.; Kowalewski, Z. Freezing of soils (*Przemarzanie gruntów*). In *Informator Instytutu Technicznego Wojsk Lotniczych*; Seria A, Nr 5/1962: Warsaw, Poland, 1962. (In Polish)
10. Zavarina, M.V. Methods of Computing Maximum Soil Freezing Depth; Translated from Voyeykov Main Geophysical Observatory (Trudy GGO). *Sov. Hydrol. Sel. Papers* **1969**, *2*, 131–139.
11. Nixon, J.F. Discrete ice lens theory for frost heave in soils. *Can. Geotech. J.* **1991**, *28*, 843–859. [CrossRef]
12. Tarnawski, V.R.; Wagner, B. On the prediction of hydraulic conductivity of frozen soils. *Can. Geotech. J.* **1996**, *33*, 176–180. [CrossRef]
13. Vermette, S.; Christopher, S. Using the rate of accumulated freezing and thawing degree days as a surrogate for determining freezing depth in a temperate forest soil. *Middle States Geogr.* **2008**, *41*, 68–73.
14. Goodrich, L.E. The Influence of Snow Cover on the Ground Thermal Regime. *Can. Geotech. J.* **1982**, *19*, 421–432. [CrossRef]
15. Grodecki, W. Analysis of Certain Real-World Ground Freezing Factors. Ph.D. Thesis, Politechnika Warszawska, Warszawa, Poland, 1971. (In Polish).
16. Gontaszewska, A. Comparison of calculated and observed depth of frost penetration in west Poland. In *Faculty of Civil and Environmental Engineering, Institute of Structural Engineers*; ASCE: New York, NY, USA, 2003.
17. Steurer, P.M. Probability distributions used in 100-year return period of air-freezing index. *J. Cold Reg. Eng.* **1996**, *10*, 25–35. [CrossRef]
18. Mageau, D.W.; Morgenstern, N.R. Observations on moisture migration in frozen soils. *Can. Geotech. J.* **1980**, *17*, 54–60. [CrossRef]
19. Żurański, J.A.; Sobolewski, A. About measurements of the soil temperature and forecasting of the depth of soil freezing. *Inżynieria Bud.* **2013**, *3*, 141–145.
20. Janiszewski, F. *Instruction for Meteorological Stations*; IMGW, Wydawnictwa Geologiczne: Warszawa, Poland, 1988. (In Polish)
21. World Meteorological Organization. *Guide to Meteorological Instruments and Methods of Observation, WMO-No. 8*; WMO: Geneva, Switzerland, 2018.
22. Gumbel, E.J. *Statistics of Extremes*; Columbia University Press: New York, NY, USA, 1958.
23. Żurański, J.A.; Sobolewski, A. *Snow Loads in Poland (Obciążenie Sniegiem w Polsce)*; Instytut Techniki Budowlanej, Monografie: Warszawa, Poland, 2009.
24. Smith, M.G.; Rager, R.E. Protective Layer Design in Landfill Covers based on Frost Penetration. *J. Geotech. Geoenviron. Eng.* **2002**, *128*, 794–799. [CrossRef]

25. Lieblein, J. *Efficient Methods of Extreme-Value Methodology*; Technical Report, Institute for Applied Technology, National Bureau of Standards: WaWashington, DC, USA, 1976.
26. Żurawski, J.A.; Godlewski, T. *Seasonal Ground Freezing in Poland*; Instytut Techniki Budowlanej, Monografie: Warszawa, Poland, 2017.
27. Jakubek, M. Statistica (in Polish). Politechnika Krakowska, Poland. 2017. Available online: https://www.l5.pk.edu.pl/~{}mj/lib/exe/fetch.php?media=pl:dydaktyka:matematyka2:skrypt_statystyka.pdf (accessed on 5 April 2017).
28. Kozłowski, E. *Mathematical Statistics. Lecture II: Point Estimation*; Politechnika Lubelska: Lublin, Poland, 2020. Available online: http://www.kozlowski.pollub.pl/wyklady/SM/sm_wyklad2.pdf (accessed on 1 February 2020).
29. Plucińska, A.; Pluciński, E. *Introduction to the Probability and Mathematical Statistics Account*; Wydawnictwa Politechniki Warszawskiej: Warsaw, Poland, 1974. (In Polish)
30. Kordecki, W. *Probability Account and Mathematical Statistics: Definitions, Theorems, Formulae*; Oficyna Wydawnicza GiS: Wrocław, Poland, 2003. (In Polish)
31. Pekasiewicz, D. Selected compliance tests for extreme statistics distributions and their use in financial analysis. *Pr. Nauk. Uniw. Ekon. Wrocławiu* **2014**, *371*, 268–276. [CrossRef]
32. Xu, X.Z.; Wang, J.C.; Zhang, L.X. *Physics of Frozen Soil*; Science Press: Beijing, China, 2001.
33. Yu, F.; Guo, P.; Yuanming, L.; Stolle, D. Frost heave and thaw consolidation modelling. Part 1: A water flux function for frost heaving. *Can. Geotech. J.* **2020**, *57*, 1581–1594. [CrossRef]
34. Ickiewicz, I.; Pogorzelski, J.A. Influence of selected factors on the freezing depts in soils. *Inżynieria Bud.* **1987**, *10–11*, 338–342.

applied sciences

MDPI

Article

Numerical Analysis of Pressure Profiles and Energy Dissipation across Stepped Spillways Having Curved Risers

Najam us Saqib [1], Muhammad Akbar [2,*], Huali Pan [2,*], Guoqiang Ou [2], Muhammad Mohsin [3], Assad Ali [3] and Azka Amin [4]

1 Faculty of Engineering, Lahore Leads University, Lahore 54000, Pakistan; najam.fraz@gmail.com
2 Institute of Mountain Hazards and Environment, Chinese Academy of Sciences, Chengdu 610041, China; ougq@imde.ac.cn
3 Department of Civil Engineering, Jiangsu University of Science and Technology, Zhenjiang 212013, China; mohsinnumani@gmail.com (M.M.); assadali041@gmail.com (A.A.)
4 Faculty of Business Administration, Iqra University, Karachi 75500, Pakistan; azka.amin@iqra.edu.pk
* Correspondence: akbarmohammad0092@mails.ucas.ac.cn (M.A.); Phl0902@126.com (H.P.)

Abstract: In this study, curved risers stepped spillways models based on the increasing angle of suspension were tested to check for improvement in energy dissipation and pressure distributions. Four fourteen-steps stepped spillway models with a slope 1:0.84 were selected, using Froude's number non-dimensional similarity. The risers of steps were made curved, based on three angles of suspensions, i.e., 30°, 60°, and 90°. The simulations were performed by FLOW 3D software and by the turbulence model Renormalization Group (RNG) for discharges between 0.020 and 0.068 m^3/s followed by the model calibration. The 3D Reynolds-averaged Navier–Stokes equations were solved, which included sub-grid models for air entrainment, density evaluation, and drift–flux, to capture free-surface flow over the stepped spillway. It was estimated that curving the risers increases the energy dissipation up to three percent for lower flow rates, whereas it has no significant impact on energy dissipation for higher flow rates. It was found that in simply stepped spillway lower steps dissipate more energy as compared to curved risers stepped where energy dissipation is shifted to higher steps. On the other hand, curved risers stepped spillways showed lower values of negative pressures as compared to the simply stepped spillway. It was seen that a higher energy dissipating step as experienced more negative pressures as compared to the lower energy dissipating step.

Keywords: stepped spillways; flow 3D; energy dissipations; pressure profiles; cavitation

Citation: Saqib, N.u.; Akbar, M.; Pan, H.; Ou, G.; Mohsin, M.; Ali, A.; Amin, A. Numerical Analysis of Pressure Profiles and Energy Dissipation across Stepped Spillways Having Curved Risers. *Appl. Sci.* **2022**, *12*, 448. https://doi.org/10.3390/app12010448

Academic Editors: Krystyna Kazimierowicz-Frankowska, Gabriele Chiaro, Jaroslaw Rybak and Małgorzata Jastrzębska

Received: 11 October 2021
Accepted: 23 November 2021
Published: 4 January 2022

1. Introduction

Spillways are the integral part of the dam as they allow safe passage of overtopping flow [1]. There are many types of spillways but stepped spillways hold prominent importance due to presence of steps, which are responsible for the loss of kinetic energy and better aeration [2]. Stepped spillways contain steps that induce the macro roughness and project the high turbulence in flow [3]. As compared to simple ogee spillways, they offer more loss of kinetic energy of flowing water [4]. Stepped spillways have gained more interest due to their compatibility with roller compacted concrete dams [5]. Flow over the steps spillway is divided into three regimes: (1) nape flow; (2) transition flow; and (3) skimming flow [6]. Nape flow usually occurs at small flow rates and total fall is divided into number of smaller free fall, while skimming flow occurs at high flow rates and flow occurs as a coherent stream over the steps with recirculation trapped within the steps [7]. Geometry of steps is very important to determine the flow parameters over the steps [8].

Energy dissipation to reduce the length of the stilling basin and promote the safe flow over the stepped spillway, and pressure variations over the steps to study the aeration patterns, are the most important flow parameters that many researchers have investigated [9,10]. The studies on spillway models comparing ogee and stepped spillways have found that stepped spillways provide more resistance to flow [11]. Studies on lower scale models have found that energy dissipation is more in stepped spillways, as compared to ogee smooth spillways. The authors in [12,13] calculated the pressure profiles along the vertical and horizontal surfaces of the steps for a model stepped spillway and found that vertical surfaces of the steps are subjected to negative pressures. The authors in [14] used the k-e turbulence model to study a model stepped spillway using FLUENT software to plot pressure profiles on the horizontal and vertical surface of the steps.

The authors in [15] used ADINA software to simulate the turbulence flow over the stepped spillways of various step configurations. The authors in [16] performed model experiments to check the aeration efficiency of stepped spillways, in particular the effects of varying chute angle and step height. The authors in [17] did an experimental study on stepped spillway, using Froude number similitude in large size experimental facilities, using 10 configurations, which included smooth steps and steps with devices to enhance the energy dissipation. The authors in [18] did model studies on steeply sloped spillways, with high velocity flows, to investigate cavitation damage, along with formulation of mathematical expression to calculate the location of the inception point. The authors in [19] did model studies to look for energy dissipation due to the production of turbulence due to steps. They concluded that energy dissipation during the skimming flow is at a maximum for low discharge as compared high flow rates. The authors in [9] investigated the stepped spillways with non-uniform heights and found that there was no improvement in energy dissipation instead there was more turbulence for smaller flow rates. The authors in [20] studied stepped spillway models with stones and gabions and found that energy dissipation improved as compared to horizontal steps. The authors in [21] did laboratory experiments in order to determine the cavitation potential during the flow across the stepped spillway, along with an investigation into the relationship between cavitation index and friction factors. The authors in [22] studied the impact of inclined steps and inclined steps with end sills on energy dissipation. It was found that energy dissipation considerably increased.

The authors in [23] used smooth particle hydrodynamics to investigate pressure distribution on steps in a non-aerated flow region of a stepped spillway subjected to different discharges in a skimming flow condition. The authors in [24] adopted the Neuro fuzzy approach to simulate the stepped spillways with different slopes and steps. They found that the neuro fuzzy approach was best for calculation of energy dissipation as compared to regression analysis. The authors in [25] used FLOW3D to evaluate the step geometry, in which steps contained Λ-shaped steps at angles of 25 degrees in terms of calculating the energy dissipation in the steps spillways.

The above mentioned experimental and numerical studies have investigated different parameters, such as energy dissipation, air entrainment, and pressure profiles for different steps geometries. The present study estimates the impact of curved risers on the stepped spillway model having skimming flows on the three stepped spillway parameters, i.e., pressure, total energy dissipation, and energy dissipation distribution. The study was undertaken by numerical solver FLOW 3D using a realizable RNG k-ε turbulence model, which is capable of accounting the smaller scales of turbulence motion. The main aims of the study are given below.

(1) To estimate the total energy dissipation, energy dissipation distribution, and pressure profiles along the curved surface of the steps of the stepped spillway models (with curved risers) and along the vertical surface of steps for simple stepped spillways models under the skimming flow regime of flow.

For validation experimental results consisting of pressure profiles along vertical and horizontal surfaces of the steps of a stepped spillway model of slope 1; 0.75 and 13 steps from a previously published paper [14] were compared with the results from numerical solver FLOW 3D. A good agreement was obtained. Furthermore, a scale down spillway model of Khanpur Dam's (Islamabad, Pakistan) spillway was selected using Froude's number dimensional analysis whose risers were made curved based on the three different angles of suspension, i.e., 30, 60, and 90 degrees. Table 1 shows stepped spillway models having curved risers that have never been studied before. The three parameters are: pressure profiles, total energy dissipation and energy dissipation distribution for both simple and curved riser models, were made for the flow rate of 0.020–0.068 m^3/s.

Table 1. Autor study contribution table.

Reference	Author(s)	Country	Type of Spillway Model	Pressure Profiles	Total Energy Dissipation	Energy Dissipation Distribution	Methodology
28	[1]	Iran	Vertical risers and Horizontal tread	×	√	×	Physical Modelling/Laboratory setting
29	[2]	Algeria	Vertical and Horizontal Tread	×	×	×	Numerical Modelling/Ansys Fluent
30	[3]	China	Sky Jump/Vertical risers and Horizontal/Inclined Treads	×	√	×	Physical Modelling/Laboratory setting
31	[4]	Vietnam	Vertical Risers and Horizontal Tread	×	×	×	Physical Modelling/Laboratory Settings.
32	[5]	Iran	Sloping steps, Separation between steps	×	√	×	Physical Modelling & Numerical Modelling.
27	[6]	Canada	Smooth and Rough Vertical and Horizontal Treads/Steps with edges	√	×	×	Numerical Modelling FLOW 3D
33	[7]	Thailand	Vertical Risers and horizontal treads	×	√	×	Numerical Modelling/Physical Laboratory Model.
34	[8]	China	Vertical Risers and horizontal treads.	√	×	×	Numerical Modelling/Physical Laboratory Model.
35	[9]	Sweden	V type of steps with vertical and horizontal risers.	√	√	×	Numerical Modelling FLOW 3D
36	[10]	China	V type of steps with vertical and horizontal risers.	×	√	×	Numerical Modelling FLOW 3D
37	[11]	Iran	Vertical Risers and Pooled horizontal treads.	×	×	×	Numerical Modelling
38	[12]	Iran	Vertical Risers and Horizontal Treads.	×	×	×	Artificial Neural Network, Support Vector Machine
39	[13]	Germany	Vertical Risers and Horizontal Treads.	×	×	×	Laboratory Experimentation
40	[14]	United Kingdom	Vertical Risers and Horizontal Treads.	×	√	×	Laboratory Experimentation
	This Research Work	Pakistan	Curved Risers and Horizontal Treads with different angle.	√	√	√	Numerical Modelling FLOW 3D, ArcGis

2. Materials and Methods

2.1. Model Validation

To carry out the confidence in the study, a calibration process is always needed. For this purpose, experimental results were taken from a previous paper [14] and were compared with the same model results from FLOW 3D. The authors in [14] presented experimental results for the pressure profiles along the horizontal and vertical surfaces of three steps, i.e., step nos. 5, 9, and 13 of the stepped spillway model, along with water surface elevation over the stepped spillway model, for a flow rate of 0.027 m^3/s. The same model geometry was constructed in SOLIDWORKS software package and was named as Model 1. The geometry file was imported to numerical solver software package FLOW 3D. Grid Sensitivity analysis was performed at step no. 13 (for which the results are discussed later). Different mesh sizes 20 mm, 15 mm, 10 mm, and 5 mm were opted to run the simulation. It was 5 mm mesh which gave the lowest error (less than 10%) when the values were compared to experimental results for step no. 13 taken from [14].

2.2. Modelling of the Spillways

Two scaled down models were adopted. The first model (Calibration model) as previously discussed was created from the information provided in the paper [14], while for the second model, dimensional analysis was performed to get its dimensions. The second model, which was further processed and simulated, was brought from spillways of Khanpur Dam Taxila, Pakistan. General features of Khanpur dam are given in Table 2. Using Froude's number analysis of model generation, a model was constructed in the ratio of 1:50. Head and number of steps were considered in a way to minimize the scale effects and present the skimming flow along the steps [26]. General features of Model 2 are presented in Table 3. Figure 1a,b presents the pictorial view of Model 2 and the magnitude of curvature induced in each riser of Model 2, respectively.

Table 2. General Features of Khanpur Dam.

Items	Description
Location	Taxila on Haro River (Pakistan)
Catchment Area	308 square miles
Design Flood discharge	166,000 cusecs
Main Dam type	earth and Rock fill
Maximum dam height	167 ft
No. of spillways	5

Table 3. General features of Model 2.

Items	Description
Type of spillway	Stepped
No of steps	14
Maximum design head	213 mm
Maximum design discharge	0.064 m^3/s

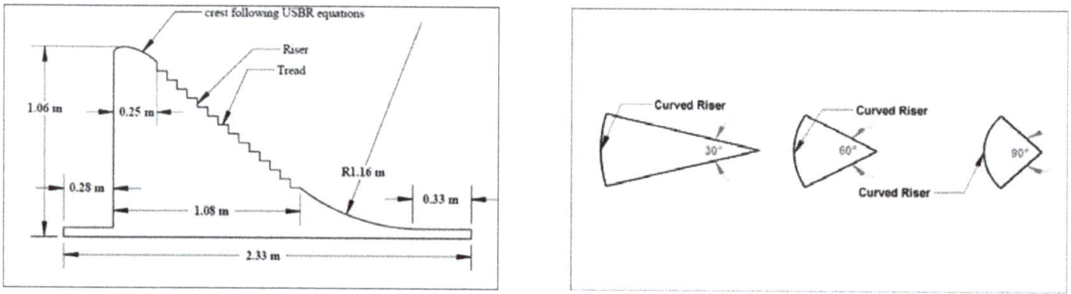

(a) (b)

Figure 1. (**a**) Pictorial View of Model 2; (**b**) Magnitude of Curvature.

2.3. *Properties of Model*

Physical properties of both models were incorporated in software package SOLID Works. The physical properties (length, width, and breadth, etc.) of the 1st model (Validation Model) (refer to Figure 2a) were documented in a previous paper where they were obtained by dimensional analysis for Model 2 (refer to Figure 2b), which has already been discussed in detail. The following Table 4 gives us the detailed description of geometrical/physical properties/features of the both the models.

Both the models were made up of standard Plexiglass material. FLOW 3D has already built-in material properties, but Plexiglass was not present in the built-in list. Moreover, general properties of Plexiglass given Table 5 were inculcated in FLOW 3D.

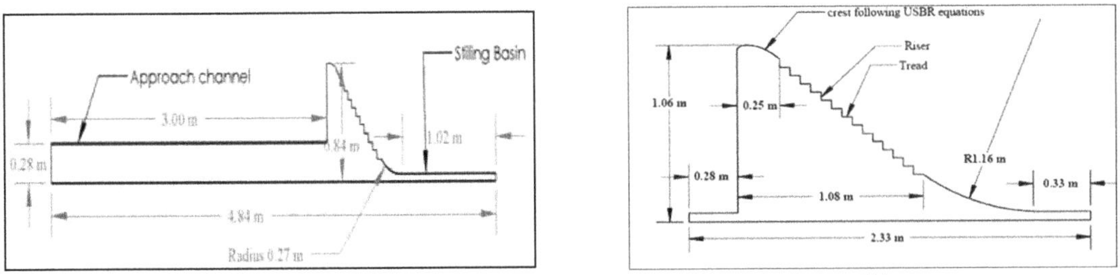

(a) (b)

Figure 2. (a). Pictorial View of Model 1; (b) Pictorial View of Model 2.

Table 4. Geometrical and physical properties of both models.

Sr No	Part	Model 2	Model 1 (Validation Model)
1	Total Height	1.06 m	78.9 cm
2	Total Width	30 cm	30 cm
3	No of Steps	14	13
4	Riser Height	6 cm	First 5 steps (2, 2.4, 3, 4, and 5 cm) others 6 cm
5	Tread length	5 cm	First 5 steps (1.5, 1.8, 2.25, 3, and 3.75 cm) others 6 cm
6	Equation of Crest	$y = 2.632x^{1.64}$	$y = 3.632x^{1.85}$
7	Tread Shape	Horizontal	Horizontal
8	Riser Shape	Vertical/Curved	Vertical
9	Slope of spillway	1 V:0.80 H	1 V: 0.75 H
10	Radius of Bottom Curve	116 mm	28 cm

Table 5. Properties of Plexiglass.

Sr No	Property	Value
1	Source Material	acrylic
2	Specific Gravity	1.18
3	Refractive index	1.49
4	Modulus of Elasticity	450,000 psi

2.4. Boundary Conditions

A boundary condition represents the type of behavior performed/exhibited by a particular boundary in a problem. This represents a constraint that helps to solve the problem. FLOW 3D incorporated with CFD calculates the values of a particular parameter at nodes of grides by using numerical approximation. Accuracy of numerical approximation depends upon many factors such as mesh size, type of boundary conditions, and model geometry, etc. In this study, a stepped spillway model was considered. As we know, the main function of the spillways is to dissipate the energy coming from upstream; therefore, at the upstream, we need to consider the discharge or the fluid height at the upstream mesh boundary of the model. In our case fluid height was considered as the boundary condition at the upstream. Similarly, as the water flow over the steps, it exits from the outlet boundary. So, the outlet boundary is considered at the x maximum.

In physical modeling, stepped spillway model made with plexiglass is placed into a flume. The flume presents the constraints/obstructions on the sides. Therefore, in our study, the wall is considered on the sides and bottom of the model. Figures 3 and 4 present the boundary conditions for Model 2 and Model, 1 respectively. On the other side, at the top of stepped spillway, atmospheric pressure is considered as the boundary condition. Moreover, fluid fraction is put equal to zero as at the top boundary of the mesh has no fluid. The commercial code FLOW 3D was used to perform this study. It is the commercial code that can solve 3D Reynolds-averaged Navier–Stokes (RANS) equations for one fluid including the RNG k-ε turbulence model numerically. FLOW 3D uses TurVOF method for interface tracking [27].

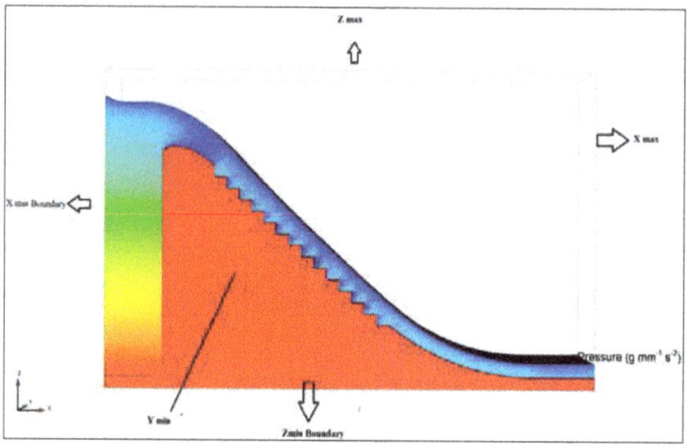

Figure 3. Boundary conditions for Model 2.

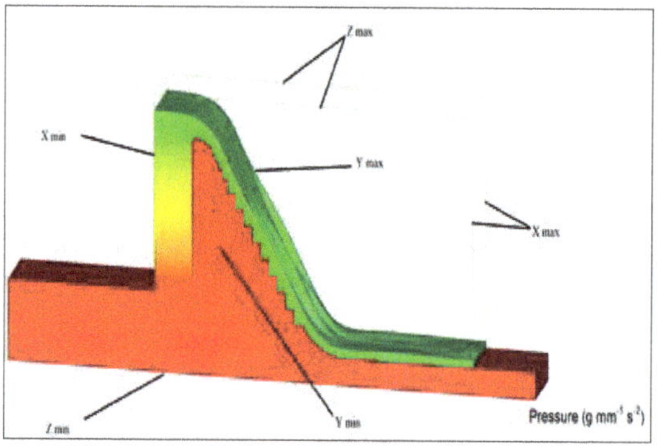

Figure 4. Boundary conditions for Model 1.

2.5. Flow 3D FEA Numerical Modelling

2.5.1. Mass Continuity Equation

As the flow over the stepped spillway is two phase flows (air water flow), therefore continuity equation in the form of volume weighted average density and velocity of corresponding phases (air and water), is given in Equations (1) and (2).

$$\frac{\partial \rho_m}{\partial t} + \nabla \times (\rho_m u_m) - \nabla \times (\vartheta \nabla \rho_m) = 0 \quad (1)$$

$$\vartheta = \frac{S_c u_m}{\rho_m} \quad (2)$$

where ρ_m and u_m are average density and velocity, respectively. The term $\nabla \times (\vartheta \nabla \rho_m)$ is the turbulent diffusion term. It accounts for turbulent mixing process in fluids with non-uniform density. On the other hand, S_c is the constant that is equal to reciprocal of turbulent Schmidt number, and u_m is volume weighted average dynamic viscosity.

2.5.2. Momentum Equation

Momentum equation for the fluid mixture is given in Equation (3),

$$\frac{\partial \rho_m u_m}{\partial t} + \nabla \times (\rho_m u_m u_m) = -\nabla P + \rho_m g + \nabla \times \tau \quad (3)$$

where τ is Reynolds's stress tensor, P is pressure, and g is gravitational acceleration. τ can be calculated from Boussinesq hypothesis as in the Equation (5)

$$\tau = -\vartheta_{\mathrm{eff}}\left(\nabla u_m + \nabla u_m^T - \frac{2}{3} I \nabla u_m \right) + \frac{2}{3} I k \tag{4}$$

where ϑ_{eff} is effective kinematic viscosity, k is turbulent kinetic energy, and I is identity Matrix.

2.5.3. RNG k-ε Turbulence Model

FLOW3D provides many models to account for turbulence. The accuracy of the model varies depending upon the type of problem being dealt with. In this work, the RNG k-ε turbulence model, which is capable of accounting the smaller scales of motion was selected. Equations (6) and (7) are the governing equation for this model.

$$\frac{D}{Dt}(\rho_m k) = \nabla \times (\rho_m D_k \nabla k) + P_k - \rho_m \varepsilon \tag{5}$$

$$\frac{D}{Dt}(\rho_m k) = \nabla \times (\rho_m D_\varepsilon \nabla \varepsilon) + \frac{C_1 P_k \varepsilon}{k} - \frac{C_2 \rho_m \varepsilon^2}{k} \tag{6}$$

where D_k and D_ε show the effective diffusivity of k and ε. P_k is the generation k due mean velocity gradients. C_1 is equal to 1.42, and C_2 can be calculated from C_2 = 1.68 k. More details about the RNG k-ε turbulence model can be found in user model of FLOW3D.

2.5.4. VOF Model

FLOW3D uses the TruVOF technique to track the interface between two non-soluble fluid (air and water in our study). It uses an indicator scalar, whose value range from 0 to 1 representing fractional volume of main fluid, i.e., is water in this study. TruVOF also applies suitable boundary condition at the fluid interface, i.e., atmospheric pressure in our case. Considering the Equation (7) for the calculation of f.

$$\frac{\partial f}{\partial t} + u_m \times \nabla \times f = \nabla \times (\vartheta \nabla f) \tag{7}$$

The term on right hand side stands for turbulent diffusion.

2.5.5. Air Entrainment Model

Air Entrainment Model in FLOW3D is based on the assumption that air entrainment in the free surface will occur. This will occur due to instabilities created by turbulence produced by flow over the complex geometry. Thus, this turbulence can overcome the stabilizing forces, caused by gravity and surface tension. As a result, air with the volume δV may be entrained into the fluid, which can be expressed by the Equations (8)–(10).

$$L_T = \frac{CNU^{\frac{3}{4}} k^{\frac{3}{2}}}{\varepsilon_T} \tag{8}$$

$$P_t = \rho_w k \,;\, P_d = \rho_w g_n L_T + \frac{\sigma_{sur}}{L_T} \tag{9}$$

$$\delta V = \left\{ k_{air} A_s \left[\frac{2(P_t - P_d)}{\rho_w} \right]^{\frac{1}{2}} \text{ for } P_t > P_d,\ 0 \text{ for } P_t < P_d \right\} \tag{10}$$

where CNU has a value of 0.09, L_T is turbulence scale length, and ε_T is turbulent dissipation. ρ_w is the water density σ_{sur} is the coefficient of surface tension. A_s is the surface area, k_{air} is proportionality constant. δV is the volume of air entrained per unit time.

2.5.6. Density Evaluation Model

As the air is entrained air, the density of water became non-uniform. There is the formation of an air–water mixture whose density can be computed as Equation (11).

$$\rho_m = (1 - C_a)\rho_w + C_a\ \rho_a \tag{11}$$

Here, ρ_m is the mixture density while ρ_a is the air density. C_a is the air concentration.

2.5.7. Drift Flux Model

According to [28] phase drag and bubble diameter can be modeled using a drift flux model. According to this, model air particles are dispersed into the air over the continuous water flow. The relative velocity between the velocity of dispersed particles and continuous water flow are considered steady. Equation (12) is referred as air transport.

$$\left(\frac{1}{\rho_w} - \frac{1}{\rho_a}\right)\nabla P = \left(\frac{f\rho_w + (1-f)\rho_a}{(1-f)\rho_w\rho_a}\right) \times Ku_r \tag{12}$$

K is cell drag coefficient, f is volume fraction of water, u_r shows relative or slip velocity. K can be computed from Equation (13) given below

$$K_p = \frac{1}{2}A_p\rho_w \times \left(C_d U_r + \frac{12\mu_w}{\rho_w\ R_p}\right) \tag{13}$$

$$K = \frac{(1-f)}{V_p}(K_p) \tag{14}$$

K_p is single drag coefficient. A_p is the cross-sectional area of the bubble. U_r Depending on u_r, μ_w is dynamic viscosity of water, C_d is user defined drag coefficient, and R_p denotes bubble diameter.

2.6. General Simulation Setup

Both models (Models 1 and 2) were constructed in SOLIDWORKS in the form of STL (Stereolithography) 3D file. Numerical solver FLOW 3D can import 3D files from SOLIDWORKS or AUTOCAD 3D to begin the simulation. To account for the turbulence in flow, due to the presence of steps, the RNG k-ε model was adopted. Different sub models (as discussed in Section 2), such as the VOF model, to capture the free surface; air entrainment model, to capture the formation of bubbles in the flow due to air; density evaluation model, to address the variation in density of the water; and drift flux model, to cope the formation of phase drag, were selected to create a virtual environment. Mesh sizes was adopted in a way to effectively capture all the features of geometry and free surface. Additionally, very fine mesh was avoided keeping in view the computing power. Same boundary conditions were adopted as discussed in Section 2.

2.7. Methodology Limitation

The study was carried out through numerical solver FLOW 3D which is the latest software incorporated with CFD coding. The use of numerical methods in hydraulic engineering is still found to be skeptical by many researchers due to the development of errors in the code. That is why the validations are always needed in case of numerical modeling. Numerical approximations are always unpredictable. Approximating a natural phenomenon of flowing water from the top of the stepped spillway sometimes presents unexpected errors and divergent solutions. Therefore, there is always a need for a sweet spot, where the validation errors are at a minimum, and based on these, further simulations are made.

3. Results

3.1. Model Validation

Model 1 was used to build up the confidence over the FLOW 3D. A simulation run was made for the flow rate of 0.02 m^3/s as used by the paper. Two parameters pressure profiles and free surface elevation were compared with results from [14] to look for the validity of the model. The same sub models that are given in Section 2 were adopted in FLOW 3D. To get a promising mesh size, grid sensitivity analysis was performed at step no. 13, as shown in Figure 5. We found that 20 mm mesh size gave the maximum error (more than 50%), while the error was less than 10 percent when the 5 mm mesh size was adopted. Therefore, for the further validation same 5 mm mesh size was adopted.

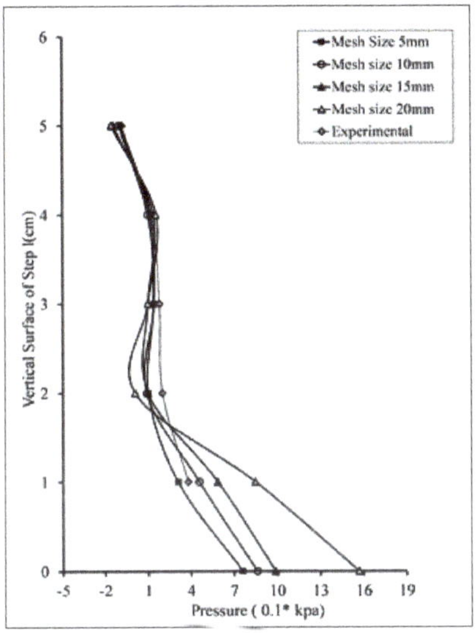

Figure 5. Grid Sensitivity Analysis.

3.1.1. Pressure Profiles

Figure 6 shows the comparisons of horizontal pressure profiles over step nos. 5, 9, and 13 of model no 1. In FLOW 3D, pressure values just at node closest to the horizontal surface were considered. It is clear from Figure 6 that numerical solver software FLOW 3D is sufficiently accurate.

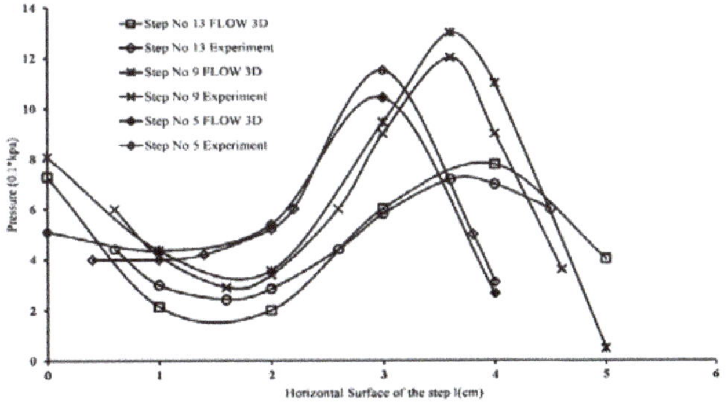

Figure 6. Pressure profiles comparison along the horizontal surface.

Similarly, Figure 7 presents the comparisons of pressure profiles over the vertical surface of step nos. 5, 9, and 13, respectively. Experimented values were taken from the paper, while for the vertical surface pressure profiles from FLOW3D, the value of pressure at the nodes near the vertical surface was considered. Similar to the case of horizontal pressure profiles, vertical pressure profiles also present a good agreement.

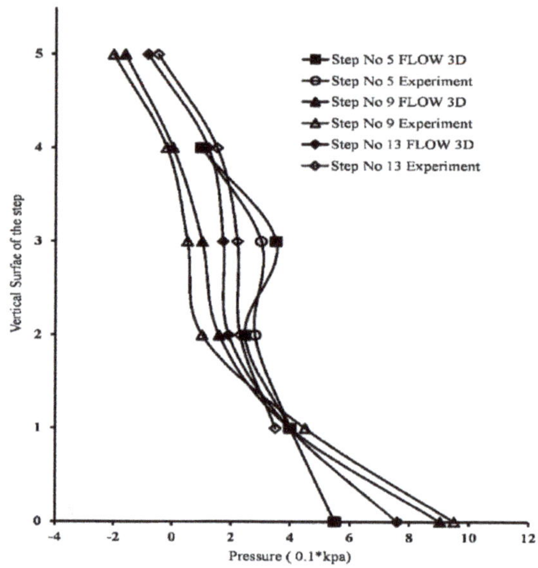

Figure 7. Pressure profiles along the vertical surface of steps.

3.1.2. Free Surface Elevation

This section presents the comparison of free surface elevation between the FLOW3D and experimented values. FLOW3D directly gives values of free surface elevation along three dimensions.

The three values along the free surface elevation were taken and then were averaged. Moreover, FLOW3D presents values of free surface elevation at every node, which is at distance of 5 mm (as the mesh size is 5 mm). A comparison is presented in Figure 8. A good agreement values can be seen.

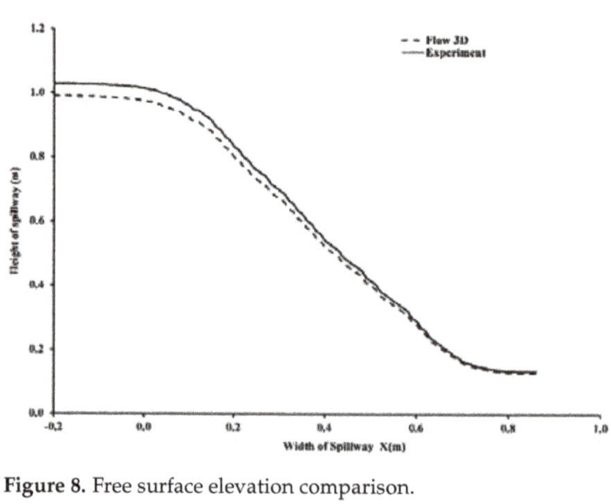

Figure 8. Free surface elevation comparison.

3.1.3. Pressure Profiles

Pressure profiles for each flow rate for each step configuration for Model 2 were plotted along the vertical surface of the steps and along the curved surfaces in the case of curved risers' steps from the flow rate 0.0208 to 0.068 m^3/s. The pressure decreases along the vertical surface of the steps in the case of vertical riser and eventually becomes negative near the tip. This produces cavitation, which in real spillway cause damages to spillway [27]. This behavior of pressure profiles along the vertical surface of the steps have been studied by many researchers, such as [23]. The reduction in the negative pressure along the vertical surface of the steps will show a reduction in cavitation. Therefore, many researchers have tried different techniques to counter that pressure reduction. Curved risers as presented in the plots from 9 to 13, encounter fewer negative pressures as compared to simple stepped spillways. This reduction in negative pressures as compared to simple stepped spillways and previous studies [16–18] will reduce the cavitation damage.

From the pressure profiles in Figures 9–13 on vertical surfaces, it is clear that vertical surfaces of steps are vulnerable to cavitation. The pressure profile gets negative while going toward the tip of the step. Although pressure increases a bit at the end, maximum negative pressure still can cause the cavitation but only for the case simple vertical riser. When pressure profiles for the simple vertical riser were compared to the curved riser, it was found that the curved riser contains larger negative pressure areas as compared to simple vertical steps. Simple stepped spillways are exhibiting a maximum negative pressure value as compared to an angled riser

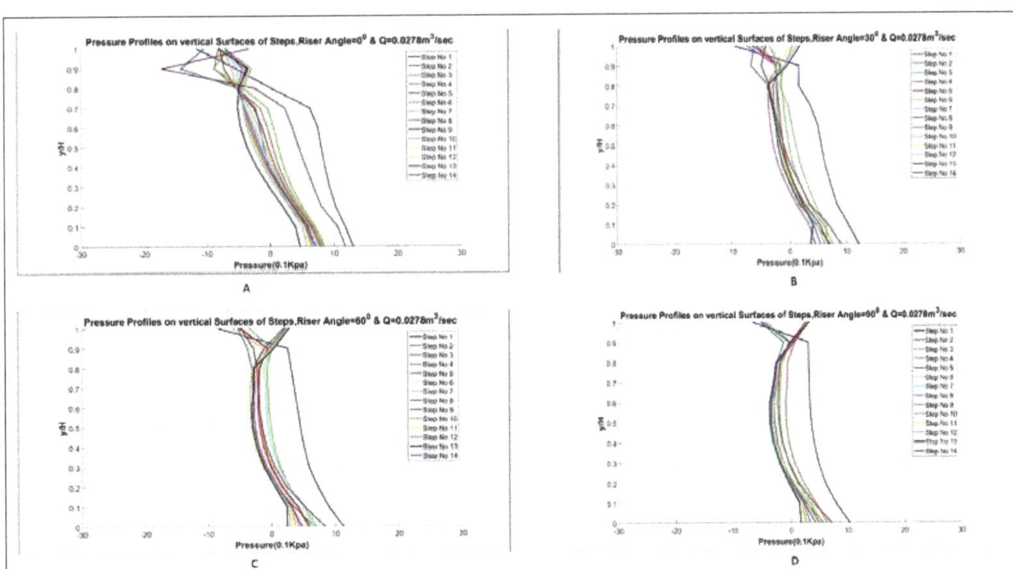

Figure 9. Pressure profiles for (**A**) simple stepped, (**B**) 30°, (**C**) 60°, and (**D**) 90° curved risers spillway for the head of 100 mm.

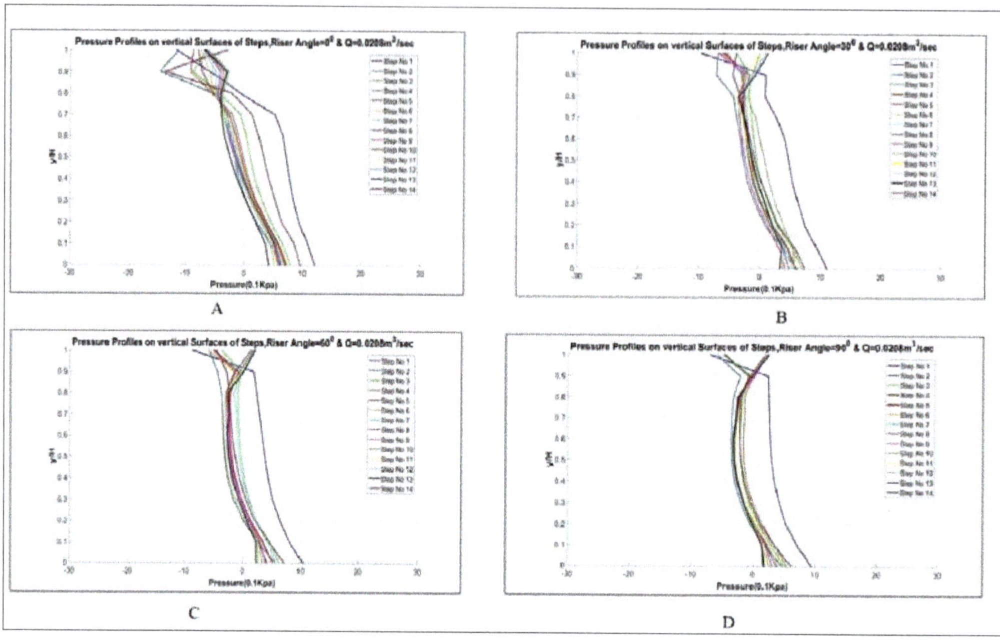

Figure 10. Pressure profiles for (**A**) simple stepped, (**B**) 30°, (**C**) 60°, and (**D**) 90° curved risers spillway for the head of 120 mm.

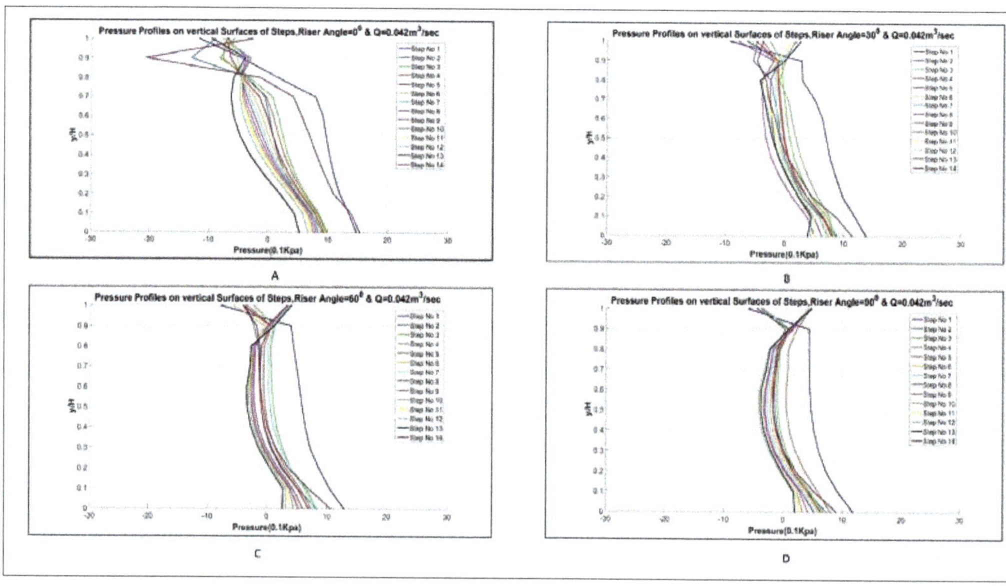

Figure 11. Pressure profiles for (**A**) simple stepped, (**B**) 30°, (**C**) 60°, and (**D**) 90° curved risers spillway for the head of 160 mm.

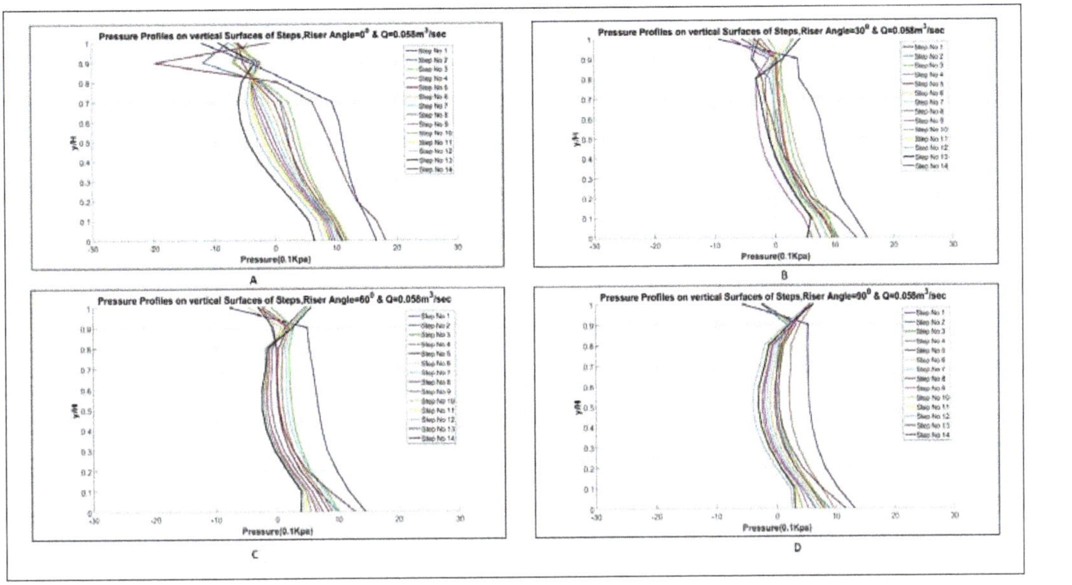

Figure 12. Pressure profiles for (**A**) simple stepped, (**B**) 30°, (**C**) 60°, and (**D**) 90° curved risers spillway for the head of 200 mm.

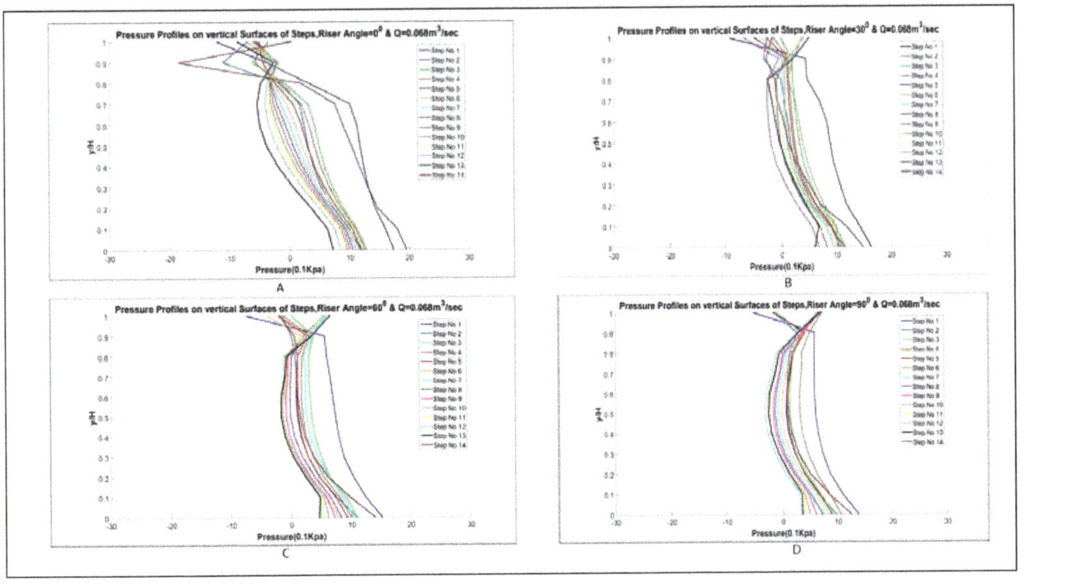

Figure 13. Pressure profiles for (**A**) simple stepped, (**B**) 30°, (**C**) 60°, and (**D**) 90° curved risers spillway for the head of 220 mm.

The reduction in the negative pressure in the curved riser models is due to the reduction in the local velocity due to the curvedness of the riser. As it is evident the fluid particular has to cover more distance in the case of curved risers, due to its curvedness as compared to simple vertical model.

3.2. Total Energy Dissipation

Previously changes in geometry were done in order to improve the energy dissipation across the steps. Many researchers, such as [9,22,25], altered the steps geometry and improved the energy dissipation to a certain limit. However, as compared to [9,22,25], changing the risers of stepped spillway do not increase energy dissipation. Hence, there was no significant increment in the energy dissipation for models with curved risers. A maximum of 3–4% energy dissipation was obtained as shown in Table 6 and Figure 14. Energy dissipation was calculated by percent head loss as compared to head at upstream. It was found that energy dissipation is dependent on the flow rate. For lower discharges, energy dissipation is more as compared to the high discharges. Details are given in the following Table 6. Q is given in m^3/s.

Table 6. Total Energy Dissipation.

Riser Type	Q = 0.027	Q = 0.042	Q = 0.059	Q = 0.068	Q = 0.0208
Vertical	59.17%	52.58%	43.81%	37.93%	61.42%
30°	59.91%	52.53%	42.08%	36.28%	64.61%
60°	59.99%	52.33%	40.12%	36.11%	65.48%
90°	59.63%	51.4%	42.27%	34.8%	65.12%

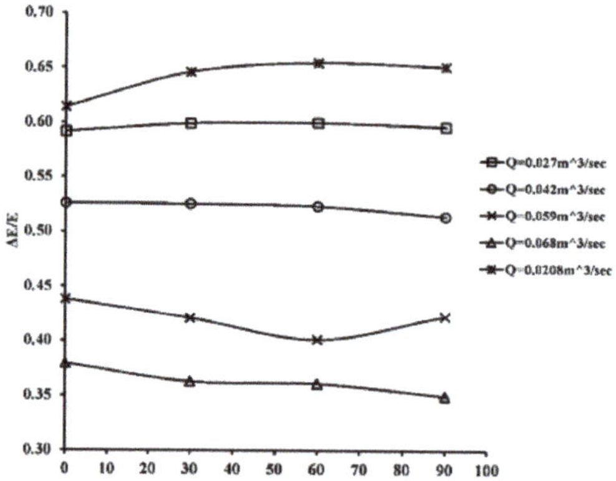

Figure 14. Graphical view of total energy dissipation.

Therefore, curved risers in a stepped spillway model do not increase in energy dissipation. In a practical sense, the steps in the stepped spillways, which were damaged due to cavitation, can be made curved in order to with stand the negative pressures. This can be done without any changes in total energy dissipation.

3.3. Energy Dissipation Distribution

EDD (Energy dissipation distribution) is the magnitude of energy dissipated by the each step. This distribution is useful in a sense that it presents individual energy/head dissipating behavior of the steps during the flow. In the literature [7–9], the authors have mostly stressed the total energy dissipation. Our study presents EDD in the form of a new variable named dissipation ratio. Energy dissipation distribution is the head or energy dissipated by each step at the steady state flow condition. In general, energy dissipation should be different at different steps due to difference in velocities. The flow velocity also changes when water flows from crest of the model to the bottom; therefore, different values of EDD for each flow rate are expected.

Dissipation ratio for a particular step = 100 × (head at the riser/head at the tread)/(head at the riser). Energy dissipation distribution across different steps is presented below.

From Figure 15a,b, it is evident that dissipation increases at the lower steps. Moreover, for the curved risers, this increase in dissipation is shifted to other steps. It can be seen in that for a 90-degree curved riser for each unit discharge, increase in energy dissipation is smooth. Ideally, a stepped spillway dissipating equal energy at each step is the best spillway as it causes less cavitation and is thus less prone to damage. A regular behavior in the nature of EDD is missing. Figure 16a,b show that energy dissipation increases while going down along the steps. Moreover, in contrast to this, for other flow rates, this shifted to higher steps in the curved risers.

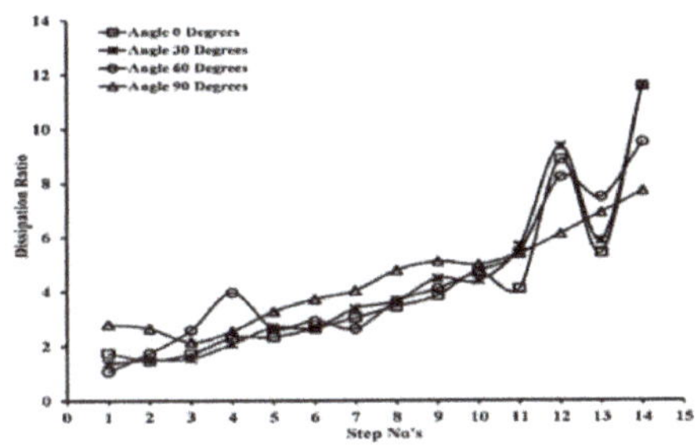

(**a**)

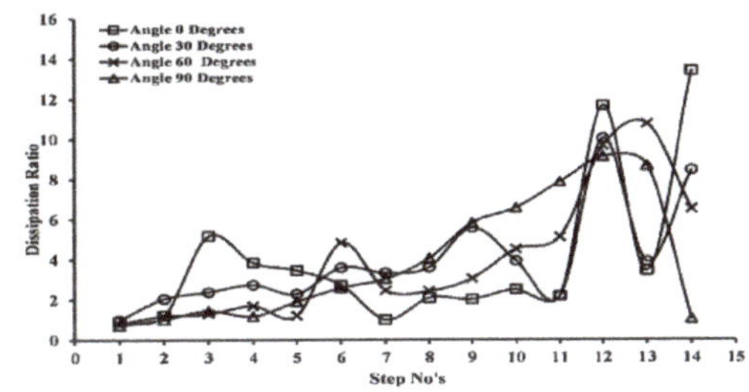

(**b**)

Figure 15. (**a**). EDD for the Q = 0.0208 m^3/s. (**b**). EDD for the Q = 0.027 m^3/s.

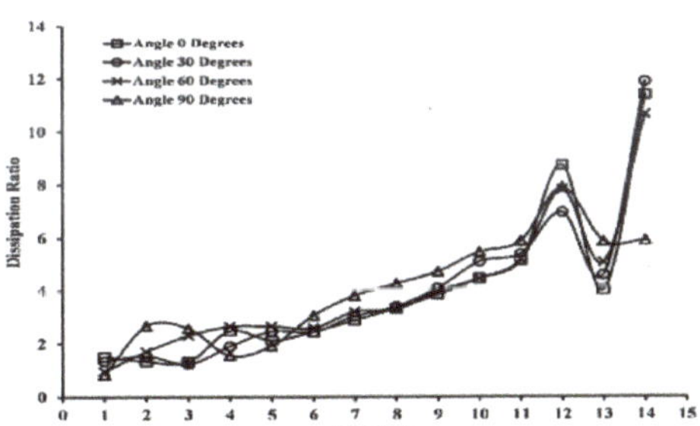

(**a**)

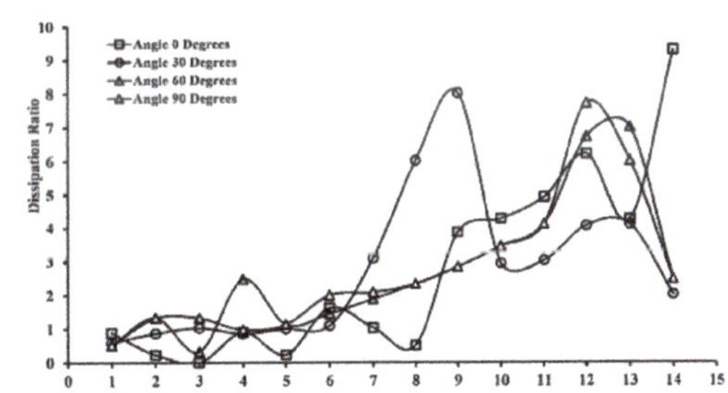

(**b**)

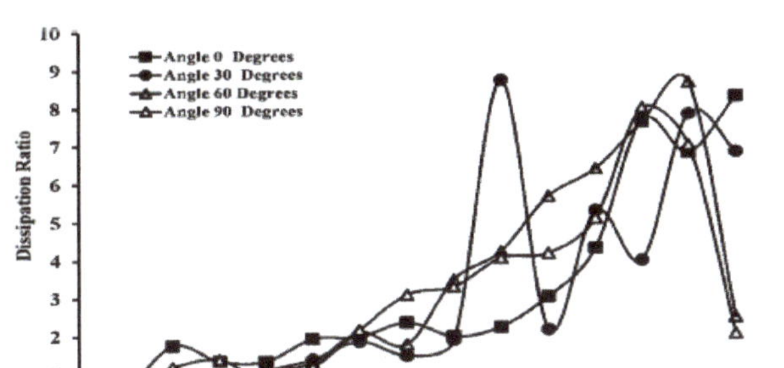

(**c**)

Figure 16. (**a**).EDD for the Q = 0.042 m^3/s. (**b**). EDD for the Q = 0.059 m^3/s. (**c**). EDD for the Q = 0.068 m3/s.

4. Conclusions

The primary purpose of the study was to evaluate the impact of alteration in step geometry on flow parameters such as pressure and energy dissipation across the stepped spillway model using the latest CFD (computational modelling dynamics) software package FLOW3D.

- Total energy dissipations and pressure profiles were calculated for each of simulation. No increment in energy dissipation was obtained expect for lower flow rates like 0.020 m^3/s where it increased to three percent in curved risers stepped spillways as compared to simple stepped spillways.
- In a simple stepped spillway, the lower steps, such as step nos. 12, 13, and 14 were dissipating more energy as compared to higher steps. Moreover, this shifted to higher steps in the case of stepped spillways having curved risers.
- Pressure profiles, along the vertical surface of steps in the case of simple stepped spillways, and along curved risers in the case of curved stepped spillways, showed that simple stepped spillway possess more negative pressure values as compared to curved stepped spillways. This can cause more cavitation in the case of simple stepped spillways as compared to curved stepped spillways, when the local pressure across the step face decreases less than the vapor pressure of water.

Police Recommendations

- This research was conducted through the latest CFD commercial code FLOW3D. Though the calibration process provides enough accuracy through the use of fine grids, there is still possibility for inaccuracies. The finer meshing and the use of second-order advection options can be utilized in FLOW3D to find inaccuracies.
- Spillway models with curved risers demonstrate an increment of three percent energy dissipation for lower flow rates. These types of spillways can be utilized for dams that are subject to continuous low emissions. Compared with a simple stepped spillway, the curved risers efficiently diminish material costs and allow more energy dissipation.
- Spillways with curved risers do not significantly increase energy dissipation for high flow rates. Compared with simple stepped spillways, they dissipate almost the same amount of energy. It is important to use these types of flow rates instead of simple stepped spillways because they can save material costs, but can dissipate the same amount of energy.
- Cavitation is an important parameter studied by dam spillway researchers. The spillways of many dams have been damaged by cavitation. Current research has identified that simple stepped spillways have more negative pressure than spillways with curved risers. The practical use of stepped spillways with curved risers can reduce the possibility of cavitation because they dissipate the same amount of energy.
- The present research shows that the steps that consume more energy are experiencing more negative pressure. Thus, specific steps that dissipate more energy are more susceptible to cavitation. The energy dissipation distribution of the steps can be used to analyze the number of steps that consume more energy and endure more negative pressure. This fact is very useful in a practical sense because more resources can be utilized to build steps that dissipate more energy and are more prone to cavitation than steps that dissipate less energy and are less likely to occur. This can help the design of the spillway in the future, and can stimulate the life of the spillway.

Author Contributions: Conceptualization, N.u.S. and M.A.; methodology, N.u.S., M.A; software, N.u.S.; validation, M.A., N.u.S.; formal analysis, N.u.S., M.A.; investigation, M.A. and N.S resources, N.u.S.; data curation, N.u.S., M.A; writing—original draft preparation, N.u.S. and M.A; writing—review and editing, A.A. (Azka Amin) and M.M.; visualization, A.A. (Assad Ali); supervision, H.P., G.O.; project administration, H.P.; funding acquisition, H.P. All authors have read and agreed to the published version of the manuscript.

Funding: This paper was supported by the Second Tibetan Plateau Scientific Expedition and Research Program (STEP) (Grant No. 2019QZKK0902), and National Natural Science Foundation of China (Grant No. 42077275). It was also supported by Youth Innovation Promotion Association of the Chinese Academy of Sciences (2018405).

Data Availability Statement: All data is available for request.

Acknowledgments: This paper was supported by the Second Tibetan Plateau Scientific Expedition and Research Program (STEP) (Grant No. 2019QZKK0902), and the National Natural Science Foundation of China (Grant No. 42077275). It was also supported by Youth Innovation Promotion Association of the Chinese Academy of Sciences (2018405).

Conflicts of Interest: The authors declare that there is no conflict of interest regarding the publication of this paper.

References

1. Tabari, M.M.R.; Tavakoli, S. Effects of Stepped Spillway Geometry on Flow Pattern and Energy Dissipation. *Arab. J. Sci. Eng.* **2016**, *41*, 1215–1224. [CrossRef]
2. Morovati, K.; Eghbalzadeh, A. Stepped Spillway Performance Study of the Pressure and Turbulent Kinetic Energy versus Discharge and Slope. *J. Water Sci. Res.* **2016**, *8*, 63–77.
3. Bung, D.B.; Valero, D. Re-aeration on stepped spillways with special consideration of entrained and entrapped air. *Geosciences* **2018**, *8*, 333. [CrossRef]
4. Christodoulou, G.C. Energy Dissipation on Stepped Spillways. *J. Hydraul. Eng.* **1993**, *119*, 644–650. [CrossRef]
5. Sarfaraz, M.; Attari, J.; Pfister, M. Numerical Computation of Inception Point Location for Steeply Sloping Stepped Spillways. In Proceedings of the 9th International Congress of Civil Engineering, Isfahan, Iran, 8–10 May 2012; pp. 2–9.
6. Chanson, H. Stepped spillway flows and air entrainment. *Can. J. Civ. Eng.* **1993**, *20*, 422–435. [CrossRef]
7. Felder, S.; Chanson, H. Energy dissipation, flow resistance and gas-liquid interfacial area in skimming flows on moderate-slope stepped spillways. *Environ. Fluid Mech.* **2009**, *9*, 427–441. [CrossRef]
8. Felder, S.; Chanson, H. Energy dissipation down a stepped spillway with nonuniform step heights. *J. Hydraul. Eng.* **2011**, *137*, 1543–1548. [CrossRef]
9. Chen, Q.; Dai, G.; Liu, H. Volume of fluid model for turbulence numerical simulation of stepped spillway overflow. *J. Hydraul. Eng.* **2002**, *128*, 683–688. [CrossRef]
10. Sorensen, R.M. Stepped spillway hydraulic model investigation. *J. Hydraul. Eng.* **1985**, *111*, 1461–1472. [CrossRef]
11. Rice, C.E.; Kadavy, K.C. Model study of a roller compacted concrete stepped spillway. *J. Hydraul. Eng.* **1996**, *122*, 292–297. [CrossRef]
12. Juny, M.S.; Pomares, J.; Dolz, J. Pressure field in skimming flow over a stepped spillway. *Hydraul. Stepped Spillways* **2020**, *8*, 137–145. Available online: http://www.flumen.upc.edu/admin/files/63.pdf (accessed on 15 November 2021).
13. Yasuda, Y.; Takahashi, M.; Ohtsu, I. Discussion of "Volume of Fluid Model for Turbulence Numerical Simulation of Stepped Spillway Overflow" by Qun Chen, Guangqing Dai, and Haowu Liu. *J. Hydraul. Eng.* **2004**, *130*, 170–172. [CrossRef]
14. Tabbara, M.; Chatila, J.; Awwad, R. Computational simulation of flow over stepped spillways. *Comput. Struct.* **2005**, *83*, 2215–2224. [CrossRef]
15. Baylar, A.; Emiroglu, M.E.; Bagatur, T. An experimental investigation of aeration performance in stepped spillways. *Water Environ. J.* **2006**, *20*, 35–42. [CrossRef]
16. Gonzalez, C.A.; Chanson, H. Hydraulic design of stepped spillways and downstream energy dissipators for embankment dams. *Dam Eng.* **2007**, *17*, 223–244.
17. Amador, A.; Sánchez-Juny, M.; Dolz, J. Developing flow region and pressure fluctuations on steeply sloping stepped spillways. *J. Hydraul. Eng.* **2009**, *135*, 1092–1100. [CrossRef]
18. Felder, S.; Chanson, H. Energy dissipation and residual energy on embankment dam stepped spillways. In Proceedings of the 19th Canadian Hydrotechnical Conference, Vancouver, BC, Canada, 9–14 August 2009; pp. 1940–1947.
19. Chamani, M.R.; Rajaratnam, N.; Beirami, M.K. Turbulent jet energy dissipation at vertical drops. *J. Hydraul. Eng.* **2008**, *134*, 1532–1535. [CrossRef]
20. Frizell, K.W.; Renna, F.M.; Matos, J. Cavitation potential of flow on stepped spillways. *J. Hydraul. Eng.* **2013**, *139*, 630–636. [CrossRef]
21. Hamedi, A.; Mansoori, A.; Shamsai, A.; Amirahmadian, S. Effects of end sill and step slope on stepped spillway energy dissipation. *J. Water Sci. Res.* **2014**, *6*, 1–5.
22. Husain, S.M.; Muhammed, J.R.; Karunarathna, H.U.; Reeve, D.E. Investigation of pressure variations over stepped spillways using smooth particle hydrodynamics. *Adv. Water Resour.* **2014**, *6*, 52–69. [CrossRef]
23. Salmasi, F.; Özger, M. Neuro-fuzzy approach for estimating energy dissipation in skimming flow over stepped spillways. *Arab. J. Sci. Eng.* **2014**, *39*, 6099–6108. [CrossRef]

24. Mansoori, A.; Erfanian, S.; Moghadam, F.K.; Mansoori, A.; Erfanian, S.; Moghadam, F.K. A Study of the Conditions of Energy Dissipation in Stepped Spillways with Λ-shaped step Using FLOW-3D. *Civ. Eng. J.* **2017**, *3*, 856–867. [CrossRef]
25. Yusuf, F.; Micovic, Z. Prototype-scale investigation of spillway cavitation damage and numerical modeling of mitigation options. *J. Hydraul. Eng.* **2020**, *146*, 100–120. [CrossRef]
26. Flow Science. User Manual: FLOW-3D® Cast 3.2. 2009, 1–49. Available online: http://www.easysimulation.com/public/flow3dcast/documentation/FLOW-3D_Cast_3.2_Manual.pdf (accessed on 15 November 2021).
27. Rajaei, S.H.; Khodashenas, S.R.; Esmaili, K. Comparative evaluation of energy dissipation over short stepped gabion and rigid spillways. *J. Hydraul. Res.* **2020**, *58*, 262–273. [CrossRef]
28. Bentalha, C.; Habi, M. Inception Point and Air-Water Flow Characteristics Over Stepped Spillway: Numerical Study. *Larhyss J.* **2020**, *31*, 91–105.

applied sciences

MDPI

Article

Monitoring the Impact of the Large Building Investments on the Neighborhood

Marian Łupieżowiec

Department of Geotechnics and Roads, Silesian University of Technology, 44-100 Gliwice, Poland; marian.lupiezowiec@polsl.pl; Tel.: +48-32-237-15-32

Abstract: The article presents the concept of monitoring buildings and infrastructure elements located near large construction investments (the construction of high-rise buildings of the Oak Terraces housing estate in Katowice and the construction of a tunnel under the roundabout in Katowice along the intercity express road DTŚ). The impacts include deep excavation, lowering of the groundwater level over a large area, and dynamic influences related to the use of impact methods of soil improvement. The presented monitoring includes observation of the groundwater level with the use of piezometers, geodetic measurements of settlement and inclinations, as well as the measurement of vibration amplitudes generated during the works involving shocks and vibrations. It was also important to observe the development of cracks on the basis of a previously made inventory of damage. The results of the monitoring allow corrections to be made in the technology of works (e.g., reduction of vibration amplitudes, application of additional protections at excavations, etc.) or the use additional safety measures. Currently, there are also monitoring systems used during the operation of completed facilities.

Keywords: impact monitoring; geotechnical influence; ground improvement

Citation: Łupieżowiec, M. Monitoring the Impact of the Large Building Investments on the Neighborhood. *Appl. Sci.* **2021**, *11*, 6537. https://doi.org/10.3390/app11146537

Academic Editor: Krystyna Kazimierowicz-Frankowska

Received: 10 June 2021
Accepted: 14 July 2021
Published: 16 July 2021

1. Introduction

Monitoring of the impact of works on the environment is one of the most important elements enabling the safe implementation of investment projects in urban development [1,2]. Due to the fact that works are carried out near existing buildings, including high-rise buildings, various types of observations and measurements must be carried out, on the basis of corrections in the field of works technology (e.g., reducing the amplitudes of vibrations, using a different type of driving element in the ground, changing the frequency of soil compaction, etc.) as well as suitable safety measures that can be introduced [3]. Currently, due to the increasingly scarce availability of land for development, investors are often forced to carry out construction works near other facilities [4] (with a risk of damaging adjacent buildings) and on poor grounds [5] (e.g., anthropogenic embankments with a high variability of granulometric composition and state of compaction, organic soils, etc.). Effective methods of ground improvement as well as technologies which allow the introduction of structural elements constituting an intermediate foundation or protection of scarps or excavation walls use shocks or vibrations [6]. They are usually destructive to nearby structures and can cause various types of damage, breakdowns, or even construction disasters. In addition, due to the need to optimize spatial development, it is often necessary to make deep excavations, which may also pose a threat to existing buildings [7–10].

Properly implemented monitoring should comprise a comprehensive program of continuous observation, which would enable a fast reaction in the event of a threat [11,12]. Such a program must be properly planned at the design stage and properly implemented during construction works. Sometimes, measurements and observations during the operation stage are also provided for, as harmful influences may also occur during this period [13,14].

One of the most destructive impacts of dense urban development involves deep excavation works. Very often, the level of the trench bottom is lower than the foundation level of the neighboring objects, and the works are carried out in close contact with another building. Often, the neighboring structures are in bad technical condition, or their structures are very sensitive to additional horizontal displacements and settlements resulting from the works being carried out, and hence very strict restrictions in this respect are imposed [15,16]. A separate problem is to ensure the stability of high scarps or walls of the excavations under high ground load. Another type of harmful effect involves a temporary lowering of the groundwater level during the excavation works [17]. Such a lowering may be from a few to a dozen or so meters and can cause additional settlements of up to several cm [18]. Especially in the case of complex ground conditions, uneven settlement may be the cause of various types of structural damage. Despite the increase in the value of mean stresses in the subsoil, which generally increases the subsoil bearing capacity, accompanying settlements under existing buildings is almost always a danger.

Particularly, destructive effects occurring during the execution of construction works are shocks and the result of vibrations propagating in the soil medium [19]. Since typical building structures, located in non-seismic areas, are not resistant to such influences, we can frequently observe damage or failures effected by dynamic influences. Some of the applied ground strengthening technologies, such as, for example, dynamic consolidation, making stone columns, or driving prefabricated piers or elements of sheet piling, may cause similar damage, as in the case of seismic or paraseismic tremors (e.g., mining tremors). Harmful vibrations can also occur during the transport of materials or other types of movement of heavy vehicles or construction machinery.

In the south of Poland, where the monitored investments described in the article were implemented, mining deformations are a significant problem accompanying construction sites. They are manifested by a lowering of the surface area, and additionally, inclinations, curvature of the surface area, and horizontal deformations. Mining deformations also adversely affect buildings and infrastructure elements within the impact range, sometimes causing damage. Sometimes, different kinds of influences overlap with the final deformation of the surface area, which translates into settlement and horizontal displacement of buildings. Hence, there is a need to take into account all factors that may affect the course of the monitored process [20–22]. Only in this case can the results of monitoring be reliable and contribute to the protection of endangered objects.

Due to the nature of the impacts caused by construction works on neighboring facilities, the standards in force in Poland impose very strict requirements in terms of permissible values (displacements and vibration amplitudes), which often prevents the use of some very effective technologies, and sometimes also puts into question the realization of some investments in dense building developments. On the other hand, predictions often differ from later observations. Additionally, the technical condition of the structures subjected to hazard or ground conditions in the place where the works are carried out has a great influence. Hence, monitoring of the impact is very often a prerequisite for safe implementation of an investment project. The aim of the article is to present the experience gained during the implementation of the monitoring ring, using the examples of two large investments in Silesia. Possible measurement techniques were tried, which can be applied in practice for many larger building investments.

2. Materials and Methods

The following part of the article presents the monitoring of two large investments carried out in Katowice in the south of Poland (Figure 1). The first one is the construction of a tunnel under the roundabout along an intercity route through the city center. The second one is the construction of a large housing estate. Both projects were carried out in the vicinity of other buildings, hence many facilities were within the range of impact of the realized construction works.

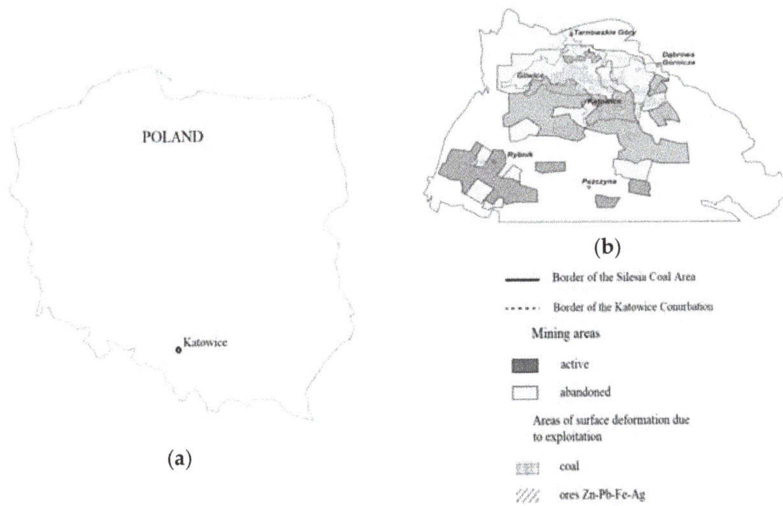

Figure 1. (**a**) Place of realization of the monitored investments. (**b**) Areas of mining activity and mining deformation in Upper Silesia.

The first element involving the monitoring of the impact of works on the structural condition of objects is the inventory of the state of objects at risk of the impact. The scope of observation and, above all, the requirements involving permissible impacts, should depend on the type of structure and its technical condition. For new structures in good condition, slightly higher impacts may be allowed. Yet, when the inventory reveals the presence of excessive damage, it may necessitate a change in the execution technology of works, or the use of additional reinforcements or protections. When assessing the technical condition of objects before starting monitoring, special attention should be paid to visible cracks, in particular on reinforced concrete and masonry structures (Figure 2b). Cracks on finishing elements (plaster, glaze, etc.) do not constitute any threat to the safety of use of the object, yet they indicate the occurrence of accelerated wear of the object and are the first sign of the object's lower resistance to future impacts.

(**a**) (**b**)

Figure 2. Applied monitoring of the impact of the investment projects on neighboring facilities: (**a**) measurement benchmark; (**b**) sealant for observation of crack development.

The monitoring should be carried out on the basis of an appropriate design project. Detailed arrangements regarding the types and frequency of observations, as well as the procedures to be followed in the event of excessive impacts, should take into account the technology used for the execution of works, the relevant safeguards, and the conclusions from the inventory of technical conditions. An appropriate monitoring program should optimize the number of tests and observations to be carried out, the frequency, and costs. It must not be forgotten that some tests may be burdensome for the contractor. Nowadays, increasingly complex automatic measurement systems are used, which enable quick detection of the risk of unfavorable impacts on objects [23,24].

A particularly important aspect of the presented monitoring involves the observation of the development of cracks noticed during the inventory. A plexiglass plate is glued to existing cracks (Figure 2b), which enables the observation of the development of cracks during the execution of works, in particular the growth of their openings. Modern monitoring and diagnostic programs of structures use crackmeters for measurements [25]. Additionally, the observation should take into account the development of a new cracks or other damage. The above observations are the most significant data to be applied for expert assessments involving the harmfulness of the impacts on a given object.

The first of the presented types of research is observation of the groundwater level, which is lowered at the stage of excavation and construction of underground elements. To determine the extent of lowering, piezometers made beforehand in the subsoil are used. They consist of a perforated pipe covered with sand. Inside it, the groundwater stabilizes on an ordinate consistent with its piezometric level in the surrounding ground. The measurement of this level is carried out by a device inserted inside the pipe into the water collected inside it. The measurement can be carried out in a traditional way with the use of devices emitting an acoustic signal when contacted with the water surface, or in an electronic way. In modern systems, we can also measure water temperature, direction, and speed of filtration, and we can monitor the quality of water and its degree of contamination [26].

It is particularly important in the applied monitoring to perform geodetic measurements of settlement and displacements in the horizontal plane. Measurements of this type have been used for a long time, and they have been now innovated by the introduction of automation in measurements, which allows for faster responses when the permissible values are exceeded [27,28]. Reports on geodetic measurements enable to analyze the impact of works on neighboring buildings at any time. The implementation of geodetic measurements requires prior installation of benchmarks (for the measurements of settlements) and measurement points (for monitoring displacements), which make up the measurement matrix (Figure 2a). The design of such a control system should be based on the assessment of the range of anticipated impacts and the inventory of damage identified before the commencement of works.

Finally, the last of the discussed methods involves measurements of vibrations generated on the structures of objects which are within the range of impact. Most often, accelerometers (Figure 3a) are used to measure accelerations in the structural elements of objects [29,30]. Less frequently, the amplitudes of displacements or velocities are measured. To be able to realistically assess the destructive impact, in addition to the amplitude values, it is also important to know the corresponding frequencies. The frequency range most dangerous for building structures is 5–25 Hz (Figure 3b). In some countries, appropriate monograms have been developed to assess the vibration risk level for certain types of buildings. In general, the permissible values should be determined on the basis of structure type and the technical conditions of a given building. Vibration monitoring is particularly helpful in verifying the safety measures used during the works.

(**a**) (**b**)

Figure 3. (**a**) Sensor for acceleration measurements. (**b**) An exemplary scale for assessing the impact of vibrations on the structure of buildings (according to the Polish standard).

While discussing various types of measurements carried out in-situ with sensors installed on the structures of monitored buildings, the accuracy of the measurements must not be overlooked. The accuracy of geodetic measurements (both settlements and deflections of objects) can reach ± 1 mm. Currently used measuring instruments can offer greater precision of measurement, however, for the purposes of monitoring, such accuracy is sufficient. In the case of groundwater level measurements in piezometers, an accuracy of ± 0.1 m is completely sufficient. It allows for a sufficiently precise assessment of the influence on the behavior of nearby objects. On the other hand, vibration measurements should already be carried out with greater accuracy. It is easiest to carry out the acceleration measurements, where most of the commonly used equipment ensures a measurement accuracy of ± 1 mm/s^2. Such accuracy allows the use of various types of scales (Figure 3b) developed to assess the impact of dynamic influences. In Polish standards, the basic quantity at which the impact of vibrations on structures is assessed is the acceleration amplitude, although in many other countries such a quantity is the velocity amplitude. Measurements of velocity and displacements of vibration would require using equipment that guarantees a much greater precision of measurements, hence such measurements are less often used in practice.

It is of great importance when designing a monitoring system to be able to define the frequency of the measurements carried out. During the realization of earthworks and geoengineering works, the intervals between measurements should be short enough to react duly to a possible threat to the safety of neighboring buildings. These intervals can be slightly extended right after the completion of the works (their effects may remain in the structures for some time). Frequently, monitoring is also carried out during the operation of the finished object, when the frequency of measurements is even lower. Sometimes the dates of tests are selected based on weather conditions (season of year). In the past, due to costs, it was necessary to limit the number of measurements and the time of monitoring. Currently, with the introduced automation of measurement, registration processes, archiving, and even automatic responses, the costs of monitoring are much lower, hence the growing popularity of its use. The monitoring system fits perfectly into the BIM system, which is becoming standard in the implementation of larger investments [31].

3. Results

3.1. Construction of a Tunnel under the Roundabout in Katowice

The following section presents examples of the monitoring carried out at two construction sites in Silesia. The first one involves the construction of a tunnel along the intercity express road (DTŚ) under the roundabout in Katowice, where underground works were carried out in 2005. The preparatory works, including the lowering of the groundwater level, started in mid-2004. As part of the investment project, a tunnel 650 m long and 30.4 m wide was constructed under an important road intersection in the center of the capital of

Upper Silesia (Figure 4). The excavations for the tunnel were carried out using the ceiling method. The structure of the tunnel consists of two floor slabs (lower and upper) with a thickness of 1.05–1.27 m supported on diaphragm walls 0.8 m thick (Figure 5). The said walls make up both the tunnel casings and limit the extent of the depression crater resulting from the temporary (for the duration of the works) lowering of the groundwater level by approximately 5 m. Additionally, a diaphragm wall was built in the middle between two parts of the tunnel, which contained traffic lanes in both directions. The length of the diaphragm wall was 17 m from the ground level, while the length of the piezometers used was 25 m. Since the ceiling of impermeable grounds (rock grounds-claystone and siltstone) was too deep, it was not possible to bring down the diaphragm walls to the impermeable grounds. Subsoil above the rocky soils was composed of medium and fine sands in a medium dense or dense state (see Figure 6). Consequently, the diaphragm walls only partially limited the range of the depression crater. Hence, the need to observe the groundwater level in the area adjacent to the construction site.

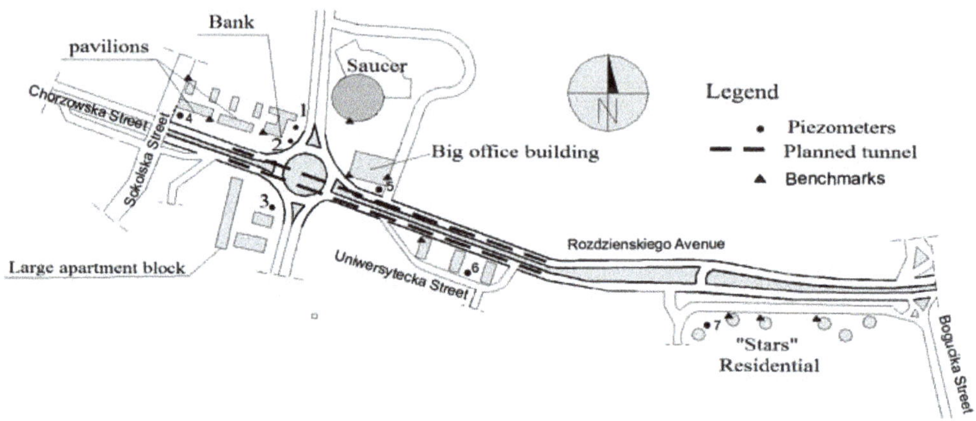

Figure 4. Schematic of the tunnel under the roundabout in Katowice.

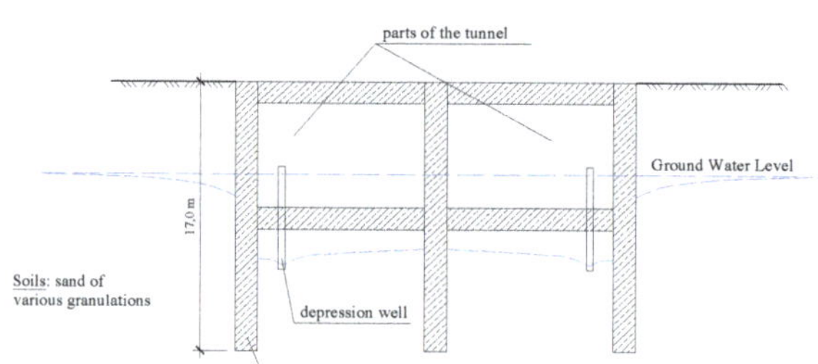

Figure 5. Cross section of the tunnel.

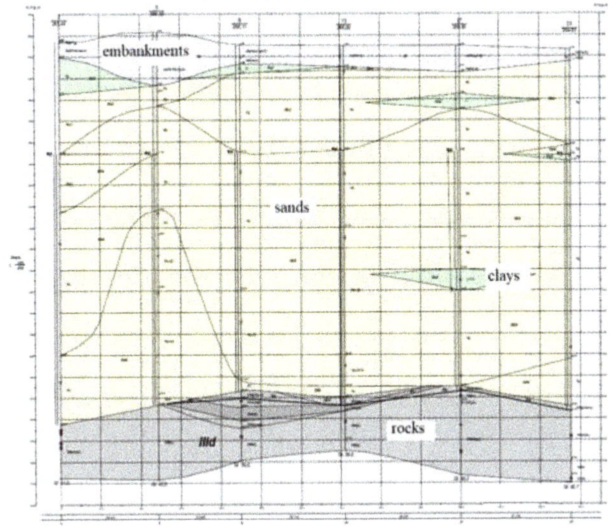

Figure 6. Exemplary geotechnical cross-section showing the ground conditions in the tunnel under construction.

The frequency of measurements as part of the monitoring carried out is presented in Table 1.

Table 1. Frequency of measurements as part of the monitoring involving the impact on the neighborhood of tunnel construction under the roundabout in Katowice.

	Number of measurement points	Frequency of Measurements					
		During the realization of investment project					During the warranty period
		X–XII 2003 Preparation works	I–XII 2004 Execution of the diaphragm walls	I–VI 2005 Excavations	VII–XII 2005 Execution of the tunnel structure	I–X 2006 Road structure of the tunnel	to X 2008 Exploration of the tunnel
Observation of crack development	123	every month	every month	every month	every month	every month	II, VI, XII, XXIV month
Settlement Inclination	79 10	every month	every quarter	every quarter	every 2 months	II, VI, X month	II, VI, XII, XXIV month
Piezometers	7	every 2 weeks	every month	every month	every month	every 2 months	II, VI, XII, XXIV month

The objects subjected to the hazard are located along the tunnel under construction (Figure 3). It was assumed that the range of the harmful influence was approximately 200 m on each side of the construction site. This resulted from the depression crater estimated in the technical project. The buildings subjected to hazard (Figures 7–10) include the sports and entertainment hall "Spodek" (Saucer), high-rise residential or office buildings (over 20 above-ground stories), and many other structures. Some of them were in poor technical condition. The results of selected measurements carried out as part of the monitoring are shown in Figures 11–15. In addition, in many cases, an increase in crack width was found (up 2 mm, mean 0.2–0.3 mm), which speaks to the real impact of the performed works on the condition of the structures. Cracks were most common on walls on the basement or ground floors (see Figure 2b) and were located near windows or ceilings.

(**a**) (**b**)

Figure 7. Execution of works as part of the construction of the tunnel under the Roundabout in Katowice: (**a**) beginning of the tunnel; (**b**) construction of diaphragm walls.

Figure 8. Objects within the impact range: sports and entertainment hall "Spodek" (Saucer).

(**a**) (**b**)

Figure 9. Objects within the impact range: (**a**) bank building with a glass elevation; (**b**) small architecture object: a monument to the Silesian Insurrections.

(**a**) (**b**)

Figure 10. High-rise buildings in the neighborhood of the works in progress: (**a**) big office building; (**b**) residential buildings of the "Stars" estate.

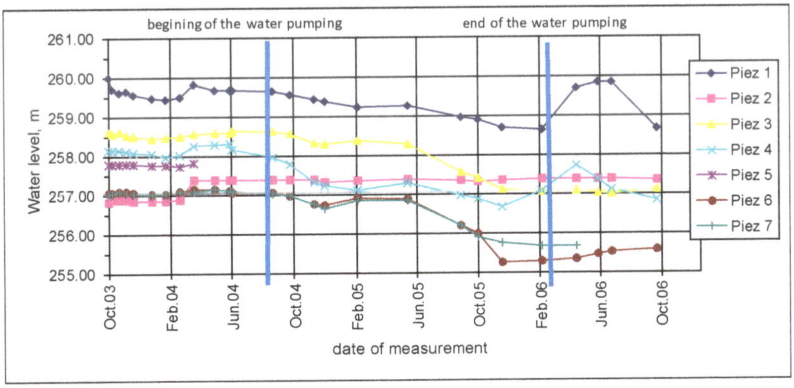

Figure 11. Change in water level measured in individual piezometers (construction of the tunnel under the roundabout in Katowice).

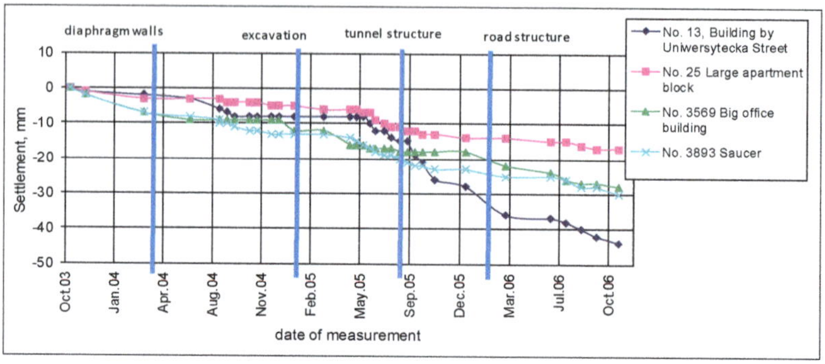

Figure 12. Settlement of selected objects measured on benchmarks (construction of the tunnel under the roundabout in Katowice—objects with significant settlements).

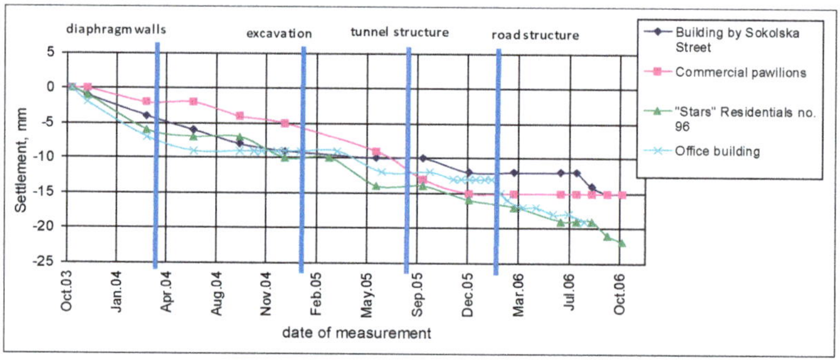

Figure 13. Settlement of selected objects measured on benchmarks (construction of the tunnel under the roundabout in Katowice—objects with small settlements).

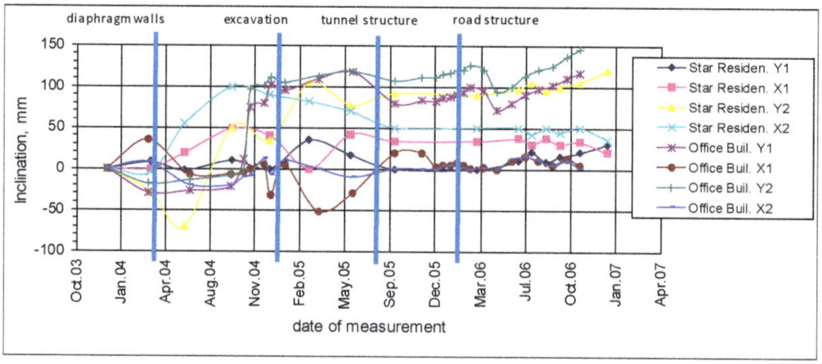

Figure 14. Inclination of selected objects (construction of the tunnel under the roundabout in Katowice—objects with significant inclinations).

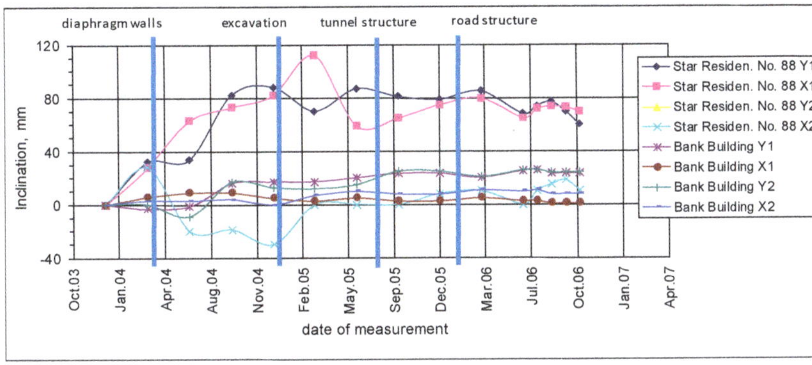

Figure 15. Inclination of selected objects (construction of the tunnel under the roundabout in Katowice—objects with small inclinations).

3.2. Construction of the Housing Estate "Oak Terraces" in Katowice

The second of the analyzed investment projects involved the construction of the housing estate "Oak Terraces" in Katowice (Figure 16). As part of the investment project realized since 2006 and divided into four stages, a total of 4 large buildings with 12 above-ground stories and 8 small 4-storey buildings are constructed (Figure 17). All buildings had one underground story. A subsoil up to a depth of up to 20 m below the ground level was formed by anthropogenic embankments composed of coal waste. The embankments were characterized by a very variable granulometric composition and state of compaction. Below there were very load-bearing and stiff rocky ground. The works that had a destructive impact on the neighborhood involved strengthening of the subsoil by means of the dynamic consolidation method [32]. In the applied technology, a heavy rammer is used for soil compaction whereby shocks are produced causing vibrations harmful to nearby objects [33]. The situation was complicated by the poor technical condition of the buildings within the impact range. The said objects were most often single-family houses with masonry structures, with Klein or wooden ceilings, without additional stiffening in the form of wreaths. Many cracks were found on the buildings, and the condition of the structures themselves was a cause of nuisance for people staying in the buildings.

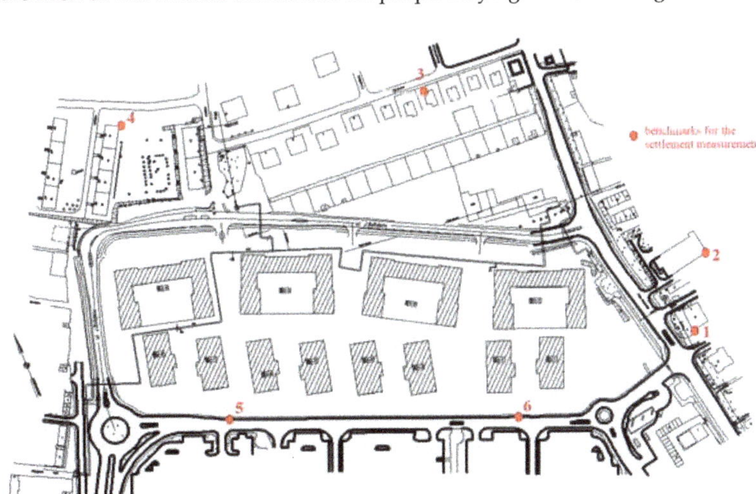

Figure 16. Construction site of the "Oak Terraces" housing estate in Katowice.

Figure 17. Buildings of the "Oak Terraces" housing estate in Katowice (project visualization, advertising materials, Tringranit).

Monitoring of the impact of the works on the neighborhood was primarily narrowed down to the measurements of settlements with previously installed benchmarks and measurement of vibration amplitudes resulting from the strengthening of the subsoil. The measurement results are shown in Figure 18 (settlements of the neighborhood) and Figure 19 (amplitudes of accelerations caused by dynamic consolidation). The latter values were used to determine the degree of dynamic damage to objects according to the scales given in the Polish standard (see Figure 3b). The settlement monitoring was carried out during the first stage of the works (from mid-2006), in which the first large building on the left (Figure 16) and the nearest two small ones were built, and during the second stage (from Autumn 2007), where there were another large and two small buildings.

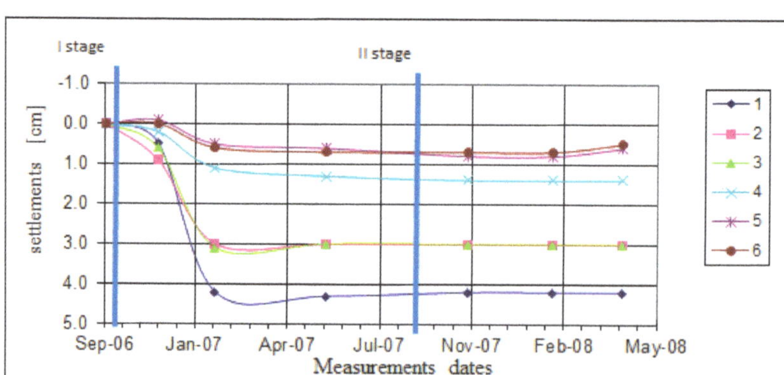

Figure 18. Settlement on selected benchmarks (near the construction site of the "Oak Terraces" housing estate in Katowice).

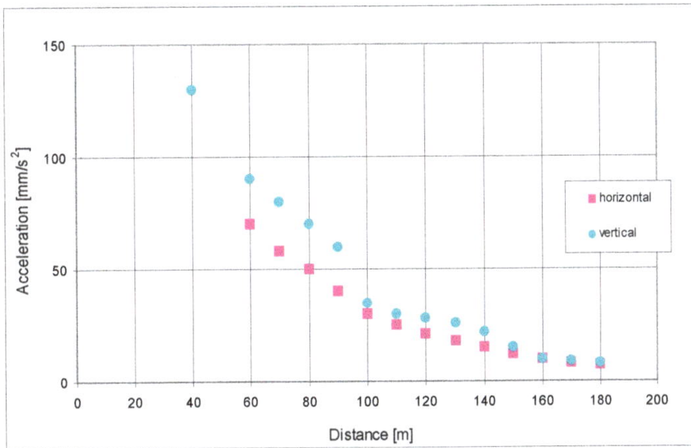

Figure 19. Dependence of acceleration amplitude on the distance from the source (neighborhood of the "Oak Terraces" a housing estate in Katowice).

4. Discussion

The monitoring results presented above in both examples (the construction of a tunnel under the roundabout and the "Oak Terraces" housing estate in Katowice), clearly indicate the considerable impact of the carried-out works on adjacent buildings. There is a visible time-based relationship between the dates of work execution and the formation of settlements or inclinations identified about 1–2 months later (see plots in Section 3, where the types of technology with the resulting displacements can be compared). The resulting additional settlements can reach even more than 40 mm, which in the case of tall objects that exert considerable pressure through the foundations on the ground, may already be regarded as a potential risk of failure. The above value of settlement is approximately equal to the value allowed by the serviceability limit state, therefore creating additional settlements of this value is often dangerous for the structure (and risk reaching the ultimate limit state). In general, it is not possible to define strict boundaries, as the safety of a specific object depends on many factors (e.g., technical condition of the building, type of structure, ground conditions, etc.). In the case of tall objects, the inclination can also be considerable; in extreme cases it can even reach 150 mm. The vast majority of settlements and inclinations remain permanent—after finishing of the works, the object does not return to its original position. Moreover, the lowering of the groundwater level has a meaningful impact on subsidence. Currently, efforts are made to prevent the depression crater from extending beyond the plot on which the investment project is being carried out. In situations where geotechnical conditions do not allow it (e.g., non-cohesive soils at great depth), observation of changes in the water level is a particularly important element of monitoring. In some objects, significantly lower settlements and displacements were found (Figures 13 and 15), which may indicate that these buildings are less endangered by the impacts caused by construction works. This may be due to the greater distance of the building from the place of works (e.g., buildings at Sokolska Street or the skyscrapers in the "Stars" estate, which are further away from the construction site, have low pressure on the ground (e.g., commercial pavilions) or a compact and rigid solid of the facility (e.g., the bank building)). Settlement is also often influenced by ground conditions, causing less deformation of the area around buildings and a better technical condition of the building's structure.

There is a noticeable coincidence in the time of construction works with the occurrence of settlements and deflection of objects within the range. It can be concluded that most often deformations measured on neighboring buildings appear about 2 months after excavation, lowering the water level or other types of works that have a large impact on the neighborhood.

It was feared that excessive and uneven settlement of buildings could damage the structure of buildings, leading to building failures. These damages are manifested in the appearance of cracks in masonry or concrete structures [34]. An example of a crack on the structure is shown in Figures 2b and 20. Hence, the need for monitoring, which allows for a quick response, consisting in changing the technology of works, execution of security or ad hoc repairs. Results of the monitoring results will also allow to determine the real impact of the works on specific damages. In the case of the two presented realizations, only a slight increase in the opening or enlargement of some cracks was found, which in no case led to a failure. Of course, there was a need to make a few repairs, but the cost of this was negligible in relation to the investment value.

Figure 20. An example of a crack was created as a result of uneven settlement of the building (at the window and under the ceiling).

In the case of smaller objects (e.g., single family houses), additional settlements and inclinations, regardless of the measured values, are not too much of a problem. This is due to the compact and rigid building structure and lower requirements (compared to large buildings) in terms of the serviceability limit state. Yet, the problem may involve the deterioration of the technical conditions of such buildings as a result of impacts from the construction site. Hence, it is essential to measure the vibration amplitudes, which in such cases are the decisive factor. It is also essential to observe the development of cracks and damage, which can be extremely helpful when choosing the technology of work execution. A detailed report on the rise of damage may also be helpful in resolving claims for compensation. It is important to have the inventory of damages prior to the commencement of works, so that the investor is not forced to pay for the damage incurred prior to the commencement of works.

The results of the monitoring presented above indicate that in the case of earthworks, in particular those carried out below the groundwater level, it is essential to measure the current water level with piezometers. It can be roughly assumed that the reduction of water level by 1 m can result in additional settlements of about 1 mm (assuming an active depth of 5 m and an average modulus of elasticity of the subsoil 50 MPa, the above value may be slightly different depending on the type of soil). With the lowering of up to a few meters, this is not a big problem for the condition of the building structure, but when the lowering is larger, additional settlements may contribute to a failure. In the case of the construction of the roundabout in Katowice, the change in ground water level did not exceed 2 m, because of the proper operation of the diaphragm walls, although they were not brought to impermeable soil.

The frequency of measurements (see Table 1) was adjusted to the work schedule. In the periods of higher intensity of construction works, measurements were carried out more often, while at the time when the impact of these works on the resulting settlement was smaller, measurements were carried out less often. Generally, in the case of large construction investments with typical soils in Poland, there is no need for measurements more frequent than once every 2 weeks. It should not be forgotten that more frequent measurements are more expensive, which can be a problem in some countries. The accuracy of the measurements of displacements (including settlements), as well as the amplitudes and frequencies of vibrations mentioned in Section 2, turned out to be sufficient to assess the impact of the works on the neighborhood.

In the case of the construction of the tunnel under the roundabout, the greatest concern was the fall of the groundwater level, which was predicted over a large area. Deep excavations could also be dangerous, but they were secured by diaphragm walls. The monitoring results showed that although some settlements and horizontal displacements were found, there was no greater threat to the safety of the structures of adjacent buildings. It also turned out to be beneficial that no mining deformation occurred during the construction works, which would constitute an additional risk of failure. The location of the investment (the center of a large city) was essential here, because there is no mining activity causing large deformations of the land surface, as well as the final phase of coal extraction in Upper Silesia in Poland.

In the case of the construction of the Dębowe Tarasy estate, the main problem was vibrations and their destructive impact on the neighborhood. The vibrations caused by use of the heavy ramming method to soil improvement. Fears were intensified by the poor technical condition of the adjacent buildings. Monitoring carried out in some places indicated the need to reduce the energy of ramming, which allowed the destructive influence of the works on the neighborhood to be minimized.

5. Conclusions

The monitoring of impacts is particularly important in the context of the influence exerted by an investment project on its neighborhood, which was demonstrated with the examples included in the article. With less and less access to investment areas, contractors often have to face the problem of limiting the impact on nearby facilities. In extreme cases, these influences may cause failure and sometimes even destruction of objects. The main objective of monitoring is to develop a capability for making quick adjustments to the applied work execution technology or to introduce additional safety measures. Nowadays, monitoring is becoming an indispensable element of the realization of larger building investment projects. The realization of monitoring begins at the design stage, when the inventory of objects within the impact range should be made. The most important aspect of the monitoring involves measurement and observation during works, when the impact on the neighborhood is the highest. Frequently, the realization of monitoring also extends to the period after the completion of construction works when the operation of the newly built facility starts. We can also indicate here such investment projects as the A1 motorway running through Piekary Śląskie, where there is a risk of significant mining deformations. In such situations, the monitoring system is inextricably linked to the operation of this motorway and is a prerequisite for users' safety.

The implementation of monitoring fits perfectly into the observation method of simultaneous design and realization of investments. This method has been included in the current Eurocode 7 standard. Monitoring, based on the two presented examples, allowed for the efficient performance of works without worrying about the safety of neighboring objects. Only a small amount of minor damage was found, which, however, is unavoidable with such large investments. It turned out that the noise from the construction site as well as dust or dirt on the roads were a greater nuisance for the residents and users of the monitored facilities. The vibrations were also annoying, but they lasted for a relatively short time. On the other hand, the carried-out monitoring of the impacts made it possible to determine that there was no threat to the condition of building structures of nearby facilities.

On the basis of the analyzes presented in the article, it is possible to present "good practices" in the field of monitoring investments carried out in urban areas or near other facilities. These principles are summarized in the below Table 2.

Table 2. Good practice rules in monitoring of structures adjacent to construction works.

Term	The Frequency of Measurements or Observations	Documentations and Projects
Before construction works (at the design stage)	1/quarter–1/half a year	Inventory of endangered objects, Inventory of existing damage, Development of a monitoring project
During construction works	1/week–1/2 months (depending on the type of works and their pace)	Observation of the development of existing damage and localization of new ones Regular reports with the results of measurements and observations along with in-depth analysis
After completion of construction works and during operation	1/2 months–1/year (more often during the warranty period, then less frequently, observation for a specified period given in the monitoring design)	Regular reports with the results of measurements and observations along with in-depth analysis

Funding: This research received no external funding.

Informed Consent Statement: Not applicable.

Conflicts of Interest: The author declares no conflict of interest.

References

1. Witakowski, P.; Kurczyński, Z.; Wójtowicz, J.; Kujawińska, M.; Dymny, G.; Gawęcki, P.; Woźniak, M. *System Kompleksowego Zarządzania Jakością w Budownictwie. Bezdotykowe Metody Obserwacji i Pomiarów Obiektów Budowlanych (Comprehensive Quality Management System in Construction. Non-Contact Methods of Observation and Measurements of Building Objects)*; Series "Instrukcje, Wytyczne, Poradniki", 443/2009; ITB Publishing: Warszawa, Poland, 2009. (In Polish)
2. Zhang, Z.; Huang, M.; Wu, B. Risk Analysis and Control Factors Based on Excavation of a Large Underground Subway Station under Construction. *Symmetry* **2020**, *12*, 1629. [CrossRef]
3. Hwang, R.N.; Moh, Z.C.; Wang, C.H. Performance of wall system during excavation for Core Pacific city. *J. Geoeng.* **2007**, *2*, 53–60.
4. Capraru, C.; Adam, D. Evaluating the influence of deep excavations on neighboring buildings by numerical analysis. *Numer. Methods Geotech. Eng.* **2014**, *II*, 729–734.
5. Leung, E.H.; Ng, C.W. Wall and ground movements associated with deep excavations supported by cast in situ wall in mixed ground conditions. *J. Geotech. Geoenviron. Eng.* **2007**, *133*, 129–143. [CrossRef]
6. Stinnette, P.; Gunaratne, M.; Mullins, G.; Thilakasiriand, S. A quality control programme for performance evaluation of dynamic replacement of organic soil deposits. *Geotech. Geol. Eng.* **1997**, *15*, 283–302. [CrossRef]
7. Hsiung, B.C.B. A case study on the behaviour of a deep excavation in sand. *Comput. Geotech.* **2009**, *36*, 665–675. [CrossRef]
8. Dmochowski, G.; Szolomicki, J. Technical and Structural Problems Related to the Interaction between a Deep Excavation and Ajacent Existing Buildings. *Appl. Sci.* **2021**, *11*, 481. [CrossRef]
9. Li, S.; Li, P.; Zhang, M.; Liu, Y. Influence of Approaching Excavation on Ajacent Segments for Twin Tunnels. *Appl. Sci.* **2020**, *10*, 98. [CrossRef]
10. Rybak, J.; Ivannikov, A.; Kulikova, E.; Żyrek, T. Deep excavation in urban areas—Defects of surrounding buildings at various stages of construction. In Proceedings of the 9th International Scientific Conference Building Defects (Building Defects), Ceske Budejovice, Czech Republic, 23–24 November 2017; Volume 146, p. 02012. [CrossRef]
11. Witakowski, P. Zdalne monitorowanie obiektów budowlanych podczas budowy i eksploatacji (Remote monitoring of buildings during construction and operation). *Tech. Trans. Environ. Eng.* **2007**, *104*, 179–189. (In Polish)
12. Xu, L.; Xu, Y.; Wang, C.; Feng, K. Data-Driven Deformation Reliability of Retaining Structures in Deep Excavations Considering Measurement Error. *Appl. Sci.* **2019**, *9*, 5466. [CrossRef]
13. Runkiewicz, L.; Sieczkowski, J. Monitorowanie stanu bezpieczeństwa obiektów budowlanych w trakcie eksploatacji (Safety condition monitoring structures during their use). *Mater. Bud.* **2015**, *11*, 131–132. (In Polish) [CrossRef]
14. Florkowska, L.; Bryt-Nitarska, I.; Gawałkiewicz, R.; Kruczkowski, J. Monitoring and Assessing the Dynamics of Building Deformation Changes in Landslide Areas. *Buildings* **2020**, *10*, 3. [CrossRef]

15. Ou, C.Y.; Hsieh, P.G.; Chiou, D.C. Characteristics of ground surface settlement during excavation. *Can. Geotech. J.* **1993**, *30*, 758–767. [CrossRef]
16. Long, M. Database for retaining wall and ground movements due to deep excavations. *J. Geotech. Geoenviron. Eng.* **2001**, *127*, 203–224. [CrossRef]
17. Adamowicz, H.; Popielski, P. Analysis of changes in groundwater and soil conditions in a highly urbanized area due to deep building foundations. *Tech. Trans.* **2015**, *24*, 4–20.
18. Howard, K.; Israfilov, R. *Current Problems of Hydrogeology in Urban Areas, Urban Agglomerates and Industral Centres*; Kluwer Academic Publishers: Dordrecht, The Netherlands, 2002.
19. Gryczmański, M.; Jastrzębska, M.; Łupieżowiec, M. A model for the forecasting of the propagation of technological impacts. *Studia Geotech. Mech.* **2008**, *30*, 59–66.
20. Juraszek, J.; Gwóźdź-Lasoń, M.; Logoń, D. FBG Strain monitoring of a Road Structure Reinforced with Geosynthetic Mattress in Cases of Subsoil Deformation in Mining Activity Areas. *Materials* **2021**, *14*, 1709. [CrossRef] [PubMed]
21. Parkasiewicz, B.; Kadela, M.; Bętkowski, P.; Sieńsko, R.; Bednarski, Ł. Application of Structure Monitoring Systems to the Assessment of the Behaviour of Bridges in Mining Areas. *IOP Conf. Ser. Mater. Sci. Eng.* **2017**, *245*, 032018. [CrossRef]
22. Rybak, J.; Kongar-Syuryun, C.; Tyulyaeva, Y.; Khayrutdinov, A.; Akinshin, I. Geomechanical substantiation of parameters of technology for mining salt deposits with a backfill. *Min. Sci.* **2021**, *28*, 19–32. [CrossRef]
23. Ornoch, L.; Popielski, P.; Olszewski, A.; Kasprzak, A. Ultrasonic Sensors Enabling Early Detection of Emergency Trends and Analysis of Structure Inclination and Stability by Means of Highly Accurate Level Measurements. *Sensors* **2021**, *21*, 1789. [CrossRef]
24. Zabielski, J.; Srokosz, P. Monitoring of Structural Safety of Buildings Using Wireless Network of MEMS Sensors. *Buildings* **2020**, *10*, 193. [CrossRef]
25. Chen, X.; Topac, T.; Smith, W.; Ladpli, P.; Liu, C.; Chang, F.K. Characterization of distributed microfabricated strain gauges on stretchable sensor networks for structural applications. *Sensors* **2018**, *18*, 3260. [CrossRef] [PubMed]
26. Radzicki, K.; Rybiański, Ł.; Popielski, P. Method of thermal detection of leakages in construction of deep excavation: A real case study in Poland. In *MATEC Web of Conferences*; EDP Sciences: Les Ulis, France, 2019; Volume 284, p. 03007.
27. Żaczek-Peplinska, J.; Kowalska, M.E.; Łapiński, S.; Grzyb, M. Multi-temporal survey of diaphragm wall with terrestrial laser scanning method. *Open Geosci.* **2020**, *12*, 656–667. [CrossRef]
28. Taghavikish, S.; Elhabiby, M. Target Based Correlation Deflection Monitoring to Analyze the Environmental Effect on Variations of Deflection on Structures. *Geomatics* **2021**, *1*, 12. [CrossRef]
29. Tomczyk, K.; Layer, E. Accelerometer errors in the measurement of dynamic signals. *Measurement* **2015**, *60*, 292–298. [CrossRef]
30. Sanayei, M.; Maurya, P.; Moore, J.A. Measurement of building foundation and ground-borne vibrations due to surface trains and subways. *Eng. Struct.* **2013**, *53*, 102–111. [CrossRef]
31. Nieto-Julian, J.E.; Lara, L.; Moyano, J. Implementation of a TeamWork-HBIM for the Management and Sustainability of Architectural Heritage. *Sustainability* **2021**, *13*, 2161. [CrossRef]
32. Pan, J.L.; Selby, A.R. Simulation of dynamic compaction of loose granular soils. *Adv. Eng. Softw.* **2002**, *33*, 631–640. [CrossRef]
33. Sękowski, J.; Kwiecień, S.; Kanty, P. The influence of dynamic replacement method on the adjacent soil. *Int. J. Civ. Eng.* **2018**, *16*, 1515–1522. [CrossRef]
34. Drobiec, Ł. Study of impact of bed joint reinforcement on load-carrying capacity and crack resistance of masonry walls made of calcium silicate units. *J. Build. Eng.* **2021**, *33*, 101841. [CrossRef]

MDPI
St. Alban-Anlage 66
4052 Basel
Switzerland
Tel. +41 61 683 77 34
Fax +41 61 302 89 18
www.mdpi.com

Applied Sciences Editorial Office
E-mail: applsci@mdpi.com
www.mdpi.com/journal/applsci

www.ingramcontent.com/pod-product-compliance
Lightning Source LLC
Chambersburg PA
CBHW061600220926
51850CB00005B/382